Praise for *The Power Makers*

"Maury Klein's *The Power Makers* allows us to step back and remind ourselves—and we *do* need reminding—that the past two centuries have been a period of extraordinary invention . . . Fascinating."

—**William Tucker, *Wall Street Journal***

"In a well-written and satisfying account, Klein makes readers aware of the magnitude of the energy, genius and tenacity of not only Edison—whose development of the world's first power station in 1881 on New York's Pearl Street was a momentous accomplishment—but also of Westinghouse and many others whose discoveries and vision made cheap electricity possible."

—***Publishers Weekly* (starred review)**

"Maury Klein's stories of heroic inventors creating the industrial revolution make the history of technology come alive."

—**Daniel Walker Howe, NBCC Award nominee for *What Hath God Wrought: The Transformation of America, 1815–1848***

"*The Power Makers* vividly and brilliantly reveals how the revolutions of steam and electricity, one facilitating the other, combined to reshape American society. Maury Klein tells a fascinating, heroic tale peopled by such giants as Thomas Edison, George Westinghouse, and J. P. Morgan, whose partnerships, subterranean deals, and marketplace battles redefined not just American commerce but the American landscape as well."

—**Edward J. Renehan Jr., author of *Commodore: The Life of Cornelius Vanderbilt***

ALSO BY MAURY KLEIN

The Genesis of American Industry, 1870–1920

The Change Makers

Rainbow's End: The Crash of 1929

The Life and Legend of E. H. Harriman

Days of Defiance: Secession, Slavery, and the Civil War

Unfinished Business: The Railroad in American Life

The Flowering of the Third America

Union Pacific: The Birth, 1862–1893

Union Pacific: The Rebirth, 1894–1969

The Life and Legend of Jay Gould

Prisoners of Progress: American Industrial Cities, 1850–1920

A History of the Louisville & Nashville Railroad

Edward Porter Alexander

The Great Richmond Terminal

THE
POWER
MAKERS
· · · ·

STEAM, ELECTRICITY, AND
THE MEN WHO INVENTED
MODERN AMERICA

MAURY KLEIN

BLOOMSBURY PRESS

NEW YORK   BERLIN   LONDON

Published by Bloomsbury Press, New York

All papers used by Bloomsbury Press are natural, recyclable products made from wood grown in well-managed forests. The manufacturing processes conform to the environmental regulations of the country of origin.

LIBRARY OF CONGRESS CATALOGING-IN-PUBLICATION DATA

Klein, Maury, 1939–
The power makers : steam, electricity, and the men who invented modern America / Maury Klein.
p. cm.
Includes bibliographical references and index.
ISBN 978-1-59691-412-4 (hardcover alk. paper)
1. Inventions—United States—History.   2. Inventors—United States—
Biography.   3. Force and energy—United States—History.   I. Title.
T20.K585 2008
609.73—dc22
2007048309

First published by Bloomsbury Press in 2008
This paperback edition published in 2009

Paperback ISBN-10: 1-59691-677-X
ISBN-13: 978-1-59691-677-7

1 3 5 7 9 10 8 6 4 2

Designed by Rachel Reiss

Typeset by Westchester Book Group
Printed in the United States of America by Quebecor World Fairfield

For Kim, with love and thanks

CONTENTS

	Acknowledgments	IX
	Introduction	XI
PROLOGUE	A Show of Power: Philadelphia 1876	1
1.	The Machine That Changed the World	14
2.	Conquering the Waters	30
3.	The Greatest Engine of All	51
4.	In Search of the Mysterious Ether	71
5.	Let There Be Light	98
6.	A Covey of Competitors	118
7.	The Light Dawns	136
8.	The Pearl Street System	159
9.	The Cowbird, the Plugger, and the Dreamer	177
10.	The Alternative System	201
11.	Eventful Currents	219
12.	Gaining Traction	239
13.	Competition and Electrocution	256
14.	Money, Mergers, and Motors	279
15.	A Show of Lights: Chicago 1893	300
16.	The Niagara Fallout	324
17.	Hard Times	345
18.	The Future Arrives	366
19.	Mastering the Mysteries of Distribution	395
20.	The Empire of Energy	417
EPILOGUE	A Show of Possibilities: New York 1939	444
	Electrical Circuits	453
	Notes	457
	Bibliography	511
	Index	521

ACKNOWLEDGMENTS

THE WRITING OF this book would not have been possible without the help of several people who lent their time and talent to assist my work. The library staff at the University of Rhode Island has once again been unfailingly helpful in obtaining resource materials. I am especially indebted to Emily Greene, who fielded an endless flow of requests with aplomb and dispatch. Sandy Sheldon and Mary Anne Sumner also did everything they could to expedite my work.

Paul Israel generously shared his vast expertise on Thomas A. Edison and helped me find my way through the Edison papers online. Bob Dischner at National Grid provided me with copies of the Tesla letters and other materials housed in the archives of his company. Jill Jonnes graciously answered several inquiries about sources used for her book. Harold C. Platt introduced me to one of his favorite Chicago restaurants and an excellent evening of discussion on Samuel Insull and things electric. Kathy Young could not have been more helpful in providing me access to and copies of materials from the Insull papers housed in the archives of Loyola University of Chicago. My colleague Steven Kay, an electrical engineer, read chapter 4 for technical bloopers, and Richard John kindly read the portion of chapter 5 dealing with the telegraph and furnished me a draft of his work on Morse's contribution. To all of these people I am deeply grateful. For any remaining errors, the buck stops with me.

Three other people deserve special mention. My editor, Peter Ginna, maintained steadfast faith in this project through all its twists and turns and provided invaluable feedback once the manuscript reached his hands. So too did my agent, Marian Young, offer constant and needed encouragement on this project as she has on so many others. Finally, my wife, Kim, sustained me through a long project and endured my often prolonged mental and physical absences with good humor and patience.

INTRODUCTION

IT IS CURIOUS that no one has put together a history of both the steam and electric revolutions. Numerous authors have written about the history of electricity or some aspect of it. A lesser number have done the same for steam power, but no one has combined the two into a single narrative. Yet they are inseparable stories at the most basic level. Electricity is the single most important technology of the twentieth century. Without it the modern world simply doesn't exist. People get at least a partial sense of this fact every time the power goes out for any length of time. However, there is a corollary that very few people realize: Without the steam engine there would be no electricity. Together they form the foundation of the modern world.

Industrialization created the modern world. It began during the eighteenth century when certain countries began increasingly to organize their economies around the manufacture of goods rather than agriculture, and to increase their productivity through the use of machinery. This transformation could not have occurred without the most basic revolution of all, the development of new sources of power—especially the steam engine and electricity. So far has this revolution advanced that today we would be utterly helpless without these sources of power.

Industrialization made the United States and the power revolution made industrialization. Similarly, technology made the electrical revolution and electricity made the technological revolution. The story of how this came about is the subject of this book. It can be fully understood only when both the steam and electricity revolutions are linked together. Theirs is an epic saga, a big-cast show filled with a variety of stars as well as supporting actors. The story involves dogged quests not only to develop startling new machines but also to explain some of science's most tantalizing mysteries: the nature of heat, light, magnetism, and electricity. In every sense the power revolution was a journey of discovery for all its players. During the story of that journey the spotlight falls sometimes on the players and occasionally on the machines themselves.

When this quest began, no one understood what heat was, or light, magnetism, and electricity. Nor did human beings have at their disposal any source of power greater than human or animal muscle, wind, water, and fire. The curious minds that explored these complex questions happened upon some of the most important scientific principles yet formulated. Their concepts, and the uses to which they were put by others, shaped our modern view of nature and its inner workings. However, they were not alone in their quest. Quite another type of explorer, the inventor blessed (or cursed) with a curiosity all his own, tinkered with the creation of new types of machines despite having only a rudimentary knowledge of the principles that underlay their workings. Gradually the paths of these two parties of explorers began to converge until by the mid- to late nineteenth century their work began at last to move in tandem. By then both the scientist and the inventor had mostly advanced from gifted amateurs to trained professionals who knew how to build on each other's insights.

There is another dimension to the power revolution that often gets overlooked. It required not only scientists, inventors, and engineers but entrepreneurs as well. No invention, however ingenious, could be worthwhile unless someone found a way to make it useful and productive. It had to serve some purpose, which means someone had to make a successful business of it. Inventors soon learned that their creations had to be commercially viable, and they were seldom good businessmen. Some, like James Watt, were lucky enough to find the perfect business partner with those talents lacking in themselves; others were not so fortunate and saw their careers flicker out in the bitter embers of what might have been. Even those who possessed both talents, most notably Thomas A. Edison and George Westinghouse, found themselves ultimately outmaneuvered in the business arena.

The story of the power revolution offers more than an interpretation of the origins of industrial America. It suggests another insight into the most elusive riddle of all: What is an American? Most answers to that question emphasize spiritual and intangible ideals and values such as liberty, freedom, rugged individualism, democracy, and religious purpose. However, one major American historian, David M. Potter, pointed to a much more material source: American abundance.[1]

Material abundance has always been present and even dominant in American life. But industrialization developed its potential to startling dimensions of productivity. A century ago Herbert Croly described the promise of American life as one "of comfort and prosperity for an ever increasing majority of good Americans . . . The general belief still is this: that Americans are not destined to renounce but to enjoy." Industrialization fulfilled that promise on a fabulous scale, thanks to its outpouring of material goods. Even more, it defined the American Dream more vigorously than ever before in material terms until increasingly material things became an end in themselves rather than a means to some larger end.[2]

This transformation could not have occurred without the power revolution. The power revolution was not merely part of the technological revolution that drove industrialization. It was in many respects both a product and the source of that revolution. Electricity became not only the dominant technology of modern times but also the basis for redefining the nature of American abundance. The American version of progress came to rest squarely on the belief that every new generation of technology would bring an array of new products and goods that promised a better life tomorrow than existed today. It rooted American beliefs and values squarely in the premise that the good life was one that revolved around material well-being, which anyone might attain with hard work, perseverance, and some good luck.

Technology rarely comes without a price tag and a learning curve. We are still toting up the one even as we struggle to master the other. The beauty of invention is that it seems like an improbable blend of magic and the obvious. Some inventions, like the can opener, seem so clear and self-evident once someone actually creates them. Others, like electricity, remain a mystery even as they become so integral to our lives that we cannot do without them. In either case the consequences of a new technology can never be predicted. It enters the world and often begins to change first the way things are done and then the way people look at their world. Above all, it creates new mind-sets imbued with an escalating scale of expectations. Few technologies did this to the extent and duration of electricity and its partner in change, the steam engine. To a larger degree than most people care to admit, we have become what our technologies made us. Those who try to reject their influence find themselves no less shaped by technologies than those who embrace the changes they foster. To know the steam and electricity revolutions is to understand how they changed in dramatic form the relationship between people and their technologies.

Maury Klein
Saunderstown, RI

A SHOW OF POWER:
PHILADELPHIA 1876

The watchword of live men—the world's reliance in all that per-
tains to the benefit of the race—is progress. There is no halting, no
supineness, but a constant, unremitting effort to press forward to
new conquests, new discoveries in science, in art, in all that is cal-
culated to prove of value to mankind.

—*PRAIRIE FARMER, MAY 27, 1876*

ONE EVENING IN SEPTEMBER 1876 a nine-year-old boy, call him Ned, got the surprise of his young life. His parents summoned him and asked if he wished to go with his father to the great Centennial Exhibition in Philadelphia. He did not have to be asked twice. Scarcely containing his excitement, Ned boarded a train in his hometown in Iowa on September 10 and watched enthralled for hours as the fields and villages flashed by his window. Next morning he ate breakfast in eastern Indiana and went on to Columbus, Ohio, where his father pointed out the statehouse and the penitentiary with its toylike guardhouses. That night they traveled through Pittsburgh and then Harrisburg until they reached Philadelphia early on September 12. They got off at the Elm Station Depot, about three miles from the entrance to the exhibition, and went at once to the Patrons of Husbandry Encampment to get their room.[1]

The encampment was a cluster of pleasant, newly furnished buildings run on behalf of the Grange for its members visiting the great show. Ned and his father dropped their baggage in their room, tidied up a bit, and hurried to board a train for the exhibition, which was located in Fairmount Park. The ride took only ten minutes and cost eight cents, or fifteen cents round-trip. At one of the entrance gates they each deposited their fifty cents and stepped inside to confront the largest building they had ever seen.

The Main Exhibition Hall was in fact reputed to be the largest building in the world, stretching 1,880 feet in length and 464 feet in width. Its vast interior enclosed more than twenty-one acres of space beneath a ceiling that soared seventy feet. Its framework of iron trusses reminded Ned's father of a gigantic railroad shed. At all four distant corners towers rose seventy-five feet into the air, with ranks of stairs that carried hardy visitors to their summits to view the surrounding sights. Rather than climb the towers, Ned and his father first bought a guide to the grounds and then walked down the left wing to the door.

"We'll come back later," Ned's father said, touching his head. "Let's go see the monster first."

Ned nodded eagerly and followed him through the door. Once outside, they crossed an open space with walkways that converged at the Bartholdi fountain and entered another enormous building. The main cavern of Machinery Hall extended 1,402 feet in length and 360 feet in width, enclosing an area of about fourteen acres. At its center a wing jutted out to the south, some 208 feet long and slightly wider. Both sides of the seemingly endless hall teemed with machinery exhibits of all kinds. The displays overflowed into clusters of annex buildings located beyond both sides of the wing. Ned and his father walked to the center of the main hall, and suddenly there it was. Looming before them stood the greatest wonder of all: the enormous Corliss steam engine. The largest engine in the world, it had been built especially for the Centennial Exhibition. Already its fame had spread throughout the world. Three weeks earlier a New York reporter inspecting the exhibits wrote that he would say little about the great Corliss engine because it had "been written about in every newspaper from Montreal to Melbourne, and is, so far as journalism is concerned, a threadbare topic."[2]

Awed by the sight of this huge yet surprisingly graceful machine, Ned and his father stood in front of it as if before an altar and watched the hypnotic rhythm of its motion. The Corliss sat on a raised circular platform fifty-five feet in diameter and reached thirty-nine feet above the floor. It was actually two separate beam engines sharing a giant flywheel thirty feet in diameter and weighing fifty-six tons. The cranks of both engines connected to the same crankshaft. Instead of using the usual flatbelt around the flywheel to transfer power, it utilized an enormous pinion gear—at thirty feet in diameter it was the largest ever made—that engaged 216 teeth on the flywheel cut so precisely that they turned with astonishing smoothness and ease. The highly polished cranks, made of gunmetal, weighed five tons each.[3]

The pinion gear resided beneath the flooring, where an amazing network of shafts extended to the ends of the building and turned eight main pulleys that were belted to overhead shafting throughout the hall. The main shaft ran 252 feet and was joined at each end and two points in between by four other shafts, each 108 feet long and 6 feet in diameter. The main belting, more than seven miles of it, passed through the hall in obscure places behind glass enclosures and became itself an exhibit of sorts. The twenty boilers needed to feed the giant engine sat in a separate house beyond the main building. They generated 70 horsepower each and fed the engine through a network of riveted, wrought-iron underground pipes. The red brick boiler house, ninety feet high with two ornamented chimneys, became an exhibit in itself with a balcony that enabled visitors to watch its operations.[4]

Through this intricate system the Corliss powered virtually every machine within the fourteen-acre confines of Machinery Hall. Each of its eight main pulleys connected to an overhead main shaft by a double belt thirty inches wide and seventy feet long. A complex network of shafting that utilized four nests of beveled gears, each one six feet in diameter, transferred power from the giant Corliss to all the other machines in the building. Like the main belting, all this lay out of sight beneath the flooring. The effect was spectacular. "Acres upon acres of machinery of every description are set in motion simultaneously," marveled a reporter, "by over seven miles of shafting, which is in its turn moved by the great central engine."[5]

The noise in the building had startled Ned when he entered, but hardly any of it came from the Corliss. The huge flywheel turned its thirty-six revolutions every minute in near silence, producing 1,400 horsepower, although some said it was capable of reaching 2,500 if needed. Novelist William Dean Howells visited the engine and conceded that its "vast and almost silent grandeur" defied description. "It rises loftily in the centre of the huge structure," he wrote, "an athlete of steel and iron with not a superfluous ounce of metal on it: the mighty walking-beams plunge their pistons downward, the enormous fly-wheel revolves with a hoarded power that makes all tremble." On the platform beneath the engine, an engineer sat "reading a newspaper as in a peaceful bower. Now and then he lays down his paper and clambers up one of the stairways that cover the framework, and touches some irritated spot on the giant's body with a drop of oil, and goes down again and takes up his newspaper."[6]

Frédéric-Auguste Bartholdi, the French sculptor who had contributed a fountain to the Centennial and less than a decade later would create the Statue of Liberty, was as enchanted as Howells at the sight. "The lines are so grand and beautiful," he wrote, "and the play of movement so skillfully arranged, and the whole machine was so harmoniously constructed, that it had the beauty and almost the grace of the human form." Poet John Greenleaf Whittier did not share these sentiments. At sixty-nine an uneasy stranger in the presence of the new industrial might, he contributed a hymn to the opening of the Centennial but could not bring himself to visit the spectacle in person. "The very thought of that Ezekiel's vision of machinery and the nightmare confusion of the world's curiosity shop appalls me," he admitted, "and I shall not venture myself amidst it." His was decidedly a minority view, however; the Corliss easily became the most popular attraction of the exhibition. "To this day," wrote historian Robert C. Post, "it remains the prime symbol of the Centennial and still evokes a whole mythology. A march was written to it. It was acclaimed the ultimate manifestation of American technological progress."[7]

And so it was. Industrialization, that most profound change in the history of man, was fast gaining momentum among Americans, a people intoxicated with the prospects for material progress and personal gain that it offered them. Power held the key to the steady march of industrialization—more power from more and bigger machines to perform ever more tasks with ever greater efficiency. The steam engine was the revolutionary source of that power, and the Corliss engine displayed in striking fashion how far its development had come in a remarkably short time.

Twice a day, at nine in the morning when the grounds first opened, and at two (later one) in the afternoon, crowds rushed into Machinery Hall to watch the Corliss start up and jolt the rest of the hall into a loud cacophony of activity. It fell silent at noon each day to allow time for whatever repairs or adjustments were needed to the other machinery in the hall. Machinery Hall also displayed thirty-five or forty stationary steam engines of all types and sizes. Near the Corliss stood an enormous blast furnace with a blowing engine second in size only to the giant that fed it. Beyond it an endless row of sewing machines assailed the eyes, none of them Singers because that company had its own building at the Exhibition. Still other brands were displayed in foreign exhibits.[8]

Beyond the sewing machines lay type-casting machines, telegraph apparatus, fire engines, locomotives, printing presses, belting and bandsaw machines, hoisting and weighing machines, dredging machines, pile drivers, a huge sugarcane mill, a cotton press, the Blake rock crusher, railroad car wheels, giant steam hammers, railway signals and switches, diamond drills, and stone saws. There were machines that cut shingles, made bricks, turned wood, made horseshoes, gummed and folded envelopes, shaped gun barrels, printed wallpaper, made paper, made needles, wound silk, made glue, curled hair, made pins, clarified sugar, made hats, and manufactured metals. Carriages, agricultural machines, files, rasps, wire and cable, screws, car wheels, hardware, and cutlery occupied other space along with other models of steam engines and a demonstration of glass-making. George Westinghouse, who had already made a name for himself in the field of railroad invention, filled six hundred feet of space with an demonstration of his pioneering air brake for railroad cars.

The sheer number of exhibits, coupled with the noise, overwhelmed Ned and his father, yet they could not resist pausing to inspect some of them. They gravitated to a long row of printing presses, at the end of which stood a dilapidated piece of goods that might have easily been overlooked except for the brass plate that identified it as a press at which Benjamin Franklin once toiled. Newer and larger versions put it to shame, especially the mighty Walter press that fed a giant roll of paper weighing nearly a thousand pounds through a series of cylinders and rollers to produce as many as seventeen thousand copies of a newspaper every hour. Invented by the proprietor of the London *Times*, it was pronounced by a *New York Times* reporter "the most perfect printing press known to man." An elderly lady from a rural village watched the Walter performing its magic and said wistfully, "La if my Tom could only see that machine printin' thousands of them papers in a few minutes, and he havin' to set up all night to print a few hundred."[9]

Just outside the southern arm of Machinery Hall stood its cluster of satellite annexes housing boilers, pumps, sawmills, and gas machines, among others. The Campbell Printing Press and Manufacturing Company had its own three-story building exhibiting a complete newspaper office and job printing operation. The Shoe and Leather Building displayed those wares, but Ned followed his father to the Hydraulic Annex, where another spectacle awaited them. Inside the building they found an enormous tub that held half a million gallons of water. Above it loomed another large tank feeding a cataract over which more than thirty thousand gallons of water crashed downward every minute.[10]

They stood watching for a time, mesmerized by the steady flow of water that drowned out all surrounding noise. The whole spectacle overwhelmed them: acres of machines performing an incredible variety of tasks that only a few years earlier had to be done laboriously by hand or not at all, and all of it made possible by the latest, most gigantic version of the first source of mechanical power in human history, the steam engine. Eagerly they turned to the guide and picked out the major buildings to visit besides Main and Machinery: Horticultural Hall, Memorial Hall (also called the Art Gallery), the Agricultural Building, and the Government Building. Then there were the separate buildings for different countries and those for many states and still others for individual companies. Ned's face scrunched into a frown.

"There's so much to see!" he cried.

"We have six days to look at it all," his father reminded him gently. "We don't have to do it all the first day."

THE CENTENNIAL EXHIBITION was conceived as a showcase not only for American pride but for American power as well. Like the Centennial itself, it boasted of the future even as it honored the past. By its hundredth birthday the United States had nearly tripled its original thirteen colonies to thirty-eight states, once Colorado joined the ranks that very year. The new nation's population of nearly four million in 1790 had mushroomed to more than forty-six million. The nation craved a birthday party for many reasons, not all of them pleasant. "The sentiment of the year . . . ," intoned the *New York Tribune,* "is twofold. It is at once jubilant and solemn, congratulatory and sternly introspective." The United States had already begun to emerge in fact, if not yet in recognition, as the largest, most powerful industrial economy in the world. Once an orphan among nations, it had become a force destined to tower above those countries that had earlier sneered at things American.[11]

Of the nation's many accomplishments Americans brayed loudly and often, but in this centennial year they also had much to hang their heads over. The economy was well into its third year of what became its longest and deepest depression—an unexpected by-product of the rapidly evolving industrial landscape. The drawn-out process of reconstructing the southern states dragged wearily on, its loftiest goals long since buried in the mire of mounting cynicism, partisanship, racism, and indifference. Political scandals afflicted government at every level. At a time when the nation lacked a single living ex-president, the incumbent of the White House entered the final year of his second term with his administration disgraced by corruption in numerous offices. State and city officials also faced a withering volley of corruption and mismanagement accusations. In New York City the notorious William "Boss" Tweed had already been hauled into court and found guilty of pilfering millions.

The nation needed and craved a major birthday celebration, but what if it gave a party and nobody came? The idea of an international exhibition was still relatively new; the first one had taken place only a quarter century earlier. The great London Exposition of 1851, which featured the magnificent Crystal Palace, ushered in a new, more ecumenical age of national showmanship. Prior to that event nations had organized expositions but guarded their manufacturing secrets like national treasures, and displays by or even visitors from other nations were rigidly prohibited. But the 1851 Exposition took a new form: "The nations of the earth were to be pitted against each other for the first time. The various countries were husbanding their strength for this great contest. They felt that if the London exhibition were successful, the great exhibitions of the future must be international in character." Despite the usual chorus of naysayers, the London show proved a smashing success and even turned a profit. Of the 13,937 exhibitors, nearly half came from outside Great Britain and her colonies. There followed a succession of expositions—Paris 1855, London 1862, Paris 1867, and Vienna 1873—that varied in style and were generally considered successful.[12]

The European fairs were usually subsidized by governments, but this approach did not appeal to Americans, who participated in the early expositions on a modest scale and without any government support. New York staged a modest international exposition in 1853 without funding or even official sanction from Congress; it flopped and cost the investors about $300,000. The proposal for a centennial exposition came not from Washington but from the mayor and city council of Philadelphia. The federal government offered the project its blessing but not a penny of monetary support. Philadelphia offered Fairmount Park as the site, and the organizing committee set out to raise $10 million through stock subscriptions from the individual states, but that source yielded less than $2.4 million. The city of Philadelphia chipped in $2.5 million, and gifts, concessions, and interest fetched another $330,000. Later Congress grudgingly authorized $1.5 million for the big show, but only as a loan.[13]

Despite the financial shortfall and the usual difficulties caused when doing something by committee, the men in charge—led by Joseph R. Hawley as president—provided capable leadership and performed the seemingly impossible task of organizing and running a large operation free of corruption or scandal. Cynics joked that the Centennial Exhibition was the cleanest operation in Philadelphia. Many visitors also found it to be the most friendly as well. William Dean Howells was moved to report that "there never was on this continent such civility and patience as that of the guards and policemen and officials of the Centennial."[14]

The exhibition ultimately had 249 buildings of all sizes, and the five major structures were open for business from the beginning: Main, Machinery, Agricultural, Horticultural, and Memorial Hall. The latter building ultimately housed the largest display of art the nation had ever seen—so large, in fact, that an annex had to be constructed for the overflow. The main building contained work from twenty nations, an astounding 3,256 paintings and drawings along with 627 pieces of sculpture and 431 of applied art; the annex featured no fewer than thirty-six galleries. Nearby stood a separate building offering the first national exhibit of an increasingly popular technology: the Photography Building, displaying 2,833 groups of photographs. For sheer beauty, however, nothing rivaled Horticultural Hall. The largest conservatory in the world, designed in the Moorish style of the twelfth century, it was "so graceful that it seemed to float above a reflecting pool and surrounding flower beds."[15]

Opening Day saw an immense crowd pile up at the gates long before the nine o'clock opening. When the gates finally parted, a sea of people rushed to fill the space set aside for the gala inaugural ceremonies. All the buildings remained closed until President Ulysses S. Grant declared the Centennial Exhibition officially open. Cash admissions totaled 76,172, but no reliable figures existed for the number of complimentary passes or exhibitors; press estimates of attendance that day ranged as high as 250,000.[16]

A glowering sky that had drenched the city in rain the day before began to break up around seven in the morning. A brisk southwest wind chased the clouds away and dried most of the flags, bunting, and other patriotic decorations that adorned nearly every public space in the city. By ten o'clock the umbrellas brought to fend off rain turned to deflecting sunlight that transformed the dismal atmosphere into a brilliant spring day. As the fair opened, the onrushing crowd swarmed through the thin line of police and occupied every inch of ground, climbed upon the statues of Pegasus and the Muses in front of Memorial Hall, and even found their way to the roofs of nearby buildings.

The crush included an army of reporters who made the exhibition the most covered event in American history. The dignitaries, who filled a platform between Main and Memorial halls, included Supreme Court justices, senators, and governors; military leaders such as General William T. Sherman; Frederick Douglass; and Dom Pedro II, emperor of Brazil, the first reigning monarch ever to visit American soil. He bowed to acknowledge the lusty cheers of the crowd. A 150-piece orchestra played as President Grant and his entourage arrived, accompanied by a grand parade of military units.

After a lengthy performance by the orchestra, including a "Centennial March" written especially for the occasion by Richard Wagner and the hymn by Whittier, Grant delivered a brief speech. Few in the crowd heard anything of what he said, but they cheered loudly after his final words: "I declare the International Exhibition now open." Grant and a gaggle of other dignitaries then made a ceremonial sweep through the exhibition halls, finally to arrive in the transept of Machinery Hall, where a nervous George H. Corliss stood waiting in front of the engine he had built.

At fifty-nine Corliss had reached the pinnacle of an already successful career in a field for which he lacked any formal training. The son of a New York physician, he revealed a flair for engineering. In 1844, at the age of twenty-seven, he went to work for a steam-engine builder in Providence, Rhode Island. Within three years he became head of the firm and launched a distinguished career as not only a builder of engines but an innovator as well. In March 1849 he patented a valve gear that proved to be a major improvement over the slide valves used on most engines. The "automatic drop-cutoff" controlled the amount of steam entering the cylinder far more efficiently than any existing type. The Corliss valve gear gradually became the standard among not only American but also British engine manufacturers. Corliss patented numerous other improvements and established himself as the premier American builder of steam engines. It was a tribute to him that several of the other engines on display at the exhibition were acknowledged copies of his design.[17]

The Corliss firm in Providence was renowned for the quality and efficiency of its engines. Corliss became a member of the Rhode Island Centennial Commission and early in the planning process offered boldly to design, build, and install a giant engine for all the machines in Machinery Hall. The officials demurred, saying it would suggest impartiality if they did not entertain other offers. Besides, they added, what would happen if such an engine suddenly stopped and brought down the rest of the machinery with it? However, when no other engine builders stepped forward with offers to drive so great an array of machines, the officials went hat in hand to Corliss to ask whether his offer still held. In only ten months the obliging Corliss designed, built, and installed the largest steam engine ever devised. It was a monumental achievement for a machine intended to be not only a source of power but a centerpiece in itself.[18]

Now the moment of truth had come. The president and the emperor separated from the rest of their party and stepped onto the raised platform with Corliss. He walked them to the twin levers and explained what they were to do. Dom Pedro took hold of his lever and gave it a turn, after which Grant did the same with his lever. "In a second the steam began to ooze and hiss," noted the *Herald* reporter. "Then came the movement of the awful thing itself, an automatic and regular climbing up and down of many tons of iron and steel to accomplish a mechanical purpose." Grant and Dom Pedro watched as the huge machine lumbered into action and sent its power through the network of shafts and belts to waiting machines.[19]

The president may have pronounced the exhibition open in his speech, but not until the Corliss brought the hall's machinery to loud and clattering life did it truly begin. The exhibition ran for 159 days, a shorter term than earlier fairs because the officers made the controversial and ultimately short-sighted decision to close the grounds on Sundays to honor the Sabbath. Despite this shortened week, the exhibition drew nearly 9.8 million visitors compared to 8.8 million for the Paris Exhibition of 1867, which ran for 217 days and held the previous attendance record.[20]

Through the sweltering summer and deep into the fall, the great show remained open. For a time several newspapers kept reporters on the scene and published almost daily accounts of the exhibits. Gradually, however, other events pushed the Centennial Exhibition off their pages. During June both political parties held their conventions and came up with Republican Rutherford B. Hayes and Democrat Samuel J. Tilden as their candidates. Late that same month George Custer led his 7th Cavalry troops into a bloody massacre by the Sioux. After the first reports trickled in early in July, the stunning news occupied the papers for weeks. The final months of the exhibition overlapped the election campaign, which culminated on November 7 in the most confused outcome yet seen. Tilden emerged with a majority of the popular vote but fell one electoral vote short. The returns from three southern states were disputed, and if the Republicans could capture all of them they would win the White House. As the Centennial Exhibition formally closed on November 10, the nation was stumbling into a bitter and dangerous political impasse.

None of this slowed the flood of visitors pouring through the Centennial gates. Those who came to Philadelphia witnessed a spectacle beyond anything ever seen on the Continent. Most of the visitors received their first and often only introduction to foreign cultures or even different regions of their own country. " 'Land sakes!' " cried the heroine in Sarah Orne Jewett's story "The Flight of Betsey Lane," as she watched a Turk stroll by in his red fez. " 'I call the Centennial somethin' like the day o' judgment!' "[21]

Besides the foreigners attached to each exhibit and the exotic foods served up by the different restaurants, visitors saw a variety of objects and events too vast to be assimilated whole. In Main Hall people could inspect everything from soda fountains to coffins and gravestones. "Every dozen yards or so," complained a *New York Times* reporter, "the visitor stumbles upon a mass of many-colored marble, labeled 'Soda water: all kinds of flavors.' " However, the ubiquitous soda fountain was, as another reporter reminded readers, a "national institution—and is all our own . . . altogether and indisputably ours." Similarly, the impressive display of American-made coffins moved a *New York Sun* reporter to observe wryly that "crops may fall, stocks may rise and fall, big bonanzas may overstock the market with silver . . . but the demand for coffins will always continue steady, with the possible chance of a brisk business growing out of war or pestilence."[22]

An impressive number of exhibits not only displayed their wares but showed them in action. Hordes of visitors flocked to the western end of Machinery Hall to view a complete demonstration of glass-making put on by a Philadelphia firm. With jaws agape they watched the glass move from a huge melting furnace through the entire process to a finished object. The newest wonders of the age beckoned to visitors at every turn. Thomas Edison contributed automatic and multiplex telegraphs as well as a curious "Electric Pen and Duplicating Press." Earlier that spring the ambitious young inventor had boldly set up shop at Menlo Park in New Jersey. On June 25 Alexander Graham Bell arrived in Philadelphia to demonstrate three of his telephonic devices to impressed judges, who pronounced them "perhaps the greatest marvel hitherto achieved by the electric telegraph." However, Bell's instruments remained on display for only two weeks, and it is doubtful whether the relatively few visitors who saw them appreciated their implications. Another exhibit featured the American Typographic machine, an early version of the typewriter, and still another a primitive version of an internal combustion engine designed by George Brayton of New Hampshire. The American Watch Company displayed 2,200 of its wares with their movements exposed, representing six days' output from the factory. The Patent Office itself put on display no fewer than five thousand patent models.[23]

And the steam engines! "Never before was so splendid a collection of steam-engines gathered together as is to be seen in Machinery Hall," marveled the *New York Times* reporter who signed himself A. P. The list began with the Corliss, of course, but it embraced many more—so many, he said, that it would require a special edition of the *Times* to describe them all. Some were on display in the hall; others, like the 65-horsepower engine in the Saw Mill Annex, provided power in other buildings. The variety of engines seemed infinite; the People's Works of Philadelphia alone had on exhibit thirty different models, and several manufacturers showed marine engines as well as pumping, traction, agricultural, motive, and stationary types.[24]

As befitted a Centennial celebration, the great show offered a few glimpses of America's past. New England contributed a log cabin built and furnished as it would have been a century earlier. At Agricultural Hall visitors could see "Old Abe," the aging bald eagle that had gone into battle with the 8th Wisconsin during the Civil War. Named in honor of the president, he had twice been wounded during fights but came home with the regiment and promptly became a celebrity in the state. At Philadelphia he perched majestically on a national escutcheon supported by a pole and delighted crowds with his fierce stare. The Art Gallery offered viewers a painting of the battle of Gettysburg that filled an entire wall.[25]

However, progress, not nostalgia, drove this celebration, and nothing blared progress to the nation or the world more loudly and insistently than the endless array of machines capable of performing thousands of tasks that people once found onerous or even impossible. A reporter, thinking back on the machines on display in London a quarter of a century earlier, dismissed them as "curious relics of a bygone age." The improvement in labor-saving machinery had been phenomenal, and most of it had come from the United States. "A very casual glance around Machinery Hall," concluded the reporter, "is quite enough to assure one of the overwhelming pre-eminence, with some few notable exceptions, of the American display over those of foreign nations." Belgium exhibited a steam engine bearing the royal shield, but it also bore the name of Corliss.[26]

The Centennial Exhibition was a show not only of rising American power but of power itself, and the power behind nearly all its displays turned out to be the greatest wonder of all. That basic power was steam, the true miracle of the age. The steam engine itself was only about as old as the United States, but like the nation it had come a long way in the past century. Steam provided the power that kept the vast acres of machinery humming and churning out a seemingly endless variety of products and materials. Americans did not invent the steam engine, but once they grasped its possibilities they embraced it eagerly and put it to more uses than anyone else. It was a very American form of power—big, bold, bumptious, loud, and utterly lacking in subtlety, at least on the surface.

What could be more American than this gargantuan show of raw power and purpose spiced with a little fun and spectacle? Patriotism infused every corner of the exposition. Howells felt it at once on his visit. "No one can now see the fair without a thrill of patriotic pride," he noted. Especially was this true in Machinery Hall, where "one thinks only of the glorious triumphs of skill and invention; and wherever else the national bird is mute in one's breast, here he cannot fail to utter his pride and content . . . it is still in these things of iron and steel that the national genius most freely speaks." But this glow of national pride was steeped not in nostalgia so much as in promise. Those who shook their head and wondered what "they" would think of next were fast coming to accept the premise that, whatever it was, "they" always *would* think of something next, that the future would bring a steady stream of more and better things.[27]

Hardly anyone, however, realized or even suspected how steady a stream it would be or what a raging torrent it would become. The acres of technology on display in Philadelphia represented nothing less than the genie of material change let loose from his bottle and free to work his magic. In this sense Philadelphia was both a salute and a valediction to America's past. Events and inventions after 1880 would accelerate the pace of change that already seemed a little too fast for some people, and hasten the world of 1876 into oblivion. In less than a generation most of the wonders that enthralled visitors to the Centennial Exhibition, an entire army of "latest things," would become curiosities better suited to a museum than to a celebration of technological wonders.

Once freed, the genie could not be put back in his bottle, which for some people came to seem more like Pandora's box.

BY THE TIME they climbed wearily aboard the train bound for home, Ned and his father had explored nearly every corner of the exhibition. The memories swirled through Ned's mind during the long train ride home. There was so much to remember and to sort out, and some of it was so fabulous as to seem unreal. Of all the wonders he had seen, however, the monster Corliss engine stuck most firmly in his mind. He felt so tiny standing before it, yet it seemed not the least threatening. He marveled at the unseen force it sent through the building, a silent perpetrator of the chaos of clanging that resounded throughout Machinery Hall, and he wondered whether he would ever see the likes of it again.

CHAPTER 1

THE MACHINE THAT CHANGED THE WORLD

England is the birth-place of the steam engine. Its invention has been a grand triumph over the material which nature has placed at our disposal. There is no limit to its sphere of usefulness, nor can anyone measure the benefits which directly and indirectly accrue to society from its employment.

—MICHAEL REYNOLDS (1880)[1]

ON A BEAUTIFUL SUNDAY AFTERNOON in May 1765 James Watt decided to escape the pressure of his work by taking a stroll across the Glasgow Green. Even as he walked along, basking in the warm sun, his mind persisted in mulling over the problem that had tormented him for months. He was intent on creating a new machine, or rather a much improved version of a machine that had already been around for half a century with little having been done to make it better. Little did he suspect that on this magnificent afternoon he would finally capture the idea that later made his name the most renowned in the history of the steam engine.

Born in 1736 in the small Scottish seaport town of Greenrock, the son of a mathematician who taught navigation, Watt was a sickly boy who loved to build models in his father's workshop. After his mother's death in 1753 James was sent to live with her relatives in Glasgow. Two years later, with his father's blessing, he went to London to learn the trade of making mathematical instruments while ducking the press-gangs that roamed the city's streets. Plagued by headaches and bouts of depression that remained his companions for years, he learned enough to return to Glasgow in 1757 and set up his own shop for making instruments. He gained permission to locate his shop within Glasgow College but found it hard going to secure enough business. Two years later, he formed a partnership with a man named John Craig, who provided enough capital to put the business on a solid footing and enabled Watt to move the shop into the city.[2]

As the shop prospered, Watt settled into the comfortable routine of a modestly successful businessman who oversaw several workmen and apprentices. He acquired an interest in a pottery company and in 1764 married his cousin Margaret Miller. By then, however, a seemingly trivial incident had already begun to turn his voracious mind in another direction. A friend at Glasgow College, Professor John Anderson, brought Watt a model of the Newcomen steam engine and asked him to repair it for his natural philosophy class. A well-known London instrument maker had already tried and failed to fix it. Watt had dabbled briefly in steam a few years earlier, and he had discussed with another friend the feasibility of using steam to power a wheeled carriage, but this request was his first close encounter with a steam engine.[3]

It took Watt little time to grasp the operation of the engine, and he not only repaired it but decided to improve its design. He already knew the basic principle that when steam hit a cold object it transferred heat and was condensed in the process. The Newcomen engine first heated the water in the cylinder, then sprayed it with cold water to condense the steam. This process perpetuated a cycle of constantly heating and cooling the cylinder—the most glaring weakness of the engine. Forty years earlier Gabriel Fahrenheit had established with his mercury thermometer that water boiled at 212 degrees. To avoid the loss of steam by condensation, the water in the cylinder should stay boiling hot; however, to create the vacuum needed to lift the beam, the cylinder water had to be cooled to atmospheric temperature, or somewhere around sixty degrees. The problem that intrigued Watt was how to reconcile these opposing demands. If it could be done, the savings in fuel and increased efficiency would be enormous.

He began by conducting experiments to determine at what temperatures water would boil under different pressures, and how much steam at atmospheric pressure was produced from a given volume of water. He calculated that a given amount of water produced 1,800 times its volume in steam, a figure impressively close to the actual one of 1,642. Then he made a new boiler for the model that showed how much steam was used at every stroke and found it to be several times the volume of the cylinder. Another experiment revealed that water, when converted into steam, could heat about six times its own weight of water from room temperature to the boiling point. This finding puzzled him until another good friend, Professor Joseph Black, clarified it.

Black had already staked his own claim to recognition as a chemist. Only thirty-six years old in 1764, he had done pioneering studies in calcium carbonate and carbon dioxide before his interest was drawn to heat, a property that still mystified scientists who still regarded it as an imponderable fluid like light or electricity. From his experiments Black developed the latent theory of heat, which posited that the *quantity* of heat was different from its *intensity*. Only the latter was measured by temperature. When he melted ice, for example, its volume shrank but its temperature remained the same. The ice absorbed some amount of what Black called "latent heat," which meant it contained more heat even though its temperature remained unchanged. He also showed that an even greater quantity of latent heat resulted from boiling water into steam. The heat taken up by water in boiling helped explain why steam at boiling-point temperature had far greater energy than the same amount of liquid water at room temperature.[4]

After talking with Black, Watt realized, as he later wrote, that he had "thus stumbled upon one of the material facts by which this beautiful theory [i.e., of latent heat] is supported." He persevered in his quest to unlock the riddle of how to eliminate the cycle of heating and cooling the cylinder. The idea came to him in a flash during that Sunday afternoon stroll. What if he condensed the steam in a separate apparatus—what he logically called a condenser—thereby enabling the water in the cylinder itself to retain its heat? Watt put it this way: "As steam was an elastic body it would rush into a vacuum, and if a communication was made between the cylinder and an exhausted vessel, it would rush into it, and might be there condensed without cooling the cylinder." It was a brilliant concept, but as with most flashes of insight it took considerable time to work out the details. In this case ten years passed before the idea produced practical results.[5]

IT BEGINS IN the simplest of ways. Someone has a task to do and must figure out how to do it. He looks around for whatever tools are available. If none is adequate or applicable, he seeks some better way than manual labor to perform the task. For all of human existence, the basic sources of energy for doing work have been human or animal muscle, wind, and water. People devised tools such as the lever, the wheel, the pulley, and many others to make tasks easier, but always the scale or scope of what they could do was limited by the amount of energy they could apply to the task. Wind and water could provide greater sources of energy, but the first could not be controlled and the second only partially. One could go with the current but not against it, and the water did not always go where one needed it to go. Fire proved to be an invaluable tool but a limited one until its properties were better understood.

All the achievements of humanity down to about the eighteenth century were constrained by the inability to find more efficient ways to do things beyond the capacity of muscle and tools that, however ingenious, still required muscle to operate them. The marvel of early civilizations is that they accomplished as much as they did within this severe limitation. If it could be overcome, if some new source of power could be created, humanity would find itself in an entirely new world of possibilities. Just as the geological history of the planet separates into different ages, so does the advent of new sources of power divide human history into separate spheres of existence.

Human existence had always been an unrelenting struggle against nature, pitting limited sources of energy against a seemingly endless series of tasks. In early America most people lived on farms that had taken them years to wrench from the wilderness. The farmer had first to clear his land of dense forest by girdling trees, chopping them, pulling up their stumps, and then cutting up the wood for use in building a home and outbuildings as well as for fuel. Rocks and boulders had to be pried up and hauled away to make a fence or simply dumped. For a house he threw up a square building of logs notched at the ends to hold it together and plastered the crevices with clay or mud to keep out the cold. Some of the stones were piled to make a chimney, its smoke wafting up through a hole left in the roof. Another hole might be cut into one side of the house for a window with a wooden shutter to cover it when needed. Most if not all of the home's crude wooden furnishings had to be knocked together by hand.[6]

Once a field was cleared, the new farmer could begin his real work. For tools he had a hoe, a plow, and a scythe—staples that would remain in use until the mid-1800s. With a horse or ox as his helper, the farmer spent an exhausting day pushing against the soil to plow maybe an acre. Then he spent the season planting his crop, cultivating it, and finally harvesting it and preparing it for storage. Afterward some crops had to be ground or milled, often by pounding the grain with a mallet. If a water-powered gristmill was available, the grain had to be hauled to it over nonexistent roads. "I had 14 miles to go in winter to mill with an ox team . . . ," recalled one weary farmer. "No roads were broken and no bridges built across streams. I had to wade two streams and carry the bags on my back . . . I got only 7 miles the first night, and on the 2nd night I reached the mill." Then came the return trip with the milled grain.

For his pains the farmer got simple meals of meat and cornbread washed down with homemade cider or milk if the family had a cow. All of it had to be prepared in tedious rituals: the animals slaughtered, butchered, and salted; the apples picked, dumped into a press, and squeezed. Water had to be hauled in a bucket from a stream or well. The family wore the simplest of garments made from wool or flax. The women of the house toiled for hours spinning, weaving, and cutting clothes for their family, using thread made by twisting short fibers together. Through the long winter months the numbing process of spinning and weaving went on seemingly without end.

Life in the one-room house revolved around the fireplace, which provided heat for warmth and cooking. It was also the only source of light after dark unless the family made or bought candles. In the modern world people are assaulted constantly by noise and light; the hardscrabble farm was a place of silence and darkness once the sun went down. Feeding the fireplace became one of the most laborious chores of all. A farmer had to cut down an acre of trees to supply enough fuel for a year, and every year the trees were farther away from him. The wood had to be chopped and split to fit the fireplace, hauled to the house, and stacked. Kindling had to be gathered and stored. By one estimate a farmer spent a third of his time during the year doing the chores that provided fuel for the house—and over time the supply around him dwindled rapidly. Here, as elsewhere, civilization always came at the expense of nature.

Everyday life cried for fresh sources of power. Its myriad of problems might be solved in two ways: new types of fuel besides wood and new ways to apply the energy created by burning fuel of any kind. Coal became the most important new type of fuel. Its use in England traced back to the Middle Ages, but it was always difficult to get at. Early methods relied on shallow pits that were abandoned as soon as drainage became a problem, and protests over the pollution caused by the burning of coal arose as brewers and smiths began to use it in quantity. The growth of the iron trade in the early eighteenth century spurred interest in coal, which had been used for smelting iron a century earlier.[7]

Quite unexpectedly the quest for coal triggered another breakthrough that ushered in a new era of human history. As increased demand required miners to dig deeper, more water began to seep into their mines. How to rid the mines of water so that the digging could go forward? Tin mines in Cornwall and Devon faced the same problem. Men carried the water out in wooden bowls or turned a windlass or experimented with hand pumps. Some mines tried waterwheels, but a single wheel could not reach deep enough to be useful. Elaborate schemes were devised to haul the water to the surface with multiple buckets attached to ropes pulled by teams of horses or mules, but these arrangements proved clumsy and ineffective. And the deeper the mine, the greater the distance the water had to be lifted. Although the use of coal went back at least to the ancient Greeks and possibly to the Chinese more than three thousand years ago, no one found a way to clear a mine of water until the end of the seventeenth century.[8]

Ironically, the water that posed the problem held the key to its solution. Many people knew that water, when heated, turned into steam, but no one understood what steam was. They knew only that if you heated water in a container, it created a pressure that must be released. If that force could somehow be harnessed during the release, it might be put to work in some way. Since ancient times active minds had toyed with the potential of steam, but these early efforts soon vanished into the mists of history.[9]

With few exceptions, the subject was not revisited until the sixteenth century, when it attracted some attention from two directions: those who developed an intellectual curiosity about the properties of steam and those who puzzled over the practical problem of how to clear mines of water. Thus arose the two broad camps of practitioners that together—though not always in tandem—were to create the greatest revolution in human history. One attacked the science of the matter, seeking to discover the principles governing the phenomena that interested them. The other dealt solely with solving the practical problems that confronted them in some activity. One concentrated on theory, the other on practicum, and the interplay between them ultimately led to the astounding developments that appeared in the nineteenth century.

During the seventeenth century several attempts were made to harness the power of steam. Most of them pursued two lines of experiment. One tried to use steam to force water up a pipe; the other used steam to push against a piston. The marquis of Worcester created the earliest version of what he called a "water-commanding engine." His device amazed a visiting historian from the court of the French king, who saw it "raise to the height of 40 feet by the strength of one man and in the space of one minute of time, four large buckets of water, and that by a pipe or tube of 8 inches." But the witness is silent on the question of whether the marquis used steam to perform this feat. The luckless marquis died impoverished and largely forgotten except by historians seeking the roots of the steam engine.[10]

The quest for a workable power for pumping hardly died with the marquis. Fifteen percent of the 127 patents granted for inventions in England between 1617 and 1642 involved devices for raising water, although only a handful conceived of fire as the means for doing so. Interest in science rose sharply in England after the restoration of Charles II in 1660 and the establishment of the Royal Society three years later. During the 1680s Sir Samuel Morland challenged the prevailing notion, which traced back to Aristotle, that the world was composed of only four elements: earth, air, fire, and water. Morland's experiments led him to infer that the air created by boiling water was somehow different from ordinary air. For one thing, it occupied a much greater space. As he put it, "The vapours from water evaporated by the force of fire demand incontinently a much larger space (about two thousand times) than the water occupied previously."[11]

In 1643 the redoubtable Galileo Galilei and Evangelista Torricelli had demonstrated that water could be drawn up a pump with a piston no more than twenty-eight feet because at that point the pressure of the atmosphere failed to sustain the column of water. They also observed that if the piston was lifted any higher, a vacuum formed beneath it. Several people promptly began to explore the nature of this vacuum. Drawing on some well-known experiments by German scientist Otto von Guericke, Robert Boyle in 1658 fashioned an air pump for creating vacuums that remains the prototype for those still in use. Using this pump, Boyle came up with the law of the pressure of gas that bears his name. Denis Papin, a French Huguenot who had sought refuge in England in 1675, created a device that used steam pressure to force a piston up a cylinder to a catch located at the top. When the cylinder cooled, a vacuum formed inside it. As the catch was released, the atmospheric pressure forced the piston down the cylinder and raised a surprising amount of weight in the process.[12]

Interesting as these and other experiments were, they had yet to result in an engine capable of doing practical work. Thomas Savery attempted that first giant step in 1698 when he received a patent for the "Raising of Water . . . by the Impellent Force of Fire, which will be of great use and Advantage for Draining Mines, Serving Towns with Water, and for the Working of all Sorts of Mills where they have not the Benefit of Water nor constant Winds." He then set up shop to offer his engines to "all Proprietors of Mines and Collieries which are encumbered with Water." In a promotional piece entitled *The Miner's Friend* Savery claimed that "consideration of the difficulties the miners and colliers labour under by the frequent disorders, cumbersomeness, and in general of water-engines, encouraged me to invent engines to work by this new force." Although Savery made improvements to his engine, he never achieved commercial success. It had several fatal limitations, one being the inability to raise water more than about fifty feet. For mines of any depth, this would require an engine every fifty feet. It was also inefficient and costly to operate.[13]

Where Savery failed, Thomas Newcomen succeeded. An ironmonger by trade, he built in 1712 what came to be regarded as the first successful steam engine near Dudley Castle in Staffordshire. Unfortunately, Newcomen remains in the shadows of history. Little is known about his life other than his birth at Dartmouth, Devon, in 1663, his trade, and his death of a fever in 1729 at a friend's house; no pictures and hardly any documents by him survive. Accounts differ on whether Newcomen knew Savery or ever saw his engine, but he had sold tools and ironworks to the mines at Cornwall and Devon, where the miners struggled vainly to haul water out of the deeper shafts. No one knows why he decided to undertake the task of creating a better pumping engine or how much scientific knowledge he possessed. There is no evidence that a mine owner or anyone else underwrote the cost of his work, but sometime around 1700 Newcomen began grappling with the problem of building a better steam engine for the explicit purpose of pumping water from the mines. What is clear is that he produced the first atmospheric steam engine capable of doing useful work.[14]

At first glance Newcomen's engine seemed little more than a combination of familiar parts, such as a cylinder, piston, pump bucket, pump handle, and boiler. But Newcomen arranged them in a new alignment and added crucial touches of his own, including a tank atop the engine house to inject cold water directly into the cylinder and a valve gear that controlled the injection. It worked this way. A boiler fed hot water to the cylinder that was attached to a beam constructed so that it was heavier on the opposite, or pumping, end. Gravity pulled the beam down to the water to be hauled away. To raise the load of water, the cylinder was filled with steam and then injected with cold water to condense the steam. This process created a vacuum that pulled the steam piston down and raised the main pump piston. At the bottom of the steam piston stroke, a valve opened to return the cylinder to atmospheric pressure. This step sent the beam back down by gravity, and the process was then repeated.[15]

Newcomen's engine pumped water successfully, and its action was self-repeating, although it needed careful adjustment. It proved successful and spread beyond England to the Continent. However, Newcomen profited little from his great invention. To his dismay, the patent on the Savery engine had in 1699 been extended an extra twenty-one years beyond the normal fourteen, and its broad terms covered Newcomen's device as well. This unhappy fact forced Newcomen to build his engines under various partnerships with the holders of the patent. Despite this handicap, he continued to build and improve his pioneering engine. By the time of his death in 1729, he had constructed more than a hundred engines in England, to say nothing of those at work abroad.[16]

It was Newcomen's fate to be a prophet without honor, at least in his lifetime. Like every inventor, he had to work within the confines of the materials and technology available to him. Savery's attempt to utilize steam at high pressure, for example, required stronger boilers and tighter seams than could be devised. Newcomen avoided this issue by keeping the steam at lower pressures, but he encountered other problems, such as the inability to bore cylinders accurately. Like artists working within the limitations of a given medium, inventors pitted their knowledge of science and/or their ingenuity against the limits of technological possibility. Here too the steam engine would effect nothing less than a revolution by expediting the development of machine tools that would transform not only what could be made but also how things were made and how many of them could be made. Newcomen's contribution to that process proved invaluable. His steam engine, while limited in many respects, remained the prototype of the only effective pumping engine for sixty years.[17]

While numerous inventors struggled to improve the steam engine, advances in textile technology were creating new demands for more and better forms of power. In 1700 the English textile industry had only three machines that required power. As the cotton industry expanded, however, inventors fueled its growth with new machines capable of increasing production. Some of these required only hand operation, but the water frame, patented by Richard Arkwright in 1769, specifically called for water power. Six years later Arkwright devised machines for carding, drawing, and roving cotton, which mechanized the preparation of cotton for his water frame. These inventions set the stage for an explosive growth of the cotton textile industry and its transformation from domestic to factory production. To get power, most mill owners simply located their factories near a river and used waterwheels. A few, however, investigated the possibility of using steam engines as a power source.[18]

Textile mills posed special problems for builders of steam engines. The engine had to be utterly reliable and run with absolute regularity to ensure a constant speed against a varying load every hour of every working day, week after week. Although some mill owners adopted versions of the Newcomen and even the Savery engines to their needs, a strong demand existed for an improved type of engine. A factory engine working long hours had to be not only reliable and durable but cost-effective as well. It had to utilize both steam and the fuel needed to create it as cheaply as possible. And, of course, it had to generate enough power to perform the task at hand.[19]

The Newcomen and its predecessors were reciprocating engines; that is, they utilized a piston connected to a beam that moved up and down. The piston strokes did not move up and down evenly because the engine did not use steam pressure directly to push the piston upward. This lack of regularity posed no problem for pumping water, but it was a fatal flaw for factory work, where evenness of motion was required. Unlike the Savery, the Newcomen was an atmospheric steam engine, so called because its greatest steam pressure stayed near that of the atmosphere. This meant that the steam cylinder of the Newcomen engine had to be heated and cooled with every stroke, which wasted fuel and also caused thermal stress. John Smeaton made several improvements that doubled the efficiency of the Newcomen engine but could not surmount its inherent limitations. Maximum economy required that the cylinder be kept hot at all times, but to get maximum power, it had to be cooled down once every cycle. How to resolve that dilemma?[20]

WHILE WATT STRUGGLED to build a model utilizing his insight, the need to earn a livelihood kept interrupting his experiments. He finally obtained a patent for his "method of lessening the consumption of steam and fuel in fire-engines" in 1769. To finance his work, Watt formed a partnership with John Roebuck, an industrialist and chemist who pioneered the manufacture of sulfuric acid for industrial use. After seeing Watt's engine, Roebuck agreed to shoulder the inventor's debts of about £1,000, mostly to Joseph Black, and advance funds to obtain a patent and continue experiments in return for a two-thirds interest in the patent. Watt happily accepted this offer because it encouraged him to devote more time to his engine. Shortly afterward Watt met another industrialist, Matthew Boulton, who grasped the potential of the engine at once and asked to take a share in it.[21]

Watt took the offer to Roebuck, who had also come to believe strongly in the engine's future. "The nearer it approaches to certainty," observed Watt of Roebuck's attitude, "he grows the more tenacious of it." He put Boulton off with an offer allowing him to build the engine in three counties. Boulton declined this pittance; he had a much grander vision. "To . . . produce the most profit," he wrote Watt, "my idea was to settle a manufactory near to my own by the side of our canal where I would erect all the conveniences necessary for the completion of engines and from which manufactory we would serve all the world with engines of all sizes." This proposal must have taken Watt's breath away; the engine, after all, had not yet been perfected. In response Watt offered to give Boulton an engine needed for grinding work at his Soho plant, but the press of other duties kept him from actually building it. Boulton kept using horses to turn his wheel at Soho, but he also continued to track Watt's progress.[22]

It came slowly. Watt could not devote full time to the engine because he still had to earn a living as a civil engineer doing surveys. He also had to cope with poor workmanship, inferior materials, and a lack of craftsmanship. Himself a newcomer to engines, Watt's only advantage, besides a brilliant mind, was a lack of preconceived notions about what an engine should be. Two years of thought and experiment passed before he decided to utilize the same type of beam engine Newcomen had made, and more time was required to conceive and implement the structural changes he deemed necessary. The condenser was far from the only innovation. To keep the cylinder hot, he used boiler steam directly by eliminating the water from the top of the cylinder and enclosing the top in a cover with a stuffing box through which the piston moved. Since the water he removed had also served as packing for the piston, he devised a substitute form. His condenser utilized a water jet similar to that in the Newcomen engine, but Watt had to adapt an air pump to clear the condenser of air and water.[23]

In slogging through these problems, Watt had to combat his own bouts of gloom and discouragement as well. "I find I am not the same person I was four years ago when I invented the fire engine and foresaw even before I made a model almost every circumstance that has since occurred," he wrote dejectedly in April 1769. "I was at that time spurred on by the alluring hope of placing myself above want without being obliged to have much dealing with mankind to whom I have always been a dupe . . . I have now brought the engine near a conclusion, yet I am not nearer that rest I wish for than I was 4 years ago; however, I am resolved to do all I can to carry on this business . . . Of all things in life there is nothing more foolish than inventing."[24]

Yet he persisted, as he resolved to do. A new opportunity arose unexpectedly when the British economy turned downward during 1772–73. Roebuck was a man of many interests who tended to take on more obligations than he could sustain. The tightening of credit came after he had invested much of his own and his wife's fortunes in coal mines at Bo'ness on the Firth of Forth. He had offered Boulton a tenth interest in the mines, but Boulton had declined. In 1773 the overextended Roebuck fell into bankruptcy. One of his creditors was the partnership of Boulton & Fothergill, to whom he owed £1,200. Boulton appointed Watt as his agent for the settlement of accounts. This seemed a curious choice since Watt hated to negotiate anything. "I would rather face a loaded cannon than settle an account or make a bargain," he admitted. "In short I find myself out of my sphere when I have anything to do with mankind."[25]

Despite Watt's reservations, Boulton came away with all rights to the engine patent because none of the other creditors saw any value in it. His partner, John Fothergill, wanted no part of it either and took only cash for his share of the debt, leaving the patent entirely in Boulton's hands. Fothergill would later regret that decision. Then, in September 1773, Watt's wife died, leaving him with two small children. Steeped in gloom and utterly discouraged, Watt threw up his surveying work and Scotland itself, saying, "I am heartsick of this cursed country." He accepted Boulton's invitation to come to Birmingham and devote himself to work on the engine. In May 1774 he took his family south to Birmingham and, at thirty-eight, entered into what he hoped would be a fresh start in life.[26]

Thus began a collaboration that was to succeed beyond either party's wildest dreams. By temperament and personality they could scarcely have been better matched. The genial, tactful Boulton possessed genuine intellectual curiosity and a broad sense of vision. He liked to surround himself with brilliant minds in all fields. His circle of friends ranged from Erasmus Darwin to Josiah Wedgwood; he had met and maintained a correspondence with Benjamin Franklin, to whom he once sent a Savery engine and asked for suggestions on how to improve it. Boulton also had a shrewd business sense, something the hardworking but often impatient Watt lacked. Plagued by ill health and bouts of depression, driven by the need to scrounge a living for his family, Watt had never had the time or the means to devote his full attention to invention. Boulton gave him that opportunity with results that profited both of them immensely. The terms called for Boulton to receive two thirds of the net profits and rights to the patents while paying all expenses for the work. He oversaw the business, and Watt tended to the design and construction of the engines.[27]

The ever practical Boulton realized that Watt's patent had only eight years to run, too little time to bring an engine to commercial success. By strenuous efforts they managed in May 1775 to get the patent extended for another twenty-five years. Nine days later they formed the partnership of Boulton & Watt to last for the same twenty-five years. In May 1776 the first two Watt engines were installed and proved successful. A month later Watt remarried. With a new wife and the congenial work he had long sought, his health began to improve. Eventually he also found a first-rate assistant in the person of William Murdock, a shy young Scot who was to play a major role in several inventions for which Watt received credit.[28]

News of the Watt engine spread rapidly, and Boulton proved masterful at the art of marketing. "I sell Here," he once proclaimed to a visitor to his works, "what all the world desires to have . . . POWER!" The pivotal innovation of the condenser improved fuel efficiency by an astounding 75 percent. Shrewdly the partners decided not to license their engines but to charge a royalty amounting to one third of the savings on fuel realized by them. Mine operators willingly paid this sum to acquire an engine that not only performed reliably but saved them money as well. They paid only when savings were realized, and the money came from the savings. Even so, they complained about having to pay a continuing royalty instead of being able to buy the engine outright.[29]

Watt had dramatically improved the reciprocating engine by adding the condenser and an improved cylinder, but it still could do little more than pump water. The need of the burgeoning textile industry for a reliable engine to drive its machines directly attracted the attention of several inventors, who worked furiously at the task of designing a rotative engine capable of the evenness of operation required for such work. In 1779 Matthew Wasborough (or

Wasbrough) received a patent for an engine that included a flywheel intended to "render the motion more regular and uniform." The concept of a flywheel had been suggested nearly forty years earlier by John Wise, but no one picked up on it until Wasborough included it in his design. Wasborough produced an engine for James Pickard of Birmingham, who found the connecting "rick-rack" mechanism unsuitable and replaced it with a crank and connecting rod. Both innovations were reputed to have been stolen from Watt by workmen in his employ.[30]

The pairing of this crank with a flywheel opened the door to the creation of a workable single-acting rotative atmospheric engine. Numerous inventors built and sold simple versions of the rotative atmospheric engine, taking care to skirt existing patents through minor design changes. By 1789 they had become so numerous that a customer inquiring about a Boulton & Watt engine complained that "we already have a great number of the common old smoaking Engines in & about the Town which I confess are far from being agreeable—& the public yet are not all inclined to believe otherwise than that a Steam Engine of *any* sort must be highly offensive." Nevertheless, other builders were designing rotative engines that got around Watt's condenser patent. Boulton saw both danger and opportunity in their doings, and he understood clearly where the market was. "The people in London, Manchester and Birmingham are *steam mill mad*," he wrote Watt in June 1781. "I don't mean to hurry you but I think . . . we should determine to take out a patent for certain methods of producing rotative motion from . . . the fire-engine."[31]

Watt still believed the strongest market lay with pumping engines but conceded that "the devil of rotations is afoot" and was already hard at work on improvements to the simple rotative engine. The result was a series of patents between 1781 and 1784 that, along with the condenser, represent the peak of Watt's creative genius. The first of these in 1781 contained five devices for doing away with Pickard's crank. The title of his application explained them as "new methods of applying the vibrating or reciprocating motion of steam or fire engines, to produce a continued rotative or circular motion round an axis or centre, and thereby to give motion to the wheels or mills or other machines." Only one of these came into general use, but that one, the so-called sun-and-planet or epicyclic gear, enabled Watt to avoid infringing on Pickard's patent and became well known in its own right. It had other advantages as well, such as allowing the output shaft to rotate at twice the engine's speed and smoothing out some variations in the piston's movement.[32]

A year later Watt received another patent for some contrivances that sought to give pumping engines a "smoother and more economical cycle." Some of them, intended to equalize the power stroke, turned up in various forms as part of the operation of valve gears in engines designed by George Corliss and Charles T. Porter more than half a century later. Watt also pointed to ways of using steam expansively in the cylinder, something made possible by the presence of his condenser that kept the cylinder hot. He did this by cutting off the steam before the piston finished its stroke and allowing expansion of the steam to complete the stroke. This arrangement produced even greater fuel economy, as Watt showed in an accompanying diagram with calculations he had made. He did not pursue his improvement because the gains in economy were modest for an atmospheric engine, but the idea became crucial later when high-pressure steam engines came along.[33]

Watt also suggested using the steam to move the piston both up and down, creating a double-acting engine. This idea actually appeared as early as 1775 in the papers Watt submitted to Parliament for extending his 1769 patent. Other inventors had proposed using two cylinders, working alternately, to create a more even turning force. To connect the piston rods of the two cylinders, a Dr. Falck in 1779 devised a rack moving a gearwheel, but it lacked sufficient evenness of operation. Watt tackled the problem of enabling the piston to move up and down smoothly "without chains or perpendicular guides or untowardly frictions, arch heads or other pieces of clumsiness." The result of his labors was cleverly buried in an elaborate patent of 1784 that, as historian Richard L. Hills noted, covered "a steam road carriage, a rotary engine and at least three other principles for obtaining straight line motions."[34]

The key element was a full parallel motion that combined Watt's three-bar motion with a pantograph. Within a small space the parallel motion allowed the piston rod to directly join the end of the beam moving in a straight line. It was flexible in that more links could be fitted easily within the existing framework of the engine. The design was both simple and elegant and did not require any advanced engineering. Hills called it "the final link in the evolution of the rotative engine." Watt himself referred to it as "one of the most ingenious, simple pieces of mechanism I have contrived" and wrote much later, in 1808, that "though I am not over anxious after fame, yet I am more proud of the parallel motion than of any other mechanical invention I have ever made."[35]

As the business of Boulton & Watt flourished, Watt continued to refine his engine while steadfastly refusing to license it to others. In 1787 he created a clever "centrifugal governor" that automatically controlled the output of steam. This ingenious device, moved by the steam itself, whirled about a vertical rod. The faster it went, the more centrifugal force flung outward the two metal balls attached to it. As the balls moved farther apart, they reduced the steam outlet. As the steam output decreased, the whirl of the governor slowed, the balls dropped, and the steam outlet widened. Steam output always stayed within the range of the ball's movements. Although no one yet realized it, Watt's governor opened the door to an astonishing future. It marked the dawn of automation in that it controlled a process by using variations in the process itself.[36]

Sometime around 1790 Watt devised an "indicator," the first crude instrument for measuring the work done inside the cylinder. His version of this crucial appendage was merely a pointer that moved over a slide; later one of his assistants, John Southern, invented an improved version that traced a diagram of the readings as they occurred. The indicator and its successors were but one more example of the process that propelled technological innovation. The appearance of Watt's engine, and every one of its components, inspired other inventors to improve upon it or come up with an entirely new design that found solutions to problems yet unsolved in the original. Watt had not merely invented an engine that made pumping more cost-efficient and galvanized the textile industry; he had launched the power revolution. In 1878 Robert H. Thurston estimated that "the total steam-power of the world is about 90,000,000 horse-power"; that figure would be dwarfed in succeeding years. Without suspecting it, Watt had set in motion forces that literally changed the world. Every material achievement that would characterize civilization during the next two centuries began with the possibilities opened by the steam engine as prime mover.[37]

Watt also left another legacy, this one of measurement. No method existed for determining the power output of a steam engine or, for that matter, knowing how much power a given textile machine needed to operate properly. Savery invoked the analogy to horses by claiming that his engine could do the work of ten or twelve horses. Marten Triewald, a Swedish engineer, and John Smeaton also tried to gauge the work done by early engines in terms of the number of horses working in shifts for continuous operation. But no one agreed on just how much power a horse put out. Smeaton thought one horse equaled five men; other writers said six or seven. Then Smeaton and John Theophilus Desaguliers calculated how much weight a horse could lift a foot high in one minute. When Watt took up the problem, he rejected both their figures. His tests showed that a horse could raise a 150-pound weight nearly four feet in one second. On this basis he defined 1 horsepower as 550 pound-feet per second, or 33,000 pounds per minute. This figure has remained the standard.[38]

The firm of Boulton & Watt endured until 1800, when it gave way to a new partnership formed by the sons of both men. By that time it had built 496 engines, of which 164 did pumping, 24 were blowing engines, and the remaining

308 drove textile machinery. Boulton's vision that the future belonged to the rotative engine proved both accurate and profitable. The partners, close friends to the end, retired wealthy and honored for their contribution. Boulton died in 1809, Watt a decade later. As Hills declared, they had created "one of the crucial machines which helped to launch the Industrial Revolution . . . a design which other manufacturers found they had to copy or else face failure." It had taken years of trials, modifications, and innovations to perfect, and many scientists as well as ordinary people thought the engine beyond improvement. However, other inventors in workshops throughout Europe and the United States toiled at the challenge of moving beyond the limitations of Watt's rotative engine.[39]

CHAPTER 2

CONQUERING THE WATERS

The time will come when people will travel in stages moved by steam-engines from one city to another, almost as fast as birds can fly, 15 or 20 miles an hour . . . A carriage will start from Washington in the morning, the passengers will breakfast at Baltimore, dine at Philadelphia, and sup in New York the same day . . . Engines will drive boats 10 or 12 miles an hour, and there will be hundreds of steamers running on the Mississippi.

—OLIVER EVANS (1813)[1]

THE ADVANCE OF STEAM POWER proceeded in stages, each one marked by its own cluster of demands and problems, the solutions to which were always confined by the existing state of technological possibility. Newcomen had created an engine capable of removing water from mines or pushing it through waterworks but little else. Watt's engine met another pressing need for a reliable source of power to operate textile machines with unflinching regularity. He had freed textile manufacturers from the need to be near running water to power their looms and thereby expedited the rise of the factory system, those "dark, Satanic mills" as William Blake called them. Both the Newcomen and Watt were atmospheric engines that relied on relatively low pressures even though this limited the amount of power they could produce. They were also large mechanisms that were installed in engine houses and not at all portable. But the very existence of these engines suggested other possible uses. If steam could power machinery, why not a vehicle moving over land or sea?

The most obvious limitation to Watt's engine was the need for so large an apparatus to produce so little power. An engine using higher pressure could develop the same amount of power in a much smaller package; it would not need a condenser or cooling water. Smaller size gave it both portability and flexibility of use even for modest tasks. It worked not by using the vacuum produced by condensing the steam but by using the steam pressure directly, at higher than atmospheric pressure, and then discharging the used steam into the atmosphere. By acting directly it did away with the ponderous beam, complex valve gear, and air pump as well as condenser. Getting rid of the vacuum meant poorer fuel economy, but superior performance and the advantages given above offset that loss.[2]

However, higher pressure increased the danger of boiler explosions at a time when the technology of boiler construction still lagged. Working parts might be slightly off line, bearings not quite true, other parts not quite square. More precision in parts awaited the development of better materials and machine tools. Because of the danger of explosion, and the fact that it could do without his condenser, Watt showed no interest in developing a high-speed engine. Neither did most inventors. "Although the dead hand of Watt was removed," wrote historian H. W. Dickinson, "his tradition, amounting among engineers almost to a fetish, held sway and the rotative beam engine continued to flourish . . . It became almost standardized." But not entirely. A few hardy souls tackled the problem of designing a workable high-pressure engine.[3]

William Murdock, Watt's loyal and ingenious assistant, had experimented with a high-pressure engine and built small working models of a steam carriage. Watt included the steam carriage in his 1782 patent to appease Murdock but took a dim view of both its prospects and the high-pressure engine. Four years later he expressed displeasure at Murdock's continued attention to the steam carriage. Another talented English inventor, Richard Trevithick, took up the challenge. The son of a Cornish mine captain, he apprenticed under Murdock before setting himself up as a consulting engineer. Aware that the Cornish mine owners were eager to get out from under their royalty payments to Boulton & Watt, Trevithick devised an early high-pressure engine for mine use. Called "puffers" because they vented used steam directly into the atmosphere, they achieved some success. But Trevithick did not stay with them; instead he turned his attention to building early versions of the locomotive.[4]

Trevithick realized that the success of any high-pressure engine depended on the design and construction of better boilers. He knew that Savery's early version had failed for want of a suitable boiler. The atmospheric steam engine had no such problem; its boiler need only be strong enough to hold the weight of the water inside it. However, the high-pressure engine, by using the steam directly, built up immense pressure inside the boiler. Steam had 1,642 times the volume of water. No one yet knew this figure, although Watt had come remarkably close with his estimate of 1,800. Such pressure required a boiler that was strong and tight, a reliable monitor to measure the temperature inside, and valves capable of releasing the pressure when needed.[5]

Few people understood how boilers functioned or what materials were best suited to keep them tight at high temperatures. The process of creating steam power began with a furnace to heat the boiler. If coal was used to heat the furnace, it mattered greatly what kind of coal and what burning properties a given type possessed. The design of the furnace had to accommodate the type of coal so as to burn it in the best manner to get maximum heat at minimum cost of fuel. Then the heat and gases had to be transferred to the water inside the boiler as efficiently as possible. From the boiler the steam moved to the engine, again with a minimum loss of heat and pressure. The boiler needed a steady water supply, and the amount had to be monitored. Safety valves were needed to permit the escape of excess steam and prevent explosions.[6]

Each one of these steps required new materials and techniques as well as better understanding of how heat actually worked. Boiler design soon became an art in itself. Solving the problems inherent in their use took the better part of the nineteenth century, which is why Newcomen, Watt, and other engine designers avoided the major problems by sticking with the low-pressure atmospheric engine. Undaunted, Trevithick attacked the problem head-on by building a cast-iron boiler in 1803 and then creating an enlarged version around 1811. However, Trevithick never patented his boiler or pursued his engine work; instead he turned to other interests. His adventures took him eventually to South America, but none of his projects proved successful, and he died impoverished in 1833. "He was ingenious, a thorough mechanic, bold, active, and indefatigable," wrote historian Robert H. Thurston, "but his lack of persistence made his whole life, as Smiles has said, 'but a series of beginnings.'"[7]

The most improbable pioneer proved to be an American whose genius and talents were exceeded only by his obscurity. Intellectual curiosity constantly drove Oliver Evans, as it did Watt, to the challenge of solving seemingly intractable technical problems. Like Watt, too, he rose from humble origins to become a leading innovator in new technologies. However, the wealth and renown heaped on Watt in England escaped Evans almost entirely. Misfortune dogged his career, yet he rivals Eli Whitney in his contributions to American technical progress. Evans was doomed to suffer the classic disappointments of the inventor whose visions far outreach the technology and materials available to realize them. He remains the earliest and most intriguing prophet of American material progress.

Little is known of Evans's early years. Born in 1755, the fifth of twelve children of a cordwainer and farmer in Newport, Delaware, he somehow gained a semblance of education at a time when public schools did not exist and only the well-to-do could afford private ones. All seven of his brothers stayed with farming for their livelihood; Oliver alone showed an interest in science and a talent for things mechanical. At sixteen he apprenticed himself to a wheelwright and probably learned the rudiments of mechanics from him. When he was seventeen, a trivial incident drew his attention to what became a lifelong fascination. One of his brothers told him that for fun a friend, the son of a blacksmith, had "stopped up the touch-hole of a gun-barrel, then put into it about a gill of water, and rammed down a tight wadding. When they put the breech-end of it into the smith's fire, it discharged itself with as loud a crack as if it had been loaded with gunpowder."[8]

The boys cackled with glee, but the story had quite a different effect on Oliver. "It immediately occurred to me," he wrote later, "that there was a power capable of powering any wagon, provided that I could apply it; and I set myself to work to find out the means of doing so." One may doubt that revelation and determination came this quickly to Evans, yet it bespeaks a pattern typical of creative minds: an unexpected epiphany that sparks both an interest and a chain of reasoning followed by a long and arduous period of experimentation to realize the vision. Evans scarcely knew where to begin. He labored fruitlessly for a time until he happened upon a book that described the atmospheric steam engine. "I was astonished to observe that they had so far erred as to use the steam only to form a vacuum, to apply the mere pressure of the atmosphere, instead of applying the elastic power of steam, for original motion," he recalled, "a power which I supposed was irresistible."[9]

Thus began Evans's quest to design a high-pressure steam engine, a quest that often seemed as futile as the search for the Holy Grail. "I . . . soon declared that I could make steam wagons," he admitted, "and endeavoured to communicate my ideas to others; but, however practicable the thing appeared to me, my object only excited the ridicule of those to whom it was known. But I persevered in my belief, and confirmed it by experiments that satisfied me of its reality." Here again is the familiar tale of an inventor who is deemed a crackpot by others because his idea flies in the face of conventional wisdom, and who brushes aside ridicule to pursue his quest (some would say obsession). For Evans, who seemed always to be at the cutting edge of new technology, it became a lifelong pattern.

What is most remarkable about Evans is how early he came to his most important inventions, or at least the concept of them. He made small working models, but without financial backing he could not construct a working engine, and he labored in vain for thirty years to get a sponsor for the work. Like Watt, he did not have the luxury of devoting full time to his creative work but had to earn a living as well. Unlike Watt, he found no Matthew Boulton to smooth his path to commercial success. Evans had to go it alone, which meant that to survive he had to turn his attention to creating devices that might find a market. The American Revolution was in full swing when, at age twenty-two, he undertook the difficult task of making fine wire from American bar iron and then building a machine to card cotton and wool. His father and brothers heaped ridicule on him for wasting his time with such harebrained schemes. Even the blacksmith he tried to hire laughed at him and refused to work until plied with rum by a mutual friend.[10]

Finally Evans found a mill owner who asked him to create a machine capable of manufacturing five hundred complete card teeth per minute. For this device he offered Evans a measly two hundred dollars; in return, Evans was to keep the machine secret for two years and could build only one other like it. Evans produced a machine that turned out not five hundred but three thousand card teeth a minute. Delighted, the manufacturer also bought the other machine Evans was allowed to construct. The inventor had no patent, the secret of his machine soon got out, and he realized nothing more on his invention. Nothing more is known of his life during the war years other than that he was enrolled in the local militia along with two of his brothers. In 1782 he turned up in a village on Maryland's Eastern Shore operating a small store with his younger brother Joseph. The store did well, and the following year Evans married Sarah Tomlinson, the daughter of a Delaware farmer. To outward eyes he seemed at last to be settling down. In fact, Evans was on the verge of perfecting his first great invention.[11]

His time in Delaware and Maryland brought Evans into contact with several gristmills. Appalled by the inefficiency of the process, he decided to invent machines that would expedite the cumbersome process at every stage. In 1782 Oliver persuaded his father to sell part of his farm to him and two of his brothers. There Evans erected a prototype mill based on no fewer than five new inventions. As early as 1783 he told several people that he intended to build a mill capable of taking grain from the sack and transforming it into flour without any manual labor. No one believed him. By September 1785 Evans had the mill up and running. Other millers saw the advantage of each improvement as it went into service and began using them. Some paid a small royalty to Evans; others did not. Since the new United States government did not yet have a patent law, Evans secured patents from Delaware, Maryland, New Hampshire, and Pennsylvania in 1787. When the federal government passed a patent law in 1790, Evans applied for and received U.S. Patent No. 3. One of his first licenses went to none other than George Washington.[12]

In 1795 Evans publicized his innovation by publishing *The Young Mill-*

Wright and Miller's Guide, a book that went through fifteen editions. Still few people realized the significance of what Evans had done. He had in fact created a fully automated mill in which all the work was done by machines of his invention. Apart from being the first to automate a complete production process, he had transformed what became the most important industry in the young country. Millers bought grain in quantity from farmers, turned it into meal and flour, and sold it to wholesalers, who sold it to markets both at home and abroad. By 1840 the nation had about twenty-four thousand gristmills in operation, and most of them utilized some or all of Evans's innovations. The total value of flour produced soared from $14 million in 1810 to nearly $250 million in 1860, making it the nation's leading industry even though, thanks to Evans, it employed fewer people than other industries. The flour-milling industry became a sign of things to come.[13]

By revolutionizing a major industry, Evans struck at the heart of a great national problem. Labor, especially skilled labor, was chronically scarce and therefore expensive. The great value of machines and mechanization was that they dramatically increased output while minimizing the number of workers required. Evans pioneered the process that would characterize the economic revolution of the nineteenth century: the substitution of machines for workers. Machines not only did the work of many hands but also required fewer skilled workers to tend them. Their proliferation threatened skilled craftsmen in many fields with extinction, but they also increased both the amount and quality of output by astronomical sums.

Once again, however, Evans profited little from his ingenuity. Although he held the patents, millers did not always honor them or pay the required licensing fees on time or at all. Like Watt, he soon experienced what became another familiar scenario for successful inventors: the wearying task of suing those who infringed on his patents and defending himself from those who claimed he had infringed on theirs. Discouraged by his struggles, Evans packed up his household and tools and moved to Philadelphia in 1792. On that larger stage he hoped to find a more prominent place, and he soon discovered a strong demand for skilled mechanics and millwrights. He also found a number of needs that rekindled his interest in a high-pressure steam engine. The potential of portability most intrigued him. Watt's engine was so heavy that it could not have been budged by the power it produced. To become portable, an engine would have to use steam directly. Thanks to Evans, the steam engine, so far a creature of mine and mill, was about to grow legs.[14]

Steam power scarcely existed in the United States when Evans took up the challenge. Dickinson claimed that only six engines could be found in the entire country in 1803 and that "mechanical construction and skill were at least fifty years behind those of England." Yet as early as 1786 Evans petitioned the Pennsylvania legislature to grant him a patent for a steam carriage, and six years later he filed specifications with the U.S. Patent Office for both horizontal and vertical reciprocating engines, a rotary engine, and a boiler enclosing a furnace. The rotary engine had apparently been conceived as early as 1773. However, he could not interest anyone in funding his work and had to develop it on his own while supporting himself by making millstones and selling plaster of Paris and milling supplies. Efforts to promote his inventions and book in England dragged on for years without success. The intense labor involved in breaking up stone with a sledgehammer for the millstones did much to inspire Evans to perfect his engine. Despite his interest in steam carriages, Evans conceded that he had to design for mills if he were to make any sales.[15]

The city of Philadelphia had two engines of the Boulton & Watt design to pump water. They passed for state-of-the-art despite their weak pressure. By 1801 Evans had produced a high-pressure engine that could "break and grind 300 bushels of plaster of Paris, or 12 tons, in twenty four hours." To prove its worth he staged an exhibition on Market Street in which his engine drove twelve saws cutting marble. He then devised a new type of internal boiler to feed his engine. Despite its obvious merits, Evans received little encouragement to continue its development. In 1804 he patented both the engine and its boiler, which he advertised as "more simple and much smaller in size, therefore can be afforded at a much less price and require less room, water, and fuel." He calculated the cost of building one at $886.60 and continued to experiment with improvements on it. In 1805 he published his second book, this one on his experience with steam engines, and titled it *The Abortion of the Young Steam Engineer's Guide.*[16]

That same year Evans completed his most spectacular creation. The Philadelphia Board of Health accepted his proposal to construct a scow capable of dredging the putrid sludge and mud that had accumulated around the wharves at the lower end of the city. Having just failed in his effort to get his automatic flour mill patent renewed by Congress, Evans decided to make this project a showcase for steam power. He built a dredging scow thirty feet long and twelve feet wide with a moving chain of buckets designed on the same principle as the grain elevator in his flour mill. To power it, Evans designed a new type of engine with four drop valves in place of the old inefficient plug cocks. The odd motion of its beam led to its being called a "grasshopper" engine. Once the engine had been installed, the scow was fitted with wheels, but under its weight of fifteen tons they broke down almost immediately. Evans's workmen offered to make new ones without being paid, and their generosity got the scow remounted and ready to move.[17]

Nothing like it had ever been seen. Evans called it the "Oruktor Amphibolos," or Amphibious Digger. He inserted notices in at least two publications that he intended to have his new creation circle Center Square for the benefit of spectators, but it is uncertain whether this display ever occurred. However, the huge scow did lumber down Market Street to the Schuylkill River and head downstream. In one stroke Evans had demonstrated both a steam carriage and a steamboat, although neither one moved very fast or efficiently. Once put to work, the Oruktor had only limited success at dredging and was finally scrapped in 1808. Nevertheless, Evans had opened the door to a new era in transportation. Earlier he had submitted a proposal to the Lancaster Turnpike Company urging it to consider using steam wagons on the road. The bewildered directors dismissed his detailed offer, and Evans never again built an engine for a locomotive or land vehicle.[18]

But he remained busy. Business improved enough that in 1806 Evans opened his own foundry so that he could finally cast his own iron. He called it the Mars Works, and during the next decade it produced nearly a hundred steam engines and boilers along with a diverse lot of products ranging from screw presses for sugar, cotton, and tobacco to machinery for spinning wool. He also experimented with anthracite coal for smelting at a time when no one thought it useful for much of anything. One of his steam engines went to power a gristmill in Lexington, Kentucky; another was installed in a Pittsburgh flour mill constructed by Evans and some partners. These became two of the first manufacturing plants in the nation to be powered by steam rather than water or draft animals. Gradually Evans expanded the Mars Works until it became a model machine shop with the finest tools available. In 1811 he took the bold step of opening a second shop, this one in Pittsburgh, and placed his son George in charge of it.[19]

Despite the ordeal of constant litigation and the press of business, Evans's fertile mind kept churning out new concepts for devices that were far ahead of their time. He explored the possibility of using the sun's rays to heat water, thus doing away with fuel altogether. "The time will come," he predicted, "when water will be raised in great quantities by the heat of the sun at very small expense, for various purposes; but the expense of such inventions cannot in many instances be borne by those who have the mental powers to design them . . . In such cases aid from government becomes necessary." He designed a central hot-air heating system, a gas-lighting system, a vapor-compression refrigeration cycle, a perpetual baking oven, and a method of heating mill buildings by using the exhaust from his steam engines. Like so many of his insights, these would not become practicable for decades.[20]

He also continued to refine his design of boilers, enabling them to use fuel more economically and withstand higher pressures. The boilers of Watt's day produced only about 5 pounds per square inch of pressure; Evans's first engine was measured at 50 psi, a figure that later models tripled. "The more the steam is confined," he wrote, "and the shorter it be shut off by the regulator, the greater will be the power obtained by the fuel. For every addition of 30 degrees of heat to the water doubles the power. So that doubling the heat of the water increases the power 100 times." Put another way, the power of his engine rose in geometrical proportion while fuel consumption rose arithmetically. It would be another forty years before engine builders fully grasped Evans's insight into the economy of high-pressure engines.[21]

The stress of pioneering in so many areas took its toll on Evans. More than once he railed against the inability of the law to protect inventors, who often wound up spending more time and energy defending their rights than exploring new inventions. Having buried his frustration for years, he could no longer contain himself. In his introduction to the 1807 edition of *The Young Miller's Guide*, Evans denounced the patent law for giving inventors only fourteen years of protection, a period "barely sufficient to mature (in this country) any useful improvements." It was in fact a snare and a delusion:

> The inventor is deluded by the name of a patent, and his hopes raised by the accounts he has heard of the success of inventors in England, and he makes great exertions and sacrifices to mature, and introduce into use, his improvements; but just as he begins to receive compensation his patent expires, his sanguine hopes are all blasted, he finds himself ruined, and conceives that he has been robbed by law, is thrown into a despair, and tempted to deem the precious gift of God . . . as a curse . . . I . . . declare, that all my study, labour and time expended during the most vigorous half of my life in making new inventions, etc. I account as lost to myself and family . . . Two years ago I totally relinquished all pursuit of new improvements . . . Had the laws been such as to ensure adequate compensation, I could . . . have invented and introduced into use other improvements that would have proved ten times as beneficial to my country, as all those which I have accomplished: but I have been forced to bury my talent with disgust.[22]

Evans followed this startling outburst with even more shocking actions. He took all his drawings, notes, and specifications, including those still unborn, bound them together, and put them in storage with a vow never to open them "until the laws make it my interest or their own to do so." The law remained as intractable as ever, and Evans's resentment continued to fester. The last straw came in May 1809 when Judge Bushrod Washington declared in a ruling that a patent right infringed on the public interest. Convinced that patent rights were no longer safe, that to leave them to his children would put them on "the same road to ruin, that had subjected me to insult, to abuse and robbery all my life," he went home and piled up all the papers he had stored two years earlier—the collected works of his life as an inventor. After summoning his family, he told them that what he was about to do was for their own good, then pitched the entire pile into the fire. His career as an inventor was finished.[23]

Although he continued to explore and tinker, Evans devoted the last decade of his life to business. Plagued by an inflammation of the lungs, he lay ill in the New York home of his father-in-law in 1819 when news came that fire had destroyed the Mars Works. Four days later he died, leaving his machine shops in the capable hands of his son George and two sons-in-law. Even in death Evans found no rest. He was buried in the small graveyard of the nearby Zion Episcopal Church. In 1854 the church property was sold and Evans's remains, unclaimed by relatives, were moved to a new church on Murray Hill. The new church was sold in 1890, and Evans's remains, along with several others, were again exhumed and moved to Trinity Cemetery at 157th Street and Broadway, where they were deposited in plot 641. In death, as in life, Evans suffered the indignity of undeserved neglect and anonymity. Nevertheless, he left the nation a legacy of invention that included the steam engine of the future. That legacy unfolded not only in industrial use but most spectacularly in transportation on land and sea.[24]

FOR AMERICANS ON the move, the grand objective was to apply steam power to conquer their vast, sprawling interior. Especially did they want to utilize the impressive network of inland waterways. The Atlantic seaboard had its bays, sounds, and tidal rivers reaching back to the fall line, which gave it a useful transportation system. Once past the Appalachian Mountains, however, the problem of moving goods and people became acute. Decent roads were one obvious answer but a slow and costly one in so vast a territory. The navigation of western rivers rendered sails all but useless. Unlike eastern rivers such as the Hudson, they tended to be narrow, meandering, and filled with snags. Thick stands of trees along the banks acted as windbreaks. During the early nineteenth century three major solutions to the transportation problem arose: steamboats, canals, and railroads. The two most important of these responses required steam power and the development of suitable engines. Oliver Evans had not only anticipated both types of travel but tried to develop vehicles for them.

The application of steam to water travel would represent an enormous advance. On the high seas it would mean that a ship need no longer be at the mercy of the prevailing wind and might steer a straighter course at a steady speed toward its destination. On rivers like the meandering Mississippi it would mean that for the first time boats might travel upstream against the current. The barges and keelboats that carried goods down the Mississippi, for example, were either broken up and sold for scrap at New Orleans or taken laboriously back home by poling or hauling with rope, a journey that took three or four months at best. The steam engine offered the tantalizing prospect of turning this journey into a two-way trip as well as the opportunity to replace exhausting overland travel with a pleasant cruise aboard a well-appointed boat. However, few inventions endured a more painful or prolonged birth than the steamboat.[25]

Legend has long credited Robert Fulton with inventing the steamboat, but the story is much more complex. As historian Louis C. Hunter observed, "There is good reason to question whether Fulton's name should be placed much if any higher than the names of John Fitch, John Stevens, or even Oliver Evans." Evans had long been intrigued by the prospect of using steam power to travel over land or water. The Oruktor had managed, however briefly, to do both in one vehicle, but it was hardly practical. Evans also had dealings with all three of his chief rivals in the quest to develop a practical steamboat. Fitch was an acquaintance who had discussed aspects of steamboat development with Evans several times over the years. Both were moody men given to bouts of depression who felt that fate had somehow robbed them of their just due. In Fitch's case, fate robbed of him of just about everything.[26]

The high-pressure engine, with its relatively light weight and portability, seemed ideal for application to water travel, although it posed the danger of boiler explosions. Evans had long viewed it as the best type for water travel. In 1802, three years before the Oruktor made its debut, he agreed to install one of his engines in a steamboat being built by some men in the West. After completing the engine, he sent it along with the boiler to New Orleans, where the boat was being constructed. Early in 1803 the steamboat was nearly ready for its trial run up the Mississippi River when heavy rains unleashed a flood that carried the boat half a mile inland to a place where it could not be moved back to the water. Thwarted once again, Evans gloomily removed the engine and installed it in a sawmill, where it did splendid work. He continued to advocate the high-pressure engine for western riverboats, but its acceptance came slowly. Once again he remained a prophet without profit.[27]

Fitch's life resembled that of a Charles Dickens character with all its improbable twists and turns. Born on a hardscrabble Connecticut farm in 1744, he got only a taste of education, although he loved school and begged for more time off from the farm to attend class. At eleven he heard about a geography book and planted an extra patch of potatoes to earn the money to buy it. Once the book arrived, he practically memorized it. Apprenticed to a clockmaker at eighteen, he was worked like a slave but taught nothing of the trade. Freed of his contract at twenty-one, he lived with his father for two years and eked out a living making small brass items until he had saved enough money to go partners in a potash works. He married a local woman older than himself who bore him a son. She nagged him constantly until Fitch, unaware that his wife was again pregnant, carried out his threat to leave despite her pleas for him to stay.[28]

For six months he wandered as a tramp, tried unsuccessfully to enlist in the British army, and then settled in Trenton, New Jersey. There he launched a successful career as a silversmith only to have it disrupted by the American Revolution. He joined the local militia but got embroiled in one tiff after another and left his unit, although he toiled long hours at repairing and refitting arms. When the British entered Trenton, they destroyed nearly all of Fitch's furniture and belongings. Discouraged, he packed up what little was left and moved to a small village in Bucks County, Pennsylvania. There he started a new career supplying beer, rum, tobacco, and other goods to the Continental army during the harsh winter of 1777–78, when Washington's men were enduring their ordeal at Valley Forge. He earned good money at the work only to see its value shrivel to nothing because of rampant inflation. After the army departed, he took up his old trade of silversmith but found little business.

Seeking yet another fresh start in life, Fitch managed to finagle a post as deputy surveyor in Kentucky. His first two trips, in the springs of 1780 and 1781, filled him with high hopes for making a fortune by buying land warrants in that western region. During the spring of 1782 he bought several barrels of flour and loaded them on a barge in Pittsburgh, hoping to take them to New Orleans and earn enough profit to buy land. On the trip downriver, however, he and his companions were captured by hostile Indians and endured several harrowing weeks culminating in a forced march to Detroit, where they were turned over to the British. After languishing in a prison near Montreal, Fitch was released and put on a ship to Boston. Ten stormy weeks at sea landed him in New York before he finally made his way back to Bucks County.

Sick and exhausted, Fitch clung tightly to his dream of western land. He formed a land company and spent the summer and fall of 1783 surveying in the Ohio Valley, but his hopes for securing large tracts of land were blasted by the Land Ordinance of 1785, which called for public land to be sold in lots. The act, he later wrote, reduced him "from an immense fortune . . . to nothing at one blow." In the spring of that same year, however, Fitch was struck by an epiphany that changed his life. Walking home from a rare visit to church, he cursed in pain as rheumatism buckled his knee. While bent over, he noticed a gentleman passing by in a horse-drawn carriage. "A thought struck me," he claimed later, "that it would be a noble thing if I could have such a carriage without the expense of keeping a hors[e]." Thus was born the insight that led him to consider steam as a way of propelling such a craft.[29]

Fitch knew nothing about steam and, like the vast majority of Americans, had never even seen a steam engine. Yet within a week he "gave over the Idea of Carriages but thought it might answer for a Boat." Two or three weeks later he visited a friend who showed him a book with an illustration of a steam engine. Chagrined at not knowing of its existence and more determined than ever to create his own version of an engine, Fitch threw himself into the task. A few months later, in November 1785, he walked to Mount Vernon plantation in Virginia and presented himself to George Washington. Dirty and unshaven, his clothes filthy from the journey, his dark eyes fired with determination, he described his project to the general in detail and asked for a letter of introduction to the Virginia Assembly. Washington declined, and Fitch went on to Richmond to experience more disappointments there. Although Washington gave no reason for his refusal, Fitch thought he knew the source: James Rumsey.[30]

In September 1784 Washington had met Rumsey, who impressed the general with his model of a boat that pushed upstream by using a complex set of paddlewheels and poles. The action was purely mechanical; no engine was involved. Nevertheless, Washington agreed to give Rumsey a written statement to help him attract investors and possible funding from the state of Virginia. He also hired Rumsey to build a house and some outbuildings for him at Bath, Virginia, where the encounter took place. The possibility of a boat capable of moving up western rivers excited Washington, who personally owned nearly fifty thousand acres of western land. Rumsey had been tinkering with his mechanical poleboat for some time. A year older than Fitch, he too was a self-taught farm boy turned inventor. Where Fitch was an intense, sometimes abrasive personality with the grim air of a prophet about him, Rumsey was handsome, charming, and well-mannered. Within a few months he managed to obtain monopolies for his boat from Virginia, Maryland, and Pennsylvania. He called the vessel a *stream*-boat, thereby setting the stage for years of confusion and controversy.[31]

That same winter of 1784–85, Rumsey began experimenting with steam as a power source for his boat. Like Fitch, he had to work in an intellectual vacuum. He knew only about the early Savery and Newcomen engines; one of the latter pumped water from an iron mine near Cranston, Rhode Island. Watt was just coming out with his more efficient double-acting engine, but only

Americans who had ventured to Europe knew anything about it. Yet Rumsey, like Fitch, managed to design and build an engine from scratch. He even came up with a pipe, or tube, boiler, which later became the standard for steamboats. For propulsion he concluded—wrongly, as it turned out—that a paddlewheel would not do and decided to use a water jet system instead. His brother-in-law built for him a six-ton boat, and on March 14, 1786, the first trial was attempted. The machinery worked poorly, and Rumsey settled in for what proved to be a long and frustrating series of tests.[32]

Fitch, too, was thwarted and frustrated by his early trials, but he had made significant progress. Despite lukewarm responses from Benjamin Franklin and the American Philosophical Society, he had won over some important backers and managed to persuade the New Jersey Assembly to grant him exclusive rights to operate his steamboat on state waters. This coup enabled him to organize a company and raise money for his work. By March 1787 Fitch had secured a monopoly from Pennsylvania. He had also encountered a bright mechanic named Henry Voigt (or Voight), who became his indispensable partner in building the boat. Fitch called him "the first Genius that ever I was acquainted with." Through trial and error Fitch and Voigt devised an engine that bore two key features of Watt's engine: a condenser (Fitch called it a "cistern") and a double-acting cylinder. Fitch designed a hull and a paddlewheel system but decided against the latter. Voigt convinced him that water jet propulsion was even more inefficient than a paddlewheel. Instead Fitch came up with an elaborate crank and paddle system.[33]

Although money problems continued to dog Fitch's efforts, his fundraising got a boost when Delaware and New York both granted him monopolies. In April 1787 the boat's hull arrived in Philadelphia and was aptly named the *Perseverance*. Fitch and Voigt built a brick furnace on the deck and installed the machinery even though it continued to pose problems. In August 1787, while the Constitutional Convention toiled away in the summer heat only a few blocks away, Fitch gave his vessel a trial run. He had invited the convention delegates to attend, and a few accepted. The engine worked well, but Fitch realized gloomily that it could not produce enough power to be commercially viable. A more powerful engine was needed, he concluded, and with it a longer, narrower hull. Nearly a year passed before Fitch was ready to undertake the next trial. By that time the race between him and Rumsey had erupted into an open war.[34]

After his early good fortune, Rumsey had endured a string of mishaps. A recommendation by George Washington obliged him to accept a post as superintendent of the Potomac Navigation Company. The job proved a disaster and led Rumsey to resign just short of a year later. It also took time away from his work on the steamboat, where boiler leaks and other problems continued to plague him. Rumsey's engine too suffered from lack of power, and he grew increasingly agitated over news of Fitch's progress and especially of his rival gaining monopolies in state after state. Not until September 1787 did Rumsey have his boat ready for another trial. Although it failed miserably when several joints in the new boiler leaked steam badly, Rumsey was encouraged. The new boiler design worked fine and needed only better workmanship to seal the joints. In November he heard that Fitch had just secured monopoly rights in Rumsey's own state of Virginia. Rumsey had not planned another trial until spring, but the news impelled him to schedule a public trial on December 3 at Shepherdstown.[35]

It was a bold and risky decision. Already some locals had labeled him "crazy Rumsey," and a public failure might doom his hopes for acceptance. A host of local dignitaries, including Revolutionary War hero General Horatio Gates, gathered along the banks of the Potomac to watch the spectacle. As the boat moved away from shore and turned into the current, the general exclaimed, "My God, she moves!" For about two hours the boat continued up and down the river, making a steady three miles an hour when going against the current. Thrilled (and doubtless relieved) by his success, Rumsey scheduled another public demonstration for December 11. This trial came off nearly as well, and the boat managed to do four miles an hour going upriver. Buoyed by these triumphs, Rumsey pondered his next step. Fitch had sought and narrowly failed to get monopoly rights from Maryland. How was he to get around Fitch? Reluctantly he realized that he had to move to Philadelphia, where both potential backers and skilled mechanics could be found even though it meant facing off against Fitch directly.[36]

As a first step, Rumsey published in January 1788 a pamphlet describing his invention and defending it against Fitch's claims that Rumsey stole his ideas about steam from him. Once in Philadelphia, Rumsey used all his charms to win financial support and endorsements from influential citizens. He formed a new company called the Rumseian Society, the investor-members of which included Benjamin Franklin, who bought one share. The American Philosophical Society (APS), which had snubbed Fitch, invited Rumsey to join. That April the APS received notices of two new inventions. One was Rumsey's imaginative tube boiler, the other a series of improvements for boilers designed by Voigt that included a tube boiler resembling Rumsey's. Rumsey immediately charged Fitch and Voigt with stealing his design, although no evidence exists that they did.[37]

Surprised by the attacks, Fitch laboriously gathered evidence for a rebuttal to Rumsey's "wicked and invidious" pamphlet. Going on foot to Frederick and

Baltimore, he collected affidavits from workmen and others who had done work for Rumsey to show that the latter's boat had not been built first. By early May Fitch had his own pamphlet ready for the printer. Before delivering it, he paid a call at Rumsey's boardinghouse and read his startled rival some passages. Rumsey left the response to his brothers-in-law, exhorting one of them to "leave no stone unturned to detect Fitch in his villainy," and sailed to England in search of patents. The resulting pamphlet presented nineteen affidavits to refute Fitch's claims. It damaged his reputation, but neither man profited from the fight.[38]

Once in England, Rumsey met with Boulton and Watt and impressed them enough that talk turned to a business arrangement between them. However, the proposal offered by the Englishmen so offended Rumsey that he dashed all hope of an alliance with the most prominent steam engine firm in the world. Although Rumsey managed to secure an English patent, his efforts to raise money for a new boat, the *Columbian Maid*, led him through one frustrating escapade after another. He narrowly escaped being thrown into debtor's prison, worked for an Irish canal project, and went into the business of improving mills with new inventions. Not until the fall of 1792 was his boat ready for another trial, which failed thanks to bad luck and careless handling. On December 15 Rumsey tested the engine at dockside with encouraging results. "I have very little doubt of success," he wrote a friend, but on December 20 he died suddenly of an apparent stroke. Although his boat received another public trial, it still needed considerable work. Without Rumsey to push the project, it soon vanished.[39]

While Rumsey struggled in England, Fitch and Voigt toiled at getting their boat in shape to test. Modifications to the boiler and oars enabled them to make a trial run in July 1788. With two of the company's stockholders aboard, Fitch and Voigt chugged smoothly upriver from Philadelphia to Burlington, about twenty miles, the longest nonstop journey by steam yet made. All went well until the tube boiler developed a leak and the boat had to drift back home with the tide. Three months later they repeated the trip successfully, this time with more than thirty passengers, but at a speed that disappointed Fitch. "Although we had made our boat go fast enough to answer a valuable purpose on the Ohio," he wrote later, "it did not go fast enough to answer a valuable purpose on the Delaware."[40]

The dilemma was painfully clear and inescapable. His balky, low-pressure engine could deliver a speed of only about four miles an hour. That might do on a winding western river where no competition existed, but on the broad Delaware a steamboat had to outdo a stagecoach running over decent roads. A stage could run the thirty-eight miles from Philadelphia to Trenton in less than five hours, or between seven and eight miles an hour. A steamboat could not compete unless it could match or beat that performance. Fitch knew this realization would lead most if not all of his investors to abandon him. To make matters worse, Voigt decided that he had sacrificed enough and left the project. Fitch fell into despair. "I frequantly [sic] wished that Heaven had rather put it in my mind to have cut my own throat, than to have put me in mind of building Steam Boats," Fitch lamented later.[41]

Yet Fitch did not give up. He formed a new company, acquired two new partners, and studied every component of the engine for ways to increase its power. Voigt relented and returned to help create a new and better engine. He brought some fresh ideas that improved the engine, but disaster struck again in September 1789 when the boat caught fire. Fitch managed to sink it and thereby save the machinery. By December the partners had contrived to raise and patch the boat, but still its speed did not exceed four or five miles an hour. During the winter months, the partners bickered over what to do next but did reach some major decisions. The cranky tube boiler was replaced with a conventional pot boiler despite Fitch's objections, and the side paddles gave way to a stern paddlewheel. The old machinery received a new condenser and air pump before being installed in March 1790. They proved to be crucial modifications.[42]

On April 12 they took the boat out for another trial. To their delight the engine worked so well that its force snapped one of the pulleys in two. They fitted the boat with a stronger chain and four days later went out again. Despite a strong northeast wind, they made steady progress upriver. "We reigned Lord high admirals of the Delaware," Fitch wrote gleefully, "and no Boat on the River could hold sway with us, but all fell a-stern." Nothing broke down on the trip, convincing Fitch that his time had come at last. He admitted to being "flushed with success, and knowing that 4 men could navigate 100 Tons up the Mississippi, we concluded that our Troubles were at an end." More trials confirmed that the boat was seaworthy. Eagerly Fitch and his partners installed cabins and prepared the boat for commercial use that summer on the run between Philadelphia and Trenton.[43]

Fitch advertised the service heavily and the boat performed admirably—going five hundred miles before suffering mechanical problems—but the passengers did not come. In desperation the partners offered free food and drink on board, but every voyage continued to lose money. What had proven to be a technological triumph turned out to be a commercial failure, and the service was abandoned after that first summer. His investors backed away, and Fitch eventually found himself nearly penniless once again. He received an offer to build a steamboat in France but arrived in the midst of the revolution and endured a struggle to get home again. Once back in the United States, he drifted about before deciding to retire to Kentucky, where he still owned 1,300 acres of land near Bardstown. There he traded a tavernkeeper 150 acres of farmland for room, board, and a pint of whiskey a day. Legend has it that he drank himself nearly to death and finally committed suicide by taking pills, but a descendant claimed after investigating the facts that he died in 1798 of natural causes.[44]

Despite his bitter disappointments, Fitch never stopped talking or thinking about steam engines and steamboats. While in Kentucky he built some models to occupy himself, and he understood clearly what his true legacy would be regardless of how poorly his contemporaries had treated him. He had created the first commercial steamboat operation in history, but even more important, he had unlocked the door to the future. "This will be of the first consequence to the United States," he predicted after his successful run of 1790, "and make our Western Territory four times as valuable as otherwise it would be. This has been effected by little Johney Fitch and Harry Voigt, one of the Greatest and most useful arts that was ever introduced into the World. And altho the World nor my country does not thank me for it, yet it gives me a heart felt Pleasure."[45]

THESE THREE MEN—Evans, Fitch, and Rumsey—became the prophets and pioneers of the steamboat only to suffer bitter personal disappointments in the end. Fame and fortune went instead to Robert Fulton, whose story has been told many times and need not be repeated in detail here. Born in 1765, the son of a Pennsylvania farmer, he took up painting before turning to invention. After a bout with consumption Fulton went to England in 1786, where he fell under the spell of painter Benjamin West. He went to France in 1797 and remained in Europe nearly ten years, working first on canals and then trying to interest the French and then the British government in his "diving boat," or submarine, and an early version of the torpedo. Five years later he met Robert Livingston, the American minister to France, and formed a partnership that was to last until Livingston's death in 1813. A wealthy, influential New Yorker who helped negotiate the Louisiana Purchase in 1803, Livingston joined with the ambitious Fulton in trying to develop a steamboat for use on the Hudson River that flowed past Livingston's vast estate, Clermont, ninety miles north of New York City.[46]

Although Fulton continued to work on a variety of projects, the steamboat gradually took priority. Livingston managed to get Fitch's New York monopoly annulled and acquired it for himself and Fulton. It called for demonstrating a boat capable of traveling four miles an hour upriver. Fulton did not invent the boat himself; it was rather, in Cynthia Owen Philip's words, "a collaboration of experts." Instead of designing his own engine, Fulton acquired one from Boulton & Watt and had it shipped to New York. Fulton himself reached New York City in December 1806 after nearly twenty years' absence from the United States and went to work on the steamboat. Already Fulton sensed that the true market for his craft was the Mississippi rather than the Hudson, but Livingston insisted that they conquer the New York river first. Working furiously, endowed with ample funds, and displaying an impressive talent for managing and inspiring his men, Fulton had his boat ready for trial by August 1807.[47]

The historic trial took place on August 17. Fulton's boat sailed upriver from New York to Albany, about 150 miles, in thirty-two hours, stopping overnight at Clermont. The same trip under sail took about four days. Although Fulton's boat was also rigged with sails, he did not use them. Despite its later fame, this triumphant journey received hardly any notice in the newspapers. Nevertheless, it marked the prelude to what became regular steamboat service on the Hudson and soon afterward on the other major eastern rivers. Rather than inventing the steamboat, Fulton had made it work in a commercially viable service. Moreover, he performed this feat using a low-pressure Boulton & Watt engine, which proved adequate for the broad eastern rivers. During 1808 the refurbished boat earned profits as high as $1,000 a week. It also brought him endless litigation over patents and other issues. But Fulton wanted more; he aspired to conquer the western rivers as well.[48]

In 1811 Fulton and Livingston obtained exclusive rights to run steamboats on the lower Mississippi River. Fulton commissioned Nicholas Roosevelt to build a boat at Pittsburgh. Roosevelt had the boat ready by September, and on October 20 the *New Orleans* began what became a historic and memorable journey down the Ohio and Mississippi rivers to its namesake city. A series of bizarre and unnatural events surrounded the trip. In April the Great Comet, one of the biggest and brightest ever known, opened a nine-month stand in the heavens. During the summer the Mississippi and other rivers spilled over their banks for no apparent reason, and thousands of gray squirrels swarmed en masse across the Ohio River as if fleeing for their lives. A solar eclipse darkened the sky in September, and in December abnormally warm weather and hazy skies turned the sun into a "glowing ball of copper."[49]

The worst was yet to come. On December 14 Roosevelt tied the *New Orleans* up near what is now Owensboro, Kentucky, to replenish his supply of firewood. He had lost a month at Louisville waiting for water high enough to pass over the Falls of the Ohio, during which time his wife had given birth to their second child. Two days later everyone on board awoke at 2:00 A.M. to a loud noise as if the boat had run aground. At daylight they found the river muddy and in turmoil, the smell of sulfur in the air, and both banks lined with twisted or fallen trees. A major earthquake had struck during the night, and around 8:00 A.M. a second, more powerful shock set the ground trembling with a deafening roar. The great earthquake of 1811 devastated the Mississippi Valley with a force so explosive that it briefly reversed the course of the river. Miraculously the boat, shielded by an island behind it, remained tied to its moorings.

When the ground stopped shaking, Roosevelt decided that the river might be the safer place and cast off. Slowly the boat threaded its way downriver through trees, whole chunks of riverbank, an occasional empty flatboat, and piles of furniture floating in the water. The aftershocks continued to come— eighty-nine of them between December 17 and 23 by one count. One night the boat's pilot took refuge by tying up to a tree at the foot of an island. Next morning everyone looked in astonishment at a bewildering sight: The tree was underwater and the island was gone, broken up during the night. At the town of New Madrid they saw massive destruction. Some people begged to be taken aboard the boat, but Roosevelt lacked both room and provisions and had to steam away. Not until they reached Natchez on December 20 did the river begin to look normal. Eleven days later, on January 10, 1812, they reached New Orleans and were greeted by a cheering crowd.

For the first time a steamboat had traversed the Mississippi River from Pittsburgh to New Orleans. Fulton put the boat into service going upriver from New Orleans to Natchez, and later added two more boats to the run. On this 300-mile deepwater stretch of the river his low-pressure engines proved adequate, but the boat lacked enough power to cover the remaining 1,700 miles to Pittsburgh. It soon became clear that the high-pressure engine held the key to future development on the river. At Pittsburgh George Evans, Oliver's son, and his partner Luther Stephens began building high-pressure engines after their works opened in 1811. Other builders, most notably Henry M. Shreve, designed their own versions of high-pressure engines. In 1815 the *Enterprise*, with Shreve at the helm, became the first boat to make the upriver journey from New Orleans to Pittsburgh. Two years later Shreve brought the *Washington*, then the largest steamboat in the West, from New Orleans to Louisville in a record twenty-five days.[50]

The reliability and performance of the *Washington* launched a trend toward use of high-pressure engines. By the 1830s a standard type of engine emerged that remained in use for sixty years without serious modification. Robert Thurston hailed it as "the simplest possible form of direct-acting engine." An 1838 survey of steamboats in the Mississippi Valley found only one of 285 boats in operation that used a low-pressure engine. Gradually there emerged two distinct types of American steamboat. The eastern version, designed chiefly for the passenger trade, featured a low-pressure engine, a broad hull, fine lines, an elaborate finish, elegant furnishings, and speed. The western type, devoted to freight, boasted a high-pressure engine, a narrow hull, little or no keel, multiple rudders, no frills, and several layers of superstructure to compensate for its lack of width. One wag described it as "an engine on a raft with $11,000 worth of jig-saw work."[51]

What the western steamboat lacked in grace or elegance it made up in practicality. Although a blizzard of litigation over water rights, patents, and other issues slowed its development, the steamboat proved crucial to the settlement and growth of the West. From modest beginnings the number of boats grew at an amazing rate. In 1817 some 17 boats were plying western rivers; the following year, more steamboats were built in the West than had been previously constructed up to that time. Their number reached 75 by 1823 and 187 by 1830, yet this marked only the beginning of the steamboat era. By 1840 some 536 boats plied western waters, and by 1860 the number reached 735. Not only did steamboats multiply in number, they grew larger as well. Between 1820 and 1860 the average tonnage of a western steamboat increased 40 percent. Steamboat construction, equipping, and repairing became a major industry in itself, especially in the river cities of Pittsburgh, Cincinnati, and Louisville.[52]

Americans and foreigners alike recognized the important role played by the steamboat in settling and developing the region beyond the Appalachian Mountains. "Steam navigation colonized the West!" declared one American observer. A visiting Frenchman agreed, noting that "without the intercourse made possible by the steamboat, Ohio, Indiana, and Illinois would today be a desert unknown to civilization." The most astute French visitor of them all, Alexis de Tocqueville, understood that the steamboat brought not only settlement but closer ties among far-flung regions of the young nation. "The discovery of steam," he wrote, "has added unbelievably to the strength and prosperity of the Union, and has done so by facilitating communications between the diverse parts of this vast body."[53]

The steam engine had conquered water and with it distance—but only for those parts of the West that had navigable rivers. Even then the steamboats could run only when the river was not frozen over or the water too low to permit passage. The other half of the vision of Oliver Evans and others had yet to be realized, and it was to prove even more revolutionary than the first. The steam engine was about to conquer land as well.

CHAPTER 3

THE GREATEST
ENGINE OF ALL

*Concede that there are now no Steam Rail-Ways anywhere in the
world. This is not to say that they will not come—and that soon.
As civilization progresses, water-carriage will prove too slow and
cumbersome to satisfy the demands of humanity. And this, too,
though it remain fairly cheap. What has been accomplished, in
comparatively few years, with Steam Boats, points . . . directly at
the Steam Carriage. Merely by developing a method of correctly
applying the same principles on land, a great saving in time and
cost will be effected.*

—JOHN STEVENS[1]

IN 1723 SEVENTEEN-YEAR-OLD Benjamin Franklin decided to go from
New York to Philadelphia, where he would live for the next sixty-seven years.
To make this ninety-mile-plus journey, he planned to take a boat to Amboy on
the north Jersey shore, ride a stagecoach across New Jersey to the Delaware
River, and then float down the river to Philadelphia. What seemed a simple
trip turned into an all too typical nightmare in an age when decent roads did
not exist even between the colonies' two largest cities and the sea proved as
treacherous as ever.[2]

On Franklin's first day out, adverse winds drove his ship dangerously close
to the shores of Long Island, which were already clogged with the hulks of
shipwrecked vessels. It took the captain thirty hours to nurse his ship to Amboy. A drenching rain soaked Franklin's stagecoach journey across New Jersey
and slowed progress enough that he missed the boat for Philadelphia. Learning that another boat would not sail for two weeks, he sighed in disgust, rented
a rowboat, and muscled his way to the city he planned to call home. Such were
the vagaries of travel in early America even on one of the most heavily traveled

routes on the continent. Franklin's hardships paled before those awaiting travelers heading west into the wilderness or even to other colonies.

Land, a rich, teeming continent of it, was the greatest asset owned by the American people, but for more than a century they had no way to get at most of it. Going by horse or wagon required roads, which scarcely existed and became impassable in winter or wet weather. Traveling by river was a one-way journey in that boats could only travel downstream. One could always go by horseback or on foot, but the sheer distances involved rendered that a long, arduous, and often perilous journey. No obstacle proved greater to the settlement and development of the nation than the lack of transportation. The steam engine and its offspring removed that obstacle in spectacular fashion.[3]

The steamboat's conquest of western waters proved instrumental in opening the region beyond the Appalachians to settlement, yet the impact even of this achievement paled before that of the railroads. No matter how extensive the river system, it could reach only select parts of the nation's vast interior. Moreover, river travel suffered from limitations that could never be overcome. In northern climates the rivers remained frozen for several months in the winter, and during the summer low water made navigation impossible for boats with any significant draft. No river ran in a straight line, and some, like the Mississippi, meandered so badly as to pose a serious challenge to travel. The system of rivers east of the Mississippi ran mostly north and south, which posed problems for the movement of goods and people from east to west and vice versa. West of the Missouri River, the rivers dwindled down to a precious few, and some of them, like the Platte, were too shallow for steamboats.

To create a truly continental transportation system, the steam engine had to walk not only on water but on land as well. Travel for any distance by stagecoach was bone-jarring, exhausting, and dangerous. "I have just finished six days and nights of this thing . . . ," wrote one weary traveler in 1865. "I shall not undertake it again. Stop over nights? No you wouldn't. To sleep on the sand floor of a one-story sod or adobe hut, without a chance to wash, with miserable food, uncongenial companionship, loss of seat on a coach until one comes empty." As late as 1877 the *Omaha Herald* offered this advice for those about to undertake the adventure:

> If a team runs away, sit still and take your chances; if you jump, nine times out of ten you will be hurt. In very cold weather abstain entirely from liquor while on the road; a man will freeze twice as quick while under its influence. Don't growl at food stations; stage companies generally provide the best they can get . . . Be sure to take two heavy blankets with you; you will need them. Don't swear, nor lop over on your

neighbor while sleeping . . . Take small change to pay expenses. Never attempt to fire a gun or pistol while on the road; it may frighten the team . . . Don't discuss politics or religion . . . Don't grease your hair before starting or dust will stick there . . . Tie a silk handkerchief around your neck to keep out dust and prevent sunburn . . . Don't imagine for a moment you are going on a picnic; expect annoyance, discomfort and some hardships.[4]

The heart of the transportation revolution lay not in the steamboat but in the railroad, which grew from a flock of fledgling local lines to a massive network that crisscrossed the entire nation. In 1830 American railroads totaled a mere 23 miles of track. The following year, in South Carolina, a locomotive called the *Best Friend of Charleston* began the nation's first regular train service. When the Philadelphia Centennial opened its gates in 1876, the country already boasted 76,808 miles of operating railroad, an impressive amount but still only prelude to the great period of expansion that took off during the 1880s. By the century's end the United States had an astounding 258,784 miles of track, 206,631 of it on main lines. In comparison, the rest of the world totaled 284,435 miles.[5]

Those early inventors like Fitch, Evans, and Trevithick who had dreamed of creating a land vehicle never realized how much more complex a technological system it would require compared to a steamboat. For the latter, power had only to be transferred to some device like a paddlewheel to propel the boat. Not only was the locomotive a more complex piece of machinery, it also required a roadbed over which to travel. Versions of a "rail road" had been around since at least the seventeenth century. To ease the hauling of coal and ore from mines, roads paved with stone were built. Around 1630 an Englishman named Beaumont got the bright idea of replacing the stone with heavy planks. These gradually evolved into rails made first of wood and then of iron. By the late eighteenth century, the wheels running over these rails had flanges to improve their grip. The road of rails had become familiar at least to miners by the early nineteenth century, but it lacked a source of power beyond the horse.[6]

The steam engine was an obvious possibility. Work on it expanded rapidly after 1800, when the Boulton & Watt patents expired. Across England, the Continent, and the United States men with a mechanical bent tinkered doggedly with versions of the steam engine, each one in search of a unique solution to whatever problem or challenge intrigued him. Some worked in the privacy of their own homes; others went into the business of building engines. Many of the men who had worked for Boulton & Watt took their expertise elsewhere or went into business for themselves. They soon realized, as did most other builders, that the Watt low-pressure engine would not do for a land carriage because it simply could not generate enough power. Yet strong fears remained about the dangers of high-pressure engines. As late as the 1820s in England the stationary steam engine was still considered a serious, even preferred alternative to the high-pressure engine for railroad motive power.[7]

Most of the pioneer builders created vehicles that ran on roads. As early as 1769 a French army officer, Nicolas-Joseph Cugnot, built a steam carriage intended to haul artillery, but it failed to perform. Watt's assistant, William Murdock, built several working models of his steam carriage in 1784 but never moved to a larger version. Oliver Evans put a giant vehicle on the road briefly with his lumbering Oruktor in 1804. Two years earlier Richard Trevithick made and patented a model steam carriage and followed it with a full-sized version that received a public trial in 1803. In typical fashion Trevithick did not develop his vehicle but turned instead to devising one that ran on rails. In February 1804 he put together a steam vehicle that traveled nearly ten miles over iron plates, making him the first to haul a load with a steam locomotive over a fixed track. Four years later he built a primitive locomotive and operated it on a circular track in London until a broken rail damaged it beyond repair. Lacking the funds to build a new one, he turned to other projects.[8]

Murdock, Evans, and Trevithick all built high-pressure engines for their vehicles. The reason was obvious: Initial high pressure was needed for any use that required high starting torque. Textile mills made do with low-pressure engines because they carefully disengaged their machinery at day's end. When the engine started up next morning, it had to deal only with its own internal friction and that of the line shafting. A locomotive at rest, however, had to gather a head of steam to move the entire weight of the train forward. The same need existed in the winding engines that lifted coal or ore out of mines and had to raise the entire load at the beginning. Early attempts to use atmospheric engines for this work ranged from frustrating to futile.[9]

George Stephenson, a self-taught English inventor, worked extensively with winding engines before turning to the steam locomotive. His quick, incisive mind mastered the Newcomen, Watt, and other versions of the steam engine at an early age. In 1815, at the age of thirty-four, he built a locomotive that included all the basic elements of later versions. In fact, he built two different models and patented the second, more efficient one. Two years later, he built an engine for the duke of Portland that successfully hauled coal for thirty years. During the next decade Stephenson studied every aspect of the railroad closely and conducted experiments to determine everything from resistance and friction to hauling power over grades. One finding especially impressed him: He calculated that a steam-powered land carriage might carry twenty to thirty passengers at ten miles per hour, while a carriage on rails could move ten times the number of people at three or four times the speed.[10]

Convinced that rail travel held the key to the future, Stephenson and a partner in 1824 went into the business of building steam locomotives. The next year he and his son, Robert, created an advanced experimental engine. In 1829 they unveiled the *Rocket*, the first locomotive to include in the same engine such basic elements as a horizontal tubular boiler, forced draft, and outside cylinders that connected directly via rods to crankpins in the driving wheels. This remarkable engine earned Stephenson wide renown as the father of the railroad. On one run it carried thirty passengers at a speed ranging from twenty-five to thirty miles an hour, the fastest that people had yet traveled on land. Two days later the *Rocket* hauled thirteen tons of freight at an average of fifteen miles an hour and reached twenty-nine miles an hour on a favorable stretch. The age of railways had dawned, and with it a new mission for the steam engine.[11]

American inventors did not lag far behind their British rivals. One man in particular had been agitating for the development of railroads since 1811 when he first asked the New Jersey legislature for a charter to construct one. Colonel John Stevens was already fifty-one years old when the nineteenth century opened, and he had been in the transportation wars for more than twenty years. At different times he had tried to beat out Rumsey, Fitch, and Evans with his own version of the steamboat. Later he partnered with Robert Fulton and Robert Livingston before waging a long and bitter campaign against them over their monopoly rights. It did not help that Livingston was Stevens's brother-in-law and that he had genuine affection for both men despite their rancorous dispute. During their fight, in 1808, Stevens had advanced an argument that was later echoed in the landmark Supreme Court decision *Gibbons v. Ogden* (1824), which eliminated the steamboat river monopolies.[12]

A year after approaching the New Jersey legislature, Stevens petitioned Congress to support the construction of a national railroad. He published a pamphlet entitled *Documents tending to prove the superior advantages of Railways and Steam-Carriages over Canal-Navigation* in which he boldly declared that "I can see nothing to hinder a steam-carriage moving on its ways with a velocity of 100 miles an hour." This at a time when not a single railroad existed except for the crude trams at collieries! Stevens admitted that such a speed was not likely but added that he "should not be surprised at seeing steam-carriages propelled at the rate of 40 or 50 miles an hour." At this remarkably early date Stevens was a voice crying in the wilderness, but he clung to his belief in the future of the railroad. In 1826, fourteen years later, he built a locomotive at his own expense and ran it around a circular track on his Hoboken estate. It was the first steam engine to travel on rails in the nation, but it went nowhere and served only to amuse his friends.[13]

Despite the lack of working models or examples, the new technology especially intrigued merchants in cities on the eastern seaboard who grew concerned at the decline in their influence. The spectacular success of the Erie Canal, which opened in 1825, marked a shift in trade routes that launched a frenzy of canal building by states, cities, and private interests in an attempt to recapture lost trade or build business where none had existed before. Despite the craze for canals, not everyone regarded them as the best way to reach hinterland markets. As early as 1826–27 Baltimore merchants began talking seriously about building a railroad to protect and extend their commercial interests. From their efforts arose the first major American railroad, the Baltimore & Ohio (B&O), which laid its first track in October 1829 and completed thirteen miles of road by May 1830.[14]

On that stretch of track Peter Cooper, a clever New Yorker who rose from humble origins to become a successful businessman and beloved philanthropist, demonstrated the feasibility of train travel by hauling some passengers behind a locomotive he had built himself. An enthusiastic supporter of the project, Cooper came to Baltimore and bought three thousand acres of land southeast of the city to develop a planned industrial community that would benefit from the railroad's presence. Although he embellished his story as the years went by, Cooper built his locomotive, named the *Tom Thumb* much later, to show that it could perform on the new track. Though small, the high-pressure engine generated about 1.5 horsepower at a pressure of 50 psi. Cooper intended his locomotive only as a demonstration model, but it became the first American-built engine to carry passengers and continued to do so for more than six months.[15]

The merchants of Charleston, South Carolina, also saw their trade slipping away and determined to stop the drain. Looking eagerly at the rich cotton country west of the city, they organized a company to build a road to the small town of Hamburg on the Savannah River, just across from Augusta, Georgia. Completed in October 1833, the 136-mile Charleston & Hamburg (later South Carolina) Railroad became the longest railway line in the world. Its first engine, the *Best Friend of Charleston*, built at the West Point Foundry in New York for $4,000, was the first locomotive built for sale in the United States and in December 1830 carried 141 passengers on the first scheduled train driven by steam in the nation. It lasted only until June 1831, when a fireman, annoyed by the steam hissing from the pop valve, tied the valve down. The explosion killed him, sent the engineer flying, and demolished the engine.[16]

Boston merchants took an even more ambitious tack, underwriting rail lines to Lowell, Providence, and Worcester. Before the latter road was even completed, a new company organized to extend it another 150 miles through Springfield to Albany. As the 1830s wore on, railroad mania began to upstage the canal craze. The two sons of John Stevens organized the Camden & Amboy Railroad and completed it in 1833. The following year a group of three roads connected Philadelphia and Pittsburgh; later they would be combined into the Pennsylvania Railroad. By 1840 only four of the nation's twenty-six states still lacked any railroad track. As rails spread across the land, opposition came from canal owners, some states that had invested heavily in canals, turnpike and bridge companies, stagecoach lines, tavernkeepers, and anyone who saw his business threatened by the railroad. Some people even raised religious and moral objections. An Ohio school board warned in 1828 that "if God had designed that His intelligent creatures should travel at the frightful speed of 25 miles an hour by steam, He would have foretold it through His holy prophets. It is a device of Satan to lead immortal souls down to Hell."[17]

Like it or not, this latest device of Satan had come to stay. During the 1850s four railroads emerged as the trunk lines connecting eastern ports to distant western waters. When the Baltimore & Ohio completed its 379-mile line from Baltimore to the Ohio River at Wheeling, Virginia (now West Virginia), it cut a journey that required several days by stagecoach to sixteen hours. The historic meeting of the Union Pacific and Central Pacific railroads at Promontory Point, Utah, in 1869 gave the nation its first rail link to the West Coast. A traveler could board a train in New York City and reach San Francisco seven days later. Prior to that, the trip took thirty-five days to cover 5,250 miles if the traveler sailed to Panama, crossed the malaria-infested isthmus, and boarded another ship. Or one could devote nearly five months to the 13,500-mile route around Cape Horn. The overland stage route from St. Louis to San Francisco reduced the journey to a mere 2,800 miles and thirty days for those willing to endure the constant jouncing, vagaries of weather, and dangers of traveling through hostile territory.[18]

Not everyone agreed that the saving of time was worth the aggravation of early train travel. A Bostonian riding to Providence lamented being "sit cheek by jowl" with "poor fellows [who] squeezed me into a corner while the hot sun drew from their garments . . . smells made up of salt fish, tar and molasses . . . The rich and the poor, the educated and the ignorant, all herded together in this modern improvement in traveling . . . and all this for the sake of doing very uncomfortably in two days what could be done delightfully in eight or ten." A wealthy Philadelphia merchant complained that "if one could stop when one wanted, and if one were not locked up in a box with 50 or 60 tobacco-chewers; and the engine and fire did not burn holes in one's clothes . . . and the smell of the smoke, of the oil, and of the chimney did not poison one . . . and [one] were not in danger of being blown sky-high or knocked off the rails—it would be the perfection of traveling."[19]

British visitors added their own flavor of disdain. The actress Fanny Kemble, while on her American tour, compared the railroad car she rode to "a long greenhouse upon wheels; the seats, which each contain two persons (a pretty tight fit, too), are placed down the whole length of the vehicle, one behind the other, leaving a species of aisle in the middle for the uneasy . . . to fidget up and down, for the tobacco-chewers to spit in, and for a whole tribe of itinerant fruit and cake-sellers to rush through, distributing their wares at every place where the train stops." Charles Dickens, in his *American Notes*, was even more emphatic in his disgust:

> There are no first or second-class carriages, as with us, but there is a gentlemen's car and a ladies' car; the main distinction between which is that in the first, everybody smokes; and in the second nobody does . . . On, on, on . . . tears the mad dragon of an engine with its train of cars; scattering in all directions a shower of burning sparks from its wood fire; screeching, hissing, yelling, panting, until at last the thirsty monster stops beneath a covered way to drink, the people cluster around, and you have time to breathe again.[20]

Nevertheless, the enthusiasm for railroads mounted steadily. By 1856, when the Rock Island Railroad built the first bridge across the Mississippi River, seven railroads had already reached its banks. Two years earlier a New York politico, Thurlow Weed, already grasped the profound shift in trade that had begun. "In a business point of view this river is beginning to run upstream!" he wrote to the *New York Tribune*. "There is a West growing with a rapidity that has no parallel [and] the railroads that are being constructed . . . are to take the corn, pork, beef, &c, &c, to a northern instead of southern markets." The reasons were obvious: Shipping by rail was from the start faster, cheaper, and more reliable. Neither the canals nor the rivers could compete with the railroads in these areas. An Iowa produce merchant recalled the convenience of the service:

> When the railroad got in operation, produce men were as thick as potato bugs. If a man could raise two hundred and fifty dollars he could begin business. That amount would buy a carload of wheat. In the morning he would engage a car, have it put where he could load it, and have the farmer put his wheat in the car. By three o'clock in the afternoon the car would be loaded and shipped.[21]

Like all new technologies, the railroad quickly became a laboratory that attracted all sorts of inventors and tinkerers to the many challenges it posed. During the 1820s the Erie Canal had served as a training ground for engineers and inventors seeking to master canal technology; for those interested in railroads, the B&O took on that role as the project where every problem had to be faced and solved for the first time. However, a canal was merely a water road; a railroad required an entire technological system to function. Like all such systems, an improvement in one area demanded upgrades in other components to be effective. A bigger, more powerful locomotive could haul longer and/or heavier trains, which in turn required heavier rails, a more substantial roadbed, and stronger bridges. Although every element was crucial to efficient operation, the process began with the locomotive. Without a good engine, nothing moved.[22]

The development of the locomotive resembled that of the steamboat. It began with crude designs and rough construction because there existed no fund of engineering knowledge on which to draw and no machinery capable of producing heavy castings and forgings. All the engine's parts had to be made and put together by hand, making each locomotive a unique creation. Few endeavors offered more challenges or more potential for growth than enginemaking. Of the many men who entered the field, Matthias Baldwin soon emerged as the most successful. Born in Elizabethtown, New Jersey, in 1795, he apprenticed to a jeweler at sixteen, became a journeyman, and showed his ingenuity early by patenting a process for gold plating that became widely used. He then went into the manufacture of tools for bookbinding, built the first hydrostatic press made in the United States, and turned out engraved cylinders for printing calico textiles. During this work he developed an interest in steam engines and built a 5-horsepower version for his factory that continued to operate for forty years. Encouraged by its success, he began to build stationary engines.[23]

Baldwin ventured into the realm of locomotives when Franklin Peale asked him to build one for his Philadelphia museum. Although he had never even seen a locomotive, Baldwin responded early in 1831 with a model version that ran on a circular track at the museum hauling two cars that carried eight people. Impressed by the model, officials of the Philadelphia, Germantown & Norristown Railroad asked Baldwin to build them a full-sized locomotive to replace the horses used on their six-mile line. In constructing the engine, Baldwin had to teach his men everything. So primitive were the tools that he bored the cylinders by using a chisel set in a block of wood and turned by hand. Despite these difficulties, Baldwin completed the engine and named it *Old Ironsides*. On its first run the engine reached a speed of twenty-eight miles an hour.

Company officials accepted the shiny new machine but at first ran it only in good weather because they didn't want it to get wet. Despite its good service— *Old Ironsides* kept running for more than twenty years—company officials noted that it weighed seven tons, two more than specified, and tried to penalize Baldwin by paying him only $3,500 instead of the $4,000 called for in the agreement. Disgusted at his treatment, Baldwin told a colleague, "This is our last locomotive."[24]

He could not have been more wrong. The success of *Old Ironsides* led Baldwin to found the Baldwin Locomotive Works, which soon became the leading manufacturer of locomotives in the world. From its crowded Broad Street site in Philadelphia came a steady procession of locomotives, an average of 15 a year during the 1830s. By 1839 he had produced a total of 136 engines; twenty-five years later he was turning out 130 locomotives a year. Most of them were the so-called American, or 4-4-0, type designed by Henry R. Campbell, who created the prototype in 1837. Well suited to a wide variety of needs, the American type remained in production for the rest of the century. It served the military during the Civil War and hauled the first transcontinental trains. As the rail system expanded, designers created bigger, heavier engines to haul longer trains. By 1850 a good American engine could be bought for $8,000 to $10,000.[25]

Once the locomotive appeared, inventors heaped improvements on every aspect of it. As Baldwin himself wrote in 1852, "Not to go ahead now days is to go behind very fast." The engine acquired a bell, a whistle, a headlamp, and a cowcatcher to keep livestock and other obstacles from derailing it. When a plague of grasshoppers brought trains to a standstill in Pennsylvania in 1836, a sandbox was added to provide better traction. By 1840 the locomotive had already acquired the horizontal fire-tube boiler with a rectangular firebox at the rear that remained standard even though modifications steadily refined it. In 1869 young George Westinghouse filed the first of what eventually became 103 patents pertaining to an invention that would transform train operations: the air brake. Three years later he introduced an automatic version that historian Alfred Bruce called "a landmark in the history of technology." Westinghouse's air brake demonstrated in fact what seemed absurd in theory: that a large train could be safely and effectively stopped by the use of compressed air.[26]

Since locomotives worked in a wide variety of terrains, designers came up with more specialized versions to supplement the popular 4-4-0 American. By 1850 they had introduced a variety of new and larger types, some of which remained in use for both passenger and freight service into the 1920s. By the late nineteenth century the basic steam locomotive had become relatively stable in both design and construction. Then came a steady stream of new types that dominated the new century, each of them designed for some specific kind of task or terrain, such as mountains. In 1941 the gallery of ever larger locomotives culminated with the two most powerful articulated steam locomotives ever built: the gigantic 4-8-8-4 Union Pacific "Big Boy," and an enormous 2-6-6-6 type built for the Chesapeake & Ohio Railroad.[27]

Historian John H. White has argued persuasively that American locomotive construction in the nineteenth century was marked by "conservatism and steadfast resistance to the acceptance of novel or 'new-fangled' designs." Builders, faced with a soaring demand, wanted above all else a model that would provide good service and not break down. The basic components for such a machine came together by the early 1830s, and the 4-4-0 emerged as the version best suited to provide a wide range of uses reliably. To increase power, builders before 1900 had only to enlarge the boiler and cylinder size or raise the steam pressure. But as railroads confronted a growing variety of terrains and service demands, designers produced new types of engines to meet more specific needs. They also engaged in a continuous quest for more economical operation and fuel economy.[28]

Although the basic elements of the steam locomotive—the boiler, frame, driving wheels, and trucks—remained nearly the same during its lifetime, other changes vastly improved its operation. For decades engines could only be reversed by the engineer pulling mightily at the reverse lever known as the Johnson bar. As the locomotives grew larger and harder to reverse, the Johnson bar finally gave way to a pneumatically operated power reverse gear early in the twentieth century. Growth also strained the ability of the firemen to shovel enough coal to generate the needed horsepower until the first mechanical stoker, using a steam-jet overhead feed system to distribute coal, appeared around 1912. This system overcame the limitations imposed by hand firing and eventually increased horsepower by 400 percent, making possible even larger locomotives. It also enabled engines to burn lower-grade coal, of which the United States had huge deposits. Compound cylinders were introduced during the 1880s only to be superseded around 1905 by the superheater, which improved heating efficiency by 45 percent.[29]

The number of steam locomotives on American railroads did not peak until 1924, when about seventy thousand engines were in service. The breed ruled the railroad world until the 1950s, when nearly every company launched programs to replace them with the much more efficient electro-diesel locomotive. Here, as elsewhere, steam gave way to electricity, but this change should not obscure the revolutionary role played by the steam locomotive. It was the most important technology of the nineteenth century after the steam engine itself.

No other device did more to reshape American life in so many areas until the coming of the automobile in the next century. The railroad literally created inland America and, to a large extent, urban America. It became the nation's first big business and for more than a century remained its biggest business. In hastening the settlement of the interior, it created cities and towns where none had existed. In the process it routinized the movement of goods and people as nothing had before. Its voracious need for capital virtually created the American securities market and did much to make New York and Wall Street the financial hub of the nation.[30]

The railroad connected rural America with distant points and enabled goods to flow to markets they had never before reached. It did this by lowering the cost of transportation and by facilitating the settlement of remote regions. Resources that were once unreachable developed into large businesses. The construction of railroads gave rise to a host of new enterprises and dramatically boosted the output of existing industries, most notably iron and steel. Construction also developed into a major industry thanks to railroad work. As the first big business, the railroads were the first to utilize the corporate form and to tackle a host of problems connected with running a large operation over a far-flung landscape. They were the first business to deal with unions and the first to undergo government regulation. They had to devise new forms of organization, new methods of accounting, new ways of raising large amounts of capital. In these and many other areas, the railroads pioneered in the development of corporate structures that later became standard in the business world. They set the stage for the coming of the corporate economy and corporate society.

The locomotive made all this possible. It was literally the power behind the throne of American economic development. Even more, it became the cultural symbol of the age for millions of Americans, the machine that best expressed the essence of their time. Walt Whitman felt its power and celebrated it in his "Passage to India":

> I see over my own continent the Pacific railroad
>   surmounting every barrier,
> I see continual trains of cars winding along the Platte
>   carrying freight and passengers,
> I hear the locomotives rushing and roaring, and the
>   shrill steam-whistle,
> I hear the echoes reverberate through the grandest
>   scenery in the world.

STEAM HAD CONQUERED both land and sea. Just as the steamboat filled the western rivers, so did new and larger steamships ply the oceans. Sea travel posed a greater challenge because of the length of the voyage and the amount of fuel required for it. In his river wars with Fulton, John Stevens devised innovations for his boats that proved to be harbingers of ocean travel as well. The steamboat he built in 1804 had not only a direct-acting high-pressure engine but also two remarkably advanced features: a tubular boiler and twin screw propellers. Both anticipated later developments; the screw propeller would not be appreciated until John Ericsson brought it into general use thirty years later. Transatlantic steam navigation began in 1819 when the *Savannah* sailed from that city to Liverpool in twenty-six days. Although the ship was outfitted with steam machinery and paddlewheels, it also carried a full rig of sails. The engines saw duty on eighteen days of the crossing, with the sails providing power the rest of the time. From Liverpool the *Savannah* went on to several Baltic ports and St. Petersburg, Russia, before heading back to New York.[31]

Despite the success of the *Savannah* and a few other ships—one steamer made it to India in 1825—many authorities continued to regard ocean travel under steam as impracticable, mostly because of the fuel dilemma. In 1838, however, two ships demonstrated the feasibility of steam-powered transatlantic travel. The *Sirius*, a 700-ton ship of 250 horsepower, left Cork on April 4, while the 1,340-ton *Great Western*, generating 450 horsepower, sailed from Bristol four days later. Both ships reached New York on the same day, April 23, arriving to a fanfare of cheering dockside crowds, dipping flags, and boats filling the waters around them. The *Great Western*, carrying 660 tons of coal and only seven brave passengers, cut the usual crossing time for a packet ship in half. After a successful return trip the *Sirius* was deemed too small for transatlantic work and moved instead to a run between London and Cork. The *Great Western*, however, made seventy passages across the Atlantic during the next six years, averaging 15.5 days heading west and 13.5 days sailing east.[32]

A new age of Atlantic travel had dawned. Other ships were designed especially for the service. The Cunard Line began life in 1840 with four ships plying the Liverpool–New York route. By the late 1870s the company had eighty vessels on the route, some of which made the crossing in as little as eight days. None of them rivaled the monster *Great Eastern*, which became the largest and most notorious ship of the century. Begun in 1854 and completed five years later, the *Great Eastern* stood 680 feet long, 83 feet wide, and 58 feet deep. It boasted four paddle and four screw engines that together generated 10,000 horsepower. Unlike horses, however, these engines had to run twenty-four hours a day for several days. The boilers supplying the paddle engines alone had more than an acre of heating surface; those for the screw engines were even larger. The ship could carry four thousand passengers, but hard luck dogged its entire career. The *Great Eastern* never earned its keep as a passenger vessel; its claim to fame rested largely on carrying and laying the huge transatlantic cable in 1866.[33]

As steam engines on land and sea alike grew larger and more powerful, they shared a common appetite for fuel. Both locomotives and marine engines first consumed wood at a voracious rate until even the seemingly inexhaustible American forests began to vanish from all the demands made on it. Increasingly users turned to coal for fuel, but coal varied widely in type, quality, and cost as well as chemical composition. The cost of fuel was a major component of overall costs, not only for ships and locomotives but for textile mills as well. In all three areas the primary goal became the quest for engines that produced the most power as efficiently as possible utilizing the cheapest available fuel. The rapid spread of the steam engine, not only in number but in varieties of use, intensified the demand for cheap, efficient fuel to the point where fuel emerged as the linchpin of the power revolution, which in turn acted as the driving force behind American industrialization. As historian Alfred D. Chandler, Jr. has shown, these relationships underwent a profound transformation with the opening up of the great Pennsylvania anthracite coal fields in the early 1830s.[34]

Americans came late to their vast stores of anthracite, or "stone coal," because it was both unfamiliar and inaccessible. Most of the great fields lay in rugged, mountainous country far from navigable waters. Unlike bituminous, or "soft," coal, it was almost pure carbon, burned with a tiny blue flame, produced intense heat with hardly any smoke, and was so difficult to ignite that many people who lived around it doubted that it could even be burned. Its use required a learning curve that consumed more than a decade before enough demand arose to make mining profitable. Then the problem became finding ways to get at it and transport it cheaply to market. The completion of three coal canals by 1832 enabled anthracite to reach northeastern markets and sent production soaring. Where output totaled only 9,000 tons in 1823, it jumped to 209,600 tons in 1830, 678,500 tons in 1835, more a million tons in 1837, and 3.3 million tons in 1847. As the supply increased, the price fell from about $7.50 a ton in 1830 to below $4.00 a ton by 1842.[35]

The availability of cheap, high-quality coal in steadily increasing quantities triggered a sea change in the structure of several key American industries. It had an immediate effect on the nascent iron industry, which still depended largely on charcoal or bituminous coal for fuel. Anthracite was far more economical and efficient than either one and had no sulfur or other fumes to contaminate the iron. During the 1830s and 1840s ironworks concentrated increasingly in central Pennsylvania to be near their source of fuel. Large mills utilizing steam power came to dominate every phase of iron manufacturing. By 1849 eastern Pennsylvania alone boasted more than sixty blast furnaces;

that figure doubled in only four years. The average Pennsylvania anthracite furnace in 1849 had eighty workers and a capital of $83,000. The price of iron dropped steadily, and cheap fuel remained the primary reason.

As the supply of American iron became cheap and plentiful, it enabled entrepreneurs to organize factories for producing finished metal goods in large quantities. These ranged from guns to pots to tools of all kinds. The growth of large metalworking firms spurred the rise of the crucial machine tool industry. So too did cheap fuel encourage other industries that required heat for their production process, most notably glass and paper, to develop and expand operations. Steam power became far more affordable for these enterprises and for the nation's dominant industry, textiles. Cheap fuel enabled textile factories to replace water power with steam engines. When Samuel Slater of Rhode Island decided to outdo the water-driven mills of Massachusetts in 1828, he became the first to use steam power for his machines. From the beginning he relied on anthracite coal to fuel his engines. Other manufacturers soon got the message, and steam-driven mills began appearing in Rhode Island and throughout New England. Rhode Island's three steam-powered mills in 1832 mushroomed to twenty-nine in six years.

Through this chain of interacting factors, the factory system that came to characterize much of American production spread beyond textiles into one new sector after another. No industry was more central to industrialization than iron and steel, which depended utterly on the steam-driven power that cheap, efficient fuel made possible. The fast-growing railroad industry relied heavily on the availability of cheap iron and steel; at the same time its large purchases spurred the growth of the iron and steel industry. In sector after sector growth was made possible by the increasing use of machines, all of them powered by steam. Their use extended even to the farm, where labor was usually as scarce as it was sorely needed. As early as 1812 that versatile if erratic genius Richard Trevithick had built small machines for threshing and grinding corn as well as sawing wood. "It is my opinion," he predicted, "that every part of agriculture might be performed by steam."[36]

In the young United States, farmers and planters starved for labor looked early to the steam engine for help. A survey in 1838 listed 1,860 stationary steam engines in the nation. Most of those in the Northeast did industrial work, while those west of the Appalachians drove sawmills and gristmills. Most of the 585 engines in the South did service in plantation mills, grinding sugar, ginning cotton, threshing wheat, and threshing or pounding rice. Once in place, plantation engines often took on other tasks such as driving gristmills, sawmills, straw cutters, corn mills, and cob crushers. The early cotton gins, operated by hand crank or foot pedal, enabled a worker to clean about 40 pounds of cotton a day. With a steam-powered gin three men could clean anywhere from 1,000 to 4,500 pounds a day. The 1838 survey noted that slaves operated most of these southern engines.[37]

As the Midwest began filling up with settlers eager to grow grain on ever larger farms, inventors like Cyrus McCormick and Obed Hussey eased their labors by creating mechanical reapers and mowers. Other inventors devised a variety of threshers, cultivators, drills, harrows, and other machines. All of these labor-saving devices relied on horses, mules, or oxen for power. Threshing machines posed especially difficult problems. To get enough power for efficient threshing, inventors came up with "sweep-power" machines that utilized eight to fourteen horses and a complicated gearing mechanism. Since these rigs were expensive to build and maintain, farmers often pooled their funds to buy one. Gradually some acquired a machine and went into the business of threshing grain for neighboring farms. Thus arose the custom thresherman who roved from farm to farm plying his specialized trade.

Although these developments eased the work of farmers, they all relied on animal power. The stationary steam engine was too cumbersome and heavy for field work. A portable engine was needed to bring steam power to the farm. In 1849 A. L. Archambault of Philadelphia built the first portable engine intended specifically for agricultural use. Within a short time other engine builders added portable versions to their line for farm use as well as for gristmills and sawmills operating away from a source of water power. Enthusiasts rhapsodized over the possibilities they offered. "Agriculture by steam!" proclaimed an 1856 Indiana report. "What a change! . . . Indeed, it looks as though the ancient curse had been lifted from man."[38]

But the use of steam came slowly to the farm. During the 1860s and 1870s new inventions like the middling purifier greatly improved the flour-milling industry. The spreading railroad network made it easier and cheaper to get large crops to market. As more and more acreage came under cultivation, the enlarged harvests strained the ability to thresh it all quickly enough. A horse-powered rig could thresh three hundred to four hundred bushels a day, not nearly enough for many crops. Farms had simply outgrown both the threshing machines and the power to operate them. As one custom thresherman put it, "These small threshing machines are wasting our time. We must have larger machines to do the job right." Manufacturers responded by producing machines that could thresh nine hundred bushels a day, but no more than fourteen horses could be hitched in teams to any machine, and they wore out quickly under this harsh labor.[39]

Portable steam engines were the obvious answer, but some farmers feared them. "Our so-called plain common sense practical farmers have so perfect a mortal dread of a steam engine for farm purposes," observed one farmer contemptuously; "they look upon it as a thing of destruction. Fire and water and the dreadful explosive effects of steam are terrors to which they are not easily reconciled." Certainly early versions of these engines, fizzing steam through leaky valves and belching black smoke, did not inspire confidence among the uninitiated. A wary farmer feared not only death or injury but a disastrous fire that might wipe out his home and farm. Even so, need bordering on desperation drove many to acquire a steam engine even though it cost about $1,000 for a 10-horsepower version. The arrival of the first engine in a rural community often became an event that drew a crowd of curious spectators.[40]

Between 1885 and 1914 steam engines transformed agriculture as they had the industrial sector. Self-propelled versions resembled miniature locomotives and eliminated the need to haul the engine from one place to another by wagon. These steam traction engines appeared just as innovative new types of farm machines like the combine came into use. Over time the engines grew larger, more refined, and more reliable. Steam power on the farm reached its peak between 1905 and 1914 and did not fade away until the mid-1920s, when the ubiquitous gas-powered tractor began to replace it. Between 1900 and 1914 J. I. Case, the largest manufacturer of agricultural steam engines, produced 24,032 of them. In 1905 no fewer than thirty-seven companies together turned out about 7,500 steam traction engines. The immense increase in agricultural output and farm productivity during the late nineteenth and early twentieth centuries owed much to the growing role played by steam power on the farm.[41]

AS THE STEAM engine spread into more areas of American life, it underwent constant refinement. The incessant demand for more power kept constant pressure on the builders of steam engines to produce bigger, more powerful, and more efficient versions. Improvements in one aspect of the engine inevitably led to refinements in related aspects. The spread of high-pressure engines with their increased boiler pressures spurred the development of improved boiler designs such as the water-tube type pioneered by George Babcock and Stephen Wilcox. By the time of the Philadelphia Centennial, the water-tube boiler had largely replaced the older fire-tube, or cylindrical, designs. Another innovation, the compound engine, utilized two cylinders to make the engine run more smoothly and efficiently. Taken together, these and other developments pointed up the weakness of one component that had not been markedly upgraded: the valve gear.[42]

Although the older slide valve had been improved over time, it grew increasingly inadequate to serve newer types of engines. Its shortcomings were several, chief among them being its slowness in opening and closing. The need was for a more efficient type of valve with an automatic cutoff that could handle fluctuations in load and boiler pressure while permitting a maximum amount of expansion under all conditions. It was here that George Corliss made his first outstanding contribution. Having completed his first engine in 1848, Corliss received in March 1849 a patent for "certain new and useful improvements in Steam Engines." The most important of them dealt with innovations in the valve gear for a beam engine. Among other things he set the governor to control the point of steam cutoff, which enabled him to regulate the engine's speed by "the amount of steam admitted at full boiler pressure and its subsequent expansion."[43]

Corliss had invented the automatic cutoff valve, but it took several years of refinement to perfect it. Acceptance came slowly in Great Britain, where many businessmen harbored a suspicion of American inventions, but gradually it became the standard on most American and many British engines. Corliss gained a reputation as the best-known engine builder in the United States, especially after the Philadelphia Centennial. A shrewd businessman, he not only manufactured engines and boilers but sold licenses to other builders as well. Inevitably other builders sought to imitate his work and embroiled him in the usual unending stream of patent litigation. When his patents expired in 1870, a host of other manufacturers built "Corliss" engines, which only added to his reputation as the dominant figure among American engine builders. By 1903 some thirty American companies manufactured their own version of the Corliss engine. Although higher boiler pressures and increasing speed eventually rendered the Corliss valve gear obsolete, Allis-Chalmers continued to advertise its heavy-duty Corliss engines as late as 1925.[44]

The Corliss engines were efficient and reliable, but they also tended to be large, heavy, and expensive. Increasing the output of some machinery required a smaller, cheaper engine running at higher speeds. Two American inventors, John F. Allen and Charles T. Porter, combined their talents to produce such an engine, which amazed visitors at the 1862 International Exhibition in London. Their single-cylinder horizontal engine drove cotton and woolen looms at 150 revolutions per minute (rpm), about three times the speed of other mill engines. It featured a new type of valve invented by Allen and two crucial innovations by Porter: an ingenious new type of governor and a rigid frame that allowed the engine to be almost entirely self-contained. Their engine found extensive use in rolling mills and other applications utilizing high-speed machinery.[45]

Like so many other American inventors, Porter began life in another profession, in his case the law. Approaching the age of thirty, settled and with a family, he abruptly quit the law and turned to engineering. The Watt governor still dominated the control of steam engines when Porter saw that it would not do for engines running at higher speeds. His version met the requisites for a governor so perfectly that *Scientific American* declared in 1858 that it left "nothing further to be desired." His manufacture of governors led him to the more daring challenge of creating a high-speed engine. This ambition brought him into association with Allen, an uneducated operating engineer. Improved versions of the engine shown in 1862 went on display in Paris five years later and in New York in 1870. Meanwhile, Porter made the rounds of machine shops searching for new ideas. For three years, 1873–76, he stopped producing the engine while he incorporated new improvements. The Porter type of engine, like the Corliss, did not reach its peak production until the first decade of the twentieth century.[46]

American inventors also pioneered in another crucial area: the transfer of power from the engine to the machines it operated. The usual way of distributing power from a waterwheel or steam engine to the rest of the machines in a textile mill was through an elaborate combination of gears and line shafts. This cumbersome arrangement was noisy and inefficient and required constant maintenance. Early American mills utilized this system until 1828, when a new Appleton Mill in Lowell, Massachusetts, substituted belting in place of the gear drive to hitch the waterwheel to the other machines. Wrapped around a drum driven by the waterwheel, the broad flat leather belt traveled along guide pulleys to every floor of the mill, where it passed around other pulleys on the line shafting. This belt or rope drive ran far more smoothly and quietly, cost less, and was much easier to install and repair. Other mills took up the new system, which soon became a primary distinction between American and English mills.[47]

THE GOLDEN AGE of steam power extended from the 1830s through the 1870s. During that time the steam engine remained the primary source of power in such basic fields as transportation and manufacturing. After 1880 the steam engine continued to dominate, but it faced competition from a new source of power that would ultimately subsume it: electricity. Even then steam power did not disappear but rather became subordinate by providing power for the generators that created electric power. It also continued to power factories and mills into the twentieth century until electricity developed the capability to replace it. Yet another challenge to steam's supremacy loomed in the form of the newfangled internal combustion engine, which Nikolaus August Otto had first introduced commercially in 1876. The superiority of the gas engine was so obvious that some engineers regarded steam as hopelessly out of date.[48]

But steam power was far from dead or even dying. At the century's end it still ruled travel over land and sea, and it still powered most of the machines that manufactured products. As the electrical age dawned, it did not push steam aside so much as redefine its role. The largest reciprocating engines ever built went to provide power for one of New York City's Manhattan Elevated railways in 1898. Edwin Reynolds of Allis-Chalmers designed the 10,000-horsepower engines in a compact layout that came to be known as the "Manhattan." They were used to power a single generating station that produced 60,000 horsepower to service an entire fleet of trains. The engines did so well that three years later the city's IRT line ordered a set to run its own trains. By that time a new and more efficient type of steam engine had begun to render the reciprocating engine obsolete.[49]

The turbine differed radically from the reciprocating engine in that it created energy by forcing steam through jets rather than by the action of steam pressure on the piston of an engine. It promised far greater thermal efficiency in an engine that was much lighter, more compact, and used less fuel. However, it was also a precision machine that frustrated inventors who lacked the tools and technique to create it. Crude versions of the turbine can be found as far back as the second century B.C., and Richard Trevithick devised a "whirling engine" among his many achievements. But Trevithick realized that he lacked the means to drive it at the high speed required for economic performance. An American inventor, William Avery, patented an early version of the turbine in 1831 and sold his engine to sawmills. One was even tried in a locomotive, but another half century of experiments passed before practical turbine engines became a real possibility for commercial use. When they did finally appear, they gave the steam engine a new life as well as a new role. It became the primary power source that made possible the breathtaking expansion of the electric revolution.[50]

CHAPTER 4

IN SEARCH OF THE MYSTERIOUS ETHER

The day when we shall know exactly what "electricity" is, will chronicle an event probably greater, more important than any other recorded in the history of the human race. The time will come when the comfort, the very existence, perhaps, of man will depend upon that wonderful agent.

—NIKOLA TESLA [1]

STEAM AND ELECTRICITY HAD ONE important element in common. Well into the nineteenth century, they remained a mystery to scientists and engineers alike. What was heat? What was electricity? These basic questions baffled and tantalized those who investigated them. Engineers were not scientists but practical, often self-taught men who learned their trade on the job and knew little if anything of theory. Scientists might have more formal schooling, and some became professors, learned or otherwise, but many tended to be gifted amateurs wealthy enough to devote their time to exploring the riddles of nature. Those who devoted themselves to unlocking the secrets of heat and electricity crossed paths on more than one occasion.

Early theories of heat fell into two distinct categories. Some considered it a distinct type of matter, while others regarded it simply as a condition that could be produced in certain bodies. In the seventeenth century Robert Hooke declared that "heat is a property of a body arising from the motion or agitation of its parts." Isaac Newton agreed. "Do not all fixed Bodies," he asked, "when heated beyond a certain Degree, emit light and shine; and is not this Emission performed by the vibrating Motion of their Parts?" One prominent follower of Newton, Wilhelm Jacob 'sGravesande [*sic*], tried to meld these opposing views by describing heat as a form of motion containing a particular kind of matter similar to that of light. His influential *Physices elementa mathematica experimentis confirmata*, published in 1720, was reprinted several times and translated into both English and French.[2]

Steam was produced by using fire to heat water, but the process by which this created energy remained unknown. Chemists also wondered why fire was needed to form metals out of ore, and how this transformation occurred. Around 1700 a German chemist, Georg Ernst Stahl, advanced the notion that combustible objects were rich in something he called "phlogiston," from a Greek word meaning "to set on fire," which diminished as the burning process continued until none remained in the ashes or residue. For nearly a century the notion of phlogiston dominated the thinking about heat until Antoine Lavoisier challenged it. Through a series of experiments Lavoisier cast doubt on the phlogiston theory by demonstrating that combustion did not result in a loss of mass but merely in a shifting of it from one form to another. His concept of the conservation of mass helped establish him as the father of modern chemistry. Unfortunately his work was literally cut short in 1794 when he was guillotined during the horrors of the French Revolution.[3]

Although Lavoisier dismissed the theory of phlogiston, he still viewed heat as some sort of fluid that moved from one substance to another. He called this fluid the "caloric," which became a mystery in itself. It was described as a "subtle elastic fluid which permeated the pores of bodies and filled the interstices between the molecules of matter. It could not be created, neither could it be destroyed, nor yet had it weight." Although the caloric theory actually explained little, professors and practitioners alike clung leechlike to it well into the nineteenth century. As investigators learned more about the properties of heat, they tried to stuff their findings into the caloric theory until it acquired a potpourri of meanings. One researcher thought of the caloric as an "atmosphere" engulfing atoms; another regarded it as atoms tinier than ordinary particles that repelled each other while being attracted to the latter. In 1834 a third scientist depicted it this way:

> Caloric is uniform in its nature; but there exist in all bodies two portions of caloric very different from each other. There is one called "sensible heat," or free caloric, the other "latent heat," or combined caloric . . . "Latent caloric" is that portion of the matter of heat which makes no sensible addition to the temperature of the bodies in which it exists. Caloric, as it penetrates bodies, frequently forms a chemical composition with them, and becomes essential to their composition.[4]

Not everyone subscribed to the ubiquitous caloric. As early as 1798 Benjamin Thompson (later Count Rumford) came to a radically different conclusion. A brilliant though arrogant and thoroughly unpleasant American, Rumford sided with the British during the American Revolution and found himself in exile after the war. He took refuge first in England and then, after wearing out his welcome, in Munich serving the elector of Bavaria. While boring some cannon for the elector, he forced a boring tool under ten thousand pounds of pressure against the head of a cannon immersed in water. To his astonishment, the water began to boil after two and a half hours. Why? He realized that caloric theory could not explain what was happening before his eyes.[5]

The answer, Rumford decided, must be that somehow the mechanical motion of the borer was being converted to heat. If so, this meant that heat must be some form of motion. Others before him, most notably Francis Bacon, Robert Boyle, and Robert Hooke, had approached this concept, but Rumford spelled it out in a paper submitted to the Royal Society. "What is heat?" he asked. "Is there any such thing as an igneous fluid? Is there anything that, with propriety, can be called caloric?" The only explanation, he declared, and the only one needed to explain the phenomenon, was motion itself. Impressive (and correct) as this conclusion was, it did not persuade most scientists to abandon caloric theory.[6]

Practical engineers grappling with early versions of the steam engine found a useful analogy that seemed to reinforce the caloric theory. They understood hydraulic power, the way a waterwheel derived its energy from the fall of water to a lower level. It was easy to equate steam pressure to this fall of water. James Watt, who accepted the caloric theory, formulated a law that any given amount of saturated steam contained the same amount of caloric at all temperatures. Gifted French scientist Nicolas L. S. Carnot drew a different analogy from the way water generated energy: Just as water required a lower level to generate power, so must the steam engine use the fall in temperature to create energy. In 1824 he published a book arguing that the steam engine did work only when heat fell from a higher to a lower temperature. The *change* in temperature created energy.[7]

From this insight flowed staggering implications. It explained, for example, the superiority of high-pressure steam engines over low-pressure ones: The high-pressure engine utilized a greater fall of the caloric. Later it spawned the second law of thermodynamics, which said that a self-acting machine could not transfer heat from one body to another at a higher temperature. Carnot still reasoned in terms of the caloric, even as his main point discounted it, by saying that only the fall in temperature, not the agent itself, produced work. A good steam engine, he declared, "should not only employ steam under heavy pressure, but *under successive and very variable pressure, differing greatly from one another, and progressively decreasing.*" But Carnot's work went unheeded for nearly two decades even as scientists toiled furiously at a wide range of experiments on the nature of heat.[8]

One gifted amateur, the son of a brewer, outshone most scientists in the precision and range of his experiments. James Prescott Joule was so devoted to his work that on his honeymoon he devised a special thermometer to measure the temperature of the water at the top and bottom of a scenic waterfall. Joule became fascinated with electromagnetic machinery, which led him to ponder the problem of energy conversion. For a decade he carefully measured the heat produced by every device he could contrive. Like Rumford, he found a close relationship between the amount of work each performed and the amount of heat it produced. Scientists spurned his first published efforts, possibly because he was not an academician, but Joule persisted. In 1847 he read a paper at a scientific meeting that impressed no one until a young man in the audience rose to praise it. William Thomson (later Lord Kelvin), who became a key figure in unraveling the mysteries of electricity as well as heat, recalled the moment this way:

> I . . . felt strongly impelled to rise and say it must be wrong. But as I listened on and on I saw that though Carnot had vitally important truth not to be abandoned, Joule had certainly a great truth and a great discovery, and a most important measurement to bring forward.[9]

Two years later Joule presented his "On the Mechanical Equivalent of Heat" before the Royal Society. In this paper he declared that the amount of heat needed to raise the temperature of a pound of water one degree Fahrenheit equaled the work required to raise 772 pounds one foot. This value became known as the British thermal unit (BTU) and remained unchallenged for half a century. Joule's findings confirmed the principle of conservation of energy and established the dynamical theory of heat. It also led to the formulation of the first law of thermodynamics, which said in effect that heat could be transformed into mechanical work and vice versa. The notion of heat as a force befuddled many scientists. The first law of thermodynamics directly contradicted caloric theory, which held that heat was always conserved as heat and could not change into mechanical energy. Clearly a sea change in the understanding of heat had begun.[10]

The first law of thermodynamics declared not only that heat and mechanical energy were mutually convertible but also that heat required mechanical energy for its production and in turn produced mechanical energy as it dissipated. From this revelation emerged another fundamental principle: the conservation of energy. Put simply, energy cannot be destroyed but simply assumes other forms. Any heat created by fire, for example, either has to be directed to some useful purpose or simply wasted. The trick is always to find ways to utilize as much of the energy produced as efficiently as possible and reduce the amount wasted to a minimum. Winding the spring of a clock applies all the energy produced to a desired goal; a fire raging out of control creates a large amount of energy that is wasted because it cannot be harnessed.[11]

From these insights flowed the second law of thermodynamics. When energy/heat is transferred or transformed, the total effect is directly proportional to the total amount of heat present. Equal quantities of heat produce equal effects. Heat could move from a hotter body to a colder one only with some form of compensation. That is, to make a cold body even colder by removing heat from it, one must either pass the heat removed to a still colder body or else add more energy (which is the principle used in the refrigerator). Thus emerged a startling concept that scientists and inventors alike were slow to grasp: Any form of work or energy is created by a change in temperature. It occurs only during the fall of temperature between a hot source and a colder destination. In a steam engine these became the fire in the boiler and the condensing water.[12]

The quest for a more efficient steam engine took a new direction as scientists pitched doggedly into the intricacies of thermodynamics. The work of Joule and others demonstrated that the steam engine was inefficient, but no one grasped the reason why until Thomson came up with a formula to explain it. His findings confirmed Carnot's conclusion that the greater the decline in temperature, the more efficient the engine became. In 1850 a German physicist, Rudolf J. E. Clausius, reconciled Carnot's theory with the dynamical theory by showing that both the flow of heat from a hot to a cooler body and the transformation from heat to mechanical energy and vice versa occurred. His crucial point was in demonstrating that heat itself was the vital driving force in a steam engine, with the pressure acting only in a secondary role. The message was clear: A steam engine, to be as efficient as possible, had to operate at much higher pressures than those in use at the time in order to create the higher temperatures and greater fall needed.[13]

ELECTRICITY, LIKE HEAT, had been a familiar if terrifying force even to the ancients. Nature displayed it spectacularly in the form of lightning, though no one realized the connection for centuries. Prehistoric humans also knew of another, seemingly unrelated force later called magnetism. They learned of it through a common substance called magnetite, some pieces of which were permanently magnetized and known as lodestones because of their peculiar ability to lift several times their own weight. Around 600 B.C. a Greek philosopher named Thales observed that amber, if rubbed vigorously, could pick up light objects such as grass or straw. He also experimented with the lodestone and learned of its ability to attract iron. Whether these ancient stories about Thales are true or not, the Greeks did supply the words for electricity and magnetism. The former comes from the Greek word for amber (*elektron*), and the latter probably from a district in Thessaly called Magnesia, where lodestones were found. However, the relationship between electricity and magnetism remained a mystery.[14]

Down through the centuries both phenomena continued to fascinate observers who sought to understand their nature. A major breakthrough came in 1600 when William Gilbert of England, having experimented with electricity and magnetism, poured seventeen years of research into a volume with a formidable Latin title that has been shortened to *De Magnete*. Sometime before 1200 it had been discovered that a magnetized needle, freely floating or suspended, invariably pointed north and south, but no one knew why this happened. Gilbert demonstrated that if a compass needle was given vertical room, it pointed not only northward but also downward toward the earth. He explained this behavior by describing the earth itself as a giant magnet and declared that the compass pointed not to the heavens, as was widely believed, but to the magnetic poles of the planet. He also discovered a variety of materials that possessed the same attractive force as amber when rubbed and called them "electrics." Gilbert, who became the personal doctor to Elizabeth I, was apparently the first to coin the term "electricity," and he pioneered in the use of careful experimentation as a methodology for scientific investigation.[15]

Yet electricity remained far more elusive than heat and steam. The latter were at least palpable; you could feel the one and see the other. Apart from the unexpected shock, however, electricity revealed its presence only through inexplicable phenomena. Untutored inventors had created machines to produce steam on their own, but electricity defied this approach. As one Frenchman despaired, "We are not yet in the habit of observing machines that function without apparent cause. Their occult workings baffle us. The secret of their existence escapes us." Some associated electricity with the "aether," that mysterious fifth element invented by Aristotle to explain the composition of the heavens. As late as 1745 a Swiss professor proclaimed that "electricity is a vast country of which we know only some bordering provinces."[16]

To the Greeks, "aether" (or ether) referred to the blue sky or upper air above the ordinary air of earth. Over time it was applied to space—the mysterious void between celestial bodies. Not until the seventeenth century did it become a scientific term thanks to the work of René Descartes, who sought to create a comprehensive theory of the universe. In pondering the interaction between bodies not in contact with each other, ranging from the behavior of magnets to the effects of the moon on tides, Descartes concluded that space was not a void but a medium of some sort capable of transmitting force and influencing material bodies. In his view the ether had mechanical properties, occupied the entire universe except for a tiny amount filled with ordinary matter, and consisted of particles that remained constantly in motion. Going even farther, Descartes suggested that the cosmos was in effect a huge machine, the workings of which could be predicted by mathematical calculation.[17]

Once the ether became a scientific concept, researchers devised all sorts of ingenious experiments to probe its mysteries. In 1660 Otto von Guericke, the mayor of Magdeburg, created the first electrostatic machine, which produced far more electricity than anything before it and made possible new types of experiments. Stephen Gray, an English dyer, demonstrated in 1729 that electricity could be transmitted and that some substances conducted it while others did not. Four years later, in Paris, Charles Du Fay showed that all bodies could be electrified and that conductors needed to be insulated. He also discovered two kinds of electric "fluids," which he called "vitreous" and "resinous." Each of them repelled itself but attracted the other one. In England William Watson wrote a number of books about electricity based on extensive experimentation and became the first to apply the terms "plus" and "minus" to electric polarities. A major discovery occurred during the 1740s when both a Dutch and a German researcher, working independently of each other, learned the hard way that an electric charge could be stored to high strength and then released with shocking results. News of the "Leyden jar," the first device capable of storing static electricity, spread rapidly and enabled still more imaginative research into the mysterious force.[18]

In the United States those interested in electricity, most notably Benjamin Franklin, knew little about what was happening in Europe. Franklin, however, did have a thorough knowledge of Isaac Newton's *Opticks*. From it he learned not only important concepts, such as the notion of an elastic fluid, but also the rigor of performing and reporting experiments carefully. After hearing a lecture on the subject by Archibald Spencer, a visiting Scotsman, Franklin bought the man's equipment and plunged eagerly into experimenting. A friend in England sent him a glass tube along with a 1745 article on the current state of electrical knowledge. Fascinated by the subject, Franklin retired from his business interests and during the winter of 1746–47 devoted nearly all his time to exploring electrostatic effects and related lines of research.[19]

From this work emerged several important discoveries. Franklin concluded that an electrical charge was "a real Element, or Species of Matter, not *created* by the Friction, but *collected* only." An ingenious series of experiments with the Leyden jar and other apparatus led him to several striking observations. To his surprise he found that "the whole force of the bottle, and power of giving a shock, is in the GLASS ITSELF; the non-electrics [i.e., conductors] in contact with the two surfaces." He decided that all bodies contained electricity, and that when two dissimilar but nonconductive substances were rubbed together, one acquired an excess of electricity and the other a deficiency. From this discovery he reasoned that there was only one kind of electricity, not two as Du Fay had argued, with two characteristics that he called positive and negative. Like Du Fay and Watson, he found that these attracted each other while repelling its own type in "a manner that I can by no means comprehend."[20]

Franklin also devised and named the first electric battery, which he described as "consisting of eleven panes of large sash-glass, armed with thin leaden plates pasted on each side, placed vertically and supported at two inches' distance on silk cords, with thick hooks of leaden wire . . . and convenient communications of wire and chain from the giving side of one pane to the receiving side of the other; that so the whole might be charged together and with the same labour as one single pane." He delighted in experiments that amused as well as informed, especially the "electric kiss," and was the first to suggest in 1753 that lightning emanated from the ground upward rather than from the clouds downward even though he began with the opposite view.[21]

Lightning especially fascinated Franklin. As early as 1749 he associated it with "electric fluid" and compiled a list of a dozen similarities between them. His writings on the subject reached an appreciative audience in Europe and made him the first American scientist to gain a reputation there. In June 1752 Franklin carried out his most famous experiment: flying a kite during a thunderstorm to prove that lightning was an electrical discharge. Using the kite string, he charged a Leyden jar as the storm passed over him. Thanks to his writings, others in Europe had already tried this experiment. Scientists in England and France did it successfully; an unfortunate scientist in St. Petersburg was electrocuted in his attempt. Franklin believed he had found a way to avert danger by using a pointed rod raised high to attract lightning and carry it to ground. From this insight came the first practical electrical device, the lightning rod.[22]

As historian I. Bernard Cohen has argued, these experiments led to several important and often overlooked results. Franklin demonstrated that electricity occurred in nature and that it was the same as the discharges produced in the laboratory. His invention of the lightning rod gave the first practical support to the prediction of Francis Bacon that the pursuit of basic scientific knowledge would eventually lead to practical and useful innovations. Franklin also struck a blow against superstition, especially the belief that lightning was the wrath of God directed against sinners. Finally, the experiments gave Franklin an international reputation that set the stage for his later career as a diplomat. He came to the courts of Europe as the only American celebrity.[23]

Since lightning was obviously difficult to work with, most scientists confined themselves to experiments with static electricity. By 1800 electrostatics had become the dominant line of inquiry among electricians—the name given those who studied electricity—although others continued to work on atmospheric (lightning) electricity, animal experiments, and the chemical nature of electricity. Several of them had observed the phenomenon of induction, in which a body became electrified when approached by another body carrying an electrical charge. However, the knowledge accumulated had all been qualitative; it could show but not measure or formulate the relationship between electrostatic charges. A brilliant French physicist, Charles-Augustin de Coulomb, pioneered in providing a mathematical basis for these relationships. Having gained a reputation through his impressive studies of friction and magnetic needles, he turned to electricity and magnetism.[24]

Unlike most earlier electricians, Coulomb knew both physics and mathematics and also had practical engineering experience. Around 1784 he invented a device known as a torsion balance to measure the mutual forces of electrostatic charges. One writer hailed it as the "first precision electrical instrument." Through careful, tedious experiments, Coulomb used it to formulate the first quantitative laws governing electric and magnetic fields. He showed that the force between two magnetic poles was proportional to the product of the pole strengths and inversely proportional to the square of the distance between them, and that the same relationship held true for two electrical charges as well. Apart from their immediate significance, these laws of inverse squares marked the beginning of quantitative work in electricity and magnetism. Coulomb did not take the next step of speculating that the two forces might in fact be one and the same; that insight would not appear for many more decades. However, he brought to physics the importance of mathematical measurement as part of the scientific method. In this respect he pointed the way to the crucial contributions of James Clerk Maxwell in the next century.[25]

By 1800 these men and others too numerous to mention had advanced electricity from a curiosity of nature to a growing field of scientific study. Yet no one knew what it was—the two-fluids-versus-one-fluid debate still raged— and no practical application had come from the work other than the lightning rod. Electricity remained a mysterious ether whose properties only began to be uncovered during the next half century. But how to get at it? Every source of electric current remained transitory, produced by friction, heat, or induction from the release of accumulated charges in a Leyden jar. Little progress could be made unless some more continuous source of electricity was found. The long and winding road of discovery in this area began with Alessandro Volta, an unassuming physics professor who had been drawn to some experiments with animal electricity conducted by another professor, Luigi Galvani.[26]

During his research with frogs, Galvani had in 1780 stumbled onto an astonishing discovery: A dead frog's leg convulsed when touched with a metal scalpel. This accidental observation lured Galvani into eleven years of research into the phenomenon before he published his findings. He concluded that the twitching resulted from a "nerveo-electrical fluid" that had accumulated in the muscle and apparently acted like a Leyden jar. Although the presence of electricity in animals had been known since ancient times thanks to the eel, Galvani's findings caused a sensation, in part because they hinted at the possible uncovering of the life force itself. Like many other scientists, Volta obtained a copy of Galvani's paper and eagerly repeated the experiment. Puzzled by the results, he found inconsistencies that confirmed his skepticism. Further research led him to a very different conclusion, one that sparked a fierce controversy and a major invention.

Born in 1745, Volta seemed so slow as a child that he was thought to be retarded. However, he blossomed into a prodigy, developed a fascination for science, and published his first research paper at twenty-four. In 1775 he invented the electrophorus, which accumulated an electrical charge much more efficiently than a Leyden jar. The first device to replenish an electrical charge through induction rather than friction, it became enormously popular. Volta became professor of physics at the University of Padua at thirty-seven, traveled widely, learned several languages, and won election to the Royal Society of London in 1791, the same year Galvani published his paper. By late 1793 Volta had rejected Galvani's claim of "animal electricity," believing that the source of electricity lay rather in the contact of two dissimilar metals during the experiment. Using one of his own inventions, the condensing electroscope, Volta systematically tested a variety of metal disks for their "electromotive force," a term applied to electric current as opposed to static electricity.[27]

By measuring the amount of charge on each disk, Volta found that contact between zinc and copper left the latter with a positive charge and the former with a negative one. Copper also came up negative in contact with lead, tin, and iron although the charge grew successively weaker. But when copper came into contact with silver or gold, it developed a positive charge, while the precious metals had a negative one. Convinced that a polarity relationship existed, Volta sorted the metals into a descending order from plus to minus: zinc, lead, tin, iron, copper, silver, gold, graphite. He concluded that when any two of these metals came into contact, the one higher on the list would be positive and the other negative. He also found that the greater the distance between them on the scale of metals, the stronger the electrical charge. What if the elements were multiplied? Volta decided to create a pile of zinc and silver disks, separating each pair with pieces of cardboard soaked in brine or other liquid because his experiments had revealed that certain fluids enhanced the electrical charge.[28]

The result amazed even Volta, who announced his invention to the Royal Society in March 1800. The pile produced not a transitory but a steady electric current that could be felt. The larger the pile, the stronger the current. Volta did not pretend to know why or how this flow occurred. "This perpetual motion may appear paradoxical, perhaps inexplicable," he admitted, "but it is nonetheless true and real, and can be touched . . . with the hands." Later Sir Humphry Davy showed that the current resulted not from the mere contact of the metals but from the electrochemical interaction between the metals and the liquids. Nevertheless, Volta had given the world its first practical electric battery, its basic elements consisting of two dissimilar metals with a liquid separator. The impact was immediate and far-reaching. For the first time researchers had a continuous source of electricity to use in their experiments.

Other scientists rushed to create better and more powerful versions of Volta's pile. Davy's proof that the voltaic cell was chemical in nature spurred its deployment in chemistry, especially electrolysis. For this work Davy constructed a huge voltaic pile in the basement of the Royal Institution. In 1809 he used its power for one of his most spectacular demonstrations. Before a large audience he held up two thin charcoal sticks, one of which was connected to the giant battery. As the current began to flow, Davy touched the sticks and produced a spark. Then, as he moved them apart, the spark increased and created a dazzling blue-white arc of light. Davy had shown the world its first arc light, and it made a lasting impression.[29]

However, the greatest impact of Volta's device lay in the opportunity it gave scientists to explore the nature of electricity and magnetism, and especially the relationship between them. Bigger and better batteries also made possible the development of the first great practical application of electricity, one that would join the locomotive in changing the world: the telegraph. For the first time men began not only to conceive but to experiment with ways to do the seemingly impossible: relay a communication over distance faster than man or beast could carry it.

* * *

THE RELATIONSHIP BETWEEN electricity and magnetism continued to tantalize scientists. Scraps of evidence pointed to their similarity, but nothing definite had been proven. Some argued that no relationship existed between them. Then, in 1819, an obscure Danish professor stumbled onto the answer. Hans Christian Oersted had prepared a demonstration of the voltaic pile for his class. He was about to use the current it produced to melt a wire. A compass happened to be lying near the wire. As the current was switched on, Oersted noticed that the compass needle swung toward the wire. He shut the battery down, and the needle returned to its normal north-south position. When the battery started up again, the needle again veered toward the wire, leading Oersted to conclude that it must be responding to the electric current. Intrigued, he spent several months experimenting with stronger batteries and discovered that a wire carrying an electric current, if allowed to move freely, was swayed by a magnet. In July 1820 he published his results, which declared that electric current from a battery produced an effect beyond the wire carrying it and that this effect was magnetic in nature.[30]

Oersted sent his four-page paper to every major university, learned society, and electrician in Europe. His announcement galvanized (note the origins of that word) scientists, who immediately grasped its significance. One of them, André-Marie Ampère, later wrote in his journal that "Oersted has forever attached his name to a new epoch. This learned Danish professor has opened, by his great discovery, a new field of research to the physicists." But that assessment came later. At the time Oersted's paper flabbergasted Ampère, who had long accepted Coulomb's hypothesis that no interaction between electricity and magnetism was possible. So formidable had the acceptance of Coulomb's theory been that for twenty years no scientist in France had even bothered to experiment with the action of a battery on a magnet. When Dominique-François-Jean Arago suggested the possibility of a relationship, his fellow scientists hooted it down as impossible. So attached was Ampère to Coulomb's hypothesis that the syllabus for his course in experimental physics included a demonstration showing "that the electrical and magnetic phenomena are due to two different fluids which act independently of each other."[31]

Like any good scientist, Ampère hastened to explore this contradiction. "Since I have heard of the beautiful discovery of M. Oersted . . . ," he wrote his son, "I have thought of it constantly." Only two months after Oersted's discovery, Ampère himself made a notable contribution. He showed that two conductors carrying two currents both behaved like magnets. Parallel wires carrying currents in the same direction attracted each other; those carrying currents in opposite directions repelled each other. His research then followed two distinct courses. One set of papers explored mathematically the mutual relationships between electric currents. In 1822 Ampère published the law that came to bear his name, which quantified the relationship between an electric circuit and a magnetic field. From this work flowed the laws that later earned him the sobriquet of the "Newton of electricity." His second line of inquiry was experimental and resulted in nine points that summarized his own theory of electrodynamics.[32]

In January 1821 Ampère first introduced his notion of molecular currents, which led him to formulate a new theory of matter in which such currents were an integral part not merely of magnetism but of all molecular processes. In devising this theory he resorted to the luminiferous ("light-carrying") ether as the agent bringing matter and electricity together. Few things baffled or divided scientists more than this mysterious substance. As described by Augustin-Jean Fresnel, a French physicist who argued that light consisted of waves, the luminiferous ether was a gaslike substance through which both light and solids somehow moved. A French mathematician, Augustin-Louis Cauchy, worked out a mathematical basis for the properties of ether that made Fresnel's theory at least plausible if not satisfying to scientists.[33]

The wave theory of light required that ether be perfectly elastic and offer no resistance to a body passing through it. To these demands Ampère added a new chemical wrinkle: The ether was not simple but compound in nature and could "only be considered, in the generally adopted theory of two electric fluids, as the combination of these two fluids in that proportion in which they mutually saturate one another." They tended to transmit vibrations as well as decompose and recombine, which for Ampère explained electrodynamic effects. It was the vibrations, for example, that accounted for the tendency of two wires carrying currents to attract or repel one another. As for magnetism, Ampère viewed it as the alignment of molecular currents rather than individual molecules. In this way his physical theory of electrodynamics rested on the presence of the ether and the existence of the molecules.[34]

By the time Ampère published a summary of his work in the fields of magnetism and electricity in 1827, he had formulated several laws describing the mechanical action between electric currents. He had also shown that a wire wound into a coil (he called this a "solenoid") acted like a magnet with north and south poles. This device became crucial to later electromagnetic experiments. From Oersted's experiments and his own, he concluded that magnetism was not a separate entity but rather some kind of electric action. His work endowed research in the field with bold new concepts and techniques as well as more precise tools for measurement. The old emphasis on electrostatics gave way increasingly to new theories about the electric circuit and its attributes.[35]

During that same year of 1827 an obscure German teacher at a school in Cologne published a book summarizing his work and detailing a law that came to bear his name once it was accepted. Georg Simon Ohm had followed the experiments in conductivity as he pursued his own research. Ampère's law provided the basis for defining an absolute unit of current strength. Ohm wanted to understand the fundamental forces in an electric circuit. He probed the relationship between an electric force, its resulting electromagnetic intensity, and the length of its connecting wire. Through a series of thought experiments followed by testing, he arrived at the formula that later became known as Ohm's law. Expressed as $I = V/R$, it declared that the current in any electric circuit was directly proportional to the voltage of the circuit, and inversely proportional to the resistance in the circuit. Critics laughed at his findings at first; one dismissed it as a "web of naked fancies." Not until the 1840s did scientists begin to grasp the implications of Ohm's contribution or recognize its importance.[36]

Decades later Ohm's name would be used to represent the standard unit of resistance. Ampère too would be honored by having his name reflect a unit of electric current, and Volta's was used in calling a unit of electromotive force the volt. Coulomb's contribution was recognized by giving his name to a practical unit of electrical charge. Joule got his due by having his name attached to the measurement of energy, Oersted by his name being used as the unit for the "reluctance" of a magnetic circuit. Even Watt joined this exclusive club by having his name represent the amount of power that gives rise to the energy of one joule in a single second. Although Watt had nothing to do with electricity, his steam engine provided the basis for mechanical power. He also worked out the early basis of horsepower, which later became the unit of measurement for electric motors. Those motors also came to be rated in watts (as did lightbulbs and other electrical devices) on the basis of 1 horsepower equaling 746 watts. This recognition spoke eloquently to the pioneering roles played by all these men. Their work set the stage for yet another pioneer whose insights finally thrust the investigation of electricity into the realm of practical development.[37]

Michael Faraday's life qualifies as a Horatio Alger story of rags to riches even though he was English rather than American. Born in South London, the son of a blacksmith, he overcame a meager education by reading voraciously. At fourteen he was apprenticed to a bookseller. Unlike John Fitch, who was bound to a harsh taskmaster, Faraday worked for a kindly man in an environment that fed his hunger for learning. One book in particular, *The Improvement of the Mind* by Dr. Isaac Watts, a celebrated hymn writer, became his inspirational self-help guide. Although he read widely, science especially fascinated him. He acquired some chemical apparatus and began experimenting with electricity. Not content with reading, he haunted the lectures on the relationship between chemistry and electricity given by Sir Humphry Davy at the Royal Institution, which had been founded in 1799 thanks to a proposal by none other than the roguish Count Rumford, who left London never to return three years later.[38]

Davy was one of England's brightest and most popular scientists. As a chemist he had already shown how nitrous oxide, later known as "laughing gas," could be used as anesthesia. Having read a letter of Volta's discussing how a chemical action could produce electricity through the mere contact of two dissimilar metals, Davy set out to prove the opposite—that electricity could be used to produce certain chemicals. His experiments along this line led him to discover sodium, potassium, calcium, barium, strontium, and magnesium. Later he isolated boron and explained the nature of iodine. A handsome, dynamic man, Davy thrilled audiences at the Institution with his brilliant, carefully honed lectures delivered with the flair of a showman. His lectures brought crowds, prestige, and financial support to the Institution. Samuel Taylor Coleridge once claimed that he attended Davy's performances to renew his stock of metaphors and added that "had he (Davy) not been the first chemist, he would have been the first poet of the age."[39]

In 1812 Faraday attended a series of four lectures given by Davy. By then he had completed his apprenticeship and gone to work as a journeyman bookbinder for a man he disliked. He took copious notes on the lectures and bound them together. On the advice of a friend, Faraday showed the notes to Davy, who was impressed enough to hire the young man as his assistant the following spring. Davy had married a rich widow and decided to travel on the Continent for two years, mixing science with pleasure. Faraday, who had scarcely gone a dozen miles from London in his life, found himself part of an entourage winding its leisurely way through France, Belgium, Italy, Switzerland, and Germany. In the process he met most of the scientific luminaries of his age, including Ampère, Volta, and Rumford. Faraday returned to England in 1815 a new man, well traveled, steeped in other cultures, and equipped with two new languages, French and Italian. Impressed by the young man's talent and love of science, Davy promoted him to superintendent of apparatus at the Institution, of which Davy was president.

Faraday had found his niche in life, but no one dreamed how far he would go within its confines. Chemistry occupied most of his time, and as his reputation grew he was called upon for lectures and consulting work. In June 1821 he married Sarah Barnard, a member of his own religious sect, the Sandemanians, and settled into a happy domestic life. He remained deeply religious all his life, and his own personality mirrored the serene, kind, antisocial, and ascetic qualities of his church. As an early biographer said of him, "He drinks from a fount on Sunday which refreshes his soul for the week." Faraday lacked a background in mathematics and never overcame this weakness. Although he more than made up for it with his genius for conceptualizing and designing experiments, the lack of mathematical background remained a sore point. After one of his most important discoveries he wrote a friend and fellow chemist, "It is quite comforting to me to find that experiment need not quail before mathematics but is quite competent to rival it in discovery."[40]

In every respect Faraday seemed destined for a pleasant, comfortable life unmarred by tragedy or greatness until the October day in 1820 when Davy appeared at the laboratory with a copy of Oersted's stunning article. At once he and Faraday began repeating the experiment. Both men thought the mysterious force amounted simply to some basic attraction between the wire and the compass needle. Faraday returned to other work while Davy arranged an experiment based on a suggestion by William Wollaston. A scientist and physician, Wollaston thought that the force rotated around the wire and that a wire carrying current would rotate on its own axis if a magnet was placed close to it. Faraday had no part in setting up the experiment and apparently was not present during it, but he did overhear Davy and Wollaston discussing the outcome. The predicted rotation did not occur and Davy went on to other things, but the experiment stuck in Faraday's mind.[41]

Although busy with other projects, Faraday returned to Oersted's discovery in his spare time. One of Faraday's greatest strengths was his ability to visualize and design an experiment. In pondering the relationship between the current and magnetism, the idea struck him that it was not simply push or pull but something else. Perhaps the wire was actually trying to circle around the magnet. He devised a clever contraption to test this hypothesis, and on September 3, 1821, he watched gleefully as the free end of a suspended wire whirled around a magnet as long as the current remained on. After rearranging his apparatus, he saw the magnet rod begin to circle the wire. From this simple if ingenious experiment flowed an astounding conclusion, the dimensions of which Faraday had only begun to grasp: The electrical energy from the battery had been converted into mechanical energy. Without knowing it, Faraday had created the first electromagnetic motor only fourteen months after Oersted's announcement.[42]

Sensing the importance of his discovery, Faraday rushed an article into print. It created a sensation throughout Europe but also aroused the wrath of certain friends of Wollaston, who accused Faraday of stealing the idea. In fact, the originality of Faraday's experiment and apparatus enabled him to succeed where Wollaston and Davy had failed, but this point got buried in the ugly controversy that followed. Wollaston eventually let the matter drop, but Davy did not appreciate that his protégé succeeded where he had failed. As the star of Faraday's reputation continued to rise, Davy saw it casting a shadow on his own. When Faraday's name was put forth as a candidate for the Royal Society in 1823, Davy cast the lone dissenting vote. Faraday gained the honor in January 1824, but his relationship with Davy did not survive the slight.[43]

Although Faraday devoted the next decade to work on a variety of other projects mostly involving chemical research, his mind never let go of Oersted's discovery. He followed every new development in electromagnetism with keen interest. One development occurred in 1824 when Dominique Arago devised what became known as "Arago's disk," an apparatus in which a compass needle moved in a circular pattern whenever the copper disk located below it was rotated. The best of Europe's scientists struggled to explain this movement but failed. Both Arago and Ampère toiled at the task that fascinated Faraday: finding a way to produce electricity from magnetism. So did an American scientist named Joseph Henry, whose life and work eerily paralleled Faraday's in many respects. Faraday himself performed some experiments when time permitted, but from 1825 to 1830 he found himself fully occupied in developing optical glass to improve the lenses used by the Admiralty. Not until 1831 did he have the breathing room to investigate at length the mysteries of electromagnetism. Between August and November of that year Faraday carried out ten days of experiments that shook the worlds of science and industry to their foundations.[44]

Ampère's theory of electrodynamics had convinced many if not most scientists, but Faraday could not swallow some of the assumptions behind it. He doubted the role of the omnipresent yet mysterious ether as the source of both positive and negative electricity. The whole notion of fluids left him dissatisfied, as did the hypothetical molecular currents. To accept Ampère's explanation meant buying into a full-blown theory of matter, which the ever cautious Faraday hesitated to do. Yet he could hardly ignore the widespread acceptance of Ampère's theory. In typical fashion he resolved his doubts by testing every one of Ampère's notions in careful experiments repeated again and again. This painstaking work enabled Faraday not only to discard Ampère's theory but also to formulate one of his own.[45]

As early as 1822 Faraday had written in his notebook, "Convert magnetism into electricity." But how? No known theory explained how this could be done. Ampère and Arago had tried to create this reverse effect but finally concluded that it could not be done. After all, some effects in nature could not be reversed. Coals could not be recovered by heating ashes; perhaps the same held true here. Between 1822 and 1831 Faraday made four futile attempts to induce the conversion. Then, in August 1831, he stumbled unexpectedly onto the answer, much as Oersted had done. He wound two coils of wire on opposite sides of an iron ring, taking care to insulate them from each other and from the ring. On one side he readied a voltaic battery but did not connect it until he first wired a galvanometer to the opposite side. The galvanometer had been invented in 1820 to detect the presence of an electric current. At the suggestion of Ampère it had been named after Galvani.[46]

When Faraday then connected the battery to the first side, a flicker of current registered on the galvanometer. Faraday disconnected the battery, and the needle reacted again. He repeated the experiment, and the galvanometer reacted whenever he connected or disconnected the battery from the first coil. Clearly this meant that a current somehow passed into the second coil whenever he put a current through the first. Disconnecting and reconnecting the battery had changed the magnetic field in the first coil, and it was this change that produced a momentary current in the second coil. The key to creating an electric current by use of a magnetic field proved to be what Faraday called the "transient effects" of changing the magnetic field itself. The surge of magnetism when the battery started up produced a surge of current in the second coil. Faraday had discovered the principle of induction, a crucial element in the understanding of electricity. From this discovery would flow a series of brilliant experiments that occupied him for nearly thirty years.[47]

During the fall of 1831 Faraday put together ten experiments designed to find the key ingredients in creating induction. All of them produced a similar intermittent current, lasting only as long as the relative motion of the conductor and the magnetic field. Clearly the magnet had to remain in motion to induce an electric current. To make this happen, Faraday contrived another ingenious device that amounted to a variation on Arago's earlier one. He placed a copper disk on a brass axis mounted between the poles of a horseshoe magnet. One wire was run from the disk to a galvanometer and another from the galvanometer to a metallic conductor held against the rim of the disk. When Faraday rotated the disk, the galvanometer registered a current that remained continuous as long as the disk kept rotating. "Here therefore," wrote Faraday, "was demonstrated the production of a permanent current of electricity by ordinary magnets." Another experiment confirmed that electricity could also be induced by a uniform as well as a changing magnetic field. The principle, Faraday concluded, was that "if a terminated wire moves so as to cut a magnetic curve, a power is called into action which tends to urge an electric current through it."[48]

From the experiments of these ten inspired days in 1831 emerged ramifications that would soon change the world. Apart from solving the mystery of Arago's disk, Faraday had, in historian Harold I. Sharlin's words, "discovered electromagnetic induction, found essentially all the laws that govern it, and built a working model of an electric dynamo." Faraday had laid the foundation for both the theory of electricity and its practical development. His concept of magnetism as curved lines of force was entirely original and later became the basis for the electric power industry. At the time, however, most physicists, including Ampère, rejected it because it could not be proven mathematically. Not until 1856 would this proof be forthcoming. Faraday had also demonstrated that change was the key ingredient in producing electricity from magnetism. In many respects this insight paralleled that of Carnot and his successors, who argued that change in temperature was the essential element in producing energy from steam.[49]

Faraday had given the world its first crude electric generator and shown all the possible ways electricity could be produced from magnetism. His iron rings wound with wire contained the essential design of what later became the transformer, which would be used to step voltage up or down from one side to the other. His work also helped make possible the first spectacular advance in communication, the telegraph. Modern forms of communication began with the principle that an electric current produces a magnetic field and through it a force. In discovering this concept, Oersted showed how electrical energy could be converted into mechanical energy. Faraday took the next crucial step by demonstrating how mechanical energy could be converted into electrical energy. Together these two principles formed the basis for the telegraph (and later the telephone) as well as the development of electric power systems.[50]

During the next twenty years Faraday piled still more achievements onto his reputation as the foremost scientist of Europe. Between 1832 and 1834 his research produced what came to be known as Faraday's laws of electrolysis, which describe quantitatively the "relationship between the extent of chemical decomposition of a conducting substance and the amount of electricity that passes through it." Here, as with his work in electromagnetism, Faraday showed no interest in creating practical applications from his results, but others used them to develop the basis for industrial electroplating and electroforming. He invented new devices such as the volta-meter, which could accurately measure the volume of both hydrogen and oxygen; did pioneering work in electrostatics; produced seminal studies in dielectrics (insulators) and magnetic induction; created a powerful new electromagnet; studied the relationship between light and magnetism; founded field theory; and tried without success to establish a clear link between gravity and electricity.[51]

Through all his triumphs and good fortune, Faraday remained the same modest, unassuming individual, as befit his deeply religious nature. Despite his retiring manner, he became an immensely popular lecturer. In 1826 he inaugurated two enduring series at the Royal Institution: the Friday evening discourses for the general public and the Christmas lectures for children. Although not as theatrical as Davy had been, Faraday enthralled the audiences that packed the amphitheater at the Institution. "His audience took fire with him," recalled one admirer, "and every face was flushed." Fame did not turn his head, and neither did the lure of fortune. Faraday showed no inclination to pursue his insights to practical ends, being content to continue his quest for an understanding of the natural forces that so intrigued him.[52]

Abiding by the tenets of his religion, Faraday declined a title and remained unimpressed by the honorary degrees and memberships in prestigious organizations heaped on him. He and Sarah continued to live in the same rooms above his laboratory that they had occupied since their marriage. In 1839 Faraday began to suffer from severe headaches that kept him from working for nearly five years. Although he returned to his research, the headaches continued to plague him for the rest of his life along with memory lapses that grew more frequent. By 1861 he dwelled in a mental haze that stripped him almost entirely of the past. Sorrowfully he resigned as lecturer, giving his last Friday evening discourse in June 1862. His health began to decline, and on August 25, 1867, he died quietly in his chair. At his insistence the headstone above his grave read simply "Michael Faraday" with the dates of his birth and death.[53]

FARADAY LIVED LONG enough to see his work vindicated in the one field he could not manage—mathematics—by a Scotsman born only two months before he began his monumental discoveries in 1831. James Clerk Maxwell left an indelible mark of genius on the fields of both heat and electric theory. A native of Edinburgh, the son of upper-middle-class parents, he lost his mother to cancer at the age of eight but otherwise enjoyed a happy childhood. Maxwell spent most of his childhood not in Edinburgh but at Glenlair, the country house built by his father. Shy and socially awkward, his odd country ways earned him the nickname of Daffy. He entered Cambridge in 1850 and graduated second in his class in mathematics. In 1856 he began teaching at Marischal College, Aberdeen, where he married the daughter of the college's principal. In 1860 he moved to King's College London, where he remained for five years.[54]

During his time in London Maxwell developed most of his work in electromagnetic theory and became friends with Faraday, who still had periods of lucidity amid his memory lapses. In addition to his superior grasp of mathematics, Maxwell shared with Faraday the ability to intuit and visualize the physical dimensions of an experiment. As Sharlin observed, "Faraday was a self-educated experimenter who understood electricity in physical terms. Maxwell was a mathematician by training and inclination, but one who depended on physical models for his insights." In seeking to demonstrate theory, the still crude laboratory equipment of the era could take physical experiments only so far. Faraday had gone that distance with incomparable genius yet still could not prove his controversial theories about electricity and magnetism.[55]

Many scientists continued to reject Faraday's notion of magnetic waves, what he called "lines of force," because it lacked mathematical proof. A host of questions remained unanswered. Controversy still raged over whether electricity was a fluid or something composed of molecular particles. What exactly was the mysterious ether? No one had convincingly defined the relationship between electrostatic and electrodynamic charges. In 1845 Faraday had noticed the peculiar effect of a magnetic field on polarized light but could not explain the phenomenon or fathom the relationship between light and magnetism. He was led to develop some kind of field theory by the observation that electrostatic charges were always found on the surface of conductors rather than in their interior, and by the fact that positive and negative charges could only be created together, never separately, and always in equal amounts. But what were these mysterious lines of force? Although Faraday could not answer that question satisfactorily, he insisted that they were real. A sympathetic scientist later described his efforts this way:

> With sublime patience and perseverance which remind one of the way Kepler hunted down guess after guess in a different field of research, Faraday combined electricity, magnetism, and light in all manner of ways, and at last was rewarded with a result—and a most out-of-the-way result . . . No wonder that no one understood this result. Faraday himself did not understand it at all . . . He was in a fog, and had no idea of its real significance. Nor had anyone.[56]

Amid the confusion and controversy William Thomson came first to Faraday's defense. In 1842, when he was only eighteen, Thomson published a paper in which he used Faraday's notion of lines of force as an analogy to the distribution of lines of motion of heat. Five years later he invoked Faraday's work more directly by showing how lines of electrostatic force could be used to create a mathematical theory of electrostatic action. "All the views which Faraday brought forward," he declared, "and illustrated or demonstrated by experiment, lead to this method of establishing the mathematical theory." Thomson made the notion of lines of force respectable to many scientists by showing that it could be reconciled with existing mathematical theories of electrostatics and magnetism. These papers pleased Faraday immensely, but Thomson never took the next step of using them as departures for further research and discovery. His interests lay more in heat than electricity, and much of his later work went in that direction.[57]

Thomson made another major contribution to Faraday's cause through the influence his work exerted on Maxwell, whose interest in the subject had been awakened by reading Faraday's *Experimental Researches in Electricity*. In 1854 Maxwell wrote Thomson asking for advice on the study of electricity, which had come to absorb his attention. "I was aware," he wrote later, "that there was supposed to be a difference between Faraday's way of conceiving phenomena and that of mathematicians, so that neither he nor they were satisfied with each other's language." Thomson's work convinced Maxwell that neither side was incorrect, and he undertook to bridge the gap between them. By September 1855 he could report to Thomson that he had "got hold of several truths which will find a mathematical expression in the electrotonic state." His first paper appeared that December; a second was published in 1861 and a third in 1864. Nine years later Maxwell published his most comprehensive account, entitled *Treatise on Electricity and Magnetism*.[58]

The sum of this seminal work not only advanced Faraday's insights from physical model to mathematical formulation but in the process unified existing knowledge about electricity and laid the basis for the next century of investigation. In his effort to explain electromagnetic induction, Faraday had invented what he called the "electrotonic state" to describe the condition of a wire when in the presence of a magnetic field. Any change in this state, he found, always created an electromotive force and with it an electric current if the circuit was complete. Maxwell used this notion as his starting point, saying that he hoped to "discover a method of forming a mechanical conception of the electrotonic state adapted to general reasoning." In his first paper (1855) Maxwell used two physical analogies, one of them borrowed from Thomson, to demonstrate how Faraday's lines of force could be expressed in mathematical language. A pleased Faraday wrote Maxwell to say the paper was "grateful to me, and gives me much encouragement to think on. I was at first almost frightened when I saw such mathematical force made to bear upon the subject, and then wondered to see that the subject stood it so well."[59]

But Maxwell had barely scratched the surface of his genius. In his second (1861–62) and third (1864) papers he ventured far beyond Faraday's insights to propose a breathtaking theory of his own explaining the relationship between electricity, magnetism, and light. The second paper consisted of four parts dealing with magnetic phenomena, electric currents, static electricity, and the effect of magnetism on polarized light. After first rebutting several theories that used some mechanical condition to explain the phenomena under investigation, Maxwell devised a clever mechanical model to help develop the mathematical relationship between magnetic and electric phenomena. The model explained what happened when a conductor moved across a stationary magnetic field, but what about dielectrics, which did not conduct electricity? Faraday had discovered that these insulators, when subjected to an electric charge, tended to strain or shift toward the electric force. He called this movement "polarization."[60]

In the third part of his paper Maxwell addressed this issue by unveiling a concept that became central to his work. Polarization amounted to some sort of internal displacement of the dielectric caused by the electric pressure. "Though electricity does not flow through them," he said of insulators, "the electric effects are propagated through them." He likened conductors to a porous membrane that provided some amount of resistance to the passage of a fluid, and insulators to an elastic membrane that, although impervious to the fluid, transmitted the pressure of the fluid on one side to the fluid on the other side. He called this a "displacement current" and concluded that the sum of the current flowing through a wire must include both the conduction and the displacement currents. Going beyond this discovery, he declared that a displacement current created a magnetic field in the same way a conduction current did.[61]

The concept of a displacement current was astonishing enough, but Maxwell went beyond it to postulate his first electromagnetic theory of light. In describing the properties needed for an electromagnetic medium, he noted that they were the same as those theorized for the transmission of light. In the process he also determined the velocity of the propagation of light and electromagnetic radiations to be identical. From his calculations he concluded that "*light consists in the transverse undulations of the same medium which is the cause of electric and magnetic phenomena.*" For the first time Maxwell showed in mathematical terms how all of these phenomena were related and how to convert the measurement of forces from one series of events to another. He had also taken the first leap toward an electromagnetic theory of light.[62]

In his third paper, entitled "A Dynamical Theory of the Electromagnetic Field," Maxwell made more explicit the themes he had hinted at earlier. Gone were the mechanical models and other illustrative scaffolding he had used to advance and clarify his argument. In their place came eight brilliant equations on the electromagnetic field that became a cornerstone of modern physics. Their breathtaking scope encompassed all modes of electromagnetism's behavior in terms of an electric field. From Maxwell's elaborate argument emerged a startling conclusion: Electromagnetic and optical phenomena were identical, both being transmitted by some ethereal medium.[63]

As Sharlin noted, "The relationship between the force exerted by an electrostatic charge of electricity and that exerted by a current flowing in a wire, Maxwell found, could be expressed as a velocity. That velocity turned out to be 186,000 miles per second, the speed of light." Either nature harbored an extraordinary coincidence or light was an electromagnetic phenomenon. Other mathematical evidence supported the conclusion that light was merely one form of electromagnetic wave. Maxwell still thought that some ethereal medium transmitted these forces, but with his work the classical ether began at last to dissipate. Its gradual departure as a concept marked an end to the physical picture on which Faraday and Maxwell alike had relied to conceive their theories. After Maxwell, mechanical explanations of electric phenomena no longer sufficed, yet theoretical knowledge of their nature remained far from complete.[64]

Stupendous discoveries and developments flowed from Maxwell's concepts. They led most directly to the invention of radio and more abstractly to Albert Einstein's theory of relativity. At the time, however, many physicists shrank from Maxwell's argument despite its solid grounding in mathematics. They could not accept the notion that insulators, which confined electric currents within prescribed paths, could also be the seat of electric action. A curious irony arose. Faraday struggled for acceptance because he could not root his theories derived from experiments in mathematics. Maxwell moved theory beyond the ability of the laboratory to demonstrate his concepts experimentally. To do so, an experiment would have to show that energy passes through an insulating medium at the speed of light. For that to happen, some way had to be found to produce, transmit, and detect electromagnetic waves.[65]

Meanwhile, other scientists, such as the German physicists Franz Ernst Neumann and Wilhelm Weber, advanced other and conflicting theories of electromagnetism. So controversial did Maxwell's theory remain that in 1879 the Berlin Academy of Science offered a prize to anyone who could design an experiment that would either prove or disprove it. The prize went unclaimed until November 1887, when Heinrich Hertz informed the academy that he had proven Maxwell's theory to be correct. Maxwell did not live to see his theory vindicated. In the spring of 1860 he resigned from King's College London and returned to Glenlair, where illness dogged him for several years. In 1871 he reluctantly became the first Cavendish Professor of Physics at Cambridge, where he equipped a new laboratory while completing his *Treatise on Electricity and Magnetism* and publishing another work, *Treatise on Heat*. In 1879 he was diagnosed with abdominal cancer and died on November 5 at forty-eight.[66]

Later Max Planck, himself a great physicist and Nobel Prize winner, wrote of Maxwell, "It was his task to build and complete the classical theory of electromagnetism, and in doing so he achieved greatness unequalled. His name stands magnificently over the portal of classical physics." Like Faraday, Maxwell became one of the giants on whose shoulders later scientists and engineers gratefully stood. Michael Pupin thought Hertz belonged with them as one of the giants. "The experiments of Hertz," he wrote in 1894, "transformed the visions of Faraday and Maxwell into a physical fact."[67]

The concept of the ether lingered into the new century as scientists continued to puzzle over its nature and composition. It survived in large part because no alternative notion had emerged to replace it. In 1889 the Institution of Electrical Engineers (which had begun life as the Society of Telegraph Engineers) heard an address from its new president, Sir William Thomson (who became Lord Kelvin in 1892), on "Ether, Electricity, and Ponderable Matter." The demand for some kind of mechanical explanation of electrical phenomena, he observed, was hardly new but had been growing in intensity. After offering some examples he concluded that "an electric current actually caused a rotation in the ether" but admitted that there were exceptions. So vast was the subject that he did not expect to see a complete mechanical explanation in his lifetime. In August 1893 *Electrical World* observed, "We know now that electricity is a phenomenon of the ether. According to Lord Kelvin's theory, matter is also a phenomenon of the ether."[68]

Later that same year, however, *Electrical World* conceded that "we may demonstrate the exact relation between electricity and magnetism, and may satisfactorily connect these with other phenomena, and even obtain a working hypothesis that will answer all scientific needs, but the ultimate solution may forever evade the human mind."[69]

Nikola Tesla, a scientist fast rising in reputation, also believed firmly in the all-pervasive ether and devised his own theories on its nature. In an 1893 lecture he described its relationship to electrical phenomena as crucial to efficient wireless transmission. While elaborating his own sometimes contradictory explanations of the ether, he clung to the concept all his life. It was, after all, a convenient way to explain the substance that filled the entire universe between physical bodies. Moreover, once scientists came to view light as a wave, the question naturally arose as to what it moved through. Sound waves moved through air, ocean waves through water; the lumiferous ether became the substance through which light waves moved. As such it was considered to be motionless, a constant through which all other bodies, including the earth as it orbited the sun, traveled. Light also traveled through the ether, which explained why distant stars were visible.[70]

However, this notion of the ether received an unexpected and fatal blow from, of all things, a failed experiment by physicist Albert Michelson. He first became absorbed in the task of measuring accurately the speed of light. Little progress had been made since the work of Danish astronomer Olaus Roemer in 1676, but Michelson managed in 1879 to come up with the figure of 186,320 miles a second, remarkably close to the 186,282 miles later accepted. Two years later, with financial backing from Alexander Graham Bell, he invented an interferometer, which could split a light beam in two, send the parts along separate paths, and bring them back together again. The ether was assumed to be an absolute reference point against which all movement could be measured. Since the earth moved through it at a constant speed, Michelson postulated, a beam of light traveling straight ahead should go at a faster speed than one perpendicular to it.[71]

The interferometer enabled him to devise an experiment using mirrors that not only split a light beam but made one perpendicular to the other. Theoretically the beam at right angle should move more slowly and fall out of phase with the one traveling directly ahead. This would create interference fringes that could be measured to show the earth's velocity compared to that of the ether. Michelson made his first attempt in 1881 in the Berlin laboratory of Hermann von Helmholtz but found no interference fringes. Six years later, having made other attempts with even more elaborate preparations, he tried again in partnership with Edward Morley. Despite their careful and repeated efforts, they found no interference fringes. The logical conclusion, which Michelson no less than many other scientists hesitated to accept, was that the experiment failed because the ether did not exist. If that was true, the demise of ether theory left a vacuum, and science, like nature itself, abhors a vacuum.[72]

Although Michelson later carved out a distinguished career that included a Nobel Prize, he remained best known for the experiment that failed. Yet it actually succeeded in quite another respect: It marked the death knell for ether theory and set in motion the emergence of a new era in theoretical physics. During 1905, in one of three epochal papers he published that year, a young and obscure physicist named Albert Einstein unraveled the mystery that had tormented scientists for centuries only to put in its place a still greater one. While trying to explain some seeming anomalies in Maxwell's equations, he concluded that the only constant in the universe was the speed of the light itself. Even more, without ether theory the universe had no body at rest against which to measure motion. Any motion in one frame of reference was relative to that in some other frame of reference, and the laws of physics applied equally to all such frames. Thus was born Einstein's Special Theory of Relativity.[73]

The new physics, centered in Max Planck's quantum theory as well as

Einstein's theories, not only buried ether theory but also led to a host of findings that went against common sense because they dealt with phenomena far outside the realm of ordinary observation. It also led increasingly to the demise of mechanical models in favor of mathematical ones as explanations for the vast reaches of the universe in one direction and the invisible world of the atom in the other. For electricians, at least, the mystery had at last been resolved. Only a few scientists, including Tesla and Michelson, refused to accept relativity theory. The ether was gone, never to return except in the minds of those who called relativity the new ether.

CHAPTER 5
LET THERE BE LIGHT

People stood overwhelmed with awe, as if in the presence of the supernatural. The strange weird light exceeded in power only by the sun, rendered the square as light as midday . . . Men fell on their knees, groans were uttered at the sight, and many were dumb with amazement.

—*WABASH PLAIN DEALER*, 1880[1]

LIKE ELECTRICITY ITSELF, THE development of practical ways to utilize it came in waves. Its application appeared first in the field of communication, then advanced during the nineteenth century into illumination, transportation, and finally power, where the biggest payoff occurred. At every stage the interaction between theory and application grew more complex and sophisticated. For electricity to move beyond the realm of theory, inventors had to find ways to utilize it in the real world and businessmen had to see enough potential in the outcome to invest in it. The traditional role of engineering is to transform scientific insights into practical uses and devices, but electricity posed a special problem. Where civil and mechanical engineers gained scientific knowledge in their field largely by working with real-world materials, theories about electricity came from scientists and amateurs curious about its nature and mostly uninterested in practical applications. Electrical theory preceded its engineers and lacked any solid basis of proof or testing in the real world.[2]

The steam engine came into existence well before scientists understood the nature of heat or how it created energy. In the case of electricity, however, the science remained ahead of any technology employing it usefully until the 1830s, when a decade of feverish invention produced the first practical electrical instrument, the telegraph. The notion of using electricity to transmit information went back at least to 1753, when a Scottish magazine suggested using a system of insulated wires, one for each letter of the alphabet, that could be charged in sequence to spell out a message. Ampère saw early that the relationship between magnetism and electric current opened the possibility of long-distance signaling since the magnetic effect accompanied the current everywhere. By using a pair of wires and magnetic needles for each letter of the alphabet, he suggested, letters could be transmitted simply by opening and closing the circuit for each one. But Ampère never pursued his idea.[3]

Although Ampère could not know it, he had touched on the first and in many ways simplest practical application of electricity. All the necessary elements could be had. Volta supplied the most crucial one: a battery that could provide a steady source of current. Conductors were needed to carry the current over long distances, and devices had to be created to transmit and receive the signals. In the eighteenth century a copper conductor had successfully sent an electric shock as far as four miles. The necessary instruments could be invented by applying existing mechanical knowledge. Oersted and Faraday laid the theoretical groundwork by showing that electrical energy could be converted into mechanical energy and vice versa. All electric communication, wired or wireless, depended on the interconvertibility of mechanical and electrical energy. The challenge was to create devices that utilized this fundamental principle.[4]

By the early 1830s inventors in several countries were struggling to perfect their version of a telegraphic device. After ten years of work, a Russian noble, Baron Pavel Schilling, exhibited his instrument in 1835 to a congress of German scientists in Bonn. It featured an alphabetic code operated by turning needles left and right. But Schilling died two years later, and the Russian government abandoned the project. In 1833 two German scientists devised a galvanometer telegraph and gave it to a colleague, Karl A. Steinheil, to develop. Steinheil added enough improvements to produce the first automatic recording electromagnetic telegraph that actually worked. The Bavarian government used it for signaling on some railroads, but it proved too expensive and even Steinheil conceded that other versions of the telegraph were better and cheaper than his own.[5]

In England Francis Ronalds built a working electric telegraph as early as 1816 only to have the Admiralty assure him smugly that "telegraphs of any kind are now wholly unnecessary." Twenty years later an amateur inventor, William Cooke, witnessed a demonstration of Schilling's telegraph and was transfixed. "I was so much struck with the wonderful power of electricity," he wrote later, "and so strongly impressed with its applicability to the practical transmission of telegraphic intelligence, that from that very day I entirely abandoned my former pursuits, and devoted myself . . . to the practical realization of the Electric Telegraph." Cooke devised his own version of a telegraphic instrument that utilized three needles. Through a friend he persuaded Faraday to examine the apparatus. The great man declared the principle to be correct and thought Cooke was on the right track, but he was too busy to offer any practical help.[6]

Cooke had gone as far as his limited scientific knowledge could take him. Encouraged by Faraday's response, he took his creation to a well-known physicist, Charles Wheatstone, who had made important discoveries in acoustics along with inventing the concertina and the stereoscope, which became a favorite Victorian toy. Working together, Cooke and Wheatstone in June 1837 patented their "chronometric telegraph." Unlike Cooke's earlier models, it featured a diamond-shaped board with five magnetic needles serving as arrows that combined to indicate letters when activated by underlying wires. A year later they managed to reduce the number of needles to two, and in 1839 they unveiled a different type of telegraph that operated by pointing to letters on a dial. The British railways took up this latest system and by 1852 installed four thousand miles of wire. The Cooke-Wheatstone telegraph worked and was simple to operate, but it was expensive because it required six wires between the transmitter and receiver.[7]

In improving Cooke's original instrument, Wheatstone had benefited in 1837 from advice by the foremost American scientist after Franklin, Joseph Henry. Born in 1797, the son of a day laborer in Albany, New York, who died of alcoholism when Joseph was fourteen, Henry drifted into science from this unpromising beginning through a series of accidents as improbable as those that marked the career of his friendly rival abroad, Michael Faraday. Before his father's death, Henry was sent to live with relatives in the village of Galway. Where Faraday found the doorway to his destiny in the service of a bookseller, Henry discovered his while crawling beneath a church in pursuit of a pet rabbit that had escaped. He happened on a crack in the floor, squirmed through it, and found himself in a place completely foreign to him: the village library. Enthralled by the books, he became a voracious reader, secretly at first and then through the front door. The discovery of a book on chemistry aroused his interest in science; later he called it "the first book I ever read with attention."[8]

From this modest start in life Henry rose to become a professor first at his hometown Albany Academy in 1826 and then, six years later, at Princeton. Along the way he developed an enthusiasm for electromagnetism that came to dominate his research. The field seemed wide open; for all of Franklin's accomplishments, he had not inspired a new generation of his countrymen to pursue research in electricity. In 1827 Henry visited New York City and attended a demonstration of the wondrous new electromagnet devised by an English physicist, William Sturgeon. It was a simple device: a bar of soft iron bent into the shape of a horseshoe to put the north and south poles on the same plane, with a copper wire wound loosely about it. The ends of the wires were dipped in cups of mercury attached to a voltaic battery, and the iron bar was varnished to insulate it. With the circuit closed, the magnet could lift a weight of nine pounds.[9]

Henry went home fired with the ambition to build a better magnet. His first crucial insight was to coil more wire around the iron bar to strengthen the magnetic field, but doing so caused the wires to touch and short-circuit. To remedy that problem, Henry ripped up one of his wife's silk petticoats and laboriously wrapped the silk around each coil of wire as insulation. One can only speculate on how many hours he spent at this mind-numbing task for each of his magnets, but it was time well spent. Henry built his first magnet in 1827; four years later he astonished the scientific world with a larger version that lifted 750 pounds, or thirty-five times its own weight. In an article describing his experiments, Henry called it "probably . . . the most powerful magnet ever constructed." Later that year, at the request of Yale University, he produced a larger magnet that hoisted an astounding 2,300 pounds.[10]

Ultimately Henry created two different types of electromagnets. The first, which he called a "quantity magnet," had multiple coils arranged in parallel to maximize the current; it worked best with a single battery that had a large plate area. A second type, the "intensity magnet," used only a single coil with many turns around the iron bar. It was useful for operating at the end of a long line, and it required a battery with multiple cells operated in series to produce a high voltage. Each type worked better for different kinds of use. The development of a telegraph, for example, required an intensity type of battery to move the current over a long distance and minimize the transmission loss.[11]

Like Faraday, Henry was a brilliant experimentalist who did not utilize mathematics. Like Faraday, too, he married a quiet, supportive wife who made his home a safe haven. Unlike Faraday, he did not rush into print with the results of his research but preferred to wait until he had exhausted a given line of investigation. On several occasions this hesitancy cost him dearly. His pioneering work on electromagnets got scooped by Dutch physicist Gerrit Moll, who published his refinements on Sturgeon's magnet ahead of Henry. A more painful example came in 1831–32, when Henry's work closely paralleled that of Faraday. During the fall of 1831 Henry's own experiments led him to discover induction at about the same time as Faraday. However, other work took him away from his experiments, and he did not return to them until the following spring. Neither did he publish his findings, preferring to wait until he had completed the research.[12]

In May 1832 Henry came across an article briefly describing Faraday's discovery of induction. Chagrined at losing a race he did not even know was being run, he returned to the experiments in June and the following month rushed a short article into print recounting his own findings. Despite this belated effort, Faraday would forever be honored as the discoverer of induction. Ironically, his experiments relied on the same type of multiple-coil magnet Henry had pioneered. Both men used powerful magnets that, unlike regular magnets, prompted a sharp rise or fall in magnetic force when the battery was activated or deactivated. Faraday never gave Henry explicit recognition for the design, but the slight was unintentional. Neither man knew what the other was doing, and a disappointed Henry conceded that his own dawdling had allowed Faraday to publish first. But Henry did claim the honor of being first in a related discovery made by both men, that of self-induction. Henry described the phenomenon in his July 1832 article. Apparently Faraday never saw this piece, for in 1834 he announced the same observation as if it were new. [13]

Nor was this all. In an 1831 *American Journal of Science* article Henry described what amounted to the first electromagnetic motor that repeated the reciprocating motion of an armature arm seventy-five times an hour as long as the battery power lasted. The movement depended on a continual reversal of an electromagnet's polarity, a principle later crucial to the development of direct-current motors. Despite a flood of inquiries, however, Henry dismissed the device as a "philosophical toy" useful only for demonstrating principles that might find later, more practical use. Here too, like Faraday, he left the practical exploitation of his discoveries to others. Later he attached a flywheel to the device and gave it a rotary motion, but he did not deem these improvements worthy even of publication. The steam engine, after all, could do the same thing on a much larger scale. [14]

This same attitude governed Henry's approach to an even more impressive invention, the telegraph. Most of his experiments were devised as demonstrations to impress his students. During that same busy year of 1831 Henry strung a large reel of wire around the walls of his classroom until it measured nearly a mile in length. He connected a battery to one end of the wire and a magnet to the other, then placed a permanent magnet on a pivot so it could swing freely. Henry then closed the circuit to energize the floating magnet, which repelled the permanent magnet and caused one end to swing against a small office bell placed near it. Henry had sent the first electromagnetic telegraph signal over long distance, using not a needle but an audible receiver. Although he recognized the possibilities offered by his device, he did not apply for a patent or publish the results of his demonstration. [15]

Henry's more powerful magnet made possible this demonstration and later development of the telegraph. Once he moved to Princeton, he constructed a magnet that dwarfed even his mighty Yale creation; the new one lifted a weight of 3,500 pounds. By 1835 he had strung wires across the campus from his classroom to the library and grounded their ends in two wells. This improved version enabled him to ring bells at a distance and introduced a prototype of what became a crucial element in electrical systems: the electromagnetic relay. It also demonstrated that electromagnetic waves moved through walls, trees, and other obstacles. Henry had put together the components for a telegraph system, but he made no effort to develop one. Asked late in life why he did not patent his telegraph, he replied, "I did not then consider it compatible with the dignity of science to confine benefits which might be derived from it to the exclusive use of any individual."[16]

This observation was true but not the whole truth. All his life Henry had labored uphill against long odds to pursue his scientific research. Never did he allow the potential for practical or commercial gain to distract him from it. A dedicated teacher, he had to work around the constant interruptions imposed by the classroom, lack of adequate laboratory facilities, and the demands of other projects. But Henry also had a weakness, as he well knew. He was a man of many enthusiasms in research who tended to start more projects than he could finish. And, like many other scientists, he often did not comprehend the full meaning of his discoveries until someone else pointed them out to him. As he confided to a close friend, he tended to underrate his findings until another scientist "brings them forward and develops them in a more popular or a more definite manner." Moll and Faraday had done this; so did Samuel F. B. Morse.[17]

For another decade Henry pursued his electromagnetic research intermittently, interrupted by teaching responsibilities and other projects. In 1837 he traveled to England, where he met and was befriended by Faraday and Wheatstone among other prominent scientists. For six and a half months he basked in the sights and sensations of Europe, enhancing his own reputation as a scientist in the process. Yet the fame and recognition he thought due him seemed always to elude his grasp, which grew more eager in later years. In 1846 Henry became the first secretary of the newly created Smithsonian Institution and thereafter devoted his energies to the building up of that ambitious enterprise. Despite his many successes in that work, he grew increasingly bitter over what had been denied him through the years. The example of Faraday, who had become a national hero with international fame, loomed always before him. "The maxim given by Faraday to a young friend," he admitted ruefully, "was 'finish and publish.' I have, myself, lost much credit for various researches by not adopting this rule."[18]

*   *   *

HENRY'S WORK ON the telegraph did not go unnoticed by other inventors eager to create a workable system. However, the man who did so was ignorant of Henry's work until told about it by a colleague. Samuel F. B. Morse was not even an inventor but a painter of some repute who had developed a fascination for the instrument and its potential. In this endeavor he played a role similar to that of Robert Fulton with the steamboat: He did not invent the telegraph but rather created a system that made it commercially feasible. Unlike Fulton, however, he did contribute one original element that proved crucial to its success: a practical language or code for transmitting messages. Without the Morse code it is unlikely that the telegraph would have spread so rapidly across the United States as it did.

Born in 1791, the son of a Charlestown, Massachusetts, minister who gained national attention through the writing of geography texts and travel books, Finley Morse graduated from Yale in 1810 and startled his parents by declaring his intention of becoming a painter. After resisting at first, his father acceded to this desire and underwrote a prolonged stay by Finley in England to learn the craft. Finley returned to the United States in 1815, fell madly in love the following year, and launched into a perilous career as an artist. Moving to Charleston, South Carolina, he established himself as a portrait painter and married his beloved Lucretia Pickering Walker in September 1818. When the Charleston business petered out in 1820, he moved to New Haven and struggled to support his family. Discouraged by one setback after another, crushed by the death of his young wife early in 1825 and of his father in June 1826, Morse drove himself relentlessly to gain a reputation as a painter in a nation still too raw to support its artists. In 1829 he sailed to Europe and did not return until the fall of 1832. [19]

On the voyage home Morse fell into a dinner conversation at which the question was asked whether the length of a wire retarded the flow of electricity over it. A learned Bostonian replied that electricity could travel over any length of wire and that breaking the circuit would reveal its presence at any point on the line. This remark, Morse later wrote, led him to an epiphany: "If this be so and the presence of electricity can be made visible in any desired part of the circuit," he blurted out, "I see no reason why intelligence might not be instantaneously transmitted by electricity to any distance." Absorbed in the idea, he began jotting notes and drawings for a possible system in one of his sketchbooks. From that inspiration flowed more than a decade of hard work to realize his vision. [20]

If this eureka moment seems too pat, a convenience of memory carefully arranged, it came at a critical juncture in Morse's life. He was forty-one years old, and his painting career had reached a dead end. He had lost the wife he

loved deeply and scarcely had contact with his three children. He had no training or experience as a scientist and only rudimentary knowledge of electricity
drawn from hearing lectures at Yale. He knew nothing of current research on
telegraphy and was later surprised to learn that others were already working
on the problem. But he was ready for a change of profession, and, like Faraday
and Henry, he had a marvelous capacity for visualizing and designing an experiment or the components of a system. A month after his return from Europe, he landed an appointment as professor of painting and sculpture at the
newly opened New York University. The job paid no salary; his income depended on fees from his art students. But it gave him a home base as his interest turned increasingly to designing and building a telegraph system.[21]

For five years Morse toiled at his creation. By 1837 he had arrived through
trial and error at a basic apparatus consisting of a transmitter and a receiver.
The former consisted of a port rule—a horizontal wooden rod three feet long
with metal blanks set inside it. Each of the blanks had one to nine V-shaped
notches cut into it. As a composing arm moved across the blanks, it completed
an electric circuit in short and long bursts corresponding to the notches. For a
receiver Morse nailed one of his canvas stretchers against the side of a table
and attached an electromagnet to a bar hung across it. A lever suspended from
the top of the frame held a pencil, the point of which rested on a roll of paper.
When the magnet was electrified, it pulled the lever toward it, moving the pencil. As rollers moved the paper slowly along, the pencil marked on it a series of
V-shaped lines with wide or narrow bases, depending on how the circuit was
completed and broken. These V's comprised a code that proved to be the key
element of the entire operation.[22]

It was the code that enabled the conversion of electric current into intelligi-
ble messages. This early version used series of V's as a code for numbers that
represented words; for example, "VV [space] V [space] VVVVV" stood for
215, which meant "successful" in Morse's code book. In 1838 Morse refined this
into the famous dot-dash system that translated directly into letters. He included this code in an 1840 patent. It was an elegant solution that neatly fit the
limitations of the telegraph, although Morse did not abandon his earlier
dictionary-based code until 1844. In later versions of the telegraph, beginning
in 1839, he also replaced the type-and-port-rule system with a finger key that
tapped electrical impulses directly. This innovation also eliminated the paper
roll on the receiver because a trained operator could decipher a message directly from hearing the clicks as the circuit was opened and closed.[23]

In October 1837 Morse applied for a patent on what he called the "American Electro Magnetic Telegraph." He had just solved the major problem
of how to extend the transmission range. Ignorant of both the science of

electromagnetism and the research on the subject, he turned to a colleague at the university, Professor Leonard D. Gale, author of a text on chemistry. Gale examined Morse's setup and saw the solution at once. He had read Henry's 1831 article describing an intensity battery and told Morse to wind hundreds of turns of wire around the electromagnet and use a battery of many cells rather than only one. Morse gave him a puzzled look but made the changes. Before long he could send a message a thousand feet. Then, in November, Gale strung reels of wire around his own classroom and sent a message ten miles.[24]

Elated by the demonstration, Morse launched what proved to be a six-year campaign to get federal funding for a trial telegraph system. In March 1837 Secretary of the Treasury Levi Woodbury had issued a circular seeking to gather information "in regard to the propriety of establishing a system of telegraphs for the United States." Woodbury had in mind some version of what was called an optical telegraph; Morse had something very different in mind. He responded late in September with a detailed description of his fledgling system, which he described as "an entirely new mode of telegraphic communication," along with a vision of how a nationwide system could be constructed. Gale worked with him, as did Alfred Vail, a recent graduate of the university whose family owned the Speedwell Iron Works. He brought to the project an inventive mind as well as the resources of his family, which proved a boon to the cash-strapped Morse. Vail agreed to construct a finished telegraph for exhibition and obtain a patent for it, all at his own expense, in return for a quarter share of the profits and a half share of foreign rights.[25]

To bolster the signal, Morse came up with a wonderfully simple device, the relay. Henry had discovered the ability of a weak magnetic force to act as a distant trigger for a more powerful one, and Wheatstone had created a crude version of a relay. Morse knew nothing of these developments when he devised a relay that went far beyond earlier efforts. It solved the problem of repeating a signal over long distances, and it also persuaded Vail to join forces with Morse. As the endless rounds of experiments, trials, modifications, and demonstrations rolled on, Vail came to view Morse as an arbitrary and vain man who exploited Vail and his family's fortune while taking all the credit himself, but he remained an ardent believer in the project. In 1838 Morse journeyed to Europe in search of information and patents. He met Wheatstone, who proved helpful, but eleven months of lobbying for patents brought only disappointment. Back home again in May 1839, Morse visited Joseph Henry, who responded freely to all his questions and encouraged his work.[26]

But Congress dawdled over the bill to appropriate funds for erecting a demonstration line. Discouraged and despondent, feeling his age and the loneliness that had engulfed his life since the death of his wife, Morse plodded doggedly on with improvements to his system and playing the political game. On March 3, 1843, the final day of the session, the bill finally passed. With an appropriation of $30,000 in hand, Morse put his team to work building a telegraph line along the forty-mile right-of-way of the Baltimore & Ohio Railroad between Baltimore and Washington. The original plan called for wrapping the wires in cotton thread, coating them with varnish, and burying them in lead tubing. An itinerant plow salesman, Ezra Cornell, devised an ingenious plow that could open a trench, lay the pipe, and close the trench in one operation. But some of the lead tubes proved defective and shorted the wires despite the insulation.[27]

With time and money running short, Morse cast about desperately for an alternative. He decided on a bold new approach: Instead of burying the wires, he would string them aboveground along a line of poles. Where Morse got this idea remains unclear, but it proved the key to completing the project. Cornell shifted his operation from trenching to digging holes for the poles. As the line marched toward Baltimore, Morse tested his instrument at every village along the route. By May 1, 1844, the line reached Annapolis Junction, twenty-two miles from Baltimore, where the Whig national convention had just opened. When it nominated Henry Clay for president and Theodore Frelinghuysen for vice president, the news sped toward Washington by train. At Annapolis Junction the nominees' names were given to Vail, who relayed it by telegraph to Morse. The news arrived long before the train and created a sensation. Anyone might have guessed that Clay would head the ticket, but few would have picked Frelinghuysen as his running mate.[28]

As the line drew nearer Baltimore, Morse repeated the demonstration with minor items of news. Messages flew back and forth with ease, yet Morse could not escape his bouts of moodiness. "Professor Morse is so unstable and full of notions," Vail observed. "He changes oftener than the wind, and seems to be exceedingly childish sometimes. Now he is elated up to the skies, and then he is down in the mud . . . It requires the utmost patience to get along with him." By May 24 the line was completed and in working order. The question arose as to what the fateful first message to be sent over an electromagnetic telegraph should be. Morse had promised Annie Ellsworth, daughter of the commissioner of patents, that she could choose the first communication since it was she who first brought him the news that Congress had approved the appropriation. Annie consulted her mother and decided on a line uttered by the ancient soothsayer Balaam in Numbers 23:23: "What hath God wrought?"[29]

The message could not have been more apt. Although this first transmission drew little attention, Morse got nationwide publicity for his system a few days later by reporting news directly from the Democratic convention in

Baltimore. A new age had dawned, and pundits pondered its meaning earnestly. "What has become of space?" asked the *New York Herald* shortly after the convention. "The magnetic telegraph at Washington has totally annihilated what there was left of it by steam locomotives and steam-ships." Many hailed it as "the most wonderful climax of American inventive genius."[30]

Like the railroad and steamboat, the telegraph changed American life forever in a host of unexpected ways. It moved information the way the train moved people and goods, faster and more regularly than ever before. It ushered in the age of media, as newspapers grasped the advantages of getting in minutes news that once took hours or days or even months to receive. It was the telegraph that gave rise to the Associated Press and other joint press organizations. Within two decades a network of wire spread across the land, paralleling that of railroad track and developing a symbiotic relationship with trains that greatly improved the latter's efficiency. Trains that once took days to travel a given route covered the same distance in a single day with the help of telegraphic dispatching. Once the railroads realized the advantages offered by the new technology, telegraph wires were strung alongside railroad tracks and moved west with or sometimes ahead of the lines under construction. A federal subsidy underwrote completion of the first telegraph line to California in October 1861, nearly eight years ahead of the first transcontinental railroad.

In 1866 most of the disparate telegraph companies combined into one giant firm, Western Union, through a process that foreshadowed the pattern of local and regional railroads merging into large systems. The telegraph changed the way Americans did business. Transactions that once took weeks or even months by letter could be completed in days or even hours by parties located in distant cities. This ability to move information quickly changed the way bankers, brokers, merchants, and speculators did business. It fostered the rise of New York City as the nation's commercial hub and of Wall Street as the center of financial markets. Later it enabled growing companies to centralize their operations in one city while maintaining close touch with plants or stores or other facilities across the nation. In this way the telegraph helped create the rise of regional and national markets. The rise of big business could never have occurred without the communication ability offered by the telegraph and later by the telephone.

The telegraph also changed American culture. Like the railroad, it became one more instrument speeding up the pace of American life, which in turn pressured language to become more succinct as well. Gradually the leisurely, flowery flow of Victorian prose gave way to the terse, snappy vignettes of the telegram, where more words meant higher cost. The general public gradually learned to rely on telegrams rather than letters to move urgent news, and in most towns the telegraph office was located in the train station. A decade after the creation of Western Union, the telephone arrived and launched the next phase of the communication revolution. By then Maxwell had published his brilliant theories on electricity and the quest for an incandescent lightbulb was under way.

As for Morse, like other inventors he had to endure years of litigation challenging his patent. In all he defended no fewer than fifteen suits. The worst of them revolved around an ugly wrangle with Joseph Henry over the latter's role in the invention of the telegraph. Despite these troubles, Morse lived long enough to gain both fortune and fame at home and abroad. He married his deaf cousin, Sarah Griswold, in 1848, had three more children, and raised them on a hundred-acre farm he bought near Poughkeepsie, New York. On a trip to Europe in 1856 he made a pilgrimage to the laboratory of Hans Oersted and sat reverently in the inventor's chair. Afterward, at the Porcelain Museum, he bought a bust of Oersted and happened to meet the great man's daughter. Elsewhere on the four-month trip he reveled in what he called "the kindest and most flattering attentions." The happiness that fate had squeezed from his earlier years came at last to him. He died rich and revered in April 1872.[31]

By that date the nation had more than 180,000 miles of telegraph wire. As a newsman observed, "More than half of the business of the world is transacted through its agency, and most of the news of the universe is transmitted over its wires . . . All business would stop in the absence of regular telegrams."[32]

THE TELEGRAPH PROVED a marvel not only as an invention but as a commercial triumph as well. It succeeded in part because the instrument required only a modest amount of electrical power. Once the challenge moved from communication to illumination, however, the inadequacy even of improved batteries became obvious. The demand for better lighting was enormous and growing. For inventors the payoff could be breathtaking if they could create a practical and efficient lighting system. But the obstacles were imposing. Such a system needed not only new kinds of technology in the form of electric lights but also a means of powering them. Two different types of lighting were needed—one to illuminate outdoor areas and one for indoor use—and they posed very different problems for the inventors who took up the challenge. Outdoor use demanded a bright, unvarying, reliable source of light, while indoor use required a softer, more variable source of illumination.

Throughout human history people had created light only by burning something until it was gone. Once the sun went down, Americans relied on the glow cast from the fireplace or resorted to candles or lamps fueled with whale or lard or cottonseed oil to provide light. None of them gave off a bright, clean, steady light, and the best of them, whale oil, was growing expensive by the 1850s. Only the well-to-do could afford to light their homes every evening; most of America went to bed with the coming of darkness or soon afterward. Shortly after the discovery of Pennsylvania oil in 1859, kerosene appeared on the market as a new source of fuel for lamps, but it too was costly at first.

Earlier in the nineteenth century a new source of energy appeared in the form of coal gas, which provided five times the light of candles at the same cost. In 1816 the first gas company was organized in Baltimore, and the business spread quickly to other cities. Although many companies failed, improvements in the technology solidified the industry. The newfangled gas companies provided lights for streets, businesses, industrial use, and private homes. By 1870 the number of companies reached 390, nearly all of them in larger cities. Although a major innovation, gas lamps posed problems of their own. They cast a relatively weak light and had to be turned on and off individually, a need that gave rise to the profession of lamplighter. The light also flickered, and production efficiency remained elusive. Indoors, gas lamps sucked oxygen out of a room, left a residue on walls or wallpaper, generated heat, and always posed the threat of fire.[33]

Any type of electric light required something more than a battery to power it. Oersted and Faraday had provided the key for developing such devices by showing that mechanical and electrical energy were entirely convertible and that a single device could operate in either direction on command. From their insights flowed the creation of the generator, or dynamo, which converted mechanical energy into electrical energy. The generator became the basic power source for electrical systems. However, when run in reverse, it converted electrical energy into mechanical energy and became a motor capable of using the current produced to operate a machine. Faraday had discovered all the ways electricity could be produced from magnetism and emphasized change, or "transient effects," as the key to inducing electricity into a conductor. These effects could be created by changing the field (stopping and starting the current), moving the magnet, or moving the conductor through the magnetic field. All three methods provided the crucial action of using a conductor to cut the lines of magnetic flux.[34]

Faraday provided not only the theoretical basis but also several useful insights for the development of generators; however, inventors were slow to grasp the full meaning of his research. The main exception came in 1832, only a year after Faraday's dramatic demonstrations, when Hippolyte Pixii, a Paris instrument maker, built the first generator. Driven by hand, it consisted of a rotating horseshoe magnet and a stationary coil. Its moving conductor (called the armature) produced a steady alternating current. Later a generator that produced alternating current would be called an alternator. Acting on a suggestion by Ampère, Pixii created a commutator to change the current into direct current at the terminals of his machine. At this early date no one even remotely understood the superiority of alternating over direct current for electric power systems. The voltaic battery produced direct current, and for decades generator builders remained wedded to it for their machines.[35]

Half a century passed before inventors translated Faraday's fundamental principles into practical and efficient machines. They had little incentive to pursue the creation of generators until the 1840s, when the development of electroplating and electrotyping gave rise to new industries in England that overwhelmed the ability of batteries to supply power for them. An American inventor, Joseph Saxton, had built a crude generator in 1833 to feed the medical fad for electric therapy—sending a weak current through the body—but he soon got bogged down in a patent dispute with an English inventor. Although England maintained its supremacy in electroplating for most of the nineteenth century, its inventors made only slight progress in the creation of generators.[36]

Then came the surge of interest in using electricity for illumination. At first attention went chiefly to creating intense lights for use in lighthouses. As early as 1850 a Belgian professor, Floris Nollet, had been at work on creating light by heating a block of lime to incandescence. He built a generator and started a company for this purpose, but death terminated his project. An English colleague, Frederick Hale Holmes, who had worked with Nollet, improved on the Belgian's generator and proposed to use it for powering a carbon arc light to replace the oil lamps in lighthouses. Holmes persuaded Faraday to serve as scientific adviser for his trials, which took place in 1857. His generator, with its thirty-six permanent magnets, stood five feet high, was nearly as wide, and weighed two tons. The test results delighted Faraday, who gave the project his hearty endorsement. On December 8, 1858, at the South Foreland high lighthouse, electric light illuminated the sea for the first time. An impressed observer was moved to write that "thus were magnets serving not only in the compass to direct the mariner on his course, but also in producing a most intense light to warn him of danger and guide him on his path."[37]

Despite steady improvements in the lights, only ten lighthouses in the world utilized electric illumination by 1880, five of them in Great Britain. However, between 1860 and 1880 the generator underwent major upgrades. The most crucial step came during the 1860s when inventors began switching from permanent magnets to electromagnets, which were much more efficient. Soon workers discovered that they needed no auxiliary source of power because the generator itself supplied the current needed to excite the magnets. The "self-excited dynamoelectric machine" marked a major advance over earlier generators. In 1860 an Italian physicist, Antonio Pacinotti, built a version that improved both the armature design and magnetic circuits, but his work went unnoticed for a decade. Belgian inventor Zénobe-Théophile Gramme, who had earlier worked for Nollet, created the first commercially viable generator by incorporating Pacinotti's ring winding into his own generator in 1870 and adding an improved commutator. Two years later German inventor Friedrich von Hefner-Alteneck created the drum winding, an improved armature design.[38]

Gramme had given the world its first dynamo of a modern type. It had a circular armature that preserved its relation to the magnetic field while rotating, and a continuous ring winding with leads at intervals to a commutator. It produced a smooth, unbroken current similar to that obtained from batteries. Unlike the luckless Pacinotti, Gramme associated himself with businessmen who publicized it and found markets for it. His name became closely identified with the dynamo much as Fulton's had with the steamboat, and his machine became the standard power source for electroplating and arc lights in lighthouses and factories.[39]

Together these advances, and later modifications of them, created a generator that was less crude but still pitifully inefficient. They were creatures of trial and error; much of the science behind them remained a mystery. But at least they brought the dynamo to a stage capable of supporting the first type of outdoor illumination, the carbon arc light. Such a light in itself was not new. Sir Humphry Davy had dazzled audiences as early as 1809 with his celebrated demonstration in which he used two pieces of charcoal connected to a battery. When placed near each other, the charcoal generated a spark and began turning white. Drawn apart, they produced "a most brilliant ascending arch of light, broad and conical in form in the middle." The spectacle made for a good show, but no practical use came of it until the generators of the 1870s made possible a more powerful and reliable supply of current. Among the inventors who toiled at making something practical from the arc light, Charles F. Brush emerged as the pioneer.[40]

The problem involved a number of variables. An arc-lighting system required a reliable dynamo to power it, a lamp that would burn consistently with little care, and a wiring scheme that would allow several lamps to burn at once without having the entire group fail when one of their number burned out. Light resulted from running relatively high voltage across two carbon electrodes close enough to each other that an arc jumped the gap between them. The light created by the arc was brilliant—some said blinding—and unvarying. It had a bluish, unworldly—some said ghastly—hue, but it produced more illumination than anything short of the sun. However, when the arc filled the gap, it oxidized and burned away the carbons. This tendency created other problems. Some way had to be found to move the carbons closer together as they burned, and they had to be replaced regularly.

Brush belonged to that long line of American inventors who did not create something from scratch but took some existing device and made a far superior version of it. At least two dozen inventors had devised some version of the arc light before 1860. A Russian engineer, Paul Jablochkoff, startled the world in 1876 with his "Jablochkoff candle," a simple contrivance of two long, closely spaced carbon rods parallel to one another and separated by an insulating material that refracted light. The candle required nothing more to operate it than a generator, but it used alternating current. Gramme designed a special generator, or alternator, for this purpose, and the candle burned for as long as the carbon lasted—about two hours. The system became popular in Europe, especially Paris, for several years. Some hailed it as the first practical electric light even though it consumed carbon at a rapid rate. To ease this problem, Jablochkoff arranged groups of candles so that a new pair started automatically when one set fizzled out.[41]

Jablochkoff had created the first commercial arc light, and it brought mixed reviews. One visitor to Paris rhapsodized that "the whole street, to the tops of the loftiest houses, is ablaze with a flood of beaming light which makes the streets seem like the scenes of some grand play at the opera." A less enthralled Robert Louis Stevenson called the lights "horrible, unearthly, obnoxious to the human eye; a lamp for a nightmare! . . . To look at it only once is to fall in love with gas, which gives a warm, domestic radiance." The Jablochkoff candle also caught the attention of Thomas A. Edison, who tucked a report on it into his scrapbook. For all the excitement it generated, Jablochkoff's electric candle proved too expensive and inefficient to endure. At the same time his invention made its splash in Europe, Charles Brush was perfecting a better, more sophisticated arc light in the United States.[42]

Born in 1849 on a farm outside Cleveland, Brush developed a fascination for things scientific in early childhood. He devoured books about the subject and on his own patched together crude telescopes, microscopes, and other apparatus, going so far as to grind his own lenses. Once in high school he turned his attention to electrical apparatus, creating his own static machine, Leyden jar, batteries, electromagnets, induction coils, and small motors. Accounts of experiments with arc lights, going back to Davy's original show, thrilled Brush; when in 1865 he finally succeeded in making a small one of his own, it filled him with "joy unspeakable." So too did he follow closely reports on Gramme's generator and its refinements. Inspired by news of Englishman Henry Wilde's crude generator and arc light, Brush wrote a graduation essay entitled "The Conservation of Force." Two years at the University of Michigan earned him an M.E. degree, after which he sought work as a chemist but found no takers. Instead he partnered with an old friend to broker the sale of Lake Superior ores and pig iron.[43]

Like so many other inventors, Brush kept his day job for years while he toiled at his labor of love; unlike many of them, his inventions proved successful enough for him to devote full time to the work. Brush possessed a rare blend of intellectual acumen and keen business sense. His close study of electricity convinced him that arc lights could be a commercial success if the system were simple and reliable enough. He turned to another friend, George W. Stockly of the Cleveland Telegraph Supply Company, and got permission to use that firm's foundry, shops, and even workmen for his experiments. He began with a generator, realizing that any system of affordable lighting required above all else a superior generator. By 1876 he had developed a new type known as the "open-coil" dynamo, which proved well suited to the needs of arc lights. The generator impressed Stockly enough that he formed a partnership with Brush in which his company agreed to produce and sell the latter's dynamos and any other electrical devices he might invent.[44]

Buoyed by the success of the open-coil generator, Brush devoted all his time to invention after 1877. His fertile mind raced on to new products and improvements of existing ones. He had not intended at first to invent an arc lamp, but when none of the existing versions met his needs, he designed one of his own. His "ring-clutch" lamp was simple, cheap to make, and reliable. "Its salient features," Brush wrote late in life, "have been embodied in nearly all arc-lamps ever since." In January 1878 the Telegraph Supply Company sold its first dynamo and lamp to a doctor in Cincinnati. That same winter good fortune came to Brush in the form of a competition staged by the Franklin Institute in Philadelphia, which had decided to buy a dynamo and invited the makers of several versions to submit their machines to a comparative trial. The entrants included a Wallace-Farmer dynamo made famous by its exhibition at the Centennial Exposition, a Gramme dynamo, and a Brush generator.[45]

Since no precedent existed for such tests, the institute devised its own procedures and kept careful records of the results. One of Brush's arc lamps was used in testing all the generators. After several months the institute crowned Brush's machine as far superior to the others. The results gave Brush not only widespread publicity and a competitive advantage but also data from the tests that could be used to improve his generator. During his visits to Philadelphia Brush also made the acquaintance of Elihu Thomson, a schoolteacher who would soon become a formidable competitor in the electric industry. Despite their business rivalry, Brush and Thomson remained close friends for half a century. From his single lamp Brush quickly moved to creating two- and four-lighters in which two or four separate currents were adapted to operate one of the lamps.[46]

More improvements soon followed. Brush came up with a design that wired the lamps in series with an automatic shunt around each one so that if one lamp went out, the rest kept burning. The four-lighters advanced quickly to six- and then sixteen-lighters; by 1880 he was producing forty-lighters thanks to his improved dynamo that generated high voltage and a more constant current. During 1878 Brush created an arc light with a regulating shunt coil that made possible the commercial use of the lamps from central power stations. He devised an improved regulator and a technique for using petroleum coke to produce carbon, which greatly reduced both costs and the amount of ash. To slow the burning of the carbon, Brush hit upon a method of coating it lightly with copper. This last improvement was, he admitted modestly, "the only easy invention that it was my privilege to make; and it paid well, considering its seeming simplicity." In the process he drove the selling price of the carbons he manufactured from twenty-four cents each steadily downward to a penny apiece and still profited handsomely because of swelling volume.[47]

These innovations enabled Brush to build a growing clientele for his products. In 1878 John Wanamaker bought twenty lights for his Philadelphia department store. That December a Boston clothier, having seen Brush's lights on display at the Mechanics' Fair that autumn, bought the first series (a six-lighter) run by a central plant. One of his lights illuminated the street in front of the store and became the city's first electric streetlight. In February 1879 a Providence, Rhode Island, worsted mill acquired a sixteen-light series and later bought more units until it had eighty lights in all. Other mills became customers along with some New York dry-goods houses and a San Francisco hotel. By 1880 Brush had installed more than six thousand lights. The following year the Cleveland Telegraph Supply Company became the Brush Electric Company with a capitalization of $3 million.[48]

Pleased by his progress, Brush aimed for a more ambitious market—the city streets of America. His first chance came in April 1879 when he erected twelve lamps on tall ornamental poles around his native Cleveland's public square. The 2,000-candlepower lamps made an impression but burned too quickly to last the night, making them expensive. To remedy that defect, he invented a "double-carbon" lamp that burned all night without attention. Its success later gave rise to a blizzard of litigation against imitators. During the summer of 1879 Brush installed what became the world's first central power station in San Francisco, catering first to private business firms and then to street lighting under a city contract. What began as a modest installation of two small dynamos powering a total of twenty-two lamps soon tripled in size as customers flocked to the new technology. New York, Boston, Philadelphia, Baltimore, Cleveland, Montreal, and other cities followed suit. The largest of these enterprises, the Brush Electric Light Company in New York, lit a stretch of Broadway from Fourteenth to Thirty-fourth streets at a cost to the city 20 percent below that of the five hundred gas lamps they displaced.[49]

The New York project began modestly, with twenty-two lamps, each giving 2,000 candlepower of light. The central station, featuring a Corliss engine and boiler, was installed on West Twenty-fifth Street. It turned out to be the camel's nose under the tent. Brush had every reason to be confident. In April 1880 William H. Preece, the distinguished British electrician, declared in a public address that "the performance of the Brush light are [sic] certainly the most advanced form the electric light has yet taken." In an obvious slap at Thomas Edison, who liked to feed stories to reporters, Preece added that the Brush light had "quietly crept into existence without the aid of the ubiquitous and omnipresent newspaper correspondent, or the transmission of any sensational telegrams, to the detriment and discomfort of gas shareholders."[50]

Displacing gas lamps, especially in large public spaces, was precisely what Brush had in mind. The gas companies fought his encroachment on their business bitterly, but Brush's meticulous record keeping enabled him to defend his patents successfully. He argued that the gas companies actually benefited from his work because the public, once accustomed to brightly lit streets, would burn more gas at home. A demonstration in London in 1880 led to the formation of a British affiliate to exploit the European market. Brush lights went up in India, Australia, and other corners of the British Empire. In 1882 the Brush Electric Company mounted a demonstration in Tokyo and won contracts there. That same summer the first central power station in the Far East went online in Shanghai. Throughout this rapid expansion Brush kept improving his system. The forty-light dynamos of 1880 gave way to sixty-five-lighters and then machines that powered 125 lights, which became the Brush standard and a widely used central-station unit. In 1882 Brush invented an improved storage battery and added it to the company's product line.[51]

Brush excelled at solving technical problems, but half his task consisted of educating customers who could not resist tinkering with products they did not understand. Convinced of the need to design a foolproof lamp, he tried to create an impenetrable one. "The mechanism was locked together like a Chinese puzzle," he recalled, "and difficult to get apart. It was entirely devoid of screws." Still the determined customer could get at it. One man complained that a lamp was defective. "Why I've had that lamp all to pieces four times," he protested, "and still it won't work." Poor installation by men inadequately trained for the job led to short circuits, grounds, and other accidents. On one occasion Brush journeyed fifteen hundred miles to solve a problem by removing a simple staple tack from the bottom of a dynamo, where it short-circuited a magnet.[52]

During the early years of his company, Brush operated as what he called a "'one-man' laboratory." He had no assistant and no equipment beyond a Wheatstone bridge and galvanometer. To remedy that problem, he designed and built his own instruments for shop testing. He made all the working drawings, wrote up patent specifications, tested and adjusted all dynamos and lamps. He also supervised installation and busied himself with marketing and sales. "No time was wasted," he observed wryly, "in superfluous sleep or recreation." Despite the drains on his energy, he remained the same upbeat, determined person who was always eager to learn something new and try something different.[53]

Although he was upstaged by the more flamboyant Edison in the public eye, Brush harbored little resentment. It was he and not Edison who laid the foundation for the electrical industry and for its commercial growth. Like Fulton and Morse before him, Brush had been the right man in the right place at the right time with the right set of talents. He understood electricity as well as anyone, but even more he saw that the time was right for public acceptance of a practical operating system if he could come up with one. Brush not only created such a system but kept improving it. His invention of a differential control magnet that enabled lamps to burn steadily without attention but created a short circuit when it failed proved a crucial turning point in the commercial acceptance of the arc lamp. It greatly reduced the cost of lamps in public places and also made possible the powering of several lights by a single dynamo. For this and other reasons he deserves recognition as a pioneer in the age of illumination.[54]

CHAPTER 6

A COVEY OF
COMPETITORS

The critical student of affairs perceives that, however wonderful or
however unexpected an invention may appear, it is seldom that it
is not found to be a necessary sequence of a long series of other dis-
coveries and inventions which have preceded it.

—FRANKLIN L. POPE [1]

CHARLES BRUSH DID NOT CLIMB the ladder of success in isolation.
A host of other inventors had been diligently plugging away at their own versions of electrical lights and dynamos for as long as or longer than Brush.
Some had made impressive progress and might have beat Brush to market, but
they lacked one or more elements of his remarkable blend of inventive and en-
trepreneurial talent.

Foremost among these rivals was Edward Weston, an Englishman whose
boyhood fascination with electricity survived every attempt by his parents to
make him a doctor. A devout admirer of Faraday, Weston left home in 1870 at
the age of twenty in hopes of apprenticing himself to Professor John Tyndall,
who had succeeded Faraday at the Royal Institution. On the train to London,
however, he chanced to meet an American who convinced him that greater
opportunities existed in New York than could ever be found in London. After
failing to find a place at the Institution or anywhere else in London, Weston
bought a steamship ticket to New York and went home to face the wrath of his
parents. Unmoved by their protests, he packed his few belongings and left to
catch the steamer. In May 1870 he arrived in New York with only some books,
his Bible, some pieces of apparatus he had made, and a few English pounds. [2]

At first Weston met only disappointment. Weeks of constant effort passed
before he finally landed a job as a helper at a chemical company. An explosion
at the plant ended that position a few weeks later. On his next try Weston got

lucky; he landed a job at the American Nickel Plating Company, which desperately needed someone knowledgeable in electrochemistry. Both his education and his boyhood love of chemistry and electricity prepared Weston for his entry into the field of electroplating. Moving from firm to firm in the frenzied business arena of the 1870s, he formed a partnership in a company and quickly made it a success despite the onset of economic depression in 1873. Once his business stabilized, Weston turned to a project that had long nagged at him: designing a dynamo suitable for replacing the batteries used in electroplating.[3]

Electroplating dated back to the work of Luigi Brugnatelli, a friend of Volta's. By 1805 Brugnatelli had refined a method of covering an object with a thin layer of metal by use of an electric current. The coating could serve to strengthen the object, make it more resistant to corrosion, or improve its looks. By 1840 the Elkington brothers in Birmingham, England, had adapted the process for gold and silver plating. Within another decade electroplating methods were commercialized for bright nickel, brass, tin, and zinc and developed into a major industry. English manufacturers produced gold- or silver-plated jewelry and table utensils at prices the middle class could afford.[4]

Both Gramme and Henry Wilde claimed that their dynamos could be used for electroplating, but they could not convince practitioners who believed that their trade secrets rested on their personal manipulation of current from a battery. Weston's shrewd mind saw that no existing generator could budge their prejudices, and he resolved to go his own way. His first models proved suitable for electroplating but were not cost-efficient. To obtain a necessary level of efficiency, Weston devoted nearly a decade to his experiments. His research yielded a crucial anode patent in 1875 and with it the first of many suits contesting his work. Unlike some inventors, Weston relished the legal combat of patent suits. That same year he moved from New York to Newark and set up a workshop on the top floor of his home. Absorbed in his work, he became a stranger to his wife and child. He managed to put together a dynamo to exhibit at the Philadelphia Centennial, but it dwelled in the shadow of machines built by Gramme and William Wallace, who carried off the awards.[5]

Despite this slight, the machine caught on. By 1877 Weston had sold a hundred of his dynamos for electroplating, and the orders kept coming. In June he and his partners organized the Weston Dynamo Electric Machine Company, capitalized it at $200,000, and moved into an abandoned Newark synagogue. Weston set up his own laboratory in an empty building next door, but he also went on the road to visit customers and push sales. In his spare time he built an electric motor thought by some to be the first used in an American factory, but it proved unsuccessful. He patented several improvements to his dynamo and was the first manufacturer to develop a laminated armature—a feature Brush did not adopt until 1884. Having conquered the electroplating industry, Weston turned his attention to a potential bonanza market, arc lighting.[6]

In developing a lighting system, Weston found himself at a disadvantage. He knew electroplating intimately, could fully understand its needs and visualize designs that would meet them. Arc lights were not as familiar to him; he did not grasp either their technical requirements or the market for them. Unlike Brush, who had devoted himself entirely to arc lights, Weston began designing arc lamps without a clear sense of the uses to which they would be put. But he did have an efficient dynamo, and he had installed his first arc lamp in his own shop as early as 1874. In 1877 he began work on a commercial lamp. Using trial and error to find a current value, Weston fixed on twenty amperes. The science was still inexact; no formula existed for finding the most efficient level. Brush had chosen ten amperes, which turned out to be ideal, while Elihu Thomson decided on eleven. Brush therefore got more lumens of light per watt and a more efficient lamp than Weston, who clung stubbornly to his own figure.[7]

Like Brush, Weston had to sell the public on the value of electric lighting. He got permission to install a lamp atop one of the Newark Fire Department's watchtowers and put up lights on his own factory. The usual crowds gathered to stare at the strange illumination. During the summer of 1878 Newark ordered one large light for Military Park. Orders began to come in, most of them from amusement parks. Late in 1879 a fire ruined Weston's factory and forced a move to new facilities. Shortly afterward the partners changed the name of the firm to Weston Electric Light Company. As business increased, the plant expanded into nearby vacant land and a New York office was opened. By 1880 Weston had firmly established himself in a fledgling field that was growing faster than anyone had expected. Brush dominated the industry, but Weston had become a formidable competitor, as had a third firm headed by Elihu Thomson.[8]

The scholarly, mild-mannered Thomson was another Englishman, who had come to America in 1858 at the age of five. Mechanical aptitude ran in the Thomson family; Elihu's grandfather had been a mill mechanic. His parents settled in Philadelphia, where his father found work as an installation engineer for a Philadelphia sugar factory—a job that took him to Cuba several months a year. His absence drew Elihu much closer to his mother, a strong, intelligent woman who drilled in him the virtues of hard work, perseverance, and ambition. The move to Philadelphia proved crucial in another respect: The city had one of the earliest free public school systems in the nation. Mary Ann Thomson made sure that Elihu and her other six surviving children took full advantage of it.[9]

A tough, wiry, but not robust boy, Elihu developed a passion for learning in grammar school and never lost it. So quickly did he learn that he raced through the primary grades and was ready for high school by the age of nine. Although he passed the entrance examination at eleven, school regulations prevented him from entering until he turned thirteen. Disappointed and impatient, Elihu used the time to read voraciously in science and perform his own experiments. Like Faraday and Henry, he ran across one book that changed his life. Entitled *The Magician's Own Book*, it proved to be not merely a collection of parlor tricks but also a basic introduction to physics and chemistry complete with experiments.

Enchanted by the book, Elihu plunged eagerly into its mysteries, which included instructions on how to build Leyden jars, galvanic cells, an electrostatic device, and even an electromagnet. He built the electrostatic machine as well as a camera, a microscope, an electromagnet, and a telegraph instrument. For the experiments, he found and adapted his own materials to the need at hand, developing a talent that he would utilize all his life. By the time Thomson entered Central High School in 1866, he possessed a knowledge of science far beyond most of his peers'. In the process he had also acquired a more practical grasp of technology. On one occasion he set up a miniature iron furnace in his backyard and tried to cast a cylinder for a toy steam engine.

A formidable three stories high with a blue marble front, Central was a demanding institution for the brightest boys of working-class families eager to make their way into the middle class. The curriculum ranged from Latin and German to physics, calculus, chemistry, astronomy, and physiology to composition, drawing, bookkeeping, and political economy. Every student had to pass every course to graduate. The school's challenging entrance exam routinely washed out around half the applicants. Elihu walked three miles to his classes and the same three miles home again, lugging his books and paraphernalia. He barely got through the languages but excelled at the sciences. In particular, he thrived in the natural philosophy course taught by Edwin J. Houston, who introduced Thomson to the principles of induction, Ohm's law, Faraday's laws of electrolysis, and background on the practical uses of electricity. Thomson learned the relationship between science and mathematics and that between concepts and the experiments that demonstrated them.[10]

Not all of his education came in school. In the evenings after supper Thomson hurried to the laboratory he had set up in the bathhouse behind his home. He stocked it with chemicals from the drugstore and managed to acquire a small foot-powered lathe. Using materials scrounged from anywhere he could find them, he tried everything from chemical experiments to glassblowing to an electrostatic machine, the components of which were a wine bottle, a spindle, a crank, a leather pad, and a piece of silk. In 1866 he visited the Franklin Institute to watch a demonstration of a Holtz electrostatic machine. Later, after seeing a second Holtz device, he determined not only to build one himself but also to figure out the theory behind it. Even at thirteen Thomson displayed the peculiar blend of scientist and inventor that would characterize much of his career.

As a senior Thomson and some friends organized their own scientific society, the Scientific Microcosm, which met every month in the evening, heard a number of papers, and discussed the scientific questions raised. The audience soon spread beyond students to include professors and members of the public. The experience gave Thomson a sense that he never lost of science as a community endeavor grounded in the collegial work of societies and professional organizations. Thomson finished fourth in his class of eighteen. At his graduation in February 1870 he spoke on the topic of "The Manufactures of Philadelphia" and emphasized the important role of science and scientists in the business of manufacturing. His educational career was all the more remarkable in that changed family circumstances forced Thomson to work nights as a clerk in a telegraph office during his last two years of high school.

After graduating, Thomson worked for a few months in an ironmaster's laboratory. In September 1870 he left that post to take a modest position at Central High School as assistant in the department of chemistry. At seventeen he began a career in teaching that lasted ten years. During this time he became friends with Edwin Houston, whose encyclopedic knowledge of physics and electricity was equaled only by his lack of talent at experimentation. Himself a graduate of Central, Houston had spent a year studying at the universities in Berlin and Heidelberg. He could be flamboyant as well as erudite and in the classroom had the flair of an actor. Small wonder, then, that he became Thomson's favorite teacher and mentor.

The exact relationship between Thomson and Houston remains a mystery. Thomson's biographer, and Thomson himself in his later years, portrayed it as a parasitic one in which Houston rode the talents of his former pupil to fame and fortune. However, W. Bernard Carlson concludes that Houston did in fact make important contributions to the partnership, at least in its early years. Houston was, after all, older and more experienced. He knew his way around the professional organizations and understood how to shape articles for their publications. His knowledge of electricity and other sciences was exceptional. By contrast Thomson was still a raw talent—bright, eager, full of ideas, ambitious, devoted to hard work, but untutored in the ways of the scientific world. The abilities of the two men complemented each other well in the beginning.

What is known is that Houston persuaded his young colleague to join forces in experimental work and share whatever benefits resulted. Thomson happily accepted the offer and plunged into a wide variety of experiments in the little laboratory he created in the attic of his home. Thus began the pattern that defined his life for several years: the long walk to Central, a full day of teaching, the long walk home, and work on his experiments late into the night, using tools he had to design and build himself. As with so many other inventors, the rigors of a grueling schedule seemed to invigorate Thomson rather than wear him down. He was never happier than when immersed in experiments where the hours passed unnoticed.

Another less happy pattern also developed. Houston soon realized that he had stumbled onto a good thing. Thomson did most of the experimental work and came up with important results, which Houston wrote up for scientific magazines or journals. In the published articles Houston claimed sole credit for himself, barely mentioning Thomson's contribution. At first Elihu let it slide; he was happy enough just to be doing the work, and his fertile mind fed on his discussions with Houston. Thomson was the most pleasant and trusting of souls, a shy, unassuming man who, his biographer claimed, never had an enemy in the world. Houston was a glib charmer who always talked a better game than he could play. He had the stronger personality, and doubtless it took time for Thomson to gain the self-confidence needed to move beyond what had always been a teacher-student relationship.

However, Thomson had a capacity for growth utterly lacking in Houston. During the decade he toiled in his attic laboratory, Thomson built a pipe organ and a telescope. He set up a darkroom, ground his own camera lenses, and became expert in photography. He collaborated with a friend on a telephone based on the same principles Bell later employed but laid it aside in favor of other pursuits. Soon honors began flowing in his direction. In 1875 Central awarded him an M.A. degree and in the following year made him professor of chemistry. All his life Thomson cherished the title of "professor" above all other honors. Houston helped him get into two key organizations, the Franklin Institute in 1874 and the venerable American Philosophical Society two years later. Professionally, Thomson considered himself a chemist, but electricity remained a powerful interest. In 1875 Houston gave him an unexpected opportunity in the form of publicity surrounding what Thomas A. Edison claimed to be an exciting new discovery.[11]

While working on some telegraph experiments during the fall of 1875, Edison and his assistants noticed a curious spark similar to one they had witnessed in other experiments. They rearranged their apparatus and found they could induce the same spark from metal surfaces anywhere in the room. Yet it did not register on a galvanometer, indicating lack of a charge, positive or negative. Although he had noticed similar sparks in the past, Edison uncharacteristically jumped to the conclusion that the cause of the spark must be an unknown force, one that was nonelectric. After some more experiments Edison demonstrated the phenomenon to several people and then announced to the press his discovery of what he believed to be a new fundamental force that he called "etheric." The *New York Herald* of December 2 gave a glowing account of the find, explaining:

> Under certain conditions heat energy can be transmitted into electric energy, and that again, under certain conditions, into magnetic energy, this back again into electric energy, all forms of energy being interchangeable with each other. It follows that if electric energy under certain conditions is transformed into that of magnetism under other conditions it might be transformed into an entirely unknown force, subject to laws different from those of heat, light, electricity or magnetism. There is every reason to suppose that etheric energy is this new form.[12]

If the phenomenon actually was a new force, the *Herald* reporter observed, the possibilities were staggering. "The cumbersome appliances of transmitting ordinary electricity, such as telegraph poles, insulating knobs, cable sheathings and so on may be left out ... and a great saving of time and labor accomplished ... The existing methods and mechanisms may be completely revolutionized." Another quarter of a century would pass before this prophecy would be realized in quite a different way than anyone yet imagined. For the moment the *Herald* noted editorially that "the discovery seems to promise greatly increased facility for the transmission of signals. All that the world has learned in the short life of modern science indicates that we are only at the threshold of the great secrets of nature that are yet to be opened to us."

Here was an echo of the mysterious ether, that elusive force that had not yet been entirely banished. Neither a scientist nor a mathematician, Edison knew little about previous experiments with similar findings. Always eager to make a splash, he had simply rushed into print. Controversy arose at once over the news and quickly spread from the newspapers to *Scientific American* and other scholarly forums. Houston read these accounts and recognized the phenomenon as one he and Thomson had observed four years earlier. Neither of them had known what to make of it. After reading Edison's claim, they decided to repeat the earlier experiment, this time using better apparatus designed by Thomson. In the experiment Thomson showed that the sparks could be drawn from metallic objects not only in the classroom but throughout the building and even on the roof.[13]

Houston wrote up the results in December 1875 not for the newspapers but for the journal of the Franklin Institute. In rebutting Edison, he argued that the sparks could be explained by the known principles of induction. Edison countered by saying that Houston did not understand the properties of low-resistance electromagnets. As other scientists added their voices to the controversy, Thomson eagerly designed an ingenious new experiment. Houston's second paper, published in April 1876, demonstrated that the spark resulted not from a mysterious new force but from two opposing induced currents, which left it without polarity. Similar findings by a British physicist convinced most scientists that Edison was wrong, and the inventor promptly dropped the subject and moved on to other work.[14]

In fact, Edison, Thomson, and Houston had made a very important discovery without knowing what it was, although Thomson speculated that the ability of electricity to move through space might bode a new form of communication. They had stumbled onto high-frequency magnetic waves that would later be utilized in radio. In his second experiment Thomson had even displayed a method of tuning such waves but did not recognize it as such because he had no context in which to view the phenomenon that way. He was looking for one kind of result keyed to the purpose of the experiment and found it. Only later would the full significance of his findings become clear to him.

By 1877 Thomson seemed comfortably settled into his career as a teacher-scientist. He had become a familiar face in Philadelphia scientific circles, earned a decent salary, and gained some recognition for his talents. One might have expected him to enjoy a pleasant middle-class life, yet during the next three years he abruptly shifted course toward what became an entirely new career as inventor and businessman. The reasons for this sudden change are not clear. Thomson always relished a new challenge, and the electric light business certainly provided one. However, his life had also taught him to value security, which he already had and was about to surrender. He may also have grown tired of his work at Central, where conditions were undergoing some unpleasant changes. Unlike anything else he was doing, the field of electric lighting offered the possibility of achieving greatness if he proved a worthy competitor.[15]

What is clear is that in 1876 Thomson and Houston began collaborating on several kinds of inventions. Thomson patented his first invention, a fastener for street-railway rails that saw little use. He and Houston came up with other inventions, most notably a cream separator, which gained some success, and a microphone relay to improve Alexander Graham Bell's telephone. The latter vanished after a dispute with Edison, who claimed prior invention. At the same time Thomson's attention turned toward that new wonder, the dynamo. He had first seen one in the laboratory of a physics professor at the University of Pennsylvania; it was based on the model created by Henry Wilde and William Ladd in England. Intrigued by the machine, Thomson created a smaller version in his attic laboratory sometime between 1873 and 1875. A visit to the Centennial in 1876 introduced him to the Gramme dynamo; he and Houston got Central to purchase a small Gramme model to use for classroom demonstrations. After studying it, Thomson built one himself at home. "I could, by use of a foot lathe, get a small arc light out of it," he recalled. "By working very hard and perspiring a great deal I could keep the light going for a minute or so."[16]

An unexpected opportunity arose in the fall of 1877 when the Franklin Institute, wishing to buy a dynamo, decided to sponsor a competition among the available commercial models. Although invitations went out to manufacturers at home and abroad, only Brush and William Wallace offered two models each to the institute. Disappointed, the institute borrowed from a professor at Purdue the Gramme dynamo that had been displayed at the Centennial. Thomson and Houston were asked to be part of a nine-man committee in charge of testing the dynamos. Lacking any guidelines for such tests as well as good measuring instruments, they devised and conducted a painstaking series of tests for measuring the voltage, current, and resistance of the machines when operating. Their results found the Gramme to be the most efficient, but the larger committee recommended the Brush machines for their overall excellence.[17]

From this experience Thomson and Houston gained invaluable insights into dynamos and their weaknesses. Later they produced a paper that suggested several laws governing the links between arc resistance, current strength, and illuminating power. Thomson also got to know Brush, who became a good friend and who was about to seize the lead in the budding arc light industry. In the summer of 1878, shortly after the tests concluded, Thomson left for an extended vacation in Europe. While there he visited the Universal Exposition in Paris, which featured not only exhibits of dynamos by Gramme and Siemens but also some brilliant displays of electric lighting. Two systems especially drew his close attention: the celebrated Jablochkoff candles and a less publicized arrangement by Dieudonné-François Lontin. In each case he came away with a possible solution to thorny problems.[18]

For years argument had raged over whether light could be subdivided— that is, whether several arc lamps could be run on a single circuit, because each light drew a different amount of current and thereby made the load unstable. Jablochkoff's system intrigued Thomson because it utilized alternating current, which he had not seen in the Franklin Institute tests, and because it cleverly skirted the problem of subdivision of light by using induction coils that kept the current flowing through the primary circuit even when one or more of the lamps burned out. Thomson also admired the soft, lovely glow of Jablochkoff's lights. From Lontin's more sophisticated demonstration system, installed in a Paris railway depot, Thomson gained an even more invaluable insight. Lontin too had employed alternating current to power lights that featured a unique differential regulator. This ingenious device enabled the operation of several lamps from a single dynamo instead of requiring a separate circuit for each light.

Even with his improved dynamo Gramme could operate only one light, and Brush too ran every light on a separate circuit, which was inefficient. Jablochkoff found a way around the problem with his induction coils, but he did nothing more with the idea. Lontin's differential regulator dealt with the problem in a more elegant way; it kept the resistance constant and the carbons in constant adjustment with each other. Both of these ideas, the induction coil and the differential regulator, came home to America with Thomson in August 1878. Convinced that their development could make arc lights a commercial success, Thomson shelved his other interests and plunged into designing his own lighting system. Drawing Houston into the project, he concluded that to succeed they would have to come up with a system in which a single generator powered several lights at once economically.[19]

For their system Thomson and Houston decided to emulate Jablochkoff's alternating-current dynamo and induction coils. The latter had become familiar to them during the etheric-force experiments, but no other commercial system used alternating current for its dynamos. In place of Jablochkoff's candle they came up with a vibrating arc lamp that allowed them to place several lights on a single circuit. They created a patent application for a dynamo that gave them plenty of leeway to test different designs, then turned to devising a suitable induction coil. All this work took place during the fall of 1878; in December they displayed their induction coil at the Franklin Institute. Having completed this show-and-tell, Thomson and Houston tackled the tough task of building a bigger and better dynamo to power their system. Most of this work fell to Thomson, who as usual had few tools and little money for it. Nevertheless, he had a new machine ready for display at a January 15, 1879, meeting of the Franklin Institute.

The demonstration impressed the audience, among them a photographic dealer named Thomas McCollin. He brought Thomson together with his cousin, George S. Garrett, a Quaker banker and businessman who happened also to be an agent for the Brush arc-lighting system. Garrett asked Thomson and Houston whether they could build a direct-current dynamo that could compete with the Brush system. Doing so meant abandoning their work on an alternating-current dynamo, but Garrett and McCollin promised financial support for the DC project. Since the Patent Office had just rejected their application for the induction coils, Thomson and Houston chose to pursue the bird in hand. They entered into an unwritten partnership with Garrett and McCollin to develop a lighting system. But Thomson never lost his fascination with the potential of the AC dynamo.[20]

In effect Thomson and Houston set out to create a clone of the Brush system, which was selling well at the time, but they had to avoid infringing on any Brush patents. They also had to do it at night, since both men were still teaching at Central High School. With his usual blend of insight and tenacity Thomson improved on Brush's dynamo by developing an open-coil design that produced a steadier current. Completed in March 1879, the dynamo was installed in a bakery that owned two steam engines and was willing to trade their use for free lighting. During the summer the dynamo, located in a corner window where passersby could view it, ran every evening. Sweltering in temperatures that reached 140 degrees, Thomson kept at the work of finding and fixing problems. At Garrett's request he doubled the number of lights to eight and managed to operate them on a single series circuit.[21]

Satisfied with his dynamo, Thomson turned next to improving on Brush's patented lamp. After several tries he came up with a satisfactory model that he patented in June 1879. When the bakery decided in September not to buy the system, the operation moved to new quarters rented by Garrett, who began searching out potential buyers. However, he managed only two sales—one to a brewery in South Philadelphia and the other to a saloon. Small firms could not afford them, and only one in five Philadelphia manufacturing firms had a steam engine. Although an arc light gave off two thousand candlepower compared to only 10 to 15 candlepower for a gas lamp, Thomson's dynamo could run no more than eight lights. Brush had redesigned his system to accommodate sixteen lights, but Thomson found it difficult to adapt his system that way. Clearly the system needed a broader market, but resources were limited.[22]

During these discouraging months Thomson kept at the task of improving the components of his system. In October 1879 he and Houston applied for a patent to cover an improved version of the dynamo. That done, Thomson took up the issue of a regulator to keep the line current from increasing and possibly burning out the dynamo when a lamp burned out or was switched off. An efficient regulator was crucial to arc lighting; it kept the arcs the same length when operating in series to equalize the amount of energy on each lamp. Brush had devised a regulator that increased resistance to offset the loss of a lamp, but it was wasteful and inefficient. Thomson took up the problem and devised an automatic regulator far superior to that of Brush. It not only handled the loss of a lamp flawlessly but also allowed the dynamo to generate less current and so draw less power from the steam engine running it. The automatic regulator became a key component of the Thomson-Houston lighting system.[23]

In all this work Garrett and McCollin remained steadfast supporters, but Houston had little to contribute. He lacked manual skills and so was useless in the shop. More important, the work itself was fast passing him by. Always more at home in the realm of theory, he watched helplessly as Thomson's fertile mind devised one innovation after another for the system. As Thomson grew more sure of his own ability, he became less patient with Houston's lack of contribution and increasingly advanced the work on his own. During the spring of 1880 Thomson applied for a patent to cover a newly designed dynamo with a spherical armature tucked inside a cylindrical magnetic field. In both form and function it represented what Carlson called "the culmination of Thomson's thinking about dynamo design." Elegant in design yet simple to manufacture, it came to life just as Thomson's fortunes were about to undergo a major change.[24]

Early in 1880 Thomson had been approached by Frederick H. Churchill, a lawyer from New Britain, Connecticut, who wanted to form a new company for manufacturing and marketing electric lighting systems. Churchill had access to capital and offered Thomson the prospect of greater output for a broader market. Garrett and McCollin welcomed the idea; they lacked the resources to reach beyond the Philadelphia area. In fact, Garrett subscribed $6,000 for stock in the new American Electric Company in return for exclusive rights to sell the system in Pennsylvania, New Jersey, Delaware, and Maryland. An elated Thomson agreed in July to become chief engineer for the new company even though it meant leaving Central High School and moving to New Britain. Houston did not want to move and took cash and stock for his share in the patents as well as the position of consulting electrician in Philadelphia.[25]

With this move Thomson cut his ties to the past and set sail into an uncharted future. Electricity was no longer an interest but his life's work, and the new dynamo represented an auspicious start. Like other inventors, Thomson had long been groping for ways to make a dynamo more efficient. The principles behind the machine had become well known, but no settled theory behind the principles yet existed. Put another way, inventors understood the "what" but not the "why." They had to rely on trial and error, and especially on mental models that they could play with in their heads. That is why the ability to visualize proved so crucial in the work of men like Faraday and Morse. Thomson had this ability as well, but for all of them the mental models served as a starting point from which endless experiments had to be performed under close and constant observation lest some crucial element or factor be overlooked.

Thomson had the patience and shrewd, careful eye for this work. He had another talent as well—the uncanny ability to take someone else's invention and improve on it. He had, after all, come late to the race for creating a commercial arc-lighting system, and he was still a young man. However, he also followed this same pattern of emulation in later stages of his work. In effect Thomson became the great cowbird of the electrical industry, letting other inventors create innovative nests and then moving in to lay his own eggs there. It was a talent not to be despised in a field where change came so rapidly.

IT WAS A young man's game, as new technology so often is. In 1880 Charles Brush was thirty-one, Edward Weston thirty, and Elihu Thomson only twenty-seven. Thomas A. Edison was the old man of the group at thirty-three. However, one old-timer had made a brief but important splash before any of them. Moses G. Farmer was a New Hampshire Yankee whose work pointed the way toward an alternative route to the lighting revolution. Born in 1820, Farmer attended Andover and Dartmouth College only to have his college career cut short by a bout of typhoid fever. For a time he earned his living as a schoolteacher and principal while indulging his keen interest in mechanics, mathematics, natural philosophy, and music.[26]

A chance meeting with a man who manufactured window shades sent Farmer's life in a new direction. On his own initiative Farmer invented a machine that printed shades on paper, which could be sold for a fourth the cost of linen shades. By 1845 he was selling forty thousand of his printed shades a year. He also learned of Morse's telegraph and contracted what proved to be an incurable fascination for electricity. With scarcely a look back, he abandoned both teaching and the shade business and turned to building a miniature electric railroad consisting of an engine powered by wet batteries and a passenger car. During the summer and fall of 1847 Farmer exhibited the little train in his yard and in Dover and other towns. That December he took a job as wire examiner for a new telegraph company.

For a few years Farmer worked as a telegrapher while experimenting at home with his electrical projects. He devised the nation's the first electric fire-alarm system and in 1851 got the city of Boston to install it. Four years later he gave a paper on multiplex telegraphy, having discovered the possibilities of duplex and quadruplex transmission. But his interests shifted elsewhere, and in 1856 he started a promising electroplating business only to see it fail during the panic of 1857. This reversal forced him to take a job as superintendent in a tobacco factory to support his family and his experiments. Illumination next seized his attention. As early as 1859 Farmer invented a crude early version of the incandescent light and actually used two of them to light his parlor that summer. It was reputedly the first American home to use electric lights.

Farmer's lights used a platinum wire and a wet battery. Aware that no battery could run a lighting system, he devoted several years to designing and building a dynamo. In 1866, the same year that Thomson entered high school and Brush graduated from it, Farmer patented a newfangled self-exciting dynamo. Two years later he used it to power no fewer than forty incandescent lamps in his Cambridge, Massachusetts, home. At this early date Farmer had within his grasp a crude version of a lighting system that he might have developed, but he did not. Instead he accepted an assignment in 1872 as electrician at the Naval Torpedo Station in Newport, Rhode Island. The job was supposed to last for six months; it stretched into nine years.

While in Newport, Farmer continued to work on his own inventions. He managed to persuade William Wallace, a well-known Connecticut manufacturer, to build two improved versions of the Farmer dynamo for display at the Philadelphia Centennial. The exhibit proved a hit and encouraged Wallace to push the dynamo to market, but the Franklin Institute tests of 1878 revealed its weaknesses. Wallace persisted in offering the machine for several years before finally withdrawing it. Meanwhile, Farmer resumed work on his incandescent lamp, introducing a new version in 1877 and patenting an improved model two years later. He also suggested connecting incandescent lamps in parallel rather than in series so that each one could be switched on and off individually. In all these areas Farmer was a pioneer who never succeeded in developing a commercially viable product. Yet his dynamo left one very important if entirely unexpected legacy: its influence on Thomas A. Edison.[27]

In July 1878 Edison had joined Professor George Barker of the University of Pennsylvania and others on a western expedition to view an eclipse. During the two-month sojourn Barker told Edison about some exciting new electrical work being done in the factory of William Wallace. After their return Edison went with Barker to Wallace's factory in Ansonia, Connecticut, to see for himself. Wallace gave Edison a hearty greeting and proudly showed him a series of eight 500-candlepower arc lights powered by a Wallace-Farmer dynamo. The sight excited Edison and aroused his competitive juices. As he later told a reporter, "I saw for the first time everything in practical operation. It was all before me. I saw the thing had not gone so far but that I had a chance. I saw that what had been done had never been made practically useful."[28]

"Edison was enraptured . . . ," reported an observer. "He fairly gloated . . . He ran from the instrument to the lights and then again from the lights back to the electric instruments. He sprawled on a table and made all sorts of calculations . . . the power of the instruments and the lights, the probable loss of power in transmission, the amount of coal the instrument would use in a day, a week, a month, a year." When he had finished, Edison shook hands with Wallace and said, "I believe I can beat you making the electric light. I do not think you are working in the right direction." They made a friendly wager, and Edison departed with a whole new course of action churning in his mind. The visit could not have come at a better time; he was between major projects and pondering what big thing to do next. Now he knew: It would be a new kind of electric illumination.[29]

By any standard Edison was a different kind of person who thrived on doing things differently. Born in Milan, Ohio, in 1847, he had been a sickly but mischievous child who once drove his normally good-natured father to whip him in the public square for some wrongdoing. He grew close to his mother, who encouraged his love of learning—but always in his own way. Like many people who achieved greatness, he saw what he cared to see and ignored most other things. As biographer Matthew Josephson observed, "No one ever *taught* him anything; he taught himself." He grew up in the lumber town of Port Huron, Michigan, where his father failed at a number of occupations, some of them dubious. Most of the boy's education came from his mother, although at ten he developed a passion for chemistry and spent hours doing experiments in the cellar of the house. At twelve Edison went to work selling newspapers and goodies on the train between Port Huron and Detroit. An accident that year severely damaged his hearing and began his descent toward deafness. Although he made light of the problem, he mourned later in his private diary, "*I haven't heard a bird sing since I was twelve years old.*"[30]

During his long layovers in Detroit before the return trip to Port Huron, Edison haunted the city's public library, where he read books randomly, one shelf after another. Ever industrious, he started his own newspaper to sell as well and ran his own crude telegraph line between his house and that of a friend. The telegraph pleased him because his impaired hearing could pick up the click of the key without being distracted by other sounds. After apprenticing with a man who taught him telegraphy, Edison got a job in 1864 as an operator. For five years he knocked around, plying this nomadic trade. Although a competent telegrapher, he lost jobs because of inattention and lack of discipline. Never good at obeying orders, Edison continued to read avidly and began filling notebooks with ideas and designs for new inventions. By 1869 he had landed in Boston and decided what to do with his life. He resigned his post as operator and placed a notice in *Telegrapher* magazine that hereafter he would "devote himself to bringing out his inventions."[31]

During the next nine years Edison scrambled to earn his keep as an inventor. His first patent, for an electric vote recorder, found no market and taught him the principle he would follow the rest of his life: To be successful, an invention must not only work but be commercially feasible. Having defined himself as an inventor, Edison proceeded to make it a business in his raw manner that was a strange brew of organization and utter lack of discipline by ordinary standards. Unlike most of his peers and rivals, he did not keep his day job but devoted all his time to invention. Although not a scientist, he pored over technical and scientific literature in whatever area he happened to be working. His research was as systematic as his personal life was disorderly. Having achieved some success as an inventor, Edison in 1876 brought his family to the obscure hamlet of Menlo Park, New Jersey, and erected what became the first laboratory devoted entirely to the work of invention.[32]

Hamlet may be too ambitious a description. A reporter visiting the site wrote that "Menlo Park is not a park. It is not a city. It is not a town. Although it is on the Pennsylvania Railroad, it is not even a stopping place, except when the station agent flags the train to take on waiting passengers. It is composed wholly of Edison's laboratory and half a dozen houses where his employes [*sic*] live." Near the railroad depot could be found a saloon run by an old Scot where the boys sometimes gathered in rare leisure hours to play billiards. Edison and his family occupied the best of the village's seven houses; the bachelors on his staff mostly stayed at a boardinghouse run by Mrs. Sally Jordan, but they spent little time there. Life at Menlo revolved around the laboratory at all hours of the day or night. Much if not most of the work got done at night, when there were no visitors to interrupt them. Edison himself was a night owl who needed little sleep and expected his men to follow his example.[33]

Edison called Menlo his "invention factory" where he proposed to turn out "a minor invention every ten days and a big thing every six months or so." As biographer Paul Israel has shown, it was hardly the only such facility in the country. Other inventors had impressive shops, but Edison's new laboratory dwarfed them in size and scope of activity. To it he recruited a phalanx of bright young men eager to conduct exciting work leading to great discoveries and products under Edison's direction. Teamwork became the key to research at a laboratory created by the most individualistic and maverick of inventors. Edison also found incredibly talented acolytes like Charles Batchelor, John Kruesi, and Francis Upton, all of whose talents supplemented his own by filling in the gaps of his own abilities. Upton, for example, was a physicist whose skill in mathematics earned him much teasing and even more respect from Edison.

Menlo became very much a man's world where the inmates worked intensely for long hours and relaxed over beer, cigars, and jokes. "Midnight lunch" became a welcome break from the work, usually accompanied by singing or practical jokes. Edison drove his staff hard, but most of them endured the grind cheerfully because no one worked longer hours than the boss. He kept no regular hours or schedule, needed little sleep—which he usually got in periodic catnaps—and showed an uncanny ability to focus on the problem at hand. His devotion to work was matched only by his neglect of his wife and children. He preferred the loose, amiable company of "the boys" at the shop, where work and play moved seamlessly from one to the other. Slovenly and coarse in his personal habits, Edison personified the idiosyncratic genius, an image he not only embodied but did much to cultivate as well.[34]

His early inventions had been electrical devices like the vote recorder, an improved stock ticker, an electric pen, and a host of improvements to the telegraph, most notably the duplex and quadruplex. His telegraph work had won him a contract with Western Union that did much to support the Menlo laboratory, but he had also branched out into other, related fields. In 1877 he had improved Bell's telephone with several innovations of his own, especially the carbon transmitter that had trumped Elihu Thomson's version. Western Union acquired the rights to Edison's telephone patents and began to market the machine vigorously. Once the inevitable patent litigation commenced, however, Western Union lawyers evidently concluded that Bell's patent would be sustained. In November 1879 Western Union chose to sell its telephone rights and patents to the Bell company and retire from the field.[35]

Edison's telephone research led him to a variation on voice reproduction that thrust him into the public eye in a new role as the "Wizard of Menlo Park." Paul Israel has called the phonograph "the most important [invention] of Edison's career. It made his reputation as the 'Inventor of the Age.' " When the talking machine made its debut in 1878, it astonished the world and made Edison a celebrity at home and abroad. He took the machine to Washington and showed it to President Rutherford B. Hayes, members of Congress, and the National Academy of Sciences. Visitors began flocking to Menlo to see the great man in action. Although pleased at the attention, Edison found more and more of his time devoted to entertaining the crowds with his version of parlor tricks. The loss of research time grew as annoying as the attention was gratifying. Even more, fame thrust on him the burden of living up to it. Having become America's inventor and having embraced that role willingly, he realized that the public expected new and more amazing inventions from him on a regular basis.[36]

It was during this same year of 1878 that Edison paid his fateful visit to William Wallace's factory and observed his dynamo running the battery of arc lamps. In seeking to build a better light, Edison chose a different path than that taken by most inventors in the field. Ever the practical man with an eye toward commercial applications, he saw that the arc lamp could never be adapted to reach the biggest market of all: indoor lighting. A different kind of lamp altogether was needed, an incandescent rather than arc light that produced a softer, more versatile illumination; one that could be turned on and off at will. To the invention of such a light Edison turned the full resources of his invention factory as well as his own creative genius.

CHAPTER 7

THE LIGHT DAWNS

*Discovery is not invention, and I dislike to see the two words con-
founded. A discovery is more or less in the nature of an accident. A
man walks along the road intending to catch the train . . . on the
way his foot kicks against something and . . . he sees a gold bracelet
imbedded in the dust. He has discovered that—certainly not invented it. He did not set out to find a bracelet, yet the value is just
as great.*

—THOMAS A. EDISON [1]

LIKE THE ARC LIGHT, INCANDESCENT light made its public debut in the
hands of that master scientist and showman Humphry Davy. He had simply
passed an electric current through a platinum wire or carbon rod and watched
it glow until oxidation consumed it. In the arc lamp, the arc between the rods
produced the light, while incandescence involved using electricity to heat a
substance to the point where it glowed. Davy did not pursue either phenomenon, but others began to experiment with incandescence as well as the arc
light, especially as bigger and better batteries were devised. During the 1840s
several inventors placed their burning element inside glass globes; one of
them, an American named J. W. Starr, obtained a British patent in 1845 for his
lamp. A few even tried to create a hybrid between the arc and incandescent
lamps, but their efforts proved futile. [2]

The incandescent lamp, if one could be devised, promised to revamp existing notions about artificial light. Candles, torches, fireplaces, oil lamps, and gas
fixtures always involved smoke, flickering, excess heat, consumption of oxygen, and the ever present danger of fire. Arc lights represented an improvement, but they still flickered a bit, they gave off tremendous heat, and their
carbon elements burned down like candles. An incandescent light, enclosed
in a globe or bulb, severed the link between light and fire, but it was also

fiendishly difficult to devise. Moreover, even if it came into existence, no system existed for manufacturing or delivering it.

By the late 1870s at least twenty inventors had tried or were still trying to create an incandescent lamp, but none had come up with anything near to being commercially viable. Moses Farmer had come closest in the United States; in England a chemist named Joseph W. Swan had devised a carbon lamp in 1860, but it worked poorly, and he abandoned the project for seventeen years. Many scientists thought the obstacles involved were too formidable to overcome. The conviction that light could not be subdivided remained widespread. The right material for burning had to be found; it required a high melting point and a high—but not too high—resistance to electric current. It had to be enclosed in a transparent vessel to enhance the light and slow the rate of oxidation. A vacuum would be ideal but hard to create inside a small globe. A reliable source of power was needed along with some way to switch small lights on and off individually.[3]

One by one these daunting obstacles diminished. The work done on arc lights suggested that the subdivision of light was not impossible after all. In 1865 Hermann Sprengel invented a superior mercury vacuum pump; a decade later Sir William Crookes perfected a method of using it to create a vacuum in glass bulbs. Advances in the work on dynamos promised better sources of power. The most stubborn obstacle remained the choice of material for a filament. Platinum and carbon fit the requirements better than most, but platinum was expensive and carbon tended to vaporize or combine with other atmospheric gases. Even if one solved all these and other problems, a still larger issue loomed above them: Could such a lamp be commercially viable? It was one thing to invent an incandescent lamp and quite another to make it both cheap and reliable enough to compete with gas lighting or kerosene lamps.

The potential market for a practical incandescent light was enormous. By 1880 the nation had more than five hundred gas companies representing an investment of some $150 million. The industry had a long history tracing back to Belgian chemist Jan Baptista van Helmont, who in 1609 noticed the "wild spirit" that emanated from some coal he burned and gave it the name "gas." Experiments during the next century led to the development of techniques to utilize coal gas for lighting. In England William Murdock, the brilliant engineer for Boulton & Watt, pioneered in creating a practical process to create and store gas and made it a commercial product by lighting cotton mills. By 1815 one company had installed twenty-six miles of gas mains in London; that same year a leading expert, Friedrich Christian Accum, published *A Practical Treatise on Gas-Light*.[4]

A new industry, based on the relatively simple process of collecting, storing, and distributing the vapors derived from heated coal, had been born and spread quickly to the United States. The city of Baltimore took the lead. The first known gas lamp demonstration took place there in 1802; fourteen years later the first commercial gas company was organized to provide the city with street and residential lights. By 1850 some fifty cities were manufacturing coal gas to light streets, stores, residences, and even railroad passenger cars. Then came the discovery of petroleum at Titusville, Pennsylvania, in 1859. With the oil came large quantities of natural gas, which provided the pressure that enabled oil to flow to the surface. For some years the gas was considered a nuisance and allowed to escape into the atmosphere. Not until the 1880s did entrepreneurs succeed in utilizing natural gas on a large scale. In this, as in so many other areas, the city of Pittsburgh took the lead.[5]

The arc lamp posed a threat to only about 10 percent of the coal gas industry's market, but a successful indoor lighting system would compete for the other 90 percent. Bringing an incandescent system to market, therefore, would require not only inventive genius but the financial resources to wage a long and bitter fight for market share. Most inventors were innocent of these kinds of considerations but aware of the possibility of a huge payoff if they succeeded. Five inventors had taken the lead in the competition: Farmer, William E. Sawyer, and Hiram S. Maxim in the United States and Swan and St. George Lane-Fox in England. Farmer obtained a patent for his version in March 1879; Maxim applied for one in 1878. Sawyer's work gained momentum when he got financial backing from Albon Man, a Brooklyn lawyer who helped him incorporate the Electro-Dynamic Light Company in July 1878. Lane-Fox patented his first lamp that same year, and Swan had resumed work on an incandescent light in 1877.[6]

No inventor had succeeded in creating an incandescent lamp that burned more than a few minutes, and none was even remotely close to having a commercial product. The basic problem still eluded everyone: how to heat an element to the point of glowing while stopping short of melting it down. Edison came to this fray a newcomer with an entirely different approach than any of his rivals; as Francis Upton put it, "Mr. Edison came to the investigation unhampered by the blunders of his predecessors." After only a few days of research he hit on what he called a "big bonanza." To the press he poured out a torrent of enthusiasm and optimism. "I have it now!" he told a reporter shortly after returning from the Wallace factory and conducting some experiments, "and, singularly enough, I have obtained it through an entirely different process than that from which scientific men have ever sought to secure it. They have all been working in the same groove, and when it is known how I have accomplished my object, everybody will wonder why they have never thought of it, it is so simple."[7]

What had he found? He had solved the riddle of subdivision. "With the process I have just discovered, I can produce a thousand; aye, ten thousand, from one machine. Indeed, the number may be said to be infinite." Once the public realized how cheap and bright his lights were, Edison continued, gas lights would be discarded. Using fifteen or twenty Wallace dynamos powered by a 500-horsepower steam engine, he could light the entire lower part of New York City. Wires would be laid in the ground like gas lines and reach every building. The same wires that bring electric light would also carry power and heat. "With the power you can run an elevator, a sewing machine or any other mechanical contrivance that requires a motor, and by means of the heat you may cook your food." In this one astounding interview Edison predicted the creation of an entire new industry that would change life forever, and he did it at a time when none of the infrastructure or apparatus existed. It was an incredibly wild boast bordering on fantasy, yet the paper took it seriously. If nothing else, it made good copy.

But if Edison exaggerated his progress to reporters, he was careful not to deceive himself. As his research progressed, he found the problem to be far more complex and frustrating than he had imagined. Although he underestimated the difficulty of the task, he well knew that it would require deep pockets to carry out. The unabashed boasting to the press was designed to impress not only the public but potential investors whose cash Edison sorely needed. To that end he dispatched his lawyer, Grosvenor P. Lowrey, to shake the money tree, saying, "All I want at present is to be provided with funds to push the light rapidly." Lowrey promptly went to some of the tallest trees around—J. P. Morgan, Hamilton McKown Twombly, son-in-law of William H. Vanderbilt, and other members of the board of Western Union, for whom Edison had done so much work. It helped that Lowrey's office was in the same building as Morgan's firm, Drexel, Morgan.[8]

Lowrey did his work well. Later Samuel Insull called him "just as much the inventor *of*, as Mr. Edison was the inventor *for* the Edison Electric Light Company." On October 15, 1878, thirteen men gathered to incorporate the Edison Electric Light Company with a capitalization of $300,000. Their number included three Morgan partners, Egisto Fabbri, Anthony Drexel, and J. Hood Wright, Twombly, and Norvin Green, president of Western Union. The sole purpose of the company was to support Edison's research and control whatever patents resulted. A month later Edison signed a contract with the new company giving him 2,500 of the 3,000 shares and $30,000 in cash; in return he was to assign the company all inventions and improvements in the electric light field for five years. Once the new company had earned $50,000, it would pay Edison a lump sum of $100,000 and royalties on all electric lamps sold if net income warranted it.[9]

This was a remarkable arrangement, the likes of which no inventor had ever seen. Some of the most important bankers and businessmen in America had put their money behind a project to develop a technology that did not yet even exist and that many prominent scientists said could not possibly be created. Even Edison, the man of grand visions, was slow to grasp the scale of the enterprise. Lowrey had to prod the inventor to upgrade his concept of what was needed. The work wanted not a simple workshop but a two-story building with a separate office and library richly furnished to accommodate the stream of important visitors and investors that would be received there. An underground vault was also needed to store important documents and other valuable things. The machine shop too should house the best and latest equipment. Here was a new approach to research and development. As Francis Jehl observed, "The laboratory was based upon business principles, and . . . was the first one in which *organized research* had its inception on that principle." In October 1878 it became the first to have ample capital behind the pursuit of its goal.[10]

Edison understood his mission clearly. He had started Menlo Park to make invention a business, and he never lost that objective. On one occasion he complained to an assistant that one of his colleagues tended to stray from his assignment whenever some new phenomenon that intrigued him turned up. "We can't be spending time that way!" Edison stressed. "We have got to keep working up things of commercial value—that is what this laboratory is for. We can't be like the old German professor who so long as he can get his black bread and beer is content to spend his whole life studying the fuzz on a bee!" Now he was pursuing what could be the biggest payoff of all.[11]

Two weeks after the signing J. P. Morgan himself wrote that he had secretly been at work on "a matter which is likely to prove most important to us all . . . Edison's Electric Light." He noted that gas stock prices had dropped 25 to 50 percent on news of Edison's announcement. "This matter needs careful handling if anything comes of it," he added. "It is not entirely certain—I shall do nothing until it is—but when that time comes . . . we must be prepared to strike." Although usually cautious to an extreme about new enterprises, Morgan took a personal interest in Edison's work from the start and remained a staunch supporter of it. The support of the financiers gave Edison resources unavailable to any other inventor, but it also put pressure on him to produce tangible results to keep his investors convinced that he was making progress. This goal sometimes proved as difficult as the search for a practical electric lamp itself.[12]

And skeptics abounded. Many English scientists scoffed at the idea of a lighting system based on subdivision. A book on electric lighting published in 1879 declared that any subdivision of electric current was "incompatible with the well-proven law of conservation of energy." The distinguished William Preece, in a February 1879 lecture, offered a detailed explanation as to why "a subdivision of the electric light is an absolute *ignis fatuus.*" Hippolyte Fontaine, author of *Electric Lighting,* insisted that "the subdivision of the electric light is impossible of attainment." Silvanus Thompson, the scientist whose findings had supported Elihu Thomson's refutation of Edison's "etheric" theory, assured his listeners that "any system depending on incandescence will fail," and that Edison's prattle about subdivision of currents betrayed "the most airy ignorance of the fundamental principles both of electricity and dynamics." In the United States, Edison's rival Sawyer sneered that his scheme of subdividing light would fail. [13]

To his detractors, Edison had brought this criticism on himself through his incessant boasting to the press. Reporters loved him because he welcomed their visits, answered their questions, and showed them everything. The *Herald* reporter called him "Professor Edison." So too was he hospitable to the steady stream of visitors who flocked to Menlo Park to view the Wizard at work on his latest creation, whatever it might be. With the electric light, however, Edison was more secretive, saying that he could not display it in public until he had in hand both the American and English patents. But on October 11 he talked of a system utilizing central power stations serving a radius of half a mile in New York City, with wires buried underground in lead pipes and run through existing gas fixtures at each residence. He also demonstrated a lamp for the reporter, who found its light much softer and whiter than the yellowish tint of a gas lamp. [14]

Five days later the American Gaslight Association convened in New York for its annual meeting. Hounded by reporters for reactions to Edison's new light, most delegates denounced it as either improbable or not competitive, although one conceded that the electric light was "sure to be generally adopted some day or other." One thought it was all a scheme to drive gas stocks down. Since the price of gas was a live topic, a reporter asked the president of Brooklyn Gaslight how much the price of gas could be reduced should it have to meet competition from electricity. "Not a cent lower than we sell now," came the terse reply. The editor of *American Gaslight Journal* declared that he had been to Paris to examine the heralded Jablochkoff lighting system and saw nothing that threatened the supremacy of gas. The president of a gas company made the same trip and affirmed the same verdict. [15]

Several delegates echoed the mantra "You cannot divide the light of electricity." The views of two distinguished professors reinforced this view and assured the gas executives that they had nothing to fear from electricity. "Undoubtedly," said one, "he will wind up . . . at the insurmountable obstacle—the enormous loss due to subdivision of the current . . . The electric light will not bear subdivision except at enormous sacrifice of illuminating power." Others noted that electric light was far too bright and harsh ever to be used indoors. Their confusion was understandable. Few of the delegates had any experience with electric lighting, and those who did knew only arc lights. The same held true for the professors. Told of their remarks, Edison merely smiled and repeated what became his theme: "Those who are denouncing it as impossible are predicating on what they have not seen and do not understand . . . In a little while all who doubt it may see for themselves." [16]

The *Herald*, which had been chastising the gas industry for what it deemed extortionate prices, lined up firmly behind Edison. "Parties interested in threatened businesses," it reminded readers, "have always been remarkable for the stoniest blindness." This press support only hardened the views of the professors against Edison. "I am told he does not talk to scientific people in the same strain he does to newspaper reporters," one professor told the gas executives. "No end of daily announcements as to what Edison is going to do are daily made, but the world wags on and nothing more is heard from them. This last piece of newspaper buncombe will last a few days longer, as every one hates gas bills heartily." [17]

Scientists, unimpressed by or perhaps envious of Edison's growing fame as an inventor, tended to dismiss him as a mere mechanic, a shameless self-promoter devoted to chasing publicity and making money. They seldom understood or appreciated his close concern for the practical and commercial value of his inventions or his ability to view their application in a broader economic context. Neither did they grasp his talent for organizing a research enterprise, staffing it with capable men, and inspiring them to do their best work. The scale of his approach utterly confounded them. To develop the incandescent lamp ultimately consumed $50,000 in cash and $78,225 in loans from the stockholders—a substantial sum of money in those days. Besides labor and materials, Edison added three new buildings to Menlo Park: an office-library, a steam plant and machine shop, and a glassblowing facility. He bolstered his staff with a chemist, a glassblower, a mathematician, and some machinists. [18]

The new company was formed a mere five weeks after Edison paid his visit to Wallace's factory. Lowrey had told the investors that "Edison has discovered the means of giving us an electric light suitable for every day use, at vastly reduced cost as compared with gas." This assurance lay somewhere between a tall order and a tall tale, and it put tremendous pressure on Edison to deliver. He undertook an intensive study of the gas industry and decided to use it as a model for his own system. Earlier he had told a reporter, "I can make the electric light available for all common uses, and supply it at a trifling cost, compared with that of gas." So potent was his reputation as an inventor that these words sent gas stocks plunging in both the United States and England. Amid the hoopla, however, Edison understood clearly that whatever system he devised had to perform better than gas and cost less in the bargain. An entry in his notebook makes this point clear:

> Object: E. to effect exact imitation of all done by gas, to replace lighting by gas by lighting by electricity. To improve the illumination to such an extent as to meet all requirements of natural, artificial and commercial conditions. Previous inventions failed—necessities for commercial success and accomplishment by Edison. Edison's great effort—not to make a large light or a blinding light, but a small light having the mildness of gas. [19]

Having defined his task, Edison harnessed all of Menlo Park's talent to achieve it. No other inventor commanded the resources Edison had at his disposal, and he used them freely. He surrounded himself with talented lieutenants who complemented his own strengths. Foremost among them was Charles Batchelor, a transplanted Englishman whose patience, painstaking handiwork, and methodical manner offset Edison's often scattered style. Edward H. Johnson had been with Edison since the telegraph invention days and became one of his most able and trustworthy aides. John Kruesi, a Swiss machinist, gave Edison the invaluable advantage of high-quality mechanical work in a machine shop no other inventor came close to having. Francis Upton brought his gift for mathematics and theory, Samuel D. Mott his artistic flair in drafting, and the Ott brothers, John and Fred, their skill as mechanics. Together these and other assistants did much of the hard work in translating Edison's ideas into practical experiments and products. [20]

Edison's investors blanched at the ease with which he spent money on his research and facilities, but the ever loyal and persuasive Lowrey saw to it that they remained believers. The investors, like the public and the scientists, were slow to grasp the enormity of Edison's undertaking. The attempt to invent an incandescent lamp was bold enough, but he had to do much more: He had to devise nothing less than a complete lighting system at a time when none of its components existed. Those components included a commercially viable lamp, a suitable wiring scheme, regulators, switches, a meter, and a dynamo. At first

Edison thought the Wallace dynamo might be suitable, but he soon realized that he had to come up with his own design there as well.[21]

The subdivision of light, that seemingly impossible obstacle, proved easier to solve than anyone dared hope, thanks to Edison's clear understanding of a scientific principle that had been around for half a century but had yet to be fully grasped. Despite his innocence of mathematics, Edison had mastered Ohm's law at a time when most of his rivals either did not know it or did not understand its relevance to their work. Scientists and engineers viewed the subdivision problem as one of dividing a fixed amount of current among a given number of lights. Edison needed a small light that could be turned on and off individually. This required a parallel circuit that allowed a constant-voltage generator to increase the units of current. Boosting the current with low-resistance lamps, however, exacted a toll in the form of either a severe loss of energy in transmission or excessive costs in using thicker copper wires. From these facts Edison deduced that he had to develop a high-resistance lamp.[22]

The problem revolved around the fact that current flowing through wires generated heat, which amounted to lost energy, and the longer the distance that current had to travel, the greater the waste. Conventional wisdom dictated, therefore, that the burner should have a lower resistance than the wires to minimize this loss. From a commercial point of view, however, the conductors (the wires) represented a major cost factor. The best conductor was copper wire, which was very expensive. A low-resistance burner or lamp would have to use thicker wires to receive current from any distance or be very near the source of power to keep the line short. The former would make copper lines very expensive; the latter precluded the use of a central power station. A single house or plant could utilize low-resistance lamps by having its own generator, but a lighting system powered by a central station needed high-resistance lamps to be efficient. A low-resistance lamp might be easier to build, but a larger system based on low resistance would prove commercially unfeasible, and that fact mattered greatly to Edison.[23]

On the other hand, Edison understood that the heat generated by the burner in a high-resistance lamp would be proportional to the resistance times the square of the current. This conclusion, easily reached by applying Ohm's law, had eluded even his more erudite rivals and critics. When Edison hired Francis Upton for his mathematical ability, he asked him to confirm his notions regarding the use of high-resistance lamps in a parallel circuit. Skeptical at first, Upton later admitted to being astonished at how accurate Edison's "guesses" turned out to be even though they departed from conventional wisdom. "I cannot imagine why I did not see the elementary facts in 1878 and 1879 more clearly than I did," he recalled. "I came to Mr. Edison a trained man, a postgraduate at Princeton; with a year's experience in Helmholz's laboratory; with a working knowledge of calculus and a mathematical turn of mind." Upton waded through hundreds of equations in his effort to translate Edison's ideas about networks of conductors, current losses, voltage drops, and other intricate problems into charts and tables, which Edison analyzed carefully.[24]

Earlier Edison had thought the key to solving the subdivision riddle lay in devising a thermal regulator to keep the incandescent material from melting. This reasoning led to a round of experiments that produced a usable regulator. Having devised it, Edison did little to improve it because he thought the basic obstacles to his goal lay elsewhere. His conclusion proved correct, although the reasons surprised him. Eventually he abandoned the regulator because the crucial element in a high-resistance system turned out to be the burner, or filament. It had to have high resistance yet a small radiating surface, and it had to endure a temperature of more than two thousand degrees for a thousand hours without breaking. The regulator took a backseat to the search for a suitable filament, which for Edison and his men turned into the quest for the Holy Grail.[25]

Nothing frustrated Edison and his men more than the search for the right material to use as a filament—a term he was the first to employ. Most of his rivals had used platinum or iridium burning uncovered or carbon placed in a vacuum. Previous experiments in 1877 had shown Edison that carbon oxidized rapidly and was hard to protect with the existing state of vacuum technology. He proceeded to supplement his past work in chemistry with extensive reading, which convinced him that the platinum group of metals, with their high melting points and resistance to oxidation, offered the best hope. But platinum was very expensive, and Edison seldom accepted the conventional wisdom about anything. He began a tedious set of tests on a wide range of materials, always within the parameters of his own needs. Working patiently through these experiments, he eventually came back to platinum. It seemed to burn longer even though it had a lower melting point than carbon.[26]

At the same time Edison experimented with different shapes for the incandescent material. His studies led to the formulation of a new "electric light law" stating that the "amount of heat lost by a body is in proportion to the radiating surface of that body." From this insight he drew another conclusion. Since energy consumption was proportional to the radiating surface of a lamp rather than its resistance, high-resistance lamps should require no more energy to operate than low-resistance lamps. This crucial observation baffled most experts in electricity. By January 1879 Edison had put together his first version of a high-resistance lamp using a thin spiral of platinum set in a globe. It burned for a while but served chiefly to show that he still had a long way to go.[27]

While Edison toiled, his investors fidgeted. Late in October 1878 Sawyer claimed publicly that he had won the race, as he and Man sought a patent for their carbon-electric lamp suspended in a globe filled with nitrogen. At a board meeting of Edison Electric, some nervous members suggested that perhaps Edison should join forces with Sawyer and buy out his patents. Edison denounced the idea as a lack of confidence, saying that Sawyer and others were trying to ride his coattails. He would not join with anyone because his system was "entirely original and out of the rut." The suggestion soon vanished but the investors remained skittish, and Edison realized that the challenge was much greater than he had first thought. Slowly, patiently, he inched toward his goal, exhorting his men through every discouragement.[28]

While the experiments with materials continued, Edison turned his attention to two other key issues: the dynamo and the vacuum pump. In October 1878 he had acquired a large Wallace dynamo and one of Edward Weston's electroplating dynamos. His staff promptly began a grueling series of comparison tests between the machines, but already Edison had begun fiddling with a design of his own. When both generators proved inadequate for his needs, he decided to come up with his own version. The literature on dynamos was skimpy; the best source seemed to be the machines themselves. Edison took the time to master the principles governing dynamos. He got hold of a Gramme dynamo, which he and Batchelor, his most trusted lieutenant, studied intently while awaiting completion of the new building for the steam engine.[29]

Nearly everything had to be learned by trial and observation. Once Edison decided to use a drum armature, he assigned Upton the task of drawing the winding for it. Upton found himself at a loss; firms kept their methods secret, and no guides existed for such work. Frustrated, Upton filled sheet after sheet with drawings but could not get any of them to come out right. "Why don't you have Kruesi make up a few small wooden models of the drum?" Edison suggested. "Then you can take string and actually wind it round the block instead of drawing imaginary lines."

"But that wouldn't be scientific," Upton protested.

"What do you care," countered Edison, "if it does the work?"

Reluctantly Upton agreed to give the models a try. To his surprise he found they helped him enormously. Edison then proposed a "winding bee" in the laboratory one night after dinner. Six of them, including himself, competed to see who could wind one of the wooden drums first. To the winner went one of Edison's cigars. Upton won easily and, since he did not smoke, passed the cigar on to Kruesi, who had been the first to give up. The experiments that led Edison to choose a drum armature proved more fruitful than even he imagined; he had stumbled onto what became the basic pattern for nearly all future generator development of any significance.[30]

Once the new building was ready, Edison, Batchelor, and Upton conducted a grueling series of experiments that produced a design innovation remarkably similar to the one devised by Thomson that same winter. Most generators were built on the premise that their internal resistance should match that of the external circuit. Upton provided some calculations that suggested a more efficient design featuring much lower internal resistance. Edison also knew the writings of Faraday, having consumed all three of his classic works while working as a telegrapher. This knowledge may have influenced him to use unusually large bipolar magnets, creating an odd-looking machine that the laboratory staff dubbed the "long-legged Mary-Ann." The field magnets were huge, standing fifty-four inches high and weighing 1,100 pounds. By July 1879 Upton could declare that "we have now the best generator of electricity ever made and this in itself will make a business." It operated at an efficiency of nearly 90 percent, far greater than other generators. The dynamo was still crude in many respects, but it represented a significant advance over other versions, and it suited the needs of Edison's evolving system.[31]

The vacuum pump too was coming along. That winter Edison got hold of an improved type of Sprengel pump designed in England by Sir William Crookes. He also acquired a rival version, the Geissler pump, which, like the Sprengel, used mercury to trap and expel air bubbles to the outside. Edison also hired as glassblower young Ludwig Boehm, who had apprenticed with Heinrich Geissler. With help from William Baetz of Reinmann Baetz, a glassblowing firm, the staff devised a design combining the two pumps. The result was a pump that far surpassed any existing version in efficiency, and it could not have come at a better time. It emerged as a critical component because Edison had finally rejected platinum as a filament material. He had consumed an enormous amount of time and effort only to conclude that its defects outweighed its potential. Apart from its expense, Edison had discovered that heating caused it to crack and disintegrate, largely because it seemed to absorb gases while heating. He had also made an extensive survey of potential untapped sources of platinum that might lower its cost, but the search confirmed its scarcity.[32]

During the fall and winter of 1878–79 Edison continued to talk to reporters but revealed little in detail about his lamp or his system, pleading again the need for secrecy until patents could be obtained for each component. He gave the *Herald* man a peek at the platinum lamp that revealed incandescence as its type. When the reporter noted that two patents had just been granted, Edison countered with "Two patents don't amount to much when thirty or forty are needed." The *Herald* reminded its readers that Edison "has not undertaken merely to overcome one or two difficulties in the way of the use of electricity for illumination, but to construct a complete system for that purpose." However, Edison had raised the public's expectations, and its impatience grew with every passing month. "With reference to the lighting of dwellings," declared the *New York Tribune*, "the whole subject is yet in doubt . . . The public know what extravagant claims have been made. These claims may yet be justified, but up to the present time they are not."[33]

In March 1879 Edison gave a public demonstration of the platinum lamp. Despite enthusiastic press reports, his financial backers continued to grumble about the apparent lack of progress. In mid-April Lowrey prevailed on Edison to give a private demonstration of the early high-resistance platinum lamp for Morgan, Fabbri, and some of the others. Edison agreed even though he had already decided to abandon platinum. Darkness had fallen when the financiers arrived by train at Menlo Park. Ushered into the plush reception room of the library, which at Lowrey's bidding had been furnished with fine cherry furniture to accommodate Wall Street sensibilities, the visitors listened to a report from Edison on the progress made so far. Then Edison walked them across the yard to the laboratory, where he showed them the platinum lamp he had already decided to junk, pointed to the array of lights in brackets along the wall, and described the generator he hoped to install very soon.[34]

With the room already dark, Edison ordered John Kruesi to "turn on the juice slowly." As the lamps reached a cherry red glow, Edison asked Kruesi for a little more power. Suddenly there was a popping noise and a puff of smoke, and the room went dark again. Batchelor rushed forward to replace the blown lamp, and the demonstration was repeated two or three more times with similar results. The group returned to the library for further discussion before the financiers left to catch their train. One of them, Robert L. Cutting, knew of Starr's 1845 carbon vacuum lamp. "I have read Starr's book," he said sourly, "and it seems to me it would have been better to spend a few dollars for a copy of it and to begin where he left off, rather than spend fifty thousand dollars coming independently to the same stopping point."[35]

The visitors left in funereal silence. Edison had by his own admission showed them yesterday's technology: the platinum lamp, an inadequate Wallace dynamo, and the lights arranged in series rather than parallel circuit. He had reached the point of understanding what would not work but did not yet know what would work. Although the *New York Herald* remained a staunch supporter, the technical press continued to ridicule his progress. Edison's earlier prediction of completing the project in six weeks came back to mock him, and critics feasted on his earlier boasts. Edison could say little in response, if only because he was not about to give out details about his creations until his patents were approved. To the *Herald* he continued to issue reassuring but vague statements while he rallied the troops for a final push to make the doubters eat crow.[36]

By that spring of 1879 Edison had a clear sense of what he wanted. He needed a light of 8 to 20 horsepower with resistance as high as several hundred ohms stabilized by enclosure in a good vacuum utilizing a long-burning, heat-resistant substance. Parts of the puzzle had been put together, but the final pieces remained maddeningly elusive. The search for solutions concentrated on four key areas: the material, the vacuum, insulation, and a regulator, which Edison still considered an essential component. For a time research on the lamp was interrupted by Edison's need to do some work on his telephone, which was being vigorously promoted in England. This lull during the spring and summer of 1879 cost the electric lamp development some momentum. When research resumed, it advanced at a furious pace.[37]

The vacuum held the key to the next phase of research. The ability to create a superior vacuum enabled Edison to return to carbon as a possible material for the filament because he could now better protect it from oxidation. He knew carbon well from his experiments on the telephone transmitter. Moreover, if the final product was to be encased in a globe, he no longer needed the thermal regulator. His earlier experiments on the form convinced him that a thin wire spiral would serve best because it maximized the radiating surface and increased resistance. But how to make a spiral out of carbon? While one of his teams toiled at improving the vacuum pump, Edison launched another lengthy round of painstaking tests on virtually every carbonized substance he could find—different woods, papers, celluloid, flax, cork, coconut hair, even hair from the men's beards. Eventually his staff learned to make threadlike filaments down to seven-thousandths of an inch coated with carbon lampblack and tar.[38]

But the search for the most suitable substance seemed endless. For every test Batchelor had to connect a carbonized filament—which had taken hours to produce in the furnace—as thin as a hair to an equally thin wire. He might spend two or three days setting it in place on a lamp stem only to have it break and force him to start again. Once set, the filament was placed inside a fragile glass bulb Boehm had blown, whereupon Francis Jehl put it on the vacuum pump and evacuated the air. When the vacuum had been created, Edison himself slowly drove out the occluded gases—a process that consumed six to eight hours. Only then did the test begin by sending a current through the filament and recording the number of hours it continued to give off light. When it burned out, Edison broke the bulb and took the filament to his microscope to discover why it had failed.[39]

In October 1879 they began experimenting with cotton thread and found it surprisingly promising. On the night of October 21, after hundreds of failed tests and long, exhausting months of work, trial number nine of a carbonized cotton filament was inserted in a lamp and the globe emptied of air and sealed. When the current came on, Edison and his men witnessed the sight they had long hoped for: The lamp burned for more than thirteen hours—not the forty hours later enshrined in myth. The men cheered and shouted with joy. Edison exulted with them but knew they had more work to do. He needed a filament that would burn longer, much longer. After still more tests he found what he needed in common cardboard coated with carbon. "This is then sealed in a glass bulb," Upton reported, "and the air exhausted and then a current of electricity passed through it which heats it to brilliant whiteness so that it will give a light equal to that from a good sized gas burner." The new filament burned for 170 hours.[40]

This time Edison did not rush to the press. Instead he filed a patent for a carbon filament lamp on November 4, 1879, and quietly told his investors what he had done. Fabbri and Wright slipped over to Menlo Park to see for themselves. They found both Edison's and Upton's houses brightly lit by the new lamps. Edison wanted more money to erect a pilot power station at Menlo Park, but the financiers hesitated. No one yet knew whether the new lamp could be made commercially viable. Edison had allowed only one reporter, Marshall Fox of the *New York Herald*, to get the full story on condition that he not publish it until the inventor was ready. However, rumors about Menlo Park's lights began leaking out until the story could no longer be contained. On December 21, 1879, the *Herald* interrupted its usual parade of murders, scandals, calamities, sensations, and political rants to blare the story on its front page under a banner headline:[41]

EDISON'S LIGHT
THE GREAT INVENTOR'S TRIUMPH
IN ELECTRICAL ILLUMINATION

A SCRAP OF PAPER

IT MAKES A LIGHT WITHOUT GAS
OR FLAME, CHEAPER THAN OIL

SUCCESS IN A COTTON THREAD

For months people had been excited about the prospect of a revolutionary new form of light, noted the *Herald*, and for months they had been disappointed. The gas interests, rallying from their earlier panic, had dismissed Edison's lamp as a "philosophical toy," but the dramatic announcement would "reassure the public whose faith in the 'Wizard of Menlo Park' has grown feeble." The most striking feature of the new lamp was its "unexpected and remarkable simplicity." Even those devout believers in Edison's work never imagined that he would produce a light that could operate "without an apparatus so complicated that it would need a special education to enable them to take care of it."

Caught by surprise, Edison had little choice but to arrange a public demonstration. He had been ready to make the final preparations since early December, but the investors, mindful of the repercussions that would flow from a failure, urged caution. On the seventeenth he cabled his London agent, "Exhibition ready—capitalists won't allow it until about New Year's." Even after the *Herald* let loose the dogs of publicity, Fabbri reminded Edison of the stakes involved:

> I suggest to you the wisdom & the business necessity of giving the whole system of indoor & outdoor lighting a full test of continuous work for a week, day & night, before inviting the public to come & look for themselves. As long as you are trying private experiments, even before 50 people, a partial failure, a mishap, would amount to nothing, but if you were to express yourself ready to give a public demonstration of what you considered a complete success, any disappointment would be extremely damaging and probably more so than may appear to you as a scientific man.[42]

Unmoved, Edison and his men put the finishing touches on the entire laboratory complex as well as their own houses and other buildings at Menlo Park. He welcomed a friendly *Sun* reporter to Menlo Park to view the lamp and the system as it extended to houses and buildings about the grounds. Along with his usual technical explanation, Edison again talked expansively about the potential of his creation. "There is one advantage we shall have over the gas companies," he predicted. "We can sell light all night and power all day. The electricity that will run one gas jet will run a sewing machine, and will cost only four cents a day." He also thought it would be feasible to harness the power of Niagara Falls to generate electricity but said it would not likely be possible for another decade.[43]

On Christmas Eve Edison gave a little demonstration to a select group of men, including a representative of the gas interests. Three days later he arranged a demonstration for his investors, but curiosity seekers flooded Menlo Park as well. After explaining the system he lit one bulb, then another, until twenty burned as evenly and brightly as the first. He also showed his visitors a small wheel that, when turned, operated like a rheostat and varied the brilliance of the light. When he demonstrated the process of creating a vacuum in the bulbs, one bewildered spectator asked, "But Mr. Edison, how do you extract the vacuum?" Another suggested that the lamps could be improved by closing up the pores of the glass.[44]

Throngs of curiosity seekers descended on Menlo Park from every train to mill around the grounds inspecting every facility. Sometimes in their eagerness they damaged displays or equipment despite the efforts of the staff to contain them. They never tired of lighting one of the lamps by laying it between two long wires arranged for the demonstration. Scientific men came as well to examine the lamps and their support system with a more critical eye. To all of them Edison and the staff extended a warm if weary welcome and repeated their explanations until the last train departed and they could finally get back to work.[45]

Few grasped the full significance of the sight before them: The system installed at Menlo Park for experimental and demonstration purposes was nothing less than the first incandescent central power station in the world. It included three Edison dynamos, each rated at 6 kilowatts and powered by an 80-horsepower steam engine. Overhead lines carried current to the twenty-five lamps that lit the laboratory, eight that illuminated the office, and twenty others scattered about the grounds. During the last days of December, Edison and his men scrambled to complete the system and add more lights to it. Still Edison played coy with the press. When a *New York Times* reporter visited him on December 27, Edison gave him a full demonstration but said that no date had been set for the full public display. Asked directly if he planned a public showing on New Year's Eve, he replied, "Not for the public as soon as that."[46]

Clearly Edison was stalling for time; the official demonstration was indeed planned for New Year's Eve. On December 29, the day after the *Times* article appeared, the same paper reported a successful demonstration of the Werdermann lamp in Paris as improved by Dr. Cornelius Herz, formerly of California. Although the exhibition was a success, nothing came of the project. Herz proved to be a fraud whose complicated career was mired in one scandal after another. Nevertheless, the news must have weakened the resistance of the investors to a public demonstration. Skeptics continued to doubt the commercial feasibility of Edison's lamp. The *New York Times* reminded its readers that Edison was "a sanguine youth, and has more than once indulged in glowing visions of magic results of his own ingenuity which now appear to have no chance of realization, and has been in too much haste to announce success as assured before his experiments have been completed."[47]

Despite constant interruptions by visitors, Edison and his men managed to get everything ready. On New Year's Eve and New Year's Day a crowd estimated at three thousand descended on Menlo Park despite stormy weather. Extra trains had been run from both east and west of Menlo Park to handle the demand. Edison and his assistants mingled freely with the visitors, explaining the system in detail and emphasizing that it was but a small taste of what would soon follow. His new generator, he predicted, would light all of Menlo Park and the neighboring area with eight hundred lights. The gaping visitors marveled not only at the lights but also at an electric motor set up to pump water and run a sewing machine. To demonstrate its modest power demands, Edison switched the wires back and forth from the lamps to the motor; both ran equally well on the same amount of power. Everything was on display: generators, vacuum pumps, current regulator, even the making of the horseshoe filaments from cardboard to carbonizing to furnace.[48]

Edison knew how to work a crowd. Dressed in workman's clothes, he explained each element to people in clear, simple language while taking the time to answer the more technical questions of electricians and skeptics who had come to see for themselves. Some notables could be spotted in the crowd, including Senators William E. Chandler of New Hampshire and Preston B. Plumb of Kansas. To those who doubted his achievement from a distance, Edison again issued an invitation to visit Menlo Park and see the system firsthand. On New Year's Day the crowds came again to witness the show and tromp over the grounds. The demands grew more incessant, the questions more grating to a staff that had already lost a week's work thanks to the invasion. One well-dressed woman listened intently to the explanation, then said, "Oh yes, how wonderful! I understand it all now. You bottle up the light in these little globes and then sell it for so much a bottle!"[49]

Inanity was one thing, malice quite another. To Edison's dismay the thundering herd also did its share of damage, mostly from carelessness but sometimes deliberately. A total of fourteen lamps were stolen outright, and one lout caught his feet in the rubber hose attached to the vacuum pump and managed to destroy the pump as he tried to leave. Another visitor, identified as a well-known electrician, concealed a long piece of wire up his sleeve and tried to short-circuit the lamp display, evidently to show that it didn't really work, but one of the sixty employees posted as watchmen spotted the effort and escorted him off the premises. The whole spectacle angered the normally patient Edison. When large crowds came again on January 2, Edison closed the laboratory to the public so that he could get back to work. However, he made sure the lamps still burned at night so that visitors could marvel at them on the grounds, and the people kept coming.[50]

WHILE EDISON SAVORED his triumph, William Sawyer stewed in the bile of bitterness and frustration. Nothing had gone right for him, and he blamed everyone and everything except the most obvious source: himself. Whatever talent he possessed as an electrician had always been undermined by his own flawed character. As one of his obituaries observed, "His erratic and careless habits were practically at war with his talents, and continually led him into difficulty." At twenty-nine his life already seemed in ruins despite its promise of fame and fortune.[51]

Little is known about Sawyer's early years. Born in 1850, a native of Brunswick, Maine, Sawyer had first worked as a telegraphic reporter for Boston newspapers. He was working in the Washington office of the *Boston Daily Traveler* in 1875 when he jumped at the opportunity to become the electrician for a New York firm, the United States Electric Engine Company. His ambition reached as far as Edison's. In August 1877 he received a patent for an "Improvement in Electric Engineering and Lighting Apparatus and System," the object of which, he proclaimed, was to "supply the streets, blocks or buildings of a town or city , . . with any desired quantity of electricity, for the purposes of electrical illumination, electroplating, electric heating, the running of electro-magnetic engines, &c. . . . The advantages of my invention are that it enables householders to obtain a supply of electricity for any purposes . . . that it greatly reduces the cost of electricity to consumers, and, lastly, that it renders practicable the lighting of buildings by electricity." A week later he obtained a patent for his first crude electric lamp.[52]

Although Sawyer worked diligently at trying to develop his lamp, he soon succumbed to his worst weakness. In that same year of 1877 Hiram Maxim, himself an inventor, became the company's chief engineer. It did not take him long to notice a serious problem with Sawyer. "A large, clumsy, and brutal-looking fellow, clean shaven . . . ," he recalled, "he was said to be an expert electrician and telegraph operator, but he was a great drunkard." Maxim suggested that Sawyer might find more nourishment in a pint of milk than a gallon of brandy. The next day Sawyer sent out two or three times for milk and told Maxim that the change was indeed good and he felt better for it. Soon afterward, however, Maxim learned that the milk was heavily laced with brandy. "I made up my mind," recalled Maxim, "that we had better get rid of him."[53]

In October 1877 Sawyer was fired. Three months later, through the help of several acquaintances the young inventor gained an introduction to Albon Man, a New York lawyer with a keen interest in electricity. Sawyer told Man he had been working on an incandescent lamp and convinced the lawyer that he could make it burn permanently. Man agreed to finance Sawyer's research and help with the development work. During the winter of 1877–78 they came up with two versions of a lamp, the second of which used a pencil of carbon held between two metal supports. In March Sawyer gained temporary use of a dynamo and tested the lamp. "You would have supposed a small sun was shining in the vicinity," he reported to Man. That same day he rented a small room to serve as a workshop. With it he began a pattern of finding places in remote locations that discouraged visits by those underwriting his work.[54]

Sawyer came up with some important insights, such as flashing his carbons in an atmosphere of hydrocarbon gas to prepare them. He had worked out an advanced system for distributing light and power, at least on paper, but he took several wrong paths as well. The most serious of them was his insistence that resistance in the lamp should be as low as possible. At one point he told Edison that resistance was a detriment in anything electrical and that he intended to make the resistance of his carbon rod as low as possible. So too was Sawyer convinced that the best way to reduce the oxidation rate of the carbon was to fill the globe with nitrogen. He had tried using a vacuum in the lamp and never managed to make it work.[55]

On March 16, 1878, Sawyer and Man filed their patent for a lamp that used nitrogen gas to preserve its carbon filament. However, the pencil carbon was held so rigidly that it tended to break when expanding or contracting. Sawyer plunged into more experiments to remedy the problem, using different types of carbon, including carbonized paper. Despairing, he begged Man to stick with him, saying that he would produce a viable lamp in the end. "Contrary to my better judgment, I did," Man recalled later, "and have been sorry for it." In July they formed the Electro-Dynamic Light Company with Man as president. Capitalized at $300,000 but with only a pittance of that amount paid in, the firm operated on a shoestring. Sawyer was engaged as electrician, and a new workshop space was rented, but Man's attempts to market the lamp found no takers. Efforts were made to obtain foreign patents but failed for lack of funds. No one, including the founders, cared to buy the stock.

In October, shortly after the creation of Edison Electric Light Company, Man approached Grosvenor Lowrey about joining forces with Edison. Lowrey took the offer to the board members, the more nervous of whom favored the idea, and in very delicate fashion to Edison, who dismissed it scornfully. Ever loyal to Edison, Lowrey soothed the uneasy directors and sent Man a letter rejecting his proposal. In December the lamp got good publicity for its new features, but Man was growing desperate. The treasury was nearly empty, and Sawyer was again causing problems with his drinking. By March 1879 the company had on hand only $2.61 in cash with liabilities of $3,536.72, including $1,850 owed Man. The workmen were all let go, but Man allowed Sawyer to continue working at the shop at his own expense. In April Sawyer claimed he had developed a lamp that fed carbon continuously upward through the base. The lamp was ready, he asserted; it need only be manufactured and sold. He also presented a plan to pay off the company's debts. The board took no action on any of his proposals.[56]

The following month Sawyer made an arrangement with William Wallace and his son to build and sell the Sawyer lamps. Disillusioned with Sawyer and tired of dealing with his erratic behavior, Man approved the deal with Wallace. Although Wallace was well aware of Sawyer's problems, he needed the inventor's help in manufacturing the first run of lamps. In July 1879 Sawyer moved to Ansonia to help with the work. It took him less than two months to wear out his welcome. As Man put it, "Sawyer's conduct was so bad after going to Ansonia that the Wallaces would have nothing to do with him by the reason of his drunkenness and immorality." In September Sawyer was back in New York scheming to organize a rival company to Electro-Dynamic. Somehow he managed to find backers willing to put up $10,000 to underwrite a new version of the Sawyer lamp. Again he rented facilities in obscure locations and began work.

Then, in December, came the splash of publicity over Edison's new lamp and his anointment as the Wizard of Menlo Park. Sawyer had long resented dwelling in Edison's shadow and missed no opportunity to denigrate his more publicized rival. As early as January 1876 he had ridiculed Edison's etheric force theory as utterly without value, arguing that he had noticed the same phenomenon two years earlier and attributed it to "molecular magnetic vibration." Sawyer believed that electricity consisted of "nothing more nor less than a motion of the atomic particles of matter," but his theory had swayed few people. As the race for an electric lamp heated up, he nursed the grudge of one who thought himself underappreciated and his rivals overpraised.[57]

Edison's refusal to join forces with him also rankled Sawyer, as did the splashy *Herald* article on December 21, which mentioned other inventors as well, but not Sawyer. The slight sent Sawyer even deeper into the bottle. That same day his rancor spilled over into a letter to the *Sun* attacking Edison's entire reputation as an inventor. Sawyer charged that Edison had earned his fame from newspaper publicity over three inventions in particular: the quadruplex, telephone, and phonograph. The quadruplex, claimed Sawyer, had been adapted from French and German models; moreover, Edison's version was a failure until another inventor, Gerritt Smith, made it work. The telephone had been created by Bell, not Edison, and the phonograph was "of no earthly value" and had been dropped by its manufacturer. Having warmed to his task, Sawyer challenged Edison with a hundred-dollar wager to perform eight tasks; these included maintaining a vacuum in his lamps, running his carbonized paper lamp for three hours, and proving that his dynamo could run even at 45 percent, let alone 90 percent, efficiency.[58]

Edison scarcely nibbled at the bait. "Mr. Sawyer couldn't run his lamp three hours," he said when a reporter raised the question, "and I told him so. He doesn't know what I can do with mine as well as you do." To this he added delicately, "I think that Mr. Sawyer, at the time he wrote his attack on me, did not know precisely what he was doing." On the twenty-third Sawyer responded directly to the *Herald* article, claiming that he had invented an incandescent lamp earlier than Edison but never patented it because it failed. "Mr. Edison, over a year ago, began this controversy by an attack upon me," he declared. "He has received one in return, and evidently does not like it." Sawyer's original lamp still existed "just exactly as it was removed from the bracket" and could be inspected by anyone interested in knowing the facts. "So far as I am concerned," he growled, "the Wizard is welcome to patent it."[59]

But not entirely welcome. Edison had issued a blanket invitation to electricians to attend the New Year's Eve demonstration and see for themselves. According to one account, a drunken Sawyer showed up at the event and shouted curses at Edison until the crowd shut him up. Sawyer had also promised to demonstrate his own system of household illumination during the following week, but the exhibit did not take place until March 16, 1880. However, Sawyer's letter to the *Herald* caught the eye of Charles Cheever, a patent speculator, who saw in it the chance to make some money through blackmail if nothing else.[60]

After striking a deal with Cheever, Sawyer and Man realized the value of their earlier experiments with the carbonized paper conductors. In January 1880 they filed a new patent for a lamp that basically replicated the earlier Sawyer lamp with the paper conductors and hard carbon conductors added. Gradually this patent took on a life of its own, one that complicated the already Byzantine world of electric light patent wars during the 1880s. However, Sawyer never benefited from it. His drinking grew worse and his temper more violent. In March 1880 he grew so abusive to his wife that his landlady asked them to move. A quarrel with another boarder, Dr. Theophilus Steele, so enraged Sawyer that he shot the man in April 1880. While awaiting trial, he continued to work on his lamp and to rage publicly against his new benefactors. In April 1883, shortly before he was to be sentenced for shooting Steele, Sawyer died suddenly. He was thirty-three years old, three years younger than Edison.[61]

CHAPTER 8

THE PEARL STREET SYSTEM

The Pearl Street station was the biggest and most responsible thing
I had ever undertaken. It was a gigantic problem, with many ram-
ifications. There was no parallel in the world . . . All our appara-
tus, devices and parts were home-devised and home-made. Our
men were completely new and without central-station experience.
What might happen on turning a big current into the conductors
under the streets of New York no one could say . . . Success meant
world-wide adoption of our central-station plan. Failure meant
loss of money and prestige and setting back of our enterprise.
                                                    —THOMAS A. EDISON [1]

NOW THE REAL WORK BEGAN.

It was one thing to invent a workable incandescent bulb, quite another to create a support system for it. Apart from the lamps, dynamos, and wiring, Edison needed to create a method for transmitting power to homes and offices as well as devising the switches, fixtures, junction boxes, meters, and other apparatus needed to operate and maintain the power supply. No business model existed for a complete power system, and no one made the components; Edison would have to do nearly all of it himself. And the lamp itself needed more work. It burned well but not long or efficiently enough. All this work had to be organized, and organizations created for undertaking it. Careful calculations of costs—comparing them always with that of gas—had to be made. The birth of the incandescent lamp was but a prelude to the birth of an industry.[2]

It did not come easily or quickly. Critics continued to insist that Edison's light could not be put to extended use. "We do not doubt," sniffed the *Electrician* of London, "that his cardboard lamp will also prove a failure." One carper, Henry A. Mott, Ph.D., listed ten objections to the use of electricity inside homes; a medical doctor named Ezra R. Pulling took pains to rebut every one of them. From distant Paris, Count Dumoncel, an authority on electricity, dismissed Edison's lamp even though he had never seen it. Wilfred De Fonvielle, editor of a journal on electricity, defended Edison's accomplishment, as did some other French electricians. Then William Sawyer leaped into the fray with a letter to the *Herald*. From his sickbed in January 1880 he praised Dumoncel as the highest authority on electricity and one whose work had anticipated both the phonograph and the telephone. As for Edison's lamp, Sawyer claimed that it had been derived from his and Man's work.[3]

Some said that Edison's lamp could not keep burning, yet on January 9 the *Herald* reported that some lamps had continued to burn for 320 hours, and repeated tests showed that the light remained as strong as at the start. As for Dumoncel, the *Herald* dismissed him as about as "impartial and unprejudiced as those various engineers and chemists employed by gas companies who have hitherto attempted, on this side [of] the ocean, to write down the new lamp." While Edison kept his door open to any electrician or serious scientist who wished to examine the system, he ignored the controversy and concentrated on the work before him, which was formidable enough. A more serious threat lay with what Francis Jehl called the "capitalists, patent pirates and technical experts" who recognized what Edison had done and wished either to forestall his progress or steal his idea and modify it sufficiently to coat their knockoff with legal insulation. To ward off such pirates, Edison ordered his men to "treat visitors courteously, but under no circumstances will they leave their work or give information of any kind to visitors."[4]

The challenge, then, was to put together an operating electric power system in the face of skeptics who doubted it could be done. For this to happen, several problems had to be solved. The lamp had to be made sturdier and more reliable, and ways had to be found to manufacture bulbs cheaply and in quantity. A new, more powerful generator was needed along with any number of components. A location for the first system had to be determined, and some politicking would be needed to get permission for digging up city streets for laying the pipes containing the wires. Although cities had experience with the installation of gas lighting systems, no precedents existed for an electrical counterpart. And, as always, Edison would have to keep skittish investors and the usual chorus of naysayers at bay while the work went forward. Because visitors continued to pour into Menlo Park as if it were a theme park, much if not most of the work still got done after 9:00 P.M., when the last train departed.[5]

With scarcely a pause to savor his triumph, Edison mobilized Menlo Park for the huge and diverse task of creating a practical delivery system for electric power. As Paul Israel observed, he "turned Menlo Park laboratory into a true invention factory." In August 1878 Edison had fifteen employees on the payroll; by the year's end the number had increased to about twenty-four. During the late summer and fall of 1879 a hiring surge pushed the ranks to about thirty-five, where it remained until February 1880. By then the frenzy of activity following the demonstration at year's end led to a sharp expansion to sixty-four, many of them bright young men lured by the excitement of working for the great man himself. Edison drove them hard—the workweek knew no bounds, often reaching seventy or eighty hours—but he also kept things lively with jokes, sing-alongs, and horseplay. "Edison attracts the right kind of people for his work," observed a *Herald* reporter, "and his 'go' keeps them going."[6]

Good men came because they believed in the work, and they adored Edison. "He was the gentle master in everything—word, deed, and action," recalled Francis Jehl, "and we were his disciples . . . One of the most valuable elements in his character was a superb common sense that was rigidly logical." He performed every experiment in minute detail but also waged a constant war against waste. His "boys" worked a full week but usually got the weekend off, and many went into New York for amusement. Sometimes Edison donned his top hat or bowler and went with them to walk the city, eat and drink, and take in a show. On one occasion he spent half a dollar on a phrenologist. The visit produced a detailed map of Edison's cranial landscape that gave him some good laughs. He also loved walking around the "Tenderloin" district to watch the con men at work among the throngs of people going by. Perched on a street corner, free from the press of work and lost among the crowd, he found peace.[7]

These interludes refreshed Edison, who found little solace at home because he was anything but a homebody. Work was his life and the laboratory his home. In 1871 he had married Mary Stilwell, a pretty sixteen-year-old Newark girl who worked for him. Young and affectionate, Mary entered the harsh realities of adulthood quickly with the birth of three children and a husband who seldom came home and was usually dead tired when he did. Her loneliness and insecurity grew even greater after Edison moved his family to the house he bought in Menlo Park near his laboratory. Even then he seldom came home, absorbed in work or simply preferring the company of the boys. Lonely and unhappy, a city girl trapped in a rural backwater, Mary found solace in eating and extravagance. The mounting friction between Edison and his wife, coupled with the pressures of work, only kept him in the laboratory even more.[8]

Edison was oblivious to his obligations as both husband and father. Mary proved to be as inept at raising the children as their father. The two eldest, Marion and Tom Jr., called Dot and Dash by the boys at the laboratory, both grew up spoiled and untamed, as did their younger brother, William. Edison had little time for them, and Mary simply could not manage them. Everything in the household was haphazard and irregular, and Edison himself hardly set a sterling example. Usually unkempt and unshaven, he seldom bathed and rarely changed his clothes on the premise that undressing somehow altered the body's chemistry and brought on insomnia. Although he sometimes delighted the children with mechanical toys and enjoyed playing with them, he had a volatile temper and could be cruel. Most of the time, however, he was simply absent from their lives. At Menlo Park Dot and Dash became pets of the laboratory.[9]

Although Edison kept his hand in everything, his role gradually evolved into one of director more than inventor. Discovery became a collaborative act even though the patents were filed in Edison's name alone because joint patents were subject to challenge. In this role Edison revealed an extraordinary ability to funnel his concepts and insights through the talents of his subordinates. No one doubted that he was the boss, yet he managed always to remain one of the boys. He was brighter than any of them, worked longer hours and harder than anyone. His authority came not from appearance or any air of superiority but from sheer example. The first time John Ott saw Edison, he "was as dirty as any of the other workmen, and not much better dressed than a tramp. But I immediately felt there was a great deal to him." From the men he earned a respect that bordered on reverence. All of them shared the joy of being young and reaching for something great.[10]

The system Edison envisioned did not merely emulate the existing gas structure; it utilized the fixtures as well. Electric wiring could be attached in such a way as to give the user a choice of gas or electric light and enable him to compare them directly. The user would pay for the wiring but got the lamps free of charge. He would pay for electricity used as recorded on a meter. Power would be delivered by wires placed underground in special tubes to protect them and would flow from a central station housing large dynamos driven by steam engines. To test and cost every element in the system, Edison began installing a model version at Menlo Park in the spring of 1880. Then, in April, Henry Villard, one of his stockholders, asked Edison to create a lighting system for a new steamship being outfitted in New York. Edison saw the job as prime publicity for electric lights and agreed to do it. The *Columbia* project became the first commercial Edison electric light plant. In completing it he and his staff devised the first lamp socket, key switch, and safety fuse to prevent wires from overheating and starting a fire. All became key components in later systems.[11]

While this installation proceeded, Edison divided his staff into groups, each charged with pursuing a different problem. In seeking a better filament he told the men, "Now I believe that somewhere in God Almighty's workshop there is a vegetable growth with geometrically parallel fibers suitable for our use. Look for it. Paper is man made and not good for filaments." By the summer of 1880, having tested an endless parade of materials, they presented him with the winner: bamboo. To be certain of his choice, Edison went to the extraordinary length of dispatching men to Cuba, Brazil, China, Japan, Borneo—anywhere that bamboo grew—to search for something better. Their work confirmed the superiority of bamboo and also located in Japan a better form of the plant that Edison adopted. After intensive research on this madake bamboo, he pronounced the lamp ready for commercial production.[12]

The *Columbia* installation spurred Edison to find the most efficient and economical way to produce vacuum pumps. Through the spring of 1880 his men labored at making the Sprengel pump simpler, more reliable, and faster acting. The pump was but one component that had to be transformed from a workable device into a product that could be manufactured cheaply and in quantity. By May the men had come up with a "Mercury Pump" that met those demanding criteria. Edison decided to power it not with steam but with an electric motor drawing power from generators. In doing so he created one of the first uses of electricity for industrial production. The production of lamps required a host of other equipment, including annealing and carbonization ovens, shaping and milling tools, and bulbs in large numbers. After experimenting with ways to produce the bulbs, Edison finally subcontracted them to Corning Glass Works.[13]

For his central power station Edison needed a dynamo that was far more efficient that those used in arc lighting and could maintain a constant voltage rather than a steady current. He designed a new generator that he claimed would reduce the energy lost in linking it to the steam engine that drove it. Rather than use several smaller dynamos run with belts and pulleys, he developed a large 100-horsepower generator and planned to couple it directly to the steam engine. Throughout 1880 his men struggled to realize the first large dynamo they had ever attempted. To help with this design he enlisted a well-known maker of steam engines, Charles T. Porter, who came up with a special high-speed version.[14]

Distribution of power posed another challenge. Edison's vision of a central power station required large dynamos and buried wires. From his telegraph experience he knew how much havoc weather and other factors caused with overhead lines. New York and other cities had already begun to develop a thick canopy of telegraph and telephone wires. It was to avoid this problem that Edison chose to bury his lines under the street. However, any system, above or below ground, depended on containing the cost of the copper conductors that comprised the most expensive element. Underground wires also needed a superior form of insulation. During 1880 alone the indefatigable Edison applied for sixty patents, thirty-two of which pertained to his lamps and seven to his evolving system of distribution. Late that summer he devised an ingenious "feeder and main" system that reduced the amount of copper required by 87 percent and also prevented voltage drops in the lamps farthest from the generators. It consisted of feeder wires that carried current from the dynamos to the larger main conductors under the street.[15]

Edison also needed a suitable meter to measure the electricity used by customers. William Sawyer and Samuel Gardiner had patented a mechanical meter in 1872, and several inventors had devised meters for their own use. As usual, Edison came up with a different solution in the form of an electrolytic or chemical meter that used the weight of zinc or copper plates to determine how much current had moved. The meter was both simple and accurate; it remained in use for several years. In the process of finding one that suited him, Edison took out twenty-two patents for meters. Another major question arose as to the most suitable voltage for his system. No standards yet existed for voltage, which is the force or pressure that moves current through a line. Most other inventors had chosen somewhere between 40 and 70 volts because they used low-resistance lamps that were easier to make. Edison decided on the much larger figure of 110 volts to serve his high-resistance lamps. His choice proved a wise one; in time it became the standard figure for American voltage.[16]

By the fall of 1880 Edison had the components for his system well in hand and faced two critical issues: how to manufacture the parts he needed and where to install his first power station. Behind both these questions lay the even more pressing matter of money. Since spring he had been trying to pry funds from his investors for both these projects, but they showed little interest in taking on still more risky ventures. As the annual report of Edison Electric Light declared in 1882, company policy consisted of "merely paying the expenses of experiments and of taking out and holding patents and not of investing capital in the actual business of lighting." Nor did they care to invest in manufacturing. "We were confronted by a stupendous obstacle," Edison observed later. "Nowhere in the world could we obtain any of the items or devices necessary for the exploitation of the system. The directors of the Edison Electric Light Company would not go into manufacturing. Thus forced to the wall, I was forced to go into manufacturing myself."[17]

During the summer of 1880 Edison renovated an old building near Menlo Park, which had earlier been used to manufacture his electric pen, into a factory for producing his lamps. He put up most of the $10,000 capital, with small sums coming from three key assistants, Batchelor, Upton, and Edward Johnson. This became a common pattern in the companies that followed: Edison put up most of the money and allotted interests to his best men, who were willing to risk their limited funds as the financiers were not. Batchelor took the largest share, 10 percent, and also received the same cut from all inventions. Upton took charge of the lamp factory and presided over it for a decade; Charles L. Clarke replaced him as Edison's chief mathematician. By November 11 the factory was ready to begin production. Using lines run from Menlo Park's machine shop and then its own power plant, it became the first factory to draw all its power from electricity.[18]

That same summer the Menlo Park central station itself underwent a major transformation. To test his concept of a working system, Edison created a larger power station at Menlo Park that utilized buried conductors rather than the overhead wires employed in the original version. The key to this work lay in finding a superior form of insulation. Frank Upton, Charles Clarke, and Francis Jehl emerged as the brain trust assigned the task of developing a suitable insulation and testing the resistance between lines. It took them two months and a crew of twelve to come up with something acceptable: a mixture of paraffin, linseed oil, and asphalt wrapped in two or three layers of muslin. On November 2, 1880, the first underground distribution system powered up for demonstrations. It sent electricity through six miles of underground mains to about six hundred lights. Extra trains were needed to haul eager spectators to the display, which continued for nearly three months. By then Edison was satisfied that an underground system carrying large quantities of current was feasible.[19]

The sight of a fully illuminated Menlo Park thrilled those who came to view it. The bamboo lamps made their debut and performed splendidly. Edison had parceled Menlo Park out into imaginary streets and lined them with white lampposts topped with clear globes. One writer called it "Edison's fairyland of lights" and dubbed the inventor "the Enchanter." Many years later Jehl recalled it as "really a spectacle, especially when the fields were blanketed with snow and the radiant glow of the lamps was reflected from nature's white mantle like millions of tiny dancing stars."[20]

On December 17 nine of the Edison Electric Light directors, including Fabbri and Lowrey, joined Edison in forming the Edison Electric Illuminating Company of New York. Capitalized at $1 million, it became the vehicle for obtaining the right to install a power system under the streets of New York City. A quarter of its capital went to the Electric Light Company in return for a license to use the Edison patents it held; the remaining $750,000 was offered to the stockholders of the Electric Light Company. From these proceeds Edison received $100,000 in cash for his work. Fittingly enough, the first facility would be located in lower Manhattan to illuminate Wall Street and the homes of some of its denizens. Known as the First District, it encompassed about a square mile bounded by Wall Street on the south, Nassau Street on the west, South Street on the east, and Spruce and Ferry streets on the north.[21]

The day after the new company was organized, New York City received its first electric lights, but not from Edison. That day the Brush Electric Light Company powered up its three generators to illuminate Broadway from Fourteenth to Thirty-fourth streets with arc lights. The system had been installed on a trial basis to operate at no charge for six months. Brush had taken care to utilize his best generators and lights connected by wires strung overhead on telegraph poles. The lights burned from two to five that afternoon, then were shut down and the gas lamps lit in their place. "When the gas was lighted," wrote a reporter, "the contrast between the two lights was sufficient to fill the holders of gas stocks with dismay." Merchants in the area loved the new lights and the business they attracted. At the end of the six months Brush's company was rewarded with a contract that led to the installation of fifty-five Brush arc lamps by the end of 1881.[22]

From this first installation in New York came the phrase "Great White Way" to describe Broadway flooded with light. From it too came added pressure on Edison to deliver his system. For two years he had promised the public a new and revolutionary lighting system, saying over and over that success lay just around the corner. By the fall of 1880 he had stretched his credibility to its limit, and the chorus of skeptics continued to grow. In vain did he try to get people to see the enormous difficulties involved in trying to design and install a complete system. The seeming ease with which the Brush system operated successfully contrasted sharply with the absence of any evidence beyond the Menlo Park demonstrations coming from Edison. None of the criticism and hand-wringing fazed Edison, who plodded methodically toward his goal with the single-minded purpose for which he had become famous. His system was ready; the time had come to deploy it in the real world.

The shift from invention to commercial product became strikingly clear in February 1881 when Edison moved from Menlo Park to an ornate four-story mansion at 65 Fifth Avenue, just below Fourteenth Street in New York City. Menlo Park remained in operation and Edison often came out to inspect the research done there, but he parceled much of the work out to the factories of his new companies. Many of his key assistants came to the city with him, including Charles Clarke. Like Upton, Clarke had gone to Bowdoin and finished his education in Germany; like Upton, too, he was a fine mathematician. Clarke held a degree in engineering and played a key role in the design of the new dynamo. "Come on, Clarke," Edison told his engineer, "pack up at once and come with me to New York. You have been appointed chief engineer of the company. We're going to begin business right away."[23]

Two more companies soon followed: the Edison Electric Tube Company in February 1881 to construct and lay the tubes holding the wires and the Edison Machine Works in March to build the needed generators. The money for these ventures also came from Edison and his men. To get it, Edison sold much of his Edison Electric stock and borrowed heavily. He had crossed the line from inventor to entrepreneur, willing to risk most of his funds on a new venture in which he believed ardently. Only one key enterprise started up without his financial backing, although Edison did become a silent partner in it. For some time Edison had relied on Sigmund Bergmann to manufacture many of his inventions. Early in 1881 Bergmann joined Edward Johnson in forming a company to supply Edison with lamp sockets, switches, fuses, fixtures, meters, drop cords, and other small components. It proved to be so profitable and successful a venture that Edison later joked, "I got all the glory, and Bergmann all the money."[24]

To get at the dimensions of the task before him, Edison obtained several large maps of New York City and sent several men to canvass the First District area for data on potential customers. Going house to house, the men jotted down the amount of gas each one consumed, the number of lights, the type of globes and shades on the lights, other uses of gas power, insurance rates, seasonal hours of usage, complaints about gas, and other matters. From this mass of information Edison learned that the district had about fifteen hundred potential customers who used twenty thousand gas jets. Given competitive pricing and free lamps, many if not most were willing to try electricity. Already Edison had engaged Hermann Claudius, an Austrian electrician, to construct a detailed twelve-by-fifteen-foot scale model of the First District on which to plot the entire distribution system. That Claudius spoke no English posed little problem; Upton served as interpreter. It helped greatly that Claudius was an excellent mathematician who "slung his logarithms about with a facility that commanded respect."[25]

As the spring thaw approached, Edison was eager to get at the work of laying wires under the streets. First, however, he had to jump some political hurdles. Permission from the city was needed, and the board of aldermen predictably expressed concern about the plan to tear up streets. Lowrey arranged for the aldermen to visit Menlo Park for a demonstration of the system. On December 20, 1880, the aldermen dutifully arrived, under the full glare of newspaper publicity. Edison greeted them with handshakes all around and led them off on a jaunt through Menlo Park. After two hours of touring and scientific explanation, much of which passed over the visitors' heads, Edison led them back inside for a lavish meal catered by Delmonico's and washed down with a copious supply of champagne. A mood of good cheer soon prevailed, one alderman marveling that Edison looked like a "regular fellow" because he handled his cigar just like the Tammany Hall boys. Negotiations went more smoothly after the visit, and in April 1881 the board granted Edison Illuminating Company a franchise to lay conductors and erect lampposts.[26]

One last city obstacle remained. After the laying of wires began in April, Edison was summoned to the office of the commissioner of public works, who informed him that the work required the presence of five inspectors whose total salary would be twenty-five dollars a week. "I went out very much crestfallen," Edison recalled, "thinking I would be delayed and harassed in the work which I was anxious to finish, and was doing night and day." However, Edison's fear reflected only his innocence at doing business with the city. The inspectors showed up only on Saturday afternoons to draw their pay. The real challenge lay in the work itself, which took fifteen months to complete. By that time the laborers under John Kruesi's able direction had put down more than eighty thousand feet of conductors encased in iron tubes.[27]

In seeking a suitable site for his power station, Edison received another painful lesson, this one in the cost of Manhattan real estate. He picked out two dilapidated buildings on the worst slum street he could find in the First District and hoped to get them for about $10,000 each. Instead he found the asking price to be $75,000 for one and $80,000 for the other. Hastily he revised his plans and settled for buildings at 255 and 257 Pearl Street, which cost him around $65,000. The structure at 257 had four floors and a basement. Since it lacked the strength to house the giant generators, steam engines, and boilers, the insides were ripped out and replaced with structural ironwork. The ground floor housed four Babcock & Wilcox boilers, while the aptly named Jumbo generators went onto the next floor. The third floor contained a large bank of lights that served to test the dynamos and measure the station's load, leaving the basement for storage of coal and ash removal. The building at 255 housed offices, storage, testing and measuring facilities, and sleeping quarters for the workers.[28]

The Edison Machine Works set up shop at 104 Goerck Street and plunged into the task of developing a suitable generator. Edison soon realized that for his central station he needed a bigger dynamo and a better steam engine than the Porter-Allen version he had used for the Menlo Park demonstration. A Menlo Park veteran, Charles Dean, took charge of the work. For months experiments went on to improve the original large Menlo Park dynamo. The armature posed the thorniest problem; Jehl recalled that to reshape it took fifty-five men working eight solid days and nights. From their labors emerged the giant "C" model dynamo.[29]

The first tests took place in January 1881; a month later the machine succeeded in powering all 426 lamps at Menlo Park. Gleefully Edison led a late-night parade to the neighborhood saloon for a round or three of drinks. Further tests led to more improvements until Edison was satisfied. In September the dynamo was dismantled and loaded aboard a ship for exhibition at the Paris International Electric Exposition, where it created a sensation. Since the generator traveled to France on the same ship that had brought P. T. Barnum's famous elephant, Jumbo, to America, it was dubbed the Jumbo. The name suited a machine that weighed more than thirty tons and could power as many as seven hundred lamps.

The first Jumbo was but one of twenty-three that ultimately emerged from the Goerck Street shop. The next two went to London for installation at the Holborn Viaduct station, situated just west of London's financial district. There Edison set up an operating plant with the two generators working together to power 938 lights, including 164 streetlamps, hotels, restaurants, offices, shops, and other buildings. Started up in January 1882 and running full blast by April, the half-mile London installation became both a great success and a handy proving grounds for components needed at Pearl Street. In May three Jumbos went into the Pearl Street station, but problems with the linkages to the six Porter-Allen steam engines reduced the output to only one generator for the first month. In June 1882 the boilers were fired, and the first dynamo started on July 5 to begin extensive testing. With so much at stake, Edison did not want to take any chances on the system's performance. "Well, boys," he said as he often did, "now let's find the bugs." While the testing continued, connections to houses and offices went forward rapidly during the summer.[30]

CITIES HAD ALWAYS been places of darkness. At night they sank into shadows when the sun went down. Even the best avenues relieved the darkness only by measured pools of dim light from gas lamps reinforced in places by splotches of candlelight or gaslight from the windows of homes or buildings along the way. Prudent citizens did not walk the streets alone late at night, especially in marginal neighborhoods. Occasionally brighter slivers of light streamed from buildings with their own isolated power plant. Edison's most consistent booster, the *New York Herald*, had the largest such plant in the city, capable of running six hundred lights, five hundred of them at night. The American Bank Note Company, Everett's Hotel, and several retailers had smaller stand-alone units.[31]

Pearl Street had no such refuge of artificial light because it had fallen on hard times. Earlier in the century it was crowded with fine carriages heading to and from one of the fashionable taverns that lined the street. A popular song immortalized the pleasure of "Going to Live Uptown in Pearl Street." But the street had long since been relegated to downtown and swam in the squalor of a neighborhood well past its prime. That was why its real estate was so relatively cheap, and why Edison had bought there. The building he acquired was an unimpressive mishmash of wrought iron reminiscent of the stations on the Third Avenue Elevated. Long forgotten by most New Yorkers, it was about to seize an unexpected place in the nation's history.[32]

Finally all was ready, or nearly so. By September the testing satisfied Edison enough to begin service even though all the connections had not yet been made. The start-up was scheduled for 3:00 P.M. on September 4. At nine that morning Edison arrived at the Pearl Street station dressed uncharacteristically in a frock coat and white derby. "All I can remember of the events of that day," he said later, "is that I had been up most of the night rehearsing my men and going over every part of the system . . . If I ever did any thinking in my life it was on that day."[33]

Coat and collar promptly came off and sleeves were rolled up as he made final preparations for the big moment. The switch was to be thrown at the Wall Street office of Drexel, Morgan amid a group of luminaries that included Pierpont Morgan himself. Reporters soon appeared at Pearl Street but Edison refused to see them, saying he was too busy. He gave orders that precisely at 3:00 P.M. the station's main circuit breaker was to be thrown, thereby connecting one of the Jumbo generators to the distribution system. He synchronized his watch with those of the workmen and then left for Morgan's office along with Johnson, Kruesi, Bergmann, and the young Englishman who had become his secretary in March 1881, Samuel Insull. John Lieb stayed behind to throw the switch at the station.[34]

Edison well knew how important Morgan had been in bringing his work to this point. Normally the most cautious of bankers about new enterprises, he had seen the potential of Edison's work at once and uncharacteristically moved to support the research before other financiers could get into it even though his father, Junius S. Morgan, repeatedly conveyed his doubts from London. Drexel, Morgan partners dominated both the Electric Light and Illuminating companies. Morgan himself had yet to buy any of the stock personally, but he gave Edison an even stronger vote of confidence in 1881 by asking him to install an electric light system in his spacious new residence at 219 Madison Avenue. It was the first private residence to be wired, and the installation did not go well. A cellar had to be dug beneath the stable at the rear of the house to house the boiler, steam engine, and generator. A trench had to be carved in the yard to build a brick passage for running the wires from the generator to the house, where an electric bulb took the place of a burner in each gas fixture.[35]

Once the system was installed, an engineer had to be on hand to operate the generator from 4:00 P.M. to 11:00 P.M. If a party or other occasion required later hours, arrangements had to be made to keep him over his time. Neighbors complained about the noise and vibrations from the dynamo, which compelled Morgan to install heavy rubber pads underneath the machinery and pile sandbags around the walls to deaden the noise. In the winter stray cats liked to gather on the strip above the wires where heat had melted the snow. Short circuits and breakdowns plagued the system. Nevertheless, Morgan's faith in the electric light remained undiminished. He had 106 lamps installed at Drexel, Morgan and agreed to use his office as the showcase for throwing the switch for the first central power station in New York. He was there, along with some of his partners who had joined him in the venture. Edison watched as his assistants installed the safety catches. Tension filled the air, fueled by the fear of what would happen if the start-up proved a bust.[36]

"One hundred dollars they don't go on," said Johnson to lighten the mood.

"Taken," replied Edison.

At three o'clock Edison closed the switch; half a mile away John Lieb did the same, putting the Pearl Street station online. "They're on!" cried one of the Morgan men as the lamps in the office began to glow. Elsewhere the lights made little impression in the glare of the afternoon sun, but as dusk approached, the First District glowed with a soft, mellow radiance hitherto unknown to the public. For all the paper's criticism of Edison, the *New York Times* had put its building on the system with twenty-seven lights in the editorial rooms and another twenty-five in the counting rooms. Those places, a reporter conceded, became "as bright as day" with light that "a man could sit down under and write for hours without any consciousness of having artificial light about him." The lamps needed only a simple thumbscrew to turn on. "You turn the thumbscrew," added the reporter, "and the light is there, with no nauseous smell, no flicker, no glare . . . more brilliant than gas and a hundred times steadier." During the next five months the *Times* illuminated its other key areas with a total of 288 lights.[37]

Edison beamed with a mixture of relief and glee. Of the 1,284 lamps already installed, only 400 were lit that first day, but they carried the message loud and clear: The system worked and the skeptics had been routed, although some still wondered about the cost of electric light versus gas. As the *Herald* put it, "Edison was vindicated and his light triumphed." For once the Wizard himself indulged in understatement rather than hyperbole. "We have proved today that it is a success," he told the *Tribune*'s man. To the *Sun* reporter he said simply, "I have accomplished all I promised." The newspapers gave the event surprisingly thin coverage and never fully grasped its significance.[38]

The Pearl Street station showed Edison at his best: still a young man, confident, bristling with energy, exuding a gusto for work that infected all those around him, bursting with intellectual curiosity and eager to explore every untapped source for information. He had become the nation's icon for invention by creating an object that not only changed the world but became a symbol for that change. History would forever link his name to the incandescent bulb and to the first central power station. Both would usher in a new era of technology, and with it changes in lifestyle that separated themselves from the past with remarkable speed. In time the lightbulb even became a visual metaphor for the arrival of great ideas or thoughts. Edison had conquered the world of electricity. Like all conquerors he soon found that success forced him to spend ever more time defending the heights he had climbed.

From the Pearl Street station arose the beginnings of what would become a vast industry. Demand for lighting in the First District soared. By year's end Edison had 203 customers using 3,477 lamps; by October 1883 some 508 customers burned 10,164 lamps. Other newspapers besides the *Times* rushed to seek the new lighting for their offices. James Gordon Bennett had stolen a march on them by having Edison install an isolated system first on his yacht and then in the offices of both his papers, the *Herald* and the *Evening Telegram*. Edison had tempted Bennett by setting up a printer's composing case at 65 Fifth Avenue to showcase the advantages of his electric lamps. After September 4 Edison needed no such inducements. Bankers, printers, engravers, businessmen of all kinds, the New York Stock Exchange, even a dealer in the Fulton fish market eagerly became first-year customers. The problem soon became one of finding enough men to lay the wires.[39]

During the early 1880s Edison found himself stretched thin by all his responsibilities. The Pearl Street project occupied most of his time, but he also had to oversee the manufacturing companies, conduct show-and-tell sessions with potential customers at 65 Fifth Avenue, and push the development of improved components for the system. He also had to deal with requests for stand-alone systems by customers ranging from yacht owners like Bennett to a wide range of business firms that, like the newspapers, needed power but lacked access to a central system. Isolated plants for individual firms and persons offered a lucrative market that Edison was slow to appreciate. Absorbed with the Pearl Street station, he did little with this branch of the business until the central power station went online. To handle it, he set up the Bureau of Isolated Lighting within his parent company.

To his surprise, the business grew rapidly. In November 1881 a subsidiary, the Company for Isolated Lighting, was formed to install stand-alone systems in yachts, buildings of many kinds, and homes. In its first year alone the company installed 130 plants with twenty-two thousand lamps and paid a 10 percent dividend. By the end of 1882 its capital had been doubled from $500,000 to $1 million. Unlike the central station business, the isolated plants required large investments from the Edison company. Sales offices had to be maintained around the country along with inventories of dynamos, lamps, and other components. By contrast, local interests organized central station companies in large cities and came to the Edison company for a license and to Edison's manufacturing plants for the equipment they needed.[40]

Not every stand-alone installation went smoothly. Pierpont Morgan was never quite satisfied with his system and wanted it upgraded as new equipment came online. Three times he had the house completely rewired and showed infinite patience with the upheaval caused by the work. In the fall of 1883 he first summoned Edward H. Johnson to look over the original system. Johnson saw that it was already out of date and said bluntly, "If it were my own, I would throw the whole damn thing into the street."

"That," replied Morgan, "is substantially what Mrs. Morgan says."

He told Johnson to make whatever changes were needed and to find a way to light a lamp on his desk, which had to be moved occasionally. This posed a thorny problem; all the other lamps were in wall fixtures. Johnson brought wires up from the basement through the thick, handsome rug to insulated plates that made contact with pegs in one of the desk's legs. In trials it worked fine, but the first time the lamp was turned on it shorted and started a fire that ruined the desk, the rug, part of the floor, and some other objects. Morgan had been out of the house at the time and came home to a house reeking of wet, charred wood and burnt rug. Johnson was summoned and came the next morning when Morgan was at breakfast. He stared gloomily at the debris and braced for the worst, fearing that he had caused the end of Morgan's support for the electrical company.[41]

Morgan came in from breakfast and peered at Johnson over the top of his glasses. "Well?" he said. Just as Johnson started to stammer an explanation, his eye caught Mrs. Morgan behind her husband, gesturing him to hold his tongue. After an agonizing silence, Morgan said, "Well, what are you going to do about it?"

Finding no rock to crawl under, Johnson stood his ground. "Mr. Morgan," he replied, "the trouble is not inherent in the thing itself. It is my own fault, and I will put it in good working order so it will be perfectly safe."

"How long will it take to fix it?"

"I will do it right away."

"All right, see that you do," said Morgan as he walked out of the room.

Johnson got it right the next time, and Morgan liked the new setup so well that he threw a large reception to show it off to his friends. Some of them in turn hastened to order electric lights for their homes. Morgan's patience extended through later rewiring ordeals, which involved tearing through floors and walls fitted with expensive décor. Watching the workmen trash his walls on one occasion, Morgan smiled and said, "I hope that the Edison Company appreciate the value of my house as an experimental station." Even more, he began to invest heavily in Edison stock and become active in the companies.[42]

Edison was less fortunate with William H. Vanderbilt, who came to 65 Fifth Avenue and ordered a lighting system for his new mansion on the avenue. Like Morgan's house, the Vanderbilt home had been decorated by Christian Herter, whose reputation was exceeded only by his disdain for his elite clientele's poor taste. After the system had been installed, Edison joined the Vanderbilts and their daughters for a display one evening around eight o'clock. As they admired the lighting, smoke suddenly drifted upward in the picture gallery, where the metallic thread in the wall covering had somehow crossed two wires and grown red hot. Edison saw the problem at once and had the lights shut down before flames burst out. Mrs. Vanderbilt became hysterical and, learning for the first time that the cellar had a boiler in it, refused to stay in the house until it was taken away. The whole installation had to be removed.[43]

Despite these minor disasters, enough customers signed up for isolated systems to give the company more business than it could handle. Hotels, office buildings, apartments, and other public buildings appreciated the convenience and superior quality of light found in electric lamps. Some saved money in the bargain; the Boston Post Office Building cut its energy bill in half by switching from gas to electric lighting. Textile mills and furniture manufacturers welcomed the light because it enabled workers to discern color gradations more clearly. Chemical factories and flour mills liked the fact that electric light did not use any combustible material or flame in open air. One mill owner found that his savings in insurance premiums alone paid for the isolated lighting system in two years. Between May 1880 and October 1885 the stand-alone company installed a total of 494 plants powering 125,203 lamps.[44]

Manufacturing too provided a steady stream of revenue. The demand for isolated lighting fed business to both the machine works and the lamp factory. "Edison seemed to like this work of transforming laboratory methods into practical ones . . . ," recalled Francis Jehl. "The lamp factory was his pet hobby; he visioned it as a purely Edisonian affair." The challenge was first to devise an

efficient system of manufacturing and then to drive costs relentlessly downward. Every step of the process was analyzed and tinkered with until some improved way of doing it emerged. The original lamps, for example, used a
plaster of Paris base with a beveled brass ring and screw as the contact pieces.
Edison and Bergmann improved this by replacing the ring with a small brass
cap at the bottom of the base. By April 1882 the demand for lamps had outrun
Menlo Park's production capacity and labor supply. Edison moved the operation from Menlo Park to an East Newark factory with ten times the floor
space. There his men could turn out 1,200 lamps a day.[45]

Costs went down, but not quickly or easily. At first each lamp, which sold
for 40 cents, cost $1.10 to make. By the second year the cost had dropped to 70
cents, and then to 50 cents by the third year. The fourth year proved to be the
turning point; the cost per lamp fell to 37 cents, and the volume of sales repaid
all the losses of previous years. Eventually Edison got the cost down to 22 cents
for a 40-cent lamp, and the profits rolled in. Mindful of the risks he and his assistants had taken and the hard work they had put in, Edison gleefully declared
a dividend every Saturday night for the lamp factory. Later he sold the business
for $1.085 million.[46]

The machine works on Goerck Street absorbed the tools and machinery
from Menlo Park in stepping up its operation to produce a variety of dynamos. The Electric Light Company had agreed to use only generators from
the Edison shop, which guaranteed a steady flow of business. By August 1882
the plant had eight hundred men on the payroll and had already turned out
two Jumbo dynamos and 355 smaller generators, with another twelve Jumbos
in production. Charles Dean had come to New York from Menlo Park early in
1881 to serve as superintendent of the plant. Like the indispensable Kruesi,
Dean could "execute from verbal explanations or rough sketches anything
Edison wanted." The new position brought out the best and worst in Dean.
Overbearing by nature, he ran a tight, efficient shop with a brutal hand. His
rough handling of the men eventually put Edison in a dilemma that led to
Dean's abrupt dismissal. But the machine works thrived and was incorporated
as the Edison Machine Works in January 1884.[47]

In all this work Edison kept a busy hand, moving from 65 Fifth Avenue to the
machine works to the lamp factory like a whirlwind. At the lamp factory Upton
complained about "all the trials I have, for scarcely a day passes without a new
'bug' showing itself." At one point Edison spent a week at the factory to get production back on course, and he continued to visit once or twice a week. The
move to East Newark caused a new problem with electrical arcing that again required his presence. Bugs also infested the machine works, where Edison spent
considerable time supervising tests and experiments on the steam dynamos

that kept posing new problems. Eventually he concluded that much of the problem lay with the Porter-Allen steam engine as well as a governor. He ditched both in favor of a new engine designed by the Arlington & Sims company.[48]

As the business grew, the once tight-knit core of Menlo Park regulars scattered to new posts. Batchelor had gone to Europe to oversee the installation of isolated plants on the Continent. The business mushroomed after the successful Paris demonstration of 1881, and isolated lighting emerged as the fastest-growing business for a time. Johnson and Bergmann were busy with their thriving fixtures firm. Kruesi had charge of the Electric Tube Company and Upton of the lamp factory. Jehl moved from Menlo Park to the machine works, and Clarke served as chief engineer of the Electric Light Company. As Menlo Park emptied of men and equipment, some regulars realized sadly that its day had passed. When Jehl received orders in May 1881 to move himself and his testing instruments to Goerck Street, it hit him like a bombshell. "I had always thought," he recalled, "that Edison would never give up Menlo Park, that he would return when the urgencies of affairs in New York were over."[49]

More than half a century later, through the mists of memory, Menlo Park's image still evoked in Jehl a mixture of sorrow and pride. "Menlo Park with its laboratory was a shrine," he wrote, "Edison was the high priest, and we 'boys' were his followers. I had devoted all my energies in loyal obedience to the cause."[50]

Menlo Park faded into the realm between history and myth because Edison had outgrown it. Despite the demands of business, he never stopped being an inventor. Between 1881 and 1883 he produced an amazing 259 successful patents and a host of unsuccessful ones, nearly all of them related to things electrical. Even as business claimed more of his attention, he continued to seek technological solutions to the problems that confronted him. The way to beat competitors, he concluded, was to innovate better and faster than they could. Confident that no one could out-invent him, Edison consistently opposed bringing suits for patent infringement because they would "require me to give my personal attention to the matter & take me off other far more important work." As for any such charges brought against him, he declared that "if we are proved infringers on any point, I can probably take care of that myself."[51]

The Pearl Street station launched not only a new industry but a new era in American life, one that would increasingly be dominated by electricity. Edison had pioneered a revolution, and like all pioneers he faced the arduous task of maintaining his lead against a throng of ambitious rivals. Ahead lay challenges that would try his ability both as inventor and businessman.

CHAPTER 9

THE COWBIRD,
THE PLUGGER, AND
THE DREAMER

There is scarcely a day passing on which some new use for electricity is not discovered. It seems destined to become at some future time the means of obtaining light, heat, and mechanical force.
—ELIHU THOMSON [1]

ONCE THE ELECTRIC LAMP ADVANCED from an invention to a business, it followed the same well-worn path that the steam engine had and that new technologies would continue to use for decades to come. Edison had to fend off a host of competitors seeking either to improve his devices or replace them with even superior versions. He had opened a door that other inventors had been struggling in vain for years to pry ajar. The incandescent bulb and its delivery system, as demonstrated by the Pearl Street station, offered them a model glowing with lucrative possibilities if they could push the development of their own products.

The market was there; eager customers rushed to embrace the electric light once Edison had shown them a workable system. More stood waiting in the wings for improvements to the Edison system or new and better rival versions. The American love affair with technology was as old as the country itself, and it had always been grounded in the belief that today's wonder was but the prelude to tomorrow's more advanced version. The dawn of the electric age escalated the scale of these expectations to an unprecedented level, one at which the future promised inventive miracles that even visionaries could not yet imagine. Inventors were already exploring a wide range of possible uses for electricity. George R. Carey of Boston toiled at what was called "seeing by telegraph," which amounted to a method of sending a picture by electrical wire. Others, including Edison, worked on an electric railway, and advertisements trumpeted "Dr. Scott's Electric Hair Brush," which promised to cure headaches, neuralgia, dandruff, falling hair, and baldness.[2]

The arrival of Edison's system inspired some inventors to double their efforts, if only because the stakes had gone up. During 1880 Sawyer and Man, Hiram Maxim, a Russian named Tchikoleff, and a Professor André all offered new versions of incandescent lamps. Some inventors sought the easier path of simply "borrowing" Edison's work and making it their own. Francis Jehl sneered at the "combination of capitalists, patent pirates, and technical experts" who hoped to "imitate and outgeneral [Edison] by building a rival lamp with perhaps a change in the filament form which should avoid the horseshoe shape. No legal dodge or technicality was overlooked in their effort to capitalize on his patent or force him to come to terms." Some went so far as hiring detectives and spies to visit Menlo Park and glean information or even swipe a lamp or piece of equipment. To thwart these predators Edison had his men posted at all important stations during public demonstrations.[3]

Among this crowd, no one outdid Hiram Maxim in audacity. A native of Maine, he was bound to a carriage maker at fourteen, ducked military service in the Civil War, and wandered through a succession of jobs until he took a position in his uncle's engineering works in Fitchburg, Massachusetts. In 1866 he secured his first patent, an improved iron for curling hair. His entry into the electric wars came in 1878 when he became chief engineer for the United States Electric Lighting Company, which had the backing of a wealthy investor, S. D. Schuyler. It was in this post that Maxim happened upon and fired William Sawyer for his incorrigible drinking. The company made little progress in manufacturing although Maxim did some business producing and installing arc light systems. In 1880 he turned to incandescent lighting. Later he claimed that his work anticipated much of Edison's and had preceded it. Never shy about showering praise on himself, Maxim suggested that he had originated much of what was new in Edison's lamp and later improved it. Here, as elsewhere, he conveniently rearranged the past to suit his needs.[4]

After Edison gave his first public demonstrations in December 1879, Schuyler ordered Maxim to develop an incandescent lamp. During the summer of 1880 Maxim visited Menlo Park, where Edison devoted an entire day and evening to showing him the lamp and the works. It was a courtesy he showed any electrician who came to learn, but Maxim took more than the usual advantage of it. He sent an emissary back to Menlo Park to persuade Ludwig Boehm, Edison's glassblower, to visit him secretly at his New York shop. When these trips were discovered by Edison's men, Boehm abruptly left Menlo Park and turned up in the employment of Maxim. Humorless and forever the butt of practical jokes, Boehm had never been happy at Menlo Park. At United States Electric Lighting he became the invaluable informant that Maxim needed to emulate Edison's lamp.[5]

In October 1880 Maxim announced his new lamp, which bore a striking resemblance to Edison's 1879 version. Instead of using a horseshoe filament, Maxim arranged his in the shape of a Maltese cross; later he changed it to resemble the letter M. With characteristic cheek he claimed to be the pioneer by pointing to his patent of October 1878, which happened to be for an entirely different lamp. As for the new version, Jehl sneered that "it is evident to any sane person that it was but a copy of Edison's paper lamp." With the lamp Maxim also displayed a generator, said to be one of the largest in the country, and a regulator. Maxim did make one contribution by treating his filament with a hydrocarbon vapor to equalize and standardize its resistance. For this process he secured a patent even though Sawyer and Man had an earlier one covering the same method but worded more generally.[6]

Two professors who had earlier boosted Edison's work now heaped praise on Maxim. This was too much for Sawyer, who resented Edison but loathed Maxim even more. Late in November he told a reporter that "all the devices shown in Edison's patent are used by Maxim." Although admitting that he had never liked Edison, he conceded that "Edison has never shown a light or any connected appliance that did not originate in his own mind." Maxim, he added, had visited Sawyer regularly to pick his brain on the incandescent lamp. "Since early in the past summer Maxim has been running to me," he said. "Every week, and sometimes every day for a week, he has been at my shop . . . Maxim virtually acknowledged that he was infringing the Edison horseshoe lamp." Edison himself dismissed Maxim's lamp as "a clean steal."[7]

In the fall of 1880 the United States Electric Lighting Company made its first incandescent installation at the Mercantile Safe Deposit Company in New York. The reading room had ten lamps, the vault another fifty, "lighting up the whole place like daylight." For the next few years the company struggled to compete with Edison in the incandescent field but made little headway. Charles R. Flint, the company's vice president, tried to remedy this defect by hiring technical experts and easing Maxim out of the picture. Lured by the promise of being awarded the Legion of Honor, Maxim sailed to Europe and later went on to win acclaim and a knighthood for his work in guns. Nevertheless, by 1882 United States Electric Lighting was in such weak financial shape that it put its six-story factory in New York City on the market. Sigmund Bergmann managed to buy it at a bargain price by concealing the fact that Edison had put up half the money.[8]

During 1880 another rival surfaced in the form of Joseph W. Swan, the English chemist who had experimented with incandescent lamps as early as 1860. Swan had actually used horseshoe-shaped carbonized strips of paper and cardboard in his lamp, but at low resistance. Lacking a good vacuum seal, the lamps failed quickly, and Swan put the work aside for seventeen years before resuming his experiments. On New Year's Day in 1880 he announced that he had solved all problems and was "now able to produce a perfectly durable electric lamp by means of incandescent carbon." The news puzzled Edison, who had never heard of Swan. English electricians promptly rallied around their own; one wrote in November 1880 that "as usual, it turns out that Mr. Edison has merely copied, with phenomenal exactness, an invention which an Englishman made years and years ago, but strangely neglected to make public until some time after Mr. Edison's alleged invention had attracted world-wide attention."[9]

The absurdity of this charge can be seen in the contradiction that Edison was supposed to have copied an invention that had never been publicized. *Scientific American* ridiculed the assertion. In fact, Edison and Swan—and some other inventors—had worked along parallel lines but with important differences. Swan still thought in terms of low-resistance lamps that would be wired in series, and he never conceived of an entire system of distribution. However, he obtained a British patent for his lamp and organized a company in England to start production in 1880. Two years later Edison formed his first company in England, and Swan did the same in New York. Within a short time the two companies were locked in a legal war, accusing each other of infringing on patents.[10]

Legal wars also engulfed Edward Weston, who had turned his attention to incandescent bulbs. In seeking the best filament for a lamp, he had developed the technique of "flashing" and applied for a patent only to discover that Sawyer and Man had already filed a virtually identical application. While the inevitable legal battle between them dragged on, Weston toiled at finding a filament superior to Edison's version. His training as a chemist served him well. After an exhaustive quest to find the perfect substance—one that was absolutely uniform and structureless—he developed a synthetic product that he called Tamidine, which he hastened to patent in September 1882. With it Weston finally struck pay dirt; Tamidine burned for two thousand hours, far longer than any rival filament, and soon captured a large share of the market.[11]

Tamidine brought Weston not only an enhanced reputation but enough wealth to redirect his activities. In 1882 he sold his company to the United States Electric Lighting Company, which wanted both the firm and Weston's patents as well as his services. The possession of Tamidine gave United States

Electric a clear advantage in its competition with Edison. Weston devoted the next four years to developing commercial lighting as well as central power stations. During that time he contributed several outstanding inventions and innovations as chief electrical engineer. He also grew increasingly critical of the company's president, George W. Hebard, who he thought paid more attention to manipulating the firm's stock than to selling its products in a fiercely competitive market. In repeatedly confronting Hebard with this charge, Weston drove himself into a physical breakdown.[12]

Unable to change or even dent the management, Weston finally resigned in July 1886. He was only thirty-six years old and had already amassed 186 patents. For the next two years he continued to wage war on Hebard's regime in an attempt to reform the company in his new role as consulting engineer. United States Electric could not afford to lose Weston entirely since he had developed virtually all its major products. In June 1887, however, fate intervened when a fire burned down the Newark lamp factory, temporarily putting the company out of business. It went bankrupt soon afterward while Weston launched a new career. Borrowing a page from Edison, he built himself a new laboratory and plunged into the business of creating and manufacturing electrical instruments. Pundits had already labeled him "the new competitor of the great Edison."[13]

Despite his failing health, William Sawyer had involved himself in a new firm, the Eastern Electric Manufacturing Company, after the collapse of his old one. It reorganized in 1882 as the Consolidated Electric Light Company and began using the Edison-type filament even though it owned the Sawyer-Man patent for a low-resistance filament. Sawyer himself died the following spring, but Consolidated soon emerged as a major competitor in the incandescent lamp market. In the summer of 1883 the Brush Electric Company also got into the game by acquiring the patent rights for the incandescent lamp of St. George Lane-Fox. When that lamp proved to be a failure, Brush turned to a Swan-type filament and garnered a modest market share. Although the market seemed small at first—total sales for the industry reached only 70,000 lamps in 1883 and 125,000 in 1884—its prospects for growth lured more companies to the arena. During 1883 two formidable competitors entered the fray on a modest scale: Elihu Thomson and George Westinghouse. Both emerged as the most dangerous threats to the Edison system, though in very different ways.[14]

POISED AT THE edge of launching his business career, Elihu Thomson took time to track the work of other electricians. In particular he followed Edison's work closely despite his skepticism about the incandescent light. Sometime

during 1879 his curiosity brought him to Menlo Park to see not only the work
being done but the "invention factory" itself. Edison greeted him warmly and
gave him the grand tour along with hours of earnest discussion about the lamp
and the electric business in general. As Thomson prepared to leave, Edison
gave him one of the experimental lamps and asked him to test it for himself.
Thomson took it back to Philadelphia and thrilled his classes by demonstrating it for them. From these tests he concluded that the incandescent bulb had
no commercial future. It was simply too inefficient and required huge
amounts of copper to transmit over any great distance. This judgment proved
to be a rare miscalculation on Thomson's part even though it held a grain of
truth.[15]

While Edison toiled incessantly at his task, Thomson moved to New Britain
in September 1880 to help the American Electric Company manufacture his
lighting system. For starters the new company gave him and Houston $6,000
in cash and 30 percent of its stock for their patents; it also hired Thomson for
two years as resident electrician at $2,500 a year. Houston remained in
Philadelphia as consulting electrician for $700 a year. Although he ceased to be
a factor in the business, he pursued a career as scholar and consultant. The new
post became for Thomson another classroom as well as job. While designing
improvements for both the lights and dynamos, he also oversaw the manufacturing process and got his first close glimpse of the business of invention. His
tutor was the young lawyer who created American Electric, Frederick H.
Churchill, whose enthusiasm for the future of electric lighting matched
Thomson's own. Churchill introduced Thomson to the mysteries of finance
and organization, and to the intolerable pressures they often imposed.[16]

Despite high hopes and Thomson's dedication, the new venture did not go
well. It hoped to sell isolated arc light systems to individual customers, mostly
businesses and public buildings, but Churchill soon discovered that only a few
potential clients had both the need and capital for a lighting system. During its
first nine months the company sold only six systems, and its board showed little interest or energy in devising a different or more aggressive marketing
strategy. Churchill's enthusiasm soon went astray. Deciding that he preferred
to invent, he had a workshop set up at the factory and began neglecting his
other duties. He also overreached himself speculating in real estate to the point
where he found himself unable to meet the second assessment on his American Electric stock. Distraught, he tried in vain to borrow the funds from a local bank, then retreated to his father's barn and shot himself. A stunned board
of directors chose Joseph J. Skinner, a scientist then teaching at Yale, to replace
Churchill.[17]

Through these difficult times Thomson plugged away at improving his arc light system, especially the lamp mechanism and the regulator. He had brought with him as an assistant a Central High graduate, Edwin W. Rice Jr., who had planned to attend Yale but decided he could learn more about electricity by working directly with Thomson. Once again Thomson showed his gift for taking some existing component and making it better. His lamp improvements skated deftly around the Brush patents, a problem he had to confront in all his work on arc lights. He added new features, solved old problems, and produced an end product that performed more smoothly in the overall system. Although his automatic current regulator was superior to any other one, he tried to eliminate it from the system with more integrated methods of regulation before settling on a refined version of the original. He also enhanced the system by developing some useful accessories to make it operate more conveniently.[18]

All this required long hours and hard work. On occasion Thomson, Rice, and their workmen paused for a midnight lunch much as Edison and his crew so often did, but without the hijinks. Skinner proved to be more attuned to the business than Churchill had been; by 1882 the company had sold another twenty systems. But the orders trickled in slowly, and the customer base showed little sign of broadening significantly. Skinner and Thomson agreed that another approach was needed. If customers could not afford to install their own systems, the company needed to move into selling central stations to private parties in cities and towns who would then retail power to those needing it. Edison had provided a model with Pearl Street, although the arc light market was different. This approach posed several problems. Utility companies had to be organized, capital raised, and financing arranged. Thomson had to come up with larger dynamos and some other technical upgrades.[19]

In November 1881 Thomson suggested creating a power station in nearby Hartford. American Electric found enough customers for a start and put up a small system near the Hartford Steam Heating Company, which had a steam engine that could run the dynamo. However, like Edison, Thomson overlooked the politics involved in getting permission to operate a utility. It happened that Hartford's mayor and his friends had already secured a charter for an electric light company. The mayor couldn't use it while still in office, but he could block any attempt by outsiders to provide electric lights. Thomson and Skinner had no experience at fighting political wars, and their board lacked the will to challenge the mayor. The project languished. Finally, in April 1882, American Electric landed a good contract for a power station in Kansas City. Shortly afterward came an order from some eager businessmen in Lynn, Massachusetts. Although Thomson did not know it, this order was to transform his own career.[20]

Two Lynn businessmen, Silas A. Barton and Henry A. Pevear, had seen a demonstration of arc lights on Tremont Street in Boston and wanted similar equipment for the area around their firms. After learning that American Electric had provided the system, they journeyed to New Britain, ordered twenty-six lamps, and returned home to organize a utility company. Their efforts proved so successful that by October 1883 the company was powering ninety lights in Lynn. The experience convinced Barton, who owned a stationery company, and Pevear, a shoe and leather manufacturer, that profits could be made from the electric light business. By the time they reached this conclusion, the situation at American Electric offered them a choice opportunity to enter the field.[21]

For months Thomson had been brooding over American Electric's anemic performance. The board was not only sluggish about marketing but also stingy with resources to develop new products and hesitant to explore new possibilities. To Thomson's dismay, American proved to be another shoestring operation, and Thomson had grown weary of shoestrings. The directors knew little and cared less about technical matters; they had become more interested in selling out at a profit. Angry over their indifference, Thomson began searching for some way to gain control of the company himself. This urge brought him into contact with the peripatetic Charles R. Flint, who had bought control of United States Electric in 1879 and thereby owned the patents of Maxim and Sawyer-Man.[22]

Having moved into incandescent lighting, Flint wanted to add an arc-lighting company to his operation. After failing to buy Brush's company, he managed in 1882 to acquire Weston Electric, which had expanded into incandescent lamps the previous year and had the Tamidine filament. Flint was after something big; he envisioned a merger of all the leading electrical manufacturers and their talented founders. After talking individually with all of them, he succeeded in gathering around a table in his office Brush, Thomson, Weston, and an Edison representative. Nothing came of the talks. Later Flint went on to organize the United States Rubber Company, but his efforts in the electrical industry stalled out. While he and Thomson tried quietly to buy up more American Electric stock, the major holders found a buyer of their own in George W. Stockly, the president of Brush Electric Light Company.[23]

Confused by the rapid shift of events, Thomson tried first to get his patents back from American Electric. When that effort failed, he refused to sign a new contract and resigned from the company in July 1882. He warned Stockly frankly that the American Electric stockholders might be misleading him about the company's condition. On his own he tried to obtain patents for his system in Canada and Europe. He also searched for a new position. Unlike

Edison, who relished his independence, Thomson soon found that he had no stomach for the business world or even seeking backers for his work. The attempt made him ill during the summer of 1882 and eroded his usually unflappable enthusiasm. Finally he appealed to Stockly for help. Stockly tried to lure Thomson back to American Electric, aware that without him the company had little value. When Thomson refused, Stockly agreed to sell his interest in the company if the inventor could find a buyer for it—but only after what Thomson called "a most strenuous effort and expenditure of time and money."[24]

For five months Thomson wandered between New York, Boston, and other cities in search of investors. Sick with nervous exhaustion, unhappy at losing so much time from his work, he finally connected with Silas Barton of Lynn, whose syndicate agreed to buy control of American Electric. In October 1882 the company was reorganized with Pevear as president and Barton as general manager. For the vice presidency they persuaded another shoe and leather man, Charles A. Coffin, to join the firm. On November 1 Thomson happily signed a contract to serve as electrician for five years at a salary of $3,000. To protect himself, he retained rights to some of his inventions, but he sensed that the Lynn group would take a much more aggressive approach to the business. In this he proved to be more right than he dared imagine.[25]

The Lynn group showed their true colors at once. Construction began on a new, more spacious factory at Lynn, which opened in the fall of 1883. Until that time, Thomson worked at the old New Britain plant. In April 1883 the officers secured a new charter that increased the capital stock to $1 million and renamed the firm the Thomson-Houston Electric Company. More equipment was put into the New Britain factory and a superintendent hired to reorganize production around separate departments. The new regime paid its workers weekly instead of monthly and met its bills every thirty days. In short, they put what had been a rickety operation on a sound, efficient basis. To Thomson's delight, they embraced the strategy of selling arc-lighting equipment for central stations and hired more salesmen to go after orders. These vigorous efforts paid off; by October 1883, when the operation moved to Lynn, the new company had sold 1,600 lamps—four times the number American Electric had sold during the previous two years.[26]

Here at last were the kind of businessmen Thomson had long sought. Under its new leadership the company sprang to life and became a significant player in the electrical manufacturing field. Another pleasant surprise awaited Thomson in the form of Charles Coffin, who soon emerged as the firm's guiding genius. Only thirty-eight when he joined Thomson-Houston, Coffin had made a brilliant success of his shoe firm largely through imaginative marketing. He knew nothing about the electrical industry, but he had begun to outgrow the shoe trade and his quick mind grasped that electrical manufacturing was "a waiting market, one bound for swift and vast expansion." He embraced the central-station market as a doorway into offering an extensive line of electrical products.[27]

The presence of forceful managers enabled Thomson to retreat to the factory and do what he loved best. During this last year at New Britain he filed for twice as many patents as he had during 1880–82. He also learned to consult with Coffin, Barton, and others about what products were needed so that he could focus his efforts on commercially viable inventions. One key invention was a double-carbon arc lamp that could burn all night without needing attention. Brush had already produced such a lamp, and Thomson was determined to improve on the rival model. Coffin and Barton also urged him to develop an incandescent lamp to expand the company's product line. Brush had added incandescent lights to his line in April 1883, and Thomson realized that he must follow suit. Conceding that he had been wrong about incandescent lamps earlier, he filed his first two patents on his version in 1882 and 1883.[28]

Patents were fast becoming the weapon of choice in the growing competition among electricians. They protected an invention for seventeen years and could prove a difficult if not insuperable obstacle to others in the field. Thomson and other rivals could attack the problem in two ways. They could try to skirt the patent by submitting an application for one of their own with differences significant enough to distinguish it from Edison's. Or they could follow the time-honored tradition of simply "borrowing" from the Edison patent to get production of their own version up and running, then take their chances in court if a suit for infringement was filed, as it often was.

Thomson realized he had come late to the incandescent race. The company would not have his incandescent system ready until late in 1884, and Barton and Coffin did not want to wait that long. In June 1883 Barton negotiated with Sir William Crookes for American rights to his incandescent system. When that effort failed, he turned to the old Sawyer-Man patents. In 1882 those rights had migrated from Sawyer's old firm, the Eastern Electric Manufacturing Company, to the newly organized Consolidated Electric Company, which quickly became a major producer of lamps using not the low-resistance Sawyer-Man filament but an Edison-type high-resistance one. Thomson-Houston bought control of Consolidated Electric in 1884 and thereby staked its presence in both the arc and incandescent markets. In only two years Barton and Coffin had elevated the Lynn company into a major player, and their work had just begun.[29]

They had also rejuvenated Thomson, whose old enthusiasm and energy returned. At Lynn the company provided him with all the resources he needed to dive back into inventing. Like Edison at Menlo Park, he assembled a support team in a workshop called the Model Room, which became at once his haven and the command center for the design of new equipment and improvements on existing goods. An efficient division of labor soon arose. Coffin handled finance and marketing from his Boston office, and Rice oversaw the manufacturing process in the Lynn plant, leaving Thomson free to focus on invention and design.[30]

In the bargain Thomson also found a life outside of work. For years he had lived a secluded bachelor life in a boardinghouse. In May 1884 he married a New Britain woman, Mary Louise Peck, whom he had met and courted for more than a year. In two years life had come full circle for Thomson from the depths of despondency to both personal and professional bliss, and the future promised more of the same.[31]

THREE MONTHS AFTER Thomson gained a wife, Edison lost one.

Mary Edison had never adjusted to her role as wife of a famous inventor. Perpetually lonely and miserable, she consoled herself by eating and spending money faster than her husband could make it. Edison tried to placate his family by renting an expensive house in New York's exclusive Gramercy Park neighborhood, but they lived there only a year before Mary fell ill and was ordered by the doctor to give up housekeeping. They moved to the Clarendon Hotel, where Mary seemed to get better. In the spring of 1884 she improved enough that the family returned to the Gramercy Park house, but during the summer her condition worsened. An anxious Edison took her back to the Menlo Park house where, on August 9, she died unexpectedly of what her daughter called "congestion of the brain," which was mostly likely a tumor. Finding himself a widower with three children, Edison got help from Mary's mother, who had recently lost her husband. She took charge of the household, and Edison took care of her family financially.[32]

Despite their differences, Mary's death affected Edison deeply. Their daughter, Marion, remembered him "shaking with grief, weeping and sobbing so he could hardly tell me that Mother had died in the night." For the next few years "Dot" was her father's constant companion even as he continued to neglect his two sons. "For the rest of the summer," she recalled, "once a week at least, I would drive him through the beautiful countryside. I felt instinctively that he did not want to talk and had more important things to think about. It seems wonderful to me, if not at the time, that Father had so much time for me. He was interested in my clothes, diary, the novel I was going to write." For her part Dot accompanied him everywhere, bought his cigars, and even cashed checks for him though she was only twelve.[33]

Mary's death also dealt a final blow to Menlo Park. Already most of its activities had been transferred to New York, but her death made Edison reluctant even to visit the place. As the laboratory complex turned into a ghost town, the village of Menlo Park sank back into the obscurity it had possessed before Edison made it famous. In 1921 Francis Jehl, Charles L. Clarke, and William S. Andrews made a pilgrimage back to the site where they had helped make history. They found the place deserted and the old brick railway depot gone. A few of the houses still stood, but only the brick foundation remained of Edison's home; ironically, it had burned in 1917 after being struck by lightning. The laboratory, which had been used over the years for everything from a home to a chicken coop, was crumbling in the last stages of deterioration. Much of its lumber had been stripped by parties unknown. The machine shop and glass house still stood, but in rickety condition. Then, in 1928, one of Edison's closest friends and admirers, Henry Ford, bought the property, salvaged what he could, and re-created the laboratory complex at Greenfield Village in Dearborn, Michigan.[34]

Apart from his personal loss, Edison had undergone severe trials in business as well. The ordeal of overseeing his companies took a financial as well as a physical toll, and Edison had never been very good with money. Like poor Mary, he always spent more than he had and rarely paid a bill on time. During the early 1880s he needed more funds than ever to support all the enterprises he had started, and he had to devote more attention than ever to them. Like Thomson, he believed that the future of the electric light industry lay not in isolated plants but in central stations. Developing that market required time and money, but the board of the Electric Light Company showed little interest in providing either one. Like Thomson, too, Edison found himself at odds with his company; unlike Thomson, however, he had the resources and the temperament to shoulder the load himself.[35]

The key to his belief in central stations lay in yet another invention, the patent application for which was executed only two months after he had thrown the switch for the Pearl Street station. The three-wire distribution system reduced the amount of copper needed for the feeder and main system by another 63 percent without increasing the voltage on individual lamps. Together the two systems cut the cost of copper per lamp from $25 to less than $1.50. These savings encouraged Edison's belief that central stations could be profitable in small towns and villages. The idea for the three-wire system had come to him after building an experimental 330-volt central station for the village of Roselle, New Jersey, and showed he had not lost his touch. As an inventor in the electrical world Edison still had no peer; of 321 patents held by eight leading inventors in 1883, an amazing 147 belonged to Edison. Among his closest competitors Weston held 53, Maxim 39, Brush 35, and Thomson 26. However, invention alone could not solve his immediate problem. More than ever before Edison had to wade into the business arena to get what he wanted. The inventor began to morph into an industrialist.[36]

His plan was to install central stations in several small villages and demonstrate their profitability so that he could attract local investors. In the process Edison could sell equipment and lamps, but he also had to put up capital to build the stations. To do that he formed the Thomas A. Edison Construction Department and began collecting data from canvasses of potential customers, much as he had done with Pearl Street. Hermann Claudius, the old Austrian who had plotted the First District neighborhood by building a model neighborhood, took up the task again but did not last long. The miniature replicas took too long, and one of Edison's new hires had a better way. Frank J. Sprague received his education at the U. S. Naval Academy and was a trained engineer. He used mathematics to figure in a few hours what took Claudius weeks of expensive plotting. By the end of 1883 Claudius was gone. Edison also brought William Andrews from the Machine Works to be chief engineer of the Construction Department.[37]

Another employee had come to be invaluable to Edison. Young Samuel Insull had arrived from England in February 1881 to become Edison's personal secretary at the tender age of twenty-one. In short order he had made himself indispensable to the inventor, taking on responsibilities that went far beyond his position and age. By 1883 he found himself handling all of Edison's personal affairs, managing his finances and even buying him clothes. He also took charge of the shop's finances and became an adviser to Edison on financial matters. His rapid rise, growing influence, elegant airs, and imperious manner did not endear him to some of the old Edison hands. "Yes he was able," grumbled one, "but too confident in his ability to do most anything and in the correctness of his judgment, when he ought to have consulted others." Alfred O. Tate, who worked in the Construction Department and later replaced Insull as secretary, described his manner and style with a keen eye:

> His devotion to business almost constituted a religion. He permitted nothing to interfere with his duties towards the interests he was handling ... He loved power and gloried in the exercise of authority. He was highly appreciative of the aid rendered him by his subordinates but never praised them to their faces. Commendation of this nature he might express to a third person, but in his direct relationship he

seemed to think that sustained criticism was the most effective spur towards efficiency.[38]

For his first central station, Edison chose the small town of Sunbury in the Pennsylvania coal belt. The system was put together hastily under the direction of Andrews and Sprague—so hastily that some bearings burned out on the trial test because Sprague forgot to load a feeder with oil. Then the dynamos stalled until Edison himself figured out the problem. Finally, on July 4, 1883, the first system to employ three-wire distribution went into operation. Edison watched and waited expectantly. "My greatest trouble," he said, "will be to get the people to use the light." Some locals feared the current might jump the wires and burn their houses down; others talked of evil spirits and would not go near the wires. Edison remained in Sunbury a week to observe the operation, then left detailed notes for Andrews on improvements that were needed.[39]

Edison secured the patent for his three-wire system in March 1883. The next installation went into Brockton, Massachusetts, which required an underground system of lines. Opened formally on October 1, it had a capacity for 1,600 lamps and became the showcase system for Edison's company. Elihu Thomson attended the opening ceremony and doubtless inspected the system closely. Lawrence, Massachusetts, received a system in November and Fall River in December. By the spring of 1884 the Construction Department was running well. Edison personally supervised all technical work and approved any decision regarding station design, but he had of necessity to leave many technical problems for others to solve. Although his own men handled much of the work, the Construction Department had to leave wiring and some other tasks to local contractors under their supervision. Altogether it built thirteen stations in four states.[40]

Although the Construction Department seemed to be doing well, Edison found himself confronted by a number of crises. For some time he had neglected the issue of patent infringements. The Electric Light Company wanted to bring suits against Swan and Maxim but Edison balked, saying the effort was not worth the time or money. Lowrey agreed, realizing that the basis for Edison's own patent was vulnerable to attack. In England Edison agreed to a compromise with Swan that led to the creation of a joint firm, the Edison and Swan United Electric Company, but in the United States the Brush company had acquired rights to use the Swan lamp, and the Electric Light Company wanted to sue it and Maxim as well. Sherbourne B. Eaton, who had become Electric Light's president in 1882, warned that he was "loath any longer even to threaten suit for infringement unless we really intend to sue." Still Edison said no.[41]

Edison had little faith in the court's ability to protect inventions, a belief shaped by his earlier experience as a witness in telegraph suits. Later he noted in his diary, "A lawsuit is the suicide of time." But events forced his hand. In October 1883 the patent commissioner ruled unexpectedly that Sawyer's application had preceded Edison's and granted the patent for an incandescent lamp with a carbon filament to Sawyer and Man. The Electric Light Company appealed at once and headed to court, where the suit lingered for years. To make matters worse, Edison had his own fight with the Electric Light Company itself over a host of issues that had accumulated since the Pearl Street triumph. In addition, the central stations and other problems had placed a serious drain on his finances, and he was struggling with Mary's illness and eventual death. Nor did it help that the nation underwent a sharp financial panic during 1884.[42]

The rift with the Electric Light Company especially rankled Edison. The board had declined to invest in the manufacturing companies, and it had refused to underwrite the central station program. Eaton acted as agent for procuring central station customers, but he had done little to generate business. "What is Eaton doing about putting agents all over the country to get the towns started?" Edison demanded. The sarcastic Insull, who had gone on the road trying to sell central stations, described Eaton as a pompous little man of military bearing who strutted about "like a great mogul." Eaton was in fact a military man and had distinguished himself at the battle of Atlanta. He was an extremely able but conservative lawyer, hardly the type to pursue the kind of expansionist policy Edison craved. Instead he launched a campaign to merge the manufacturing companies with the parent company, complaining that the Edison firms were profiting by overcharging Electric Light for the equipment they installed.[43]

Edison bitterly opposed this plan, as did his men who had invested in the manufacturing companies. "Now that we have risked $200,000 in this place and pointed out the way, they wish to reap the profits," declared an angry Francis Upton. "I think Major Eaton and Major Lowrey are leading us a wild goose chase." It was Upton's ruthless efficiency at the plant that drove the price of lamps steadily downward into profitability. The *New York Tribune* portrayed the growing split as one between two factions, "one slow and conservative, the other, including the inventor, energetic and willing to spend money for the sake of making money." Negotiations over manufacturing licenses only aggravated tensions when Eaton refused to grant the Edison shops exclusive licenses. Edison complained that attempts to resolve differences seemed "to breed nothing but hostility and suspicion." Endless disputes also raged over charges Edison made to the company for expenses that Eaton thought excessive.[44]

Insull devised a plan to skirt the obstacles placed before Edison by the Electric Light board. However, Eaton got wind of it and gave young Insull a stern reprimand. Never one to forget a slight, Insull turned his fertile mind toward schemes to unseat Eaton. His notion suited Edison, who, like Thomson, concluded that his only recourse was to gain control of a company that moved too slowly for his needs. It would not be easy. The Electric Light board consisted of at least four factions: Edison and his allies; the Vanderbilts; the Morgan group; and the Bostonians led by Henry Villard. Having sold much of his stock to finance his companies, Edison had to scrounge for support among the others. His relations with the Vanderbilts had soured, and the Morgan partners kept their cards close to the vest. Villard had long supported Edison, but the panic had engulfed him in a financial crisis. However, Edison held one crucial trump card: He was the heart and soul of the enterprise, and without him it might wither away.[45]

Here was a classic business imbroglio in which clashes of personality got inextricably tangled with sincere differences over policy. Edison and his partners wanted a management that would move more vigorously to seek central station customers. "I have worked eighteen and twenty hours a day for five years," he declared, "and I don't want to see my work killed for want of proper pushing." He had nothing against Eaton personally, and Grosvenor Lowrey was his longtime friend and supporter, but Eaton was the wrong man to run the company and Lowrey found himself caught in a hopeless conflict of allegiances. Edison worked hard at soliciting proxies and exploiting the dissatisfaction of some shareholders, saying greater profits lay ahead if his men managed the firm.[46]

The Morgan holdings held the key to control of the board. By this time Morgan himself had invested heavily in Electric Light. Although his stake was trivial compared to his other holdings, Morgan still regarded the industry as having great potential. Drexel, Morgan kept the company's deposits, arranged its loans, and even owned the rights for some of its foreign business. In the past the Morgan men had supported the board's conservative approach, but Edison's rebellion created a different situation. The Morgan partners held a large amount of Electric Light stock but not a majority, and Edison's five-year contract with the company was about to expire. Morgan recognized that the company could not survive, let alone flourish, without Edison and concluded that, as Insull put it, "the most graceful thing to do is to give Edison what he wants." Edison showed his appreciation by secretly giving Morgan and J. Hood Wright each 155 shares of stock in the Machine Works.[47]

The Morgan support enabled Edison to win the election. Lowrey, Eaton, and the Vanderbilt men left the board, although Eaton stayed on as counsel for the company. In their place came Upton, Charles Batchelor, and Edward Johnson. Wright remained on the board, and a younger Morgan partner, Charles H. Coster, replaced Egisto Fabbri. Coster worked well with Edison and took charge of all his dealings with Morgan's firm. To the surprise of many, Edison made Johnson the chief executive even though he had told John Tomlinson, his personal attorney, that "EHJ is a telegraph operator, near sighted and generally of no account." For the first time Edison took charge of all the companies bearing his name and was free to pursue the aggressive course he had advocated. However, Morgan's influence still remained strong, and the banker did not always agree with Edison's approach to business. As the industry grew more competitive, fresh differences would arise between them.[48]

For the present, however, Edison could savor victory amid his grief. No one did more savoring than Insull, who relished the ousting of Eaton. "You see I am at last getting even . . . ," he crowed to Tate. "There is no one more anxious after wealth than Samuel Insull, but there are times when revenge is sweeter than money, and I have got mine at last."[49]

IN THAT CROWDED year of 1884 few people noticed the modest entry of George Westinghouse Jr. into the electrical field. Although he had already earned fame and fortune as an inventor, his work had been far removed from electricity. Had there been no Edison, Westinghouse might well have claimed the title of America's premier inventor. In forty-eight years of work he averaged a new patent every month and a half. During that time he created or was associated with 102 different companies, only a few of which bore his name. Like Edison, he moved easily from one field to another, his mind insatiable for new information and fresh challenges.[50]

At the young age of twenty-three Westinghouse burst onto the public scene with the most improbable of inventions: a method for stopping trains with compressed air. Hardened railroad men laughed at the idea. "Do you pretend to tell me that you could stop trains with wind?" bellowed Commodore Vanderbilt when approached with the idea. "I'll give you to understand, young man, that I am too busy to have any time taken up in talking to a damned fool." But the damned fool persevered and in 1872 added a triple valve that enabled every car to have air pressure. The air brake transformed one of the most difficult and dangerous jobs on a train even though it remained a hard sell because of the cost in equipping every car. In all Westinghouse received 103 patents on his air brakes alone, and by 1876 nearly 38 percent of all American passenger cars used them. Later, among many other inventions for the railroad, Westinghouse devised the friction draft gear, which A. J. Cassatt of the Pennsylvania Railroad considered even more important than the air brake.[51]

Born in the tiny village of Central Bridge, New York, in 1846, Westinghouse was to the manner born. His family included generations of farmers and mechanics. In 1856 his father left the farm to set up a small machinery shop in Schenectady, where his wares included small steam engines. George senior accumulated at least seven patents on his own, and his namesake took eagerly to the shop when he began work there in 1860 at the age of thirteen. His pay was 50 cents a day, which rose gradually to $1.12½ by 1863. That year he joined the Union army and did not come home until 1865. He tried Union College for three months before returning to his first love, the shop. In all Westinghouse had only about eighteen months of school after the age of thirteen. He cared little for studies, yet he spoke and wrote excellent English and had a voracious appetite for information. At nineteen he got his first patent, for a rotary steam engine. The idea stayed with him. At forty-five, a successful inventor, he was spied one day, dressed in a frock coat, tinkering with a rotary engine during a lull between a board meeting and a reception. It was for him, said an admirer, "the equivalent of a rubber of bridge, or a game of golf."[52]

A handsome, strapping six-footer, Westinghouse was shy but strong-willed. He had not been an easy child. "I had a fixed notion that what I wanted I must have," he admitted later. "Somehow, that idea has not entirely deserted me throughout my life. I have always known what I wanted and how to get it. As a child, I got it by tantrums; in mature years, by hard work." As he matured, Westinghouse tempered this willfulness with the reins of self-control, but work remained the driving force of his life. His self-assurance, like that of Edison, proved to be at once his greatest strength and most dangerous weakness. It was a decided strength when it came to marriage. Young George met Marguerite Erskine Walker by chance on a train and told his father that same evening that he had met the woman he was going to marry. They remained a devoted couple for forty-seven years. When Westinghouse died in 1914, his wife lived only another three months.[53]

Westinghouse was no stranger to electricity. As a boy he had tinkered with a Leyden jar in his father's shop, but mechanical challenges occupied his mind until late 1883, when he began to take a serious interest in direct-current lighting. What drew his attention in that direction remains a mystery. It was certainly not a lack of things to do. Ever since he had moved to Pittsburgh in 1868 and made that city the headquarters for his diverse business interests, Westinghouse had overextended himself. During 1879–80 he obtained four telephone patents with the idea of creating a system of auxiliary telephone exchanges that might connect rural areas with nearby cities. In 1881 he invested in the Westinghouse Machine Company, which his brother Henry Herman Westinghouse had organized to build a high-speed steam engine of his own design.[54]

The air brake company continued to do well. Westinghouse spent considerable time during the 1870s in England promoting it and arranging for the brake's manufacture. While there he became interested in the switching and signaling devices used by English railroads, which were far superior to those on American roads. Intrigued, he turned his attention to inventing improved versions, using both compressed air and electricity. Like Edison, he thought always in terms of both invention and commercial viability. In 1881 he bought two small companies and combined them into the Union Switch and Signal Company. It was through Union that Westinghouse entered the electrical field. By chance Henry Westinghouse met a young engineer named William Stanley, who had devised his own version of a DC generator and done some original work on an incandescent lamp. Stanley had another intriguing item on his résumé: He got his start in the business by working on Hiram Maxim's early lighting installations. Henry put Stanley in touch with George, who was impressed enough to invite the young engineer to come to Pittsburgh and develop his system at the Union Switch and Signal shops.[55]

Stanley had already assigned his patents to the Swan Electric Light Company, but Westinghouse bought them for $50,000. Stanley agreed to assign all future patents to Westinghouse for 10 percent of the profits and an annual salary of $5,000. Westinghouse also put some of his own engineers to work on the lamp. In 1884 the Union company began selling the Stanley system, which went on display at the Philadelphia Electric Exhibition that fall. Thus did Westinghouse enter the electrical field, but only on a modest scale. His attention still lay elsewhere, especially in natural gas, of all things. Late in 1883 he decided to organize a company, buy and lease gas fields, and lay pipes to provide Pittsburgh with cheaper gas. By mid-1884 the project was in full swing, and new inventions again flowed from Westinghouse's brain. Altogether he secured 38 patents for devices relating to natural gas.[56]

Amid the extraordinary range of his interests, electricity seemed a trifling affair in 1884. But Westinghouse understood the principle of convergence as few other inventors and businessmen did. Although he could not predict which of his activities would interact with which other ones, he sensed that the points of intersection among them could bring unexpected results, sometimes on a grand scale. This proved to be the case with electricity. The air brake had given Westinghouse invaluable experience in manufacturing and marketing as well as a financial base on which to build his other enterprises. Union Switch and Signal contributed the engineering expertise, facilities, and electrical devices to underwrite a move into the electrical field. His experience with natural gas also proved useful, as it did to Edison. Westinghouse developed a system of gas distribution that involved delivering the gas at high pressure and then converting it to a lower pressure at the point of use. It became a clear and valuable analogy for the later problem of converting high voltage to lower levels.[57]

DESPITE THE GLOOM cast by the financial panic and a nasty presidential race that saw the election of the first Democrat to the office since the Civil War, 1884 proved to be a decisive year in the development of electricity. Competition had shaken the manufacture of incandescent lamps down into a handful of firms: the Edison companies, Thomson-Houston (which acquired Consolidated Electric Light Company in 1884), Brush, United States Electric (which had acquired Weston in 1882), Union Switch and Signal, and two smaller firms. Edison had gained control of Electric Light, and Thomson had settled happily into the role of his choice at his new firm. Brush had extended his operation into incandescent lights, and Swan was preparing to get into the American market. Amid all this activity, no one noticed the arrival on American soil of a man destined to be one of the nation's most brilliant, versatile, and bizarre geniuses.

Nikola Tesla was anything but a typical immigrant. He hailed from Smiljan, Croatia, a farming village nestled between the mountains and the Adriatic Sea. The son of an educated Serbian Orthodox priest, he grew up steeped in the complex stew of Slavic heroic oral literature, clashing religions, and tangled ethnic heritages filled with ancient but no less bitter conflicts and resentments. He had a difficult childhood, especially after the accidental death of his older brother, Dane, a bright and gifted boy who had been his parents' favorite. For a time the distraught parents rejected Nikola. "The recollection of his attainments made every effort of mine seem dull in comparison," he recalled late in life. "Anything I did that was creditable merely caused my parents to feel their loss more keenly." The incident scarred Nikola for life; he had witnessed Dane's death and still recoiled from the vision half a century later.[58]

An acutely sensitive child, Tesla also grew up in a world teeming with superstition and myth. As a boy he was tormented by what he described as "a peculiar affliction due to the appearance of images, often accompanied by strong flashes of light, which marred the sight of real objects and interfered with my thought and action. They were pictures of things and scenes which I had really seen, never of those I imagined." Around the age of twelve he mastered the ability to banish the images but not the flashes of light, which continued all his life. He also began to develop the long list of peculiar phobias that followed him into adulthood: an aversion to women's earrings but not other jewelry. He could not bear to touch anyone's hair. Looking at a peach gave him a fever, and the presence of camphor caused him severe discomfort. Dropping little squares of paper into a dish filled with liquid gave him an awful taste in his mouth. He began to count his steps when he walked, and could not eat unless he first calculated the cubic contents of all the food before him. "All repeated acts or operations I performed," he admitted, "had to be divisible by three and if I mist [sic] I felt impelled to do it all over again, even if it took hours."[59]

Along with these eccentricities Tesla also possessed an extraordinary mind. He had a photographic memory and developed a remarkable power of visualization; later he claimed that he could imagine an invention, modify it, and even test it all in his mind. "I needed no models, drawings or experiments," he noted with satisfaction. "I could picture them all as real in my head." In school he showed an amazing gift for languages and for mathematics. Between the ages of ten and fourteen he attended the real-gymnasium (similar to junior high school) and then was sent to a higher-level gymnasium in Karlovac (Carlstadt). There he crammed four years of work into three years only to have his health break down. During his final year he caught a fever thought to be malaria; later he came down with cholera and spent nine months in bed. During all this time his father had determined to make Nikola a priest even though the boy pleaded that he wanted to be an engineer. As he lay near death from cholera, his father relented and promised to send him to the best technical institute if he recovered.[60]

When his son did recover, Milutin Tesla chose the Polytechnic School in Graz, Austria, eighty miles south of Vienna. There Nikola plunged into his work, studying twenty hours a day, mastering nine languages, poring over the works of Descartes, Goethe, Spencer, and Shakespeare as well as physics and mathematics. His obsessive approach brought him to the complete works of Voltaire, a hundred large volumes in small print. Doggedly Tesla worked his way through the set. When he had finished the last volume, he laid it down in relief and said, "Never more!" His dedication earned him high marks, the resentment of other students, and a warning from his teachers to his father that the boy was risking his health with his obsessive devotion to studies.[61]

In a physics class during his second year, Tesla got his first exposure to electricity in the form of a DC Gramme dynamo with its commutator that transferred the current from the generator to the motor. His professor, a German named Poeschl, explained its operation as he demonstrated it. Electric current, in its natural state as waves, alternated direction (AC), changing its flow rapidly. The commutator with its series of brushes transferred the current in only one direction, making it a direct current (DC). Enthralled, Tesla noticed that when Poeschl ran the dynamo as a motor, the brushes sparked badly. Perhaps, he said abruptly, the alternating current could be applied directly, dispensing with the need for a commutator. Startled by the remark, Poeschl proceeded to deliver a lecture explaining why this could not be done. "Mr. Tesla may accomplish great things," he said in conclusion, "but he certainly will never do this. It would be the equivalent of converting a steadily pulling force, like that of gravity, into a rotary effort. It is a perpetual motion scheme, an impossible idea."[62]

Unconvinced, Tesla began a dogged campaign to visualize the workings of an AC motor. Frustrated at the effort, he began to neglect his studies and turned to gambling. Unprepared for his examinations, he left school without a degree and traveled through Slovenia to avoid facing his parents. They finally located him and persuaded the boy to return home and start anew at another university. However, Milutin died a few months before Nikola enrolled at a branch of the University of Prague. There he was exposed to the work of physics professor Ernst Mach, who had, among many other achievements, constructed a wave machine to demonstrate "the mechanical theory of the ether." Mach believed that the structure of the ether was somehow linked to a gravitational attraction between everything in the universe. He also thought that every mental activity must have a corresponding physical action.[63]

Tesla's later research and his view of life closely resembled those of Mach, yet he never mentions the professor in his autobiography. It is therefore impossible to know the degree to which Mach actually influenced his thought. At the time Tesla was still obsessed with visualizing a workable AC motor. After a year at the university his funds ran out. With his father dead, Tesla saw no alternative but to leave school and find a job. He traveled to Hungary, where he managed to wheedle a position in the government's telegraph company. When an American telephone exchange opened in Budapest, Tesla rushed to get a job with it. There he first came into contact with the work of Edison. He studied many of Edison's inventions intently, mastered the principle of induction, and on the job took apart instruments so he could figure out how to improve them. He even invented an amplifier to boost transmission signals—in effect an early loudspeaker—but never bothered to patent it.[64]

He also continued to grapple with his elusive vision of an AC motor, seeking what had become his Holy Grail of eliminating the commutator from the machine altogether. The effort led to a nervous breakdown during which his always acute hearing became unbearably sensitive. Any noise or light racked him with pain. "A fly alighting on a table in the room would cause a dull thud in my ear," he recalled. A physician treated him with large doses of bromide of potassium before pronouncing his case incurable. Steeped in despair, Tesla never expected to recover. Then, as if by magic, his strength and energy began returning until he seemed stronger than ever. He attributed his recovery to a "powerful desire to live and to continue the work, and the assistance of a devoted friend and athlete." The friend was Anthony Szigeti, an old classmate who had become Tesla's confidant. Szigeti got Tesla out of bed and outdoors to exercise, which apparently did much to calm his nerves.[65]

Bristling with renewed energy, Tesla attacked the AC motor conundrum anew. "With me it was a sacred vow, a question of life and death," he insisted. "I knew that I would perish if I failed." One evening, while walking in the park at sunset with Szigeti, Tesla began reciting a passage from Goethe, one of many he had memorized. As he spoke, an idea "came like a flash of lightning and in an instant the truth was revealed." Excitedly he paused and drew a diagram in the sand with a stick. For the next two months Tesla surrendered himself to the ecstasy of picturing and refining new motors in his head. It was, he said, "a mental state of happiness about as complete as I have ever known in life." During that time he claimed to have conceived "virtually all the types of motors and modifications of the system which are now identified with my name."[66]

What Tesla visualized was the prototype for a motor based on a rotating magnetic field. By employing two circuits instead of the usual one, he generated two currents that were ninety degrees out of phase with each other. Through induction this caused the motor's armature to rotate and thereby attract a steady stream of electrons. Putting the currents out of phase staggered them so that the peak output points of one coincided with the dead points of the other. This was necessary because alternating current momentarily ceases to flow every time it reverses. A 60-cycle-per-second current, which became the American norm, had 120 of these reversal dead spots every second. The rotary movement eliminated the dead spots and allowed the motor to operate smoothly as its poles rotated. In addition to this single-phase motor, Tesla also conceived of polyphase versions that used three or more circuits of the same frequency.[67]

All of this work took place only in Tesla's mind. Meanwhile, he had been put in charge of the telephone exchange in Budapest and did so well that Ferenc Puskas, who had originally hired Tesla, offered him a position in Paris to help run the new Edison lighting company there. The "magic city" enchanted and overwhelmed Tesla and prompted him to adopt a rigid routine. Every morning he rose at 5:00 A.M., went first to a bathing house on the Seine, swam exactly twenty-seven circuits, and then walked to the large Edison factory at Ivry-sur-Seine. There he worked under none other than Charles Batchelor, who had been sent to oversee the operation. The Americans he met at work liked him because of his skill at billiards. Eagerly he explained to them his idea for a new motor. One of them offered to form a stock company to develop it. Tesla looked at him in bewilderment; he had no idea what a stock company was. In his spare time he jotted down the mathematics and specifications for his motor in a notebook and continued work on another of his brainstorms, a flying machine.[68]

Batchelor planned to install central stations all over Europe. In April 1883 he dispatched Tesla to Strasbourg to oversee the installation of 1,200 lamps that had been shipped there earlier. There had also been trouble with the generators at the station, and servicing them required a creative engineer since, as Batchelor observed, "all our plants are differently constructed." Tesla spent the next year in Strasbourg repairing the generator that had vexed earlier engineers. In his spare time he secretly built a simple AC motor using materials he had brought from Paris. When at last it was done, he rejoiced at "the satisfaction of *seeing rotation effected by alternating currents of different phase, and without sliding contacts or commutator*." Eagerly he showed it to several friends and to the city's mayor, who was impressed enough to gather some wealthy friends for a demonstration. None of them showed the slightest interest. Mortified, Tesla finished his work in Strasbourg and in the spring of 1884 returned to Paris in hopes of being paid handsomely for the good work he had done.[69]

Instead of a reward Tesla got a classic runaround from company officials. Batchelor suggested that he go to the United States and help Edison redesign his generators. Tesla agreed to go at once. Thus it was that on a spring day in 1884 Tesla stepped ashore in the New World against a backdrop of the fabulous Brooklyn Bridge nearing completion and the audacious new Statue of Liberty being hoisted aloft. Despite these wonders, his first impression was unsettling. "What I had left was beautiful, artistic, and fascinating in every way," he observed; "what I saw here was machined, rough and unattractive . . . It is a century behind Europe in civilization." Five years later he reversed his position and thought that America was at least a century ahead of Europe.[70]

When at last Tesla was introduced to Edison, he described the experience as "a memorable event in my life. I was amazed at this wonderful man who, without early advantages and scientific training, had accomplished so much." Well might Tesla have thought that the future had at last arrived. He was a new man in the New World, eager for fame and fortune and ready to serve the most famous inventor in the nation. The dreamer had found the perfect venue for realizing his most cherished obsession. Or so it seemed.[71]

CHAPTER 10

THE ALTERNATIVE
SYSTEM

Every great invention is really either an aggregation of minor in-
ventions, or the final step of a progression. It is not a creation but a
growth . . . Hence, the same invention is frequently brought out in
several countries, and by several individuals, simultaneously. Fre-
quently, an important invention is made before the world is ready
to receive it, and the unhappy inventor is taught, by his failure, that
it is as unfortunate to be in advance of his age as to be behind it.
—ROBERT H. THURSTON [1]

PIONEERS OFTEN PAY A PRICE FOR being first, and Edison was no exception. The Pearl Street station failed to earn a profit for two years and did
not reduce its rates to customers until 1890. The early version of electricity
failed to undercut gas in price, which slowed its spread and allowed the gas
companies to breathe easier. The installation of new systems was confined almost entirely to large businesses, houses of entertainment, and the homes of
the wealthy. Although the Edison manufacturing companies did well, new
customers were slow to arrive for central stations because only a few cities had
populations dense enough to offset the high installation cost of Edison's large
copper mains. Edison realized that the future of the electrical industry required new ways to reduce the cost of generating and distributing power. He
also knew that competition in the industry was becoming more intense. To
meet it, he resorted to the same formula utilized by most of the great industri-
alists of the era. "No one is safe in this cold commercial world that can't produce as low as his greatest competitor," he declared. "No matter how much
money you are making never for an instant let up on economizing."[2]

Edison also recognized that, like it or not, his role had changed. Even before
gaining control of the Electric Light Company he announced that "I'm going

to be a businessman, I'm a regular contractor for electric lighting plants and I'm going to take a long vacation in the matter of invention." True to his word, he traveled the country overseeing the installation of lighting systems. By August 1886 the company had built 58 central stations powering an average of 2,580 lamps; by October of the same year he had also installed 702 isolated plants powering an average of 258 lamps. The larger central stations thus averaged ten times the number of lights as the isolated plants, but they cost much more to build. Pearl Street had provided an important lesson. Unforeseen problems had swollen its final cost to $600,000, more than double the original estimate of $250,000.[3]

In marketing the system the Electric Light Company made a careful distinction between small cities under ten thousand in population without gas service and larger cities that had gas. For a central station the smaller towns paid cash for their equipment and a license to use it in that location only. The larger cities with gas got a license separate from the equipment giving them the right to operate a central station and to sell isolated plants as well as all Edison equipment and lamps. For that privilege they usually paid 25 to 30 percent of the central station company's stock along with 5 percent of its capitalization in cash. This arrangement gave the Electric Light Company a stake in each of the central station firms as well as a market for Edison products, which under the terms had to be used exclusively. This stipulation benefited Edison's manufacturing companies but not Electric Light, which owned no part of them. Edison's control of Electric Light eased the tension between it and the manufacturing firms but did not resolve the awkward disconnects between them.[4]

Of necessity Edison had to delegate authority to his corps of talented associates. Edward Johnson took hold of Edison Electric Light, expanded the Pearl Street station, and began construction of two new power stations at Twenty-sixth and Thirty-ninth streets. In May 1885 he issued a stern warning that anyone infringing on an Edison patent would be promptly sued and backed this threat by bringing suits against United States Electric Lighting, Consolidated Electric Light, and some other manufacturers. By October 1886 the company could report that a hundred such suits were pending. In the end, however, Edison went the distance on only one major suit, that against United States Electric. Hearings on it did not even begin until 1889. The threat of death by litigation had little effect on Edison's rivals, if only because considerable doubt remained that the courts would uphold Edison's patent. Consolidated responded by instituting several suits against local Edison companies for patent infringements of its Sawyer-Man lamp.[5]

The Machine Works, which had been incorporated in January 1884, remained in the care of John Kruesi until Batchelor returned from Europe shortly afterward to take charge. It hungered constantly for more capital to produce the large equipment needed for installations. Bergmann's company continued to thrive at manufacturing the small apparatus, and Francis Upton rode herd on the lamp company, which had moved again to a larger plant, this one in East Newark (Harrison), New Jersey. The lamp factory remained Edison's pet project. He spent most of his limited research time working on new filaments with the help of John Ott as well as prodding Upton to produce a better product at a lower cost. To everyone's relief, business began to pick up from the doldrums following the panic of 1884.[6]

During 1885, however, Edison's outlook on life began to change in ways that astounded all his cronies. While still immersed in work, he suddenly rediscovered life outside business. He was then a widower of thirty-eight with three children, two of whom scarcely knew their father. They had received little education, thanks to Edison's prejudice against formal schooling. Amid his electrical work he had turned his attention back to the telephone and his dispute with the Bell company over the fact that they paid him $6,000 a year in salary but no royalties for use of his carbon transmitter. He renewed his acquaintance with another inventor, Ezra Gilliland, who headed Bell's experimental department in Boston. A bright, talented, jolly man, Gilliland soon became fast friends with Edison and joined him in a secret venture to break the Bell patent.[7]

Edison began visiting Gilliland in Boston. There he met John Tomlinson, who soon became his trusted lawyer, and he saw again Mina Miller, a lovely, charming girl of nineteen whom he had met briefly during an earlier trip to New Orleans. She dazzled Edison with her beauty, her gentle grace, and her poise. The Gillilands invited a number of eligible young ladies into their parlor to meet Edison. "Come to Boston," Edison wired Samuel Insull that June. "At Gill's house there is lots of pretty girls." But none of them rivaled Mina in Edison's eye. His infatuation grew to the point where, under the influence of Gilliland and his vivacious wife, Edison underwent a social metamorphosis. The crude, unkempt, cigar-chomping, tobacco-spitting inventor submitted himself to the respectable world of fashionable conversation, parlor games, rides in the bay in Gilliland's steam launch, and other rituals of proper behavior.[8]

In July Edison joined Mina and some other houseguests at Woodside Villa, the Gillilands' summer house on the north shore of Boston Bay. There Edison entered willingly into the unaccustomed role of middle-class gentility. He spent hours in the parlor discussing books they had all read, listened—as best he could—to music recitals, including one from Mina, and even joined in the fad of keeping a diary for ten days. His entries reveal a sharp wit and wry, detached amusement. Finding himself in the rare situation of having to dress properly, he donned "one of those starched horrors procured for me by Tomlinson" lest Lillian Gilliland think he had "an inexhaustible supply of dirty shirts." He kept regular hours and often went to bed early—dreaming one night of "a Demon with eyes four hundred feet apart."[9]

"Saw a lady who looked like Mina," he confided one evening after a day in the city. "Got thinking about Mina and came near being run over by a street car. If Mina interferes much more will have to take out an accident policy." Mina did continue to intrude. A few weeks later he took Marion to Chautauqua to join the Gillilands. Mina and her family were also there. Her father, Lewis Miller, had a modest reputation as an inventor and manufactured farm implements, chiefly mowers. He and Edison got on well, and Mina was allowed to accompany Edison and his party on a trip to the White Mountains of New Hampshire. Edison taught Mina the Morse code so that they could communicate privately. "Even in my courtship my deafness was a help," he recalled. It enabled him to get much closer to Mina than he otherwise dared and allowed them to use pet names for each other without anyone hearing them. Hopelessly in love, Edison screwed up his courage and tapped out a proposal of marriage. Mina accepted.[10]

Once back in New York, Edison wrote Miller seeking his approval. Miller invited Edison to the family home in Akron, Ohio. Although the devout Mrs. Miller had reservations about her daughter marrying a nonbeliever ("What a wonderfully small idea mankind has of the almighty!" Edison wrote in his diary. "My impression is that he has made unchangeable laws to govern this and billions of other worlds, and that he has forgotten even the existence of this little mote of ours"), she finally relented, and the wedding was set for February 1886. Edison found a large home for them in Llewellyn Park, the country's first exclusive suburb. Glenmont had been built at a cost of $400,000 by an embezzler who was enmeshed in legal problems; Edison acquired it for about $235,000. He had come up in the world in every sense. The newlyweds would live in style at Glenmont, and Edison would have to sustain its expensive upkeep. He also had to find some way of placating Marion, who resented being replaced as her father's companion by a woman only six and a half years older than herself.[11]

After the honeymoon Mina found herself the mistress of a large household while Edison disappeared back into the laboratory. He moved his own research work from New York to the Edison Lamp Company in East Newark. In June 1886 he decided to take charge of the lamp works because Upton had failed to solve a rash of complaints concerning bad units. He prodded Upton to pay more attention to technical matters. "You are degenerating into a mere business man," he scolded. "Money isn't the only thing in this mud ball of ours." Edison and John Ott plunged into a series of experiments to increase lamp life and improve the manufacturing process. In August Mina joined them in the laboratory and recorded their results in the usual notebooks. She stayed only a brief time before retreating to Glenmont, where a predictable veil of loneliness descended on her. It was compounded by her difficulties with Edison's children, especially Marion. [12]

But Mina was stronger than Mary had been. Although she did not yet know her new husband well, she knew that he was rich and famous, that he needed guidance in things social, and that she must become mistress of Glenmont. In that role she hired a staff, planted the gardens, and filled the stables and pastures with animals. Her mother reminded her that all young wives got the blues during their first year of marriage. Although Mina never got over the melancholy induced by Edison's prolonged absences, her strong sense of duty enabled her to push past it. "You have no idea," she said later, "what it means to be married to a great man." Despite Edison's dislike of formal education, she arranged for Marion to go away to school and for the two boys to attend St. Paul's, an elite academy in Concord, New Hampshire. The real ordeal came late in 1886 when Edison fell ill with pleurisy or pneumonia. Just as he was recovering, he developed an abscess below the ear that required surgery. [13]

Edison's marriage signaled the onset of a profound shift in his life. He was about to start a new family in a new home that reflected a very different lifestyle for him. Mina had begun what became a long and arduous task of polishing her husband's rough veneer. In business no less than his personal life, 1886 proved to be a crucial dividing line for Edison. The economy continued to boom and flooded the Edison factories with more orders than they could handle. On May 1 the workers at the Machine Works surprised Batchelor by demanding new rules for pay and hours and that the shop be unionized. Batchelor relented on the rules but held the line against unionization. On the nineteenth the workers voted to strike. Since the company needed larger facilities anyway, the board decided to move the operation elsewhere. A locomotive company had erected two large buildings in Schenectady but never used them. The strike lasted only a week, but the directors bought the Schenectady property and moved the Machine Works there during the fall. [14]

Although not directly involved in the strike, Edison backed Batchelor's stand and supported the move to Schenectady. Later he declared that he wanted to "get away from the embarrassment of the strikes and communists to a place where our men are settled in their own homes." He declined to add that skilled mechanics could be had much more cheaply in Schenectady without the bother of unions. To run the operation, Edison needed someone who was efficient, ruthless, and hard-driving. He chose Insull, giving him the burden of overseeing the Schenectady plant while still handling all of Edison's affairs. When the surprised Insull had completed the move north, he asked Edison for further instructions. Edison told him only to "run the whole thing," adding, "Do it big, Sammy. Make it either a big success or a big failure." Insull made it a big success. In two years he quadrupled sales and hiked the annual return on investment to more than 30 percent. In the process he gained invaluable experience in both business and finance.[15]

That same year Edison took another step toward integrating the manufacturing companies. He had earlier folded the Tube Company and the Edison Shafting Manufacturing Company into the Machine Works and his Construction Department into Isolated Lighting. During the summer of 1886 his three major manufacturing firms, Edison Lamp Company, Edison Machine Works, and Bergmann & Company, formed the United Edison Manufacturing Company to serve as sales agency for them all. In December he merged the Isolated Lighting Company into Edison Electric Light. These moves simplified the structure of his business empire but did not remove the underlying tensions between some of the units. Amid all this work, Edison began making plans with Batchelor for an even more audacious move: the construction of a new laboratory for himself.[16]

He had mulled the idea over during his long convalescence from illness. During 1886, the same year he severed his connection to United States Lighting, Edward Weston began building an elaborate private laboratory behind his home in Newark. Completed in 1887, it was hailed as the largest and best-equipped research facility in the country. Doubtless Edison read about its progress as he began planning his own new laboratory near his home in West Orange. He and Weston had never been friendly, and Edison's ambitions had only grown greater since his success. Apart from all his electrical interests, he still had men working on the telephone, he had developed an ore-milling machine, and the phonograph unexpectedly reclaimed his attention. Absorbed in these and other projects, Edison still found time to oversee the design and construction of a magnificent new laboratory that he envisioned as a true "invention factory." Ever conscious of his status as America's inventor, he resolved to build a facility that would not only maintain but enhance that reputation.[17]

EVEN THE INCORRIGIBLY optimistic Elihu Thomson must have been surprised at how rapidly his new company emerged as a major player in the electrical manufacturing industry. Between 1883 and 1892 Thomson-Houston sales mushroomed from $426,988 to $10.3 million. The company began with

45 employees and by 1892 had 3,492 people on its payroll. By 1891 it had installed 666 central stations for lighting, more than twice the number done by
Westinghouse and three times that by Edison, who together had managed only
525. Most of the credit for this remarkable growth belonged to Charles Coffin,
whose executive and marketing savvy exceeded all expectations for someone
who was a newcomer to the industry. To gain status for its arc-lighting system,
the company entered two competitions: an industrial exposition in Cincinnati
in 1883 and a much tougher test in 1885 at the London Inventions Exhibition.
In both cases Thomson-Houston came away with the top awards, including
the only gold medal given in London.[18]

Like Edison, Thomson thought in terms of whole systems rather than individual components. When the Franklin Institute in Philadelphia hosted the International Electrical Exhibition in 1884, Thomson refused to enter his
arc-lighting system in the competition even though it was staged in his hometown. He dismissed the tests because they concentrated only on the dynamo.
"The dynamo is not the system," he declared, "nor the regulator, nor the lamp,
nor the lines & switches, but the whole entity must be compared with another
entity, under conditions which represent real, out & out commercial usage."
Thomson's voice carried considerable weight at Thomson-Houston, but his
role differed markedly from that of his rivals. Edison and Westinghouse both
dominated every aspect of their firms; Thomson did not even manage his
company, which was much more group driven. Coffin, Pevear, Barton, and
Rice all shared in the direction of Thomson-Houston. Their ability to reach
consensus and coordinate their separate spheres of activity accounted for the
firm's rapid rise to success.[19]

Thomson entered willingly into this arrangement up to a point. Nothing
pleased him more than to confine his work to the laboratory, free of most concerns about the business aspects of the company. However, Thomson no less
than Edison or Westinghouse liked to control what he invented, how he went
about it, and the uses to which his inventions were put. In these areas key differences arose between him and Coffin as the decade wore on, especially in regard to the level of activity in the Model Room. For a time, however, Thomson
enjoyed his role in the company as much as he did his newfound domestic
bliss. Like Mina Edison, Mary Louise "Minnie" Thomson had begun the difficult task of learning the role of wife to a famous man. Her mother, a devout
Congregationalist, undertook in vain to convert Elihu to the flock. He had
long resented organized religion, saying that church was a place where you had
to listen to ideas you disagreed with from a man you couldn't talk back to.[20]

Coffin orchestrated the sales and management operation from the Boston
office. A hustler in the best sense of the word, he grasped the fundamentals of

the electric business with remarkable speed. His genial, easygoing personality masked a shrewd mind and complemented his genius as a salesman. In peddling shoes, Coffin had gone out on the road to get a feel for the market in far-flung cities. From these efforts arose an extensive sales network. His experience in the shoe industry also taught him the value of owning the latest technology to reduce costs and expedite production. These principles transferred easily to the electrical manufacturing business. He and Thomson proved to be a perfect fit; Coffin knew nothing about the technology, and Thomson was no salesman. "I have as little as possible to do with the business of the Company," Thomson wrote cheerfully in 1888, "my work being in the line of development of apparatus and the production of new inventions." Coffin also had a smooth, even gentle, style of management; he tended to suggest rather than order and welcomed input from associates, whom he never called subordinates.[21]

Despite his lack of background in the electrical industry, Coffin realized that the key lay in developing central stations. Potential customers included virtually every city and town in America, which could be tapped by offering a product line of different-size dynamos capable of expanding the number of lights served by them. A varied line of arc and incandescent lights could be used for streets, shops, homes, and public areas. However, organizing local utility companies was a tricky business. Every city had its own unique problems and political gamut to run, along with the basic task of raising the funds to build a system. The process involved four stages: 1) mobilizing local capital; 2) obtaining a franchise from the city council; 3) drawing up contracts for the equipment; and 4) organizing an operating company. The contracts usually contained a clause binding the local company to buy Thomson-Houston equipment for its station.[22]

At first Coffin ducked these difficulties by giving Edward H. Goff and his American Electric and Illuminating Company exclusive rights to market Thomson-Houston equipment for central stations. Through this arrangement Goff built some sixteen central stations, mostly in New England, adding his own markup to the equipment and selling it under his own name. In 1885, however, Coffin dispensed with Goff and began putting together his own sales staff with district offices in major cities. He also poured money into advertising in electrical journals to promote the virtues of the Thomson-Houston system. Like Edison, Coffin accepted the securities of fledgling central station companies as part payment for the equipment they bought; unlike Edison, he refused stock and took only bonds. These bonds he put into a series of trusts and offered Thomson-Houston stockholders the right to buy stock in the trusts. When these trusts were finally dissolved, they returned two dollars for every dollar invested.[23]

By 1886 all Thomson-Houston equipment was being sold by its own sales-men under the company's own name. Coffin used the funds from the sale of trust stocks to enlarge the Lynn factory and to buy smaller rival firms, some of which he first intimidated by suing them for infringing Thomson's patented regulator. These purchases brought Thomson-Houston useful patents and even more useful inventors in some cases. The acquisition of Consolidated Electric in 1884 had enabled Coffin to expand the product line with incandes-cent lights. A year later Thomson-Houston was selling a complete incandes-cent system, including dynamos and an improved filament-treatment process that made its lamps superior to those of Edison. Unlike the Edison company, which sold one basic system, Coffin envisioned a steadily expanding product line that would enable Thomson-Houston to penetrate a national market. The key to that strategy lay in constant production innovation, which was music to Thomson's ears.[24]

To match products with markets, Coffin and Thomson agreed that each one should be designed as a separate system. Edison held the vision of one universal system based on DC power stations running everything from lights to motors, but he struggled to sell his system as configured because the high cost of copper restricted its use to high-density areas where the lines could reach a large number of customers within short distances. The market Edison had done so much to create was both expanding and diversifying, which meant that one universal technology could not serve it. A second stage of the industry was fast emerging in which equipment had to be devised and adapted to fit changing needs. Coffin and Thomson understood this better than Edison and were more nimble at adjusting to its demands. It was this cluster of con-siderations that brought Thomson's thinking back to one of his early interests, alternating current.[25]

Alternating current (AC) consists of electrons that continually reverse their direction from one pole (positive) to the other (negative). The number of times per second this reversal occurs is called the frequency. By contrast, the electrons in direct current (DC) move in a straight line. Although DC exists in nature in the form of lightning and static electricity, it is usually produced by creating a circuit that enables it to flow. The great advantage that AC has over DC is that it can move over a conductor (wire) at much higher voltages, which means that it can be transmitted far greater distances than DC. Edison had built his system around DC because the lights powered by it required low volt-ages and because AC high-voltage wires could be extremely dangerous.

Like other inventors, Thomson recognized the limitations of Edison's DC system. It resembled a system of water mains that required ever larger mains to pump water to more distant points at constant pressure. One common saying had it that if one attempted to light Fifth Avenue from Fourteenth to Fifty-ninth Street, the conductors would have to be as large as a man's leg. The equation was both simple and inexorable. The amount of current that could move economically over a conductor at a fixed pressure (voltage) depended on the size of the conductor. Moving a large amount of power at low voltage required large conductors, which greatly increased the amount of expensive copper. The alternative was to fix the size of the conductor and increase the voltage by using alternating instead of direct current. The latter had to be conveyed at low voltage because the end use required low voltage. Its practical use, therefore, was always circumscribed by the cost of copper. However, current sent at high voltage posed the yet unsolved problem of how to reduce it back to direct current at the user end.[26]

Put another way, the most efficient system would be one that could transmit power to lights over much longer distances than a mile or so without incurring the prohibitive cost of larger copper mains. Such a system required much higher transmission voltages to move power over long distances, but high-voltage lines would be far more dangerous to anyone coming into contact with them. They would also need some method of stepping down the voltage at the point of use. Perhaps, thought Thomson, the solution lay in some type of induction coil, a term used to describe what later became known as a transformer. Thomson and Houston had included such a device in their 1878 dynamo that used alternating current, but their patent had been rejected, and they set the work aside after George Garrett invited them to build DC systems.[27]

An induction coil changed direct current into alternating current. It consisted of a primary coil with only a few turns of heavy wire and a secondary coil, wound over the primary one, with many turns of fine wire. When some sort of vibrating contact was used to interrupt the direct current in the primary, the disturbance induced a high voltage in the secondary. In effect it stepped up the current to a higher voltage. Reversing the arrangement—using a primary with many turns of wire and a secondary with only a few turns—stepped down the voltage to a lower level. Joseph Henry actually created the latter version in 1838 but confined its use to laboratory demonstrations. Nearly half a century passed before anyone realized that the transformer held the key to harnessing alternating current.[28]

In March 1885 Thomson sketched a basic AC system. "The idea," he wrote, "is to have a constant potential reversed (alternating) current dynamo, feeding mains which remain at constant potential.* Between these mains are con-

---

*Potential is the difference in voltage between two points in a circuit. One of them is usually assumed to be ground with zero voltage.

Colossus: the Corliss steam engine at rest. (New York Public Library)

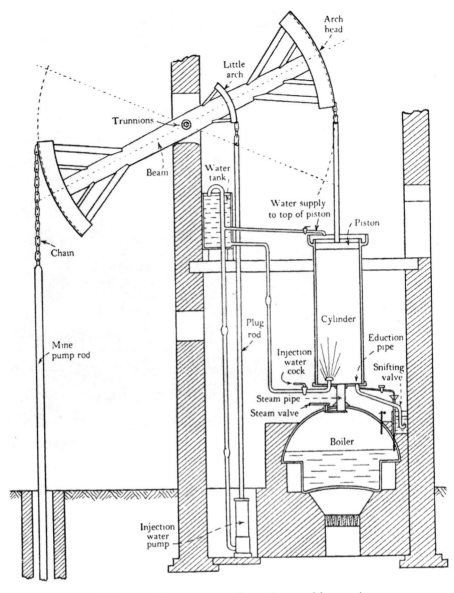

The first practical power source: Thomas Newcomen's beam engine.
(Diagram from H. W. Dickinson, *A Short History of the Steam Engine*, Frank Cass Publishers)

The progenitor of power: James Watt.
(Emmet Collection, Miriam and Ira D. Wallach Division of Art, Prints and Photographs, the New York Public Library, Astor, Lenox and Tilden Foundations)

Working out a formula: James Watt's diagram of the economy of power.

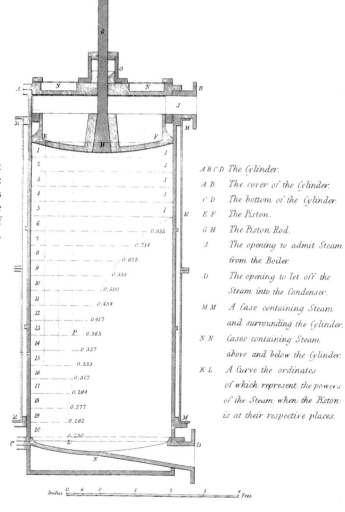

A B C D The Cylinder.

A B   The cover of the Cylinder.

C D   The bottom of the Cylinder.

E F   The Piston.

G H   The Piston Rod.

J   The opening to admit Steam from the Boiler.

D   The opening to let off the Steam into the Condenser.

M M   A Case containing Steam and surrounding the Cylinder.

N N   Cases containing Steam above and below the Cylinder.

K L   A Curve the ordinates of which represent the powers of the Steam when the Piston is at their respective places.

A drawing of Oliver Evans's incredible Oruktor Amphibolos, which was truly one of a kind.

An unsung pioneer: John Fitch's steamboat. (Library of Congress)

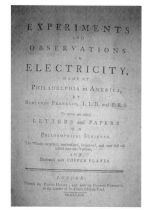

The first American electrician: the title page and some illustrations of his apparatus from Benjamin Franklin's book on electricity.

The thoughtful visage of Michael Faraday, whose brilliant experiments propelled the science of electricity into a new era. (Print Collection, Miriam and Ira D. Wallach Division of Art, Prints and Photographs, the New York Public Library, Astor, Lenox and Tilden Foundations)

This view of the National Cash Register Company factory in Dayton, Ohio, during the 1900s shows the elaborate belting required to run machines with steam power. The electric motor eliminated all this dangerous clutter by giving each machine its own engine rather than transferring power from a central power source. (Library of Congress)

This classic portrait of Samuel F. B. Morse befits a man who achieved renown as a painter as well as an inventor. (Photography Collection, Miriam and Ira D. Wallach Division of Art, Prints and Photographs, the New York Public Library, Astor, Lenox and Tilden Foundations)

The scholarly inventor: Elihu Thomson in middle age. (Library of Congress)

An elderly Charles Brush examines one of his early arc lamps. (© Schenectady Museum; Hall of Electrical History Foundation/Corbis)

The chemist at home: Thomas Edison in his laboratory. (© Bettmann/Corbis)

The light that dawned: Edison's first lightbulb.
(© Bettmann/Corbis)

The first invention factory: Menlo Park in the winter of 1880. The library/office is in front, the original building in the center, the machine shop in the rear, and the electric railroad to the right.
(Thomas A. Edison papers at Rutgers, the State University of New Jersey)

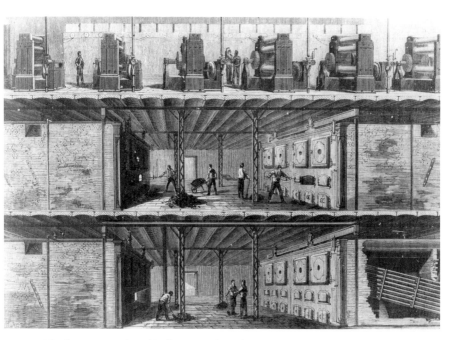

The first power station: This illustration shows the arrangement of Pearl Street station.
(Thomas A. Edison papers at Rutgers, the State University of New Jersey)

Posing for posterity: The boys at Edison's lamp factory pause in their busy workday
to gather for this group photograph. (Thomas A. Edison papers at Rutgers, the State University of New Jersey)

J. P. Morgan: the man
behind the money. This
portrait of J. P. Morgan
catches him in a posture
that is all business.

(Photograph by Edward Steichen)

A Richmond streetcar on one of the steep hills that challenged and frustrated Frank J. Sprague's
pioneering effort. (Valentine Richmond History Center)

A formal portrait of George Westinghouse, who hated publicity.
(Library of Congress)

The spider's web:
New York after
the blizzard of 1888
showing the jungle of
wires that came to
alarm people. This is
New Street looking
toward Wall Street.
(Museum of the City of New York)

Two views of the World's Columbian Exposition in Chicago. A grand enough spectacle by day, the "White City" became by night a City of Lights, the greatest display of electric illumination yet seen anywhere. More than 180,000 incandescent bulbs and thousands of arc lamps dazzled its visitors. (Chicago Historical Society)

The suave savant: Nikola Tesla, dressed impeccably as always. (© Bettmann/Corbis)

Tesla loved to shock and delight audiences with spectacular demonstrations. Here artificial lightning rages over his unconcerned head. (© Bettmann/Corbis)

This view of Montgomery, Alabama, in 1906 shows how the electric revolution had penetrated even the Deep South. Note the power poles, electric lamps, and streetcars. (Library of Congress)

The young
Samuel Insull.
(Loyola University
Archives)

The Fisk Street station, the facility built by Samuel Insull that carried central station power to new heights. (Loyola University Archives)

The Quarry Street station, built across the Chicago River from the Fisk Street station in response to the surge in demand that outstripped even the latter's generous capacity. Quarry Street featured even larger generators. (Loyola University Archives)

The future beckons: a night view of the New York World's Fair in 1939. (© Corbis)

nected induction coils with primaries of relatively high resistance, the secondaries are low resistance and feed the light. Advantage is great saving of copper
in mains and also self regulation. The potential of the mains being quite high
from 1000 to 2000 volts." Aware of the dangers posed by AC, Thomson included fuses on the primary side and ground connections on the secondary
side to deal with any short circuit between them.[29]

At the time this sketch was but one of several projects for designing new
products and not intended for commercial development. He continued working on the system, however, and in June 1886 ran an experimental AC line between two of the buildings at the Lynn plant to power incandescent lamps. For
this experiment Thomson designed an AC generator that went into the company's catalog by the year's end. Although the catalog listed four different
models of the generator, not a single one was sold. Nevertheless, Coffin and
Thomson believed that an AC system held great promise and agreed to continue work on developing one. They were not alone in this belief. In distant
Pittsburgh a newcomer to the business was hard at work developing his own
AC system. In even more distant Hungary a trio of scientists had developed a
transformer that appeared to be commercially practical.[30]

WILLIAM STANLEY WAS only twenty-five when he arrived in Pittsburgh to
work for George Westinghouse. A native of Brooklyn, he had gone to Yale intending to become a lawyer like his father until he discovered a passion for invention. It had been there since his childhood when, at the age of ten, he took
apart and reassembled a watch given him by his father. The classroom and formal academic rigor bored him. To his father's disgust, he left Yale after only a
few months, saying in a curt letter home, "Have had enough of this . . . am going to New York." Electricity aroused his interest and landed him a job first
with a manufacturer of telegraph keys and then with a nickel-plating firm. His
lucky break came when Hiram Maxim hired him for fifty cents a day to work
in the United States Electric factory. "Mr. Stanley was very young," Maxim recalled. "He was also very tall and thin, but what he lacked in bulk he made up
in activity. He was boiling over with enthusiasm. Nothing went fast enough for
him. I believe he would have preferred that each week should contain ten days,
and that each day should be forty-eight hours long. Whatever was given him to
do he did in the most thorough manner."[31]

Within a year Stanley became Maxim's top assistant and took charge of the
lamp factory and machine shop. In that position he oversaw the company's
first installation in 1881. When Maxim went to Paris and Weston took his place
as chief electrician, Stanley became his assistant but lasted only a few months.

Where Maxim had given Stanley free rein, Weston wanted things done his way. "Have had enough of this," Stanley growled, and took a job with Swan Electric Light Company. He lasted only a few months there before leaving to indulge his new passion for electrochemistry at his parents' home in Englewood, New Jersey. There in 1884 he married Lila Courtney Wetmore and settled into a family life that ultimately produced nine children. He also worked hard to develop a new kind of storage battery.[32]

Stanley knew and admired Charles Brush, who had developed a storage battery system. "At that time I was disobeying the laws of chemistry," Stanley admitted, "by attempting to produce a battery on a novel but unsuccessful plan." But he drew from Brush's achievement a spark of inspiration. Brush wanted his battery to serve as a counterforce—a variable electromotive force that could adjust instantly and accurately to varying loads and thereby maintain a constant potential. Stanley wondered why such a force could not be created by using some rotational device such as a motor. In September 1883 he sketched a possible system in his laboratory notebook. Although still crude, it became his starting point in the quest for a system that utilized alternating current to produce the force. Like so many inventors before him, Stanley found himself in uncharted waters with few books or formulas and little expertise to guide him. Moreover, conventional wisdom had already decided that there was nothing of value in alternating current. For Stanley, as for so many other inventors, that served as motivation rather than deterrent.[33]

When Stanley went to work for Westinghouse early in 1884, he found himself swamped with other work. But his mind would not let go of the tantalizing appeal of alternating current as a method for distributing power. The principle was simple: the higher the voltage, the less loss of current on the line. "It had become my major ambition by this time—my secret ambition, if I may confess it," he recalled. No one yet fully understood the laws of a magnetic circuit, and few engineers had a clear conception of the subject. Nevertheless, Stanley convinced himself that the answer lay in induction coils, which he studied intently. Lacking an alternator (an alternating-current generator), he struggled to create his own kind of induction coil that included a commutator. "It worked, some," he said ruefully, "but oh how it sparked!" Meanwhile, he soon learned that events threatened to outrun his efforts.[34]

In 1882 the Frenchman Lucien Gaulard and his English backer, John D. Gibbs, secured a patent in Europe for their version of a transformer intended to enable the high-voltage output of an AC generator to deliver current and lower voltages to local circuits. It reduced the voltage of the main to a safe level by using a magnetic connection to isolate the electric circuits from one another. Later the patent was invalidated because the system failed to operate satisfactorily. Gaulard and Gibbs never fully grasped their accomplishment. They created high voltage by utilizing connections in series, like an arc-lighting circuit, rather than changing voltages through the use of a transformer. In their system, lamps could not be operated individually. However, in 1884 three Hungarian engineers working at the firm of Ganz & Company in Budapest detected the flaw in the Gaulard-Gibbs system and improved it by connecting the transformers in parallel rather than in series. The next year they exhibited their system at a local fair; it included their transformer, which unlocked the door to the pursuit of an AC power system.[35]

None of this work made any impression in the United States. As late as October 1885 an *Electrical World* editorial compared series and parallel connections in transformers and concluded that the parallel used more copper and so was less efficient. However, that same spring George Westinghouse read an article in an English engineering periodical about an AC system on display in London. The system used Gaulard-Gibbs transformers, which were described as having no moving parts, being 90 percent efficient, and able to produce any desired voltage. The article alerted Westinghouse to the possibilities of an AC system for distributing power. One of his assistants, Guido Pantaleoni, happened to be in Italy at the time because of his father's death. Westinghouse cabled him to get an option on the Gaulard-Gibbs patents and to investigate their system thoroughly.[36]

Pantaleoni dutifully met Gaulard in Turin and examined one of his systems, which carried electricity ten miles from a waterfall to light a theater. Uncertain of his ability to evaluate such a system, Pantaleoni sought advice from Werner von Siemens of the prominent German electrical manufacturing firm Siemens & Halske. Siemens assured him that alternating current "was pure humbug and his five-wire system had rendered alternating currents useless." Unconvinced by this self-serving advice, Pantaleoni went on to consult with Ganz & Company. Their presentation, designed to trumpet the virtues of alternating current, led Pantaleoni to recommend that Westinghouse pursue the subject vigorously. During the summer of 1885 Westinghouse decided to do just that. He ordered a Siemens alternator and several Gaulard-Gibbs transformers from London. While awaiting their arrival, he came to a new arrangement with Stanley.[37]

Described by a colleague as "nervous and agile," the high-strung Stanley never worked well under others. That spring his health had broken. His doctor advised him to leave the Pittsburgh climate, where the air was heavily saturated by pollution from Andrew Carnegie's steel mills, among other industries. Stanley had become discouraged because his "surroundings were uncongenial, the work hard, and the results meagre." He agreed to sell Westinghouse a half interest in his holdings, move to the country, and conduct experiments for the company. Later he called this deal one-sided; at the time, however, he was over-joyed to exchange "the dirt of dreadful Pittsburgh" for the fresh air and green fields of Great Barrington, Massachusetts, in the Berkshires. Unable to work in the summer heat, he waited until fall to organize his laboratory in a deserted rubber mill in the village.[38]

Before Stanley left Pittsburgh, Westinghouse asked his opinion of the Gaulard-Gibbs system. Stanley replied that it would not be commercially fea-sible because of the series connections but that parallel connections might solve the problem. Gradually a way to do this dawned on him. "It seemed too simple and too easy to be true," he recalled. Stanley told his wife, who knew nothing of the subject, but she saw it as well. What he visualized was a strong similarity between a parallel-connected transformer and a shunt-wound parallel-connected motor. The trick was to build a transformer "that would regulate the energy transformed by slight variations of its induced counter electromotive force in the same manner that a shunt-wound motor regulated for energy transferred by variation of its rotational counter electromotive force."[39]

That was Stanley's idea of simple. Eagerly he designed several transformers of this type, winding them for 500 volts primary and 100 volts secondary. He described them to Franklin L. Pope, a well-known electrician, patent lawyer, and Westinghouse adviser, stressing the virtue of their automatic regulation. Pope saw the advantages but remained dubious, pointing out the dangers in-herent in a high-voltage system. Despite Pope's negative report, Westinghouse told Stanley to go ahead and develop the transformer and an AC system around it. Pope later changed his views, but his opposition foreshadowed an ironic tragedy. Ten years later he made accidental contact with wires leading to a transformer in the cellar of his Great Barrington home and was electrocuted instantly. The system had been one of Stanley's first installations.[40]

The Siemens alternator finally arrived and was dispatched to Great Barring-ton, where Stanley wound it to produce 12 amperes of current with a maxi-mum of 500 volts. Meanwhile he began constructing transformers and scouring the village for potential customers. Reginald Belfield, a young engi-neer, had been sent to help Stanley. In the February cold he began stringing four thousand feet of wire on ceramic conductors fastened to tree limbs. To conceal their presence from prying eyes, the transformers were hidden in wooden crates and installed under lock and key in half a dozen customer cel-lars. In all Stanley built twenty-six transformers at Great Barrington, ten of which went to Pittsburgh for use in a demonstration system running three miles from Union Switch and Signal's plant to East Liberty.[41]

Satisfied with his progress, Stanley left town for two weeks of vacation. In his absence an Edison system was installed in a Great Barrington mansion and dazzled the locals with the spectacle of incandescent light. Stanley was doubtless annoyed at losing the honor of being the first to bring electric lights to the town, but he still had a new system to display. On March 16 he illuminated a general store. "I believe the true way to study it is to give it a commercial test here in town," he wrote Westinghouse the next day. Stanley then connected another store, a drugstore, and a doctor's office and on March 20 gave the town a public demonstration. People thronged the streets, eager to see the lights though still wary about the danger involved. "My townspeople," reported an amused Stanley, "though very skeptical as to the dangers to be encountered when going near the lights, rejoiced with me."[42]

Great Barrington thus enjoyed the honor of being the first town to receive an AC lighting system. By the month's end Stanley had enlisted several dozen more eager customers including several stores, the post office, a barbershop, two hotels, a restaurant, and the telephone exchange. The Edison company procured some customers as well, but not as many because of the limited reach of the DC system. For the first time electricity had traveled long distances at high voltage and then been stepped down by transformers for local use. The general public, although dazzled by the light, knew nothing of the historic importance surrounding the event. Stanley could not very well discuss the difference between the two systems or enlighten the Edison people about the revolutionary new transformer.[43]

"All the converters are under lock and key so that no one knows anything about them . . . ," Stanley assured Westinghouse several days later. "I might say a great deal about the system, but briefly, it is all right." However, the Siemens alternator proved a disappointment, leading Stanley to design one of his own and dispatch the drawings to Pittsburgh. On April 6 George and Herman Westinghouse, Pope, and Pantaleoni journeyed to Great Barrington to see the system for themselves. The inspection satisfied Westinghouse that the system had great potential and that a transformer wired in parallel connection could change voltages reliably and safely. That summer a more complete test took place in Pittsburgh, using Stanley's new alternator at the Union plant and the line to East Liberty with transformers at both sites. Careful measurements of the system's efficiency produced results so favorable that Westinghouse was convinced the time had come to start yet another business.[44]

Months before these tests, in December 1885, Westinghouse had already created the Westinghouse Electric Company, which was formally organized in March 1886. He moved Union Switch and Signal to the suburb of Swissvale and gave the new company its former plant. Early in 1886 he had sent Pope and

Pantaleoni back to Europe to buy the patent rights for the Gaulard-Gibbs transformer. When Pantaleoni asked how much they should pay, Westinghouse replied tersely, "They'll tell you their price. Whatever it is, close the bargain, and I'll send the money by cable to you." The figure turned out to be $50,000. When the Gaulard-Gibbs transformers he ordered were delivered, they were accompanied by Reginald Belfield, who so impressed Westinghouse that he was persuaded to come work for the new company. Westinghouse also hired a young graduate of the U. S. Naval Academy, Oliver B. Shallenberger, to replace Stanley as engineer when the latter moved to Great Barrington. Both men joined Stanley in improving the AC system.[45]

During the fall of 1886 several transformers and four hundred lamps were set up in a building near Lawrenceville, Pennsylvania, some four miles from the dynamo that powered them. For ten days the lamps burned continuously, first with 1,000 and then 2,000 volts supplied from the dynamo. Westinghouse came daily to inspect the operation. After examining the test results, he pronounced the system ready to market even though work continued on improving it. On November 30 Buffalo, New York, received the honor of installing the first commercial alternating-current lighting system, although Greensburg, Pennsylvania, twenty miles from Pittsburgh, claimed to be the first town to obtain a complete municipal plant using the Westinghouse system. Westinghouse put Stanley to work developing an autotransformer, which used a single core to link all the circuits magnetically.[46]

"I have a very personal affection for a transformer," Stanley admitted. "It is such a complete and simple solution for a difficult problem. It so puts to shame all mechanical attempts at regulation. It handles with such ease, certainty, and economy vast loads of energy that are instantly given to or taken from it. It is so reliable, strong, and certain." Another engineer shared this sentiment. "The transformer is not interesting to look at," he wrote. "It is a mere mass of metal, dull and motionless. It is not even graceful in outline or proportion. But it is the heart of the alternating-current system. The reason . . . lies in its capabilities for simple transformation of voltage over almost any required range, from hundreds of thousands of volts down to almost nothing."[47]

George Westinghouse was fast developing his own love affair with the transformer. His shrewd mind saw at once what it could do and what it enabled him to do. Decisive as always, he moved vigorously into an entirely new field of operation and looked at once to expand the business. He understood clearly that his primary market lay in smaller towns lacking the density of population to support an Edison DC system and that he could not yet compete with Edison in large cities. The Edison people would take little notice of him at first, and with good reason. His system was new and untried, and it still lacked certain key components, most notably a meter. Westinghouse put several of his engineers to work on the problem, but it would take time. He also needed an AC motor. No one had created either of these inventions because no one used alternating current. In these areas, as in so many others, the direct-current companies held a commanding lead over Westinghouse. They held the large markets; he could only aspire to the smaller ones. But Westinghouse had made a start, and he made it along lines that, he was convinced, would dominate future development.

IT TOOK NIKOLA TESLA only a few weeks to win Edison's confidence. One of Henry Villard's fastest ships, the *Oregon*, had been fitted with lights, but both generators had broken down and delayed her sailing. The generators had been installed ahead of the superstructure and so could not be removed, leaving Edison in a quandary. One evening Tesla took some instruments aboard the ship and went to work. With help from the crew he toiled all night until he got them working satisfactorily. Walking back up Fifth Avenue at five in the morning, he ran into Edison, Batchelor, and a few of the boys heading home as well. "Here is our Parisian running around at night," said Edison. Tesla told him that he was returning from the *Oregon* and had repaired the dynamos. Edison nodded in silence, then remarked as he walked away, "Batchelor, this is a damned good man."[48]

From that day on, Edison gave Tesla free rein, and Tesla responded by working routinely from 10:30 A.M. until five the next morning. "I have had many hard-working assistants," Edison told him, "but you take the cake." Edison's other men came quickly to appreciate Tesla's talents. One of them, W. L. Dickson, called Tesla "that effulgent star of the scientific heavens" who, "like most holders of God's intrinsic gifts," was "unostentatious in the extreme, and ready to assist with counsel or manual help any perplexed member of the craft." Working at the Goerck Street shops, Tesla installed and repaired both arc and incandescent lamps, worked on Edison's DC generators, and designed some twenty-four different machines to replace those being used by Edison. He also worked on patents for several components of the DC system. What he could not manage to do was interest Edison in his new motor, which ran on alternating current.[49]

For his work Tesla received eighteen dollars a week, a meager salary even by the standards of the day. After a year's work at that pay he pressed a friend to intercede with Batchelor in hopes of getting a raise to twenty-five dollars a week. Batchelor rejected the plea out of hand. "The woods are full of men like him," he said. "I can get any number of them I want for eighteen dollars a week." To make matters worse, Edison had jokingly told Tesla he would get $50,000 for the machines he designed. Tesla took the remark seriously and asked for payment only to be refused. "When you become a full-fledged American," Edison said, "you will appreciate an American joke." Later Edison dubbed Tesla "the poet of science. His ideas are splendid, but they are utterly impractical."[50]

Tesla responded by quitting in a huff. His high estimation of Edison departed as well. Later he mocked the "Wizard of Menlo" and his style of research, saying that "if Edison had a needle to find in a haystack, he would proceed at once with the diligence of a bee to examine straw after straw until he found the object of his search . . . I was a sorry witness of such doings, knowing that a little theory and calculation would have saved him ninety per cent of his labor." Back on the street, Tesla met with a patent attorney who helped him file for his first patent, an improved arc lamp design, only to find that Elihu Thomson had preempted it. Introduced to two New Jersey businessmen, he agreed to form a lighting and manufacturing company with them in Rahway. In 1886 his arc light system went into production, but to his chagrin his partners showed no interest in his alternating-current work. Once in production, the partners forced the naïve Tesla out of the company with little more than "a beautifully engraved certificate of stock of hypothetical value."[51]

Despondent, broke, and without prospects as the winter of 1886–87 came on, Tesla endured "terrible headaches and bitter tears, my suffering being intensified by my material want." As a final humiliation he was obliged to take a job digging ditches, the sort of menial labor that utterly deflated his aristocratic pretensions. "My high education in various branches of science, mechanics and literature," he mourned, "seemed to me like a mockery." Spring brought fresh hope in the form of a meeting with Alfred S. Brown, an engineer who had worked for Western Union and held several patents on arc lamps. Brown knew well the limitations of DC current and listened readily as Tesla described his AC apparatus. He then contacted a lawyer named Charles F. Peck, who harbored the prevailing prejudices against AC current until Tesla wowed him with a demonstration that featured a rotating magnetic field. Brown and Peck agreed to create the Tesla Electric Company and provide a laboratory for the inventor to develop his work in return for a half interest in any patents developed. Here at last was the fresh start Tesla craved, and he vowed to make the most of it.[52]

CHAPTER 11

EVENTFUL CURRENTS

The business was new and presented problems which were substantially without precedent and which required new methods. People generally did not at all appreciate the need or value of electricity. They had to be educated to its use . . . Suitable manufacturing methods as well as adequate ways of distributing the manufactured product had to be devised . . . Customers did not exist; they had to be created.

—FREDERICK P. FISH, CHIEF COUNSEL, GENERAL ELECTRIC COMPANY [1]

As EDISON HAD PREDICTED, THE gas industry did not wither away once electricity arrived on the scene. Nor did anyone suppose that it would meekly accept its new competitor without a fight. The arc light cost the industry some customers, and the arrival of the incandescent lamp a few more, but gas remained competitive in terms of price and kept the lead in illumination. The gas companies had already faced down one competitor, the kerosene lamp, which became popular after the discovery of oil in Pennsylvania in 1859. They had done so largely through an innovation in the manufacture of gas known as water gas, which was produced by blowing steam through red-hot coke. Water gas was cheaper to make and created a real increase in illumination at modest cost.[2]

Electricity posed a more formidable threat than the kerosene lamp. It attacked gas on two fronts: The arc light went after the outdoor illumination market while the incandescent lamp aimed at interior lighting. The former posed little threat, if only because gas companies seldom made money on streetlights and often provided them because city councils would not grant a franchise for more lucrative private customers unless they lit the streets as well. The threat to indoor lighting was quite another matter. Once Edison's system became a reality in 1882, gas companies moved vigorously to meet the growing competition of electric lamps. Their most obvious move was to reduce the prices about which their customers had long complained. In New York City, for example, the price per thousand cubic feet dropped from $2.75 in 1878 to $1.25 a decade later. In Baltimore it fell from $2.75 to $1.00 during that same period, and in Cleveland from $2.50 to $1.00.[3]

Slashing prices was only the first step. Gas companies countered the superior electric lamp with a new technology of their own, the Welsbach mantle. Invented by Carl Auer von Welsbach of Germany in 1885, it consisted of cotton fabric impregnated with rare earths that became incandescent when placed near a gas flame. It produced a steadier, whiter light that rivaled incandescent electric lamps in quality and cost much less. It was also an economic boon, increasing the amount of light from a given quantity of gas by 600 percent. The United Gas Improvement Company of Philadelphia, which operated gas plants in more than fifty cities and built them as well, bought the American rights to Welsbach's mantle and began manufacturing it in 1890. The mantle mimicked its rival in heating a substance to produce incandescent light. Its presence slowed the spread of electric lamps but could not prevent their ultimate triumph. During these same years the technology of electric lights continued to advance as well.

Faced with a declining market, the gas industry responded by developing new ones. It began to promote the use of gas for cooking and heating. As early as 1884 half a dozen New York gas companies combined to promote the use of gas stoves in place of the coal range. It proved a hard sell because the coal range provided not only cooking but space and water heating as well. The gas companies created gas water heaters but could not heat home space until the adoption of central heating early in the twentieth century. Some gas companies also developed gas engines to replace small steam engines, and even an iron powered by gas. Another approach was for firms to redefine themselves as energy companies and sell electric as well as gas lighting. By 1889 some 266 companies, about 25 percent of all gas firms, supplied both types of lighting. Thomson-Houston took the lead in selling equipment to these companies.

Like all industries threatened by technological change, the gas interests had to innovate or perish. They managed to adapt to a rapidly changing market by developing both new products and fresh approaches. In 1878 Edison had predicted that "gas will be manufactured less for lighting as a result of electrical competition and more for heating, etc." As he suspected, the gas industry did not go away. It remained a tough competitor to electricity in several areas during the years when the electrical firms were busy fighting each other.[4]

WELL MIGHT GEORGE WESTINGHOUSE have wondered what he had undertaken in entering the electric field, but he was never one to entertain regrets about anything he did. Introspection was no more a part of his nature than indecision or hesitation. He was rather what might be called a cautious plunger. He did not make rash decisions but studied the field with measured care before acting. Once convinced to move, however, he never looked back or wrung his hands over the outcome. His calm demeanor masked an energy that knew little rest. As an associate noted, "Westinghouse was temperate in everything but work." Some problem or project always occupied his mind and demanded his attention. "His vigor and his power of work were extraordinary," said Maurice Leblanc, a French physicist and engineer. "He never took any rest." On first meeting Westinghouse, Nikola Tesla was deeply impressed by his "powerful frame, well proportioned, with every joint in working order, an eye as clear as crystal, a quick and springy step . . . Like a lion in the forest, he breathed deep and with delight the smoky air of his factories."[5]

Lion he might be, but Westinghouse neither received nor craved the lionization that Edison so coveted. He cherished his privacy as much as Edison did the limelight. An engineer who spent many years in the Westinghouse company recalled his first impression of his boss. "He did not appeal to me, even then, as being a wizard," he admitted, "but he seemed to be a plain human being with lots of initiative, with nerve to attempt difficult things, and money enough to see them through to success or failure. He met my ideas of what an engineer should be. I do not think that my earliest impressions were changed much in later years." West Orange had its wizard; Pittsburgh made do very well with its plugger.[6]

A strapping six-footer with a thatch of chestnut hair going gray and formidable sideburns, Westinghouse bolstered his commanding presence with an air of dignity that kept most people at arm's length. Despite his innate shyness, he drew the loyalty of subordinates to him like a magnet. "With his soft voice, his kind eyes, and his gentle smile," said one, "he could charm a bird out of a tree." Although the smile was not always kind or the voice soft, men did not leave Westinghouse's employ as they did Edison's. "When the old man looks at you with that smile of his," admitted one lieutenant, "there is nothing you will not do for him." They called him the "old man" not because he was old but because they were so young. Westinghouse liked having young men about him, liked their energy and capacity for growth. When a visitor to his plant asked where his older, more experienced men were, Westinghouse replied, "We have no old men; I believe in young men for a new business." His young lions developed an esprit de corps rivaling that among Edison's inner circle, but Westinghouse went one step beyond Edison: He paid his men better.[7]

"We old pioneers had to improve everything...," recalled Reginald Belfield, the young engineer who had come over with the first Gaulard-Gibbs transformers and stayed to become one of the company's brightest lights. "We all worked hard and made few mistakes, considering everything. We all were about the same age and worked very well together with only one end in view, to help G. W. We were a very loyal and happy band and for this reason we were able to accomplish so much with so little material to assist us."[8]

Even when advancing the interests of his companies Westinghouse kept himself in the background and resisted having his picture in the paper if it could be avoided. "If my face becomes too familiar to the public," he reasoned, "every bore or crazy schemer I meet in the street will insist on buttonholing me." Once he emerged as a celebrity, he went to still greater lengths to duck personal publicity. "When I want newspaper advertising," he growled, "I will order it and pay cash." Nor did he allow his name to be used in the title of any activity beyond his own business firms. When some of his shop workers organized a baseball team, he cheerfully underwrote its expense but refused to allow them to use his name.[9]

Westinghouse could be gruff and demanding, yet the men adored him because he treated them well and never held himself above them. He was never one of the boys, as Edison often was, yet he was always approachable. One young worker got the message the day he and some others stood watching a foreign worker push a wheelbarrow of copper ingots along an iron slab thrown over a muddy alley. When a wheel slipped off the slab into the mud, the spectators guffawed and shouted advice as the worker struggled to right it. Westinghouse noticed the scene and walked over to it. Although dressed in a long-tailed coat and top hat, he sloshed through the mud, removed his gloves, and, without saying a word, grabbed hold of the wheel and lifted it back onto the slab. The message was plain: Everyone had to work together, and no one was too important to do any job.[10]

Although Westinghouse lacked the easy camaraderie Edison enjoyed with his men, he took better care of his workers. As early as 1881 he instituted the half day on Saturdays at the air brake plant, the first employer in Pittsburgh if not the nation to do so. When the shop first opened, Westinghouse invited the entire workforce to Thanksgiving dinner at a local hotel. The dinner became a tradition that endured until the number of workers grew so large that

Westinghouse had to be content with giving each one a turkey. When even that practice became cumbersome because the number of workers continued to grow and often included several members of the same family, Westinghouse scrapped the turkey giveaway and put a pension plan in its place. Gradually he developed one of the largest and most extended benefits programs in the nation.[11]

Westinghouse had entered the electrical field in a modest way by deciding in 1884 to manufacture the components of a DC lighting system. Once the potential of an AC system seized his imagination, however, he faced the much larger challenge of creating and introducing a new system based on principles different from existing ones. All lighting systems required the same four basic components: lamps, generators, a conductor system, and meters. An AC system promised a striking improvement in the conductor system thanks to the transformer, and Stanley provided a workable lamp as well as an improved generator or alternator. The biggest need remained a meter, the lack of which forced the early Westinghouse central stations to sell electricity on a contract basis. No existing meter could serve an AC system, and the engineers assigned by Westinghouse to develop one struggled in vain for a year. Not until the spring of 1888 did the breakthrough occur, and it came by accident, as so many scientific and technical insights do.[12]

Oliver B. Shallenberger, Westinghouse's chief engineer, was engaged in testing an AC arc light one April day when he accidentally dropped a small spiral spring on a magnetic field at the base of the lamp. To his surprise the spring began to rotate. When an assistant reached forward to grab the spring, Shallenberger stopped him. "Let's see what makes that spring revolve," he said. After studying it for a while he exclaimed, "I will make a meter out of that." It took him less than a month to do so, and by August Westinghouse had a meter ready for market. Based on the principle that a conductor in a moving magnetic field will revolve, Shallenberger's meter measured current rather than electrical energy. Once in use, it revealed how far off contract rates were. Under that system users paid a flat rate regardless of how much or how little electricity they used; the meter charged them only for the amount of electricity actually used. By encouraging users to economize by turning off lights when not needed, the meter helped the central station save on generating capacity. Tests revealed that metered central stations required only 50 percent to 70 percent as much capacity.[13]

The alternating-current system offered another advantage as well. Use of the transformer enabled it to achieve first a ten-to-one and then a twenty-to-one ratio between the transmission and consumption voltages. This allowed the latter voltage for lamps to be set at 50 when the transmission ran at 1,000 volts. This was fewer than half the 110 volts required for Edison's lamps, for which the transmission and consumption voltages had to be virtually similar. Through his feeder-and-main system and three-wire principle Edison managed to boost his ratio between the two voltages to nearly three to one, but he could go no higher. In effect the lamp voltage always limited the transmission voltage, a shortcoming in Edison's DC system that could not be overcome. The Westinghouse lamps, with their lower voltage, were safer and more efficient, and their use of a shorter lamp filament produced a more even light pattern. It was Herman Westinghouse who came up with the idea of using shorter, thicker filaments adapted for 50 volts to make a sturdier lamp.[14]

The improved lamp enabled Westinghouse to compete with Edison in quality rather than price. To power it, Stanley's alternator underwent several refinements by Shallenberger and Albert Schmid, a Swiss engineer who had come to work for Westinghouse in 1882. An even younger hand, Lewis B. Stillwell, arrived in October 1886 and displayed his talent by devising a regulator that could raise alternating-current voltage to any desired level. An impressed Westinghouse dubbed it the "Stillwell booster," and the name stuck. Later Schmid incorporated many of the new features into a beautifully designed machine that became known as the "kodak." Asked how he could come up with a mechanical work so graceful in form, Schmid replied, "Oh, you only have to put the metal where it belongs for strength and it becomes beautiful of itself."[15]

Buoyed by the good work of his engineers, Westinghouse began to spread his wings. By September 1887 he had 68 central stations built or under contract with a total capacity of 134,000 lamps; these figures mushroomed to 350 stations and 700,000 lamps by November 1890. He decided to enter the field of arc lighting as well and in November 1888 bought a small manufacturing firm, the Waterhouse Electric and Manufacturing Company. Two months later he leased the assets of United States Electric Lighting Company, which earlier had acquired Weston Electric. These transactions enabled Westinghouse to create an arc-lighting system by combining the Weston and Waterhouse systems. Stanley thought he had found a way to power an arc-lighting system with AC current. Westinghouse promptly invested a large sum in developing and promoting Stanley's system; however, his lamp proved inferior to that used in a DC system. Reluctantly Westinghouse abandoned the project and reverted to selling a DC-powered arc-lighting system but he continued to search for a method of using AC current for arc lamps.[16]

As his production lines expanded, Westinghouse kept a wary eye on his competitors. The litigation over lamp patents remained as tangled as ever and began spilling over into other areas of dispute as well. When it came to patents,

Westinghouse was quick to protect his own. His early experience with the rail-roads over the air brake taught him bitter lessons about the need to defend his products and patents with ferocious zeal. He had managed to secure a broad patent for an AC distribution system using transformers in parallel even though it was based on the Gaulard-Gibbs system that connected them in se-ries. It pleased him immensely when in the fall of 1886 all of Elihu Thomson's applications for AC distribution were rejected, forcing Thomson-Houston to seek ways of working around Westinghouse's patent. If he wondered what Thomson would do next, he did not have long to wait.[17]

BY EARLY 1887 Thomson found himself in a quandary. Although he had far more experience with AC and had started work on his system months ear-lier than Westinghouse, the latter had already brought his product to market while Thomson still struggled with his components. Charles Coffin grew im-patient with the delays and wanted to know the reason for them. Thomson replied that he was trying to develop an entire system with matching alterna-tors, regulators, lamps, transformers, and safety devices. Such a system, he ar-gued, would excel in reliability. "When we enter this field," he stressed, "we wish to . . . be sure of success from the start with a complete and economical system." Quality, not speed, would win out in the long run. However, Thom-son offered a second explanation for his slow progress. "The development of the system," he admitted, "had outgrown the model room which is only adapted to producing small models and nothing of very large size."[18]

The obvious solution was to expand the Model Room and hire more assis-tants. Coffin understood how busy Thomson and his staff were. They had not only to come up with new products but also to deal with patent strategy and litigation, evaluate patents brought to their attention by other inventors, and provide advice on products to Coffin and the marketing people. Repeatedly Coffin suggested that Thomson hire more assistants, install more machinery, and expand the space. Outside inventors could be brought in to work on their own and other products. But Thomson steadfastly rejected these suggestions. Although he complained of being overwhelmed with work, he regarded the Model Room as his private domain and personal laboratory. He wanted full control over its operation, and he did not want to deal with other inventors who might bring rivalries and jealousy into the work. By controlling the Model Room he also shaped the direction taken by the company's technological in-novations.[19]

This attitude created a growing disconnect between Thomson's work and the needs of the manufacturing and marketing people. For Thomson, quality was paramount regardless of the problems his slow progress posed for manu-facturing schedules and marketing programs. He may also have felt a keen sense of resentment at having been wrong about both the incandescent lamp and the potential of AC power, which forced him constantly to work around the patents of Edison and Westinghouse. Thomson was, after all, the only trained scientist among the three of them, yet the others had outflanked him at every turn with their boldness, originality, and persistence. He had created in-ventions; they had created whole companies around their inventions and ex-erted control over every aspect of their operations. As Samuel Insull said of his boss, "Mr. Edison is the mainspring and ruling spirit of everything." Westing-house's men regarded their leader in the same way. No one said such things about Thomson, who could do nothing without Coffin and Edwin Rice.[20]

Rice had shown a remarkable ability as factory superintendent to convert Thomson's experimental models into manufactured products. As orders for arc lights poured in, Rice had to oversee expansion of the operation into new buildings with an enlarged workforce. His own ability grew with the company, but the onetime assistant to Thomson hesitated to intrude on his former men-tor's turf. Although the Model Room increasingly became a bottleneck to the broader operation, Rice chose to work around the problem rather than chal-lenge Thomson's refusal to expand his domain. As a result Thomson-Houston began to experience subtle but growing internal tensions as its business ex-panded, and Thomson grew progressively more unhappy with his role and the pressures put on him.[21]

The most obvious need was to get a working AC system out of the Model Room and into the marketplace. Having come late to the race, Thomson had little chance of inventing and patenting a new type of incandescent lamp. In-stead the company had bought Consolidated Electric Light to secure control of the Sawyer-Man lamp patents. It then had Consolidated form a subsidiary, the Sawyer-Man Electric Company, to manufacture lamps. Satisfied that his grounding system solved the safety issue of dealing with high voltage, Thom-son finally allowed his AC system to go into production in 1887. No sooner did it hit the market than Westinghouse slapped a suit on Thomson-Houston for infringement of his transformer patent. After a series of meetings the two sides came to an agreement in August 1887. Westinghouse agreed to let Thomson-Houston manufacture and sell its AC system in New England and to Thomson-Houston local lighting companies outside that region. In return Westinghouse received a royalty on each transformer and the right to sell the Thomson-Houston arc-lighting system throughout the United States.[22]

Even as Westinghouse staked the high ground on transformer rights, he had to concede the superiority of the Sawyer-Man lamp patents that remained in the hands of Thomson-Houston. During the fall of 1887 he agreed to stop manufacturing the Stanley lamp, sold his Pittsburgh lamp factory to Consolidated Electric, and agreed to use Sawyer-Man lamps thereafter. In payment for the factory Westinghouse took shares of Consolidated stock. Thomson-Houston also sold all of its remaining stock in Sawyer-Man Electric to Consolidated, giving the latter complete control of that lamp's patents. Westinghouse added to his holdings in Consolidated by selling it some more assets in return for stock and by buying shares in the open market. In October 1888 Westinghouse leased the assets of Consolidated from Thomson-Houston and thereby returned to the business of manufacturing lamps. Two months later he bought Consolidated outright from Thomson-Houston.[23]

This series of transactions gave Westinghouse a firm grip on the Sawyer-Man patents and strengthened his position for the looming legal battle over both the validity and scope of Edison's lamp patents. Why Thomson-Houston chose to surrender control of Consolidated remains something of a mystery. Westinghouse may have acquired enough Consolidated stock to challenge Thomson-Houston within the company or at least make their control uncomfortable. Thomson-Houston may simply have needed cash for the buying spree it launched in 1888. During the next three years it bought up seven smaller rival firms, including Brush Electric, partly to strengthen its hold on the arc-lighting market. The company may also have wished to avoid getting entangled in the bitter legal war over patents. What is clear is that the transaction did not spring from warm relations between Coffin and Westinghouse. Coffin resented the Westinghouse Company's "attitude of bitter and hostile competition," while Westinghouse regarded Coffin as a hustler.[24]

Coffin and Thomson suspected that the agreement of 1887 was more a truce than a settlement. Thomson resorted to his familiar tactic of easing around the Westinghouse transformer patents by filing patents for several different designs of transformers that would limit the broad thrust of the Westinghouse patent. Then, in April 1888, a judge undercut the Gaulard-Gibbs patent by ruling that it applied only to transformers used in series. A few months later Thomson-Houston terminated its license agreement with Westinghouse, and the latter stopped marketing the Thomson-Houston arc-lighting system. The agreement of 1887 had come unraveled. It was these moves that prompted Westinghouse to enter the arc-lighting field on his own by gaining control of the Waterhouse and United States Electric companies. These purchases, coupled with that of Consolidated, gave him control over all the major non-Edison incandescent lamp patents.[25]

While these events unfolded, Thomson continued to brood over the safety issue. "I am a believer in the establishment of all safeguards which conduce to the good working of a system," he stressed, "especially when they do not add greatly to the cost of making the installation." He viewed this approach as not only responsible but good marketing strategy. However, he reckoned without the avarice and carelessness of some Thomson-Houston customers. In their rush to get started and operate as many lamps as possible, many customers simply ignored the safety equipment and procedures—sometimes putting two dynamos on the same line even though the insulation could not handle the increased voltage. Thomson argued that "it would be better not to have the business than incur the risks which are thus involved," but he well knew that he had no control over the installation process once the equipment was sold. Nor did Coffin's group offer him much consolation. Their job, after all, was to sell systems, not lecture their customers on the need for rigid safety practices.[26]

THE NEW LABORATORY complex at West Orange was a sight to behold. Situated on fourteen acres next to the main road of the Oranges, it sat conveniently near Glenmont and the railroad line to Newark and New York. The three-story main building, 250 feet long and 50 feet wide, housed an enormous library behind huge arched windows, along with Edison's desk. At the opposite end stood the powerhouse with its boiler, steam engine, and dynamos. The building also contained Edison's private experimental rooms, a heavy machine shop on the first floor, and a precision machine shop upstairs across from the inventor's laboratory. This arrangement, he quipped, would let him build "anything from a lady's watch to a Locomotive." Four smaller buildings, each one hundred by twenty-five feet, housed electrical and chemical activity, a metallurgical laboratory, a woodworking shop, and storage.[27]

To construct and furnish the buildings with machinery, equipment, instruments, and supplies, Edison budgeted $180,000. The stockroom and storage areas bulged with materials of every kind, including $6,000 worth of pure chemicals and "every known substance on the face of the globe." A magazine reporter looked the inventory over and wrote that "the whole of nature seems to have been laid under contributions to stock these long, deep drawers." The object was to have everything on hand that might be needed in an experiment to avoid delays. Here truly was an invention factory on a scale that dwarfed Menlo Park, which was exactly what Edison intended. It was on one hand a monument to himself, intended to translate his personal visions into products, and on the other hand a well-designed machine to turn out goods on an unprecedented scale. The laboratory was to Edison as the Homestead Works were to Andrew Carnegie or River Rouge to Henry Ford.

It was the largest laboratory in the world, and it cost a lot of money not only to build and furnish but also to run. Once in full swing, Edison had to meet a payroll ranging between eighty and a hundred men headed by his main staff of twenty-five to thirty chief experimenters and assistants. Even before the laboratory was completed Edison realized what a financial load it would be to carry. "The Lord only knows where I am to get the shekels," he complained to Edward Johnson in July 1887. "Laboratory is going to be an awful pull on me." Samuel Insull, who still handled Edison's financial affairs despite his move to Schenectady, observed sourly that Edison wanted "a great deal more money than he at first anticipated, but this is simply a repetition of what has occurred so frequently before . . . Exactly how I am going to carry out his wishes and give him what he requires, I don't know." In the past Insull had always been on the scene to rein in Edison's extravagances, but no longer.[28]

To pay for his operation, Edison approached Drexel, Morgan and a Boston investment banker with plans for experimental work they might fund, but both declined his overture. He then tried to interest Henry Villard with a proposal for the banker to pay the laboratory's expenses for new research in return for a half interest in every invention produced. The idea intrigued Villard, but he decided against it. To cover part of his expenses, Edison asked the Electric Light Company, Edison Lamp Company, Edison Machine Works, and Bergmann & Company for a weekly advance to cover experiments done for the benefit of them all. The four firms responded with some funding but remained uneasy about the arrangement. He also received some money for contract research done on behalf of Bell Telephone, the American Novelty Company, and the A. B. Dick Company, which sold the Edison Mimeograph. For his own projects Edison had to come up with the funds himself. Two in particular occupied his attention: the Edison Phonograph Company and the Edison Ore Milling Company.[29]

As Edison scrambled to raise money for the laboratory, the irony of his situation became more apparent. With the new facility he hoped to elevate invention into a big business, but the need to sustain it made him more of a businessman than ever before. He needed money to experiment, but the quest for money—to say nothing of managing his complex operation—took him out of the laboratory even more. Direction for West Orange came always from the top, but Edison found himself having to move in several directions at the same time. As historian Andre Millard observed, "The policy of contract research meant that the lab had to be all things to all people." During the late 1880s Edison gave primary attention to three fields of interest: the phonograph, ore milling, and electric light and power. What he might have accomplished had he focused on one of these projects exclusively can only be imagined. His secretary, Alfred Tate, put it best. "I have often wondered," he wrote late in life, "if Edison might not have accomplished more if he had attempted less."[30]

But it was not Edison's way to attempt less. In his struggle to support the lab he plunged again into the phonograph and then into his first crude version of motion pictures. In May 1888 he broke ground for two large buildings that became his factory for manufacturing the phonograph. He also took up again his decade-long quest to perfect a process for extracting low-grade iron ore in iron and steel mills—a pursuit that became the greatest and most costly white elephant of his career. These projects occupied much of his time and thought even though things electric still held center stage in his work. Improvements in his lamp and other components still required long hours of research, as did the more mundane task of selling the Edison system. Edison hired Arthur Kennelly, a young Englishman, as his chief electrical assistant and allowed him to use the facility for his own research as well. Kennelly stayed six years and made important contributions to Edison and to electrical engineering theory.[31]

The thrust for most of the electrical work was not innovation but refinement. As competition in the industry heated up, Edison knew he had to improve his products by making them both better and cheaper. To coordinate this effort he revived the Standardizing Bureau, a committee charged with standardizing equipment throughout the companies. Under its director, W. J. Jenks, the bureau sought to "fully weigh the practical opinions of the experimenter, the manufacturer and the practitioner as to secure their combined wisdom, and avoid the friction that would otherwise be inevitable in making any necessary changes in standards." Edison and Batchelor sat on the committee along with representatives from the shops and the Electric Light Company. The object was to reduce manufacturing costs as well as improve the efficiency of lamps and other apparatus.[32]

Of necessity the work was parceled out. Kennelly took charge of the ongoing work on dynamo and motor design along with a meter suitable for the three-wire system. Edison also launched what proved a lengthy campaign to develop an insulation for house wiring that was flexible, waterproof, and fireproof. Despite large expenditures of time and money, no satisfactory solution was found. Edison himself concentrated on his specialty, lamp research. The early lamps tended to be unreliable and have short lives. They also lost illumination as carbon deposits blackened the inside of the bulb. Once again he ransacked nature in his search for a better filament and experimented with artificial fibers as well. This work took place under the shadow of the long-pending fight seeking to overturn Edison's basic carbon filament patent, and of his need to work around the Sawyer-Man patent on the flashing process. During 1888 Edison developed a process of treating filaments with a thin coating of

asphalt. This approach improved efficiency somewhat while the search continued for a more substantial leap forward.[33]

Work on the dynamos fared better. For all its progress, the electrical industry remained astoundingly imprecise in many areas. Engineers who ran central stations had only a rough notion of how much electricity they generated,
and the crucial concept of balanced load did not appear until the 1890s. Large
gaps in theoretical understanding persisted; not until November 1887 did
Heinrich Hertz prove Maxwell's theory on the nature of electromagnetic
waves to be correct. The lack of precise measuring instruments continued to
plague the industry. Both Elihu Thomson and Edward Weston were hard at
work devising better tools for the measuring trade. In 1886 it took a week to
measure a dynamo's efficiency, and even then, as Weston complained, "an accurate reading was almost an act of God." The dogged efforts to devise meters
reflected the need for even basic information; the struggle to improve dynamos betrayed a lack of basic theoretical knowledge.[34]

Few things were more crucial to dynamo development than understanding
the magnetic circuit, yet the magnetic behavior of an electromagnet's core remained largely a mystery until the mid-1880s. The concept of saturation—the
state beyond which a metal or alloy can no longer be magnetized—was still
unknown, leaving unanswered the question of whether there existed a limit on
the amount of magnetism an electromagnet could produce. Edison's Jumbo
generator offered a classic example of the notion that no such limit existed and
that field strength could be maximized by building a much larger electromagnet. Faraday had introduced the notion of a magnetic circuit, but little was
done with it until 1873, when Henry A. Rowland suggested an elementary law
governing the magnetic circuit. In 1882 and 1883 a British scientist and music
theorist, R. H. M. Bosanquet, described the general properties of a magnetic
circuit in a series of papers. Rowland, by then a renowned physics professor at
Johns Hopkins and noted authority on electricity, applied Bosanquet's formulas to the field magnets used in dynamos, thereby setting the stage for someone
to utilize his theory in the actual design of dynamos.[35]

That someone proved to be John Hopkinson, a British scientist and electrician. As early as 1879 Hopkinson had emphasized the need for some realistic
standard by which to measure dynamo performance. The usual practice was
to correlate a dynamo with a fixed number of lights, but what if a generator
built for one light was connected to two or more lamps, or vice versa? "It is
clear," he wrote, "that the attempt to examine all separate combinations of so
many variables would be hopeless, and that the work must be systematised."
Engineers needed a general law or formula to explain how much current a machine would produce under different circuit conditions, and how much power

would be needed in each case. Collaborating with his brother Edward, Hopkinson presented a paper in 1886 that illuminated the basic principles of the magnetic circuit as applied to generators. The Hopkinsons provided not only a law of magnetic circuits but an equation that enabled engineers to predict the characteristic curve of any dynamo. *Electrical World* hailed it as having "an almost incalculable effect in improving electrical designing and laying the foundations of innumerable practical applications."[36]

Arthur Kennelly doubtless read the Hopkinsons' paper. Edison certainly knew about John Hopkinson, who had invented a three-wire system at about the same time Edison had. Later Hopkinson served as a consultant to Edison's English firms. Kennelly understood from the paper that field magnets could be shortened and multiple cores done away with entirely. Redesigning the field magnets around these principles improved the magnetic circuit and increased efficiency to around 95 percent. Kennelly's new line of dynamos also reduced manufacturing costs to the point where Edison could say that the price of "all electrical machinery is entirely too high now." He was right. Kennelly was hardly the only electrician to absorb the lesson delivered by the Hopkinsons. Elsewhere new and more efficient dynamos began rolling off production lines.[37]

Despite the progress made in these and other areas, one obstacle remained unyielding: the distribution of power over distance. A sharp increase in the price of copper had made the problem even more acute. Faced with these pressures, Edison and his staff began exploring the possibilities of alternating current in 1886. That year Edward Johnson asked Frank Sprague to prepare a report on alternating current. Sprague responded by strongly urging the Edison company to adopt the AC system. In December Edison even applied for a patent covering his own version of an AC system. He also received a detailed report on the Gaulard-Gibbs system from Siemens & Halske, which licensed the Edison system in Germany. It argued that the AC system was more expensive for isolated plants, that transformers made it more difficult to maintain a constant voltage, that AC systems lacked both meters and motors, and that the cost of conversion would exceed any savings from lower copper costs.[38]

Siemens had told the same thing to Westinghouse's man, but Edison took the warning more to heart. His contempt for Westinghouse was already growing, fueled in part by the prospect of competition. "None of his plans worry me in the least," he assured Johnson; "only thing that disturbs me is that Westinghouse is a great man for flooding the country with agents and travelers. He is ubiquitous and will form numerous companies before we know anything about it."[39]

While Edison pondered the Siemens report, the Electric Light Company in November 1886 acquired the American rights to the Hungarian ZBD alternating-current system along with an option to buy the American patents when they were granted. Edison was familiar with the system. One of the inventors, Ottó Bláthy, had come to America in September 1886 to talk with Edison. Francis Upton then went to Europe to examine the system and afterward urged its purchase, if only to keep a "dangerous enemy" out of the arena. This acquisition gave Edison access to the most feasible rival AC system to that of Westinghouse, one that might have forced the latter to the wall had Edison decided to manufacture AC systems. Certainly Edison loved to compete. "I can only invent under powerful incentive," he declared. "No competition means no invention." But he could not overcome his doubts about any high-voltage system. He considered them unreliable, too complex, and entirely too dangerous, posing the twin threats of electrocution and fire. Instead of using the ZBD system as a competitive weapon, Edison insisted that the Electric Light Company drop its option on the American patent rights. He informed Henry Villard that alternating current was "not worth the attention of practical men."[40]

Edison needed to reassure Villard because he was formulating another plan that required the financier's assistance. His relationship with the various Edison companies continued to deteriorate. An ironic but predictable situation had evolved within a short time. Edison had founded the enterprise, created the companies, put some of his best boys in charge of them, and continued to control them. However, once they became public corporations, their managers had to serve broader constituencies and look more closely to their own interests. Their priorities began to diverge from those of Edison. He wanted all research contracted to West Orange, but that proved impractical. As each firm grew, it had to develop its own research capabilities in some areas. Samuel Insull put the matter clearly. "You cannot possibl[y] wipe out entirely our experimental account here," he told Edison. "There are lots of *little* things which it would be absurd for us to send to Orange to be tried, because we have all the parts here." Sigmund Bergmann said much the same thing, as did the other managers.[41]

Not surprisingly, money lay at the heart of the conflict. Edison desperately needed funds to support the new laboratory complex, and he poured a large chunk of his own money into the expensive new phonograph factory behind the laboratory buildings. He received a weekly sum from the other Edison companies and also billed them for experimental work done on their behalf, the nature of which he determined. The companies protested both the size of their payments and an arrangement that let Edison decide what experiments should be conducted. "You must remember," said Insull, "that we are running this business to show profit, that experimental expenses like other expenses have to come out of our profits." Like Edison himself, the Electric Light Company and the manufacturing firms also had to scramble for cash. In selling systems to local companies, they often had to take securities in lieu of cash for payment.[42]

Such exchanges revealed a growing breach in the ranks. Edison had conceived of the new laboratory as Invention Central, where all research and development would occur. However, the manufacturing companies soon realized that they needed their own research capabilities, and the need for financial controls led them to scrutinize bills from West Orange with a keen eye as to whether the results warranted the outlay. The bankers who loaned the companies money demanded strict control over expenditures and did not warm to supporting expensive experimental programs. Edison deeply resented this growing attitude and denounced the bankers' lack of vision and courage. Especially did he dislike the conservatism of the Electric Light Company. "With the leadened collar of the Edison Electric Light Company around me," he complained to Villard, "I have never been able to show what can be done."[43]

Henry Villard thought he had a solution to the problem. He had managed to consolidate the diverse Edison lighting firms in Europe; perhaps the same approach might eliminate the growing friction between the American companies as well. He broached the idea to Edison of organizing a syndicate to put all the firms together in one large company. Edison gave his approval on condition that Drexel, Morgan not dominate the syndicate. "When money was required for the business," he explained, "it was next to impossible to get it so long as Drexel, Morgan & Co controlled the business, or if obtainable at all the money could only be got at ruinous rates." Never again did he want to find himself in a position where those bankers "could squeeze any interest that he owned." Villard assured him that his German investors would control the syndicate and that Drexel, Morgan would have only a small share. On that basis Edison, eager for some solution to the squeeze he was already in, agreed to the negotiations.[44]

NIKOLA TESLA'S NEW laboratory was far more modest than that of Edison, but it carried an equal dimension of hope. Located at 89 Liberty Street, next to what later became the World Trade Center, it opened a new phase of Tesla's career. For the first time he could actually build the motors he had envisioned; he had become an inventor, and he was determined to produce. Working day and night, sometimes until he collapsed, he began turning out motors and on April 30, 1887, filed his first patent. Alfred Brown served as his technical assistant and his old friend Anthony Szigeti, recently arrived in New York, as his assistant. Brown and Charles Peck remained as quiet backers to Tesla for the next decade; Peck sought out other investors ranging from New York to California. In remarkably short order Tesla readied for manufacture three complete AC systems for single-, double-, and triple-phase currents. "The motors I built there were exactly as I had imagined them," he claimed later. "I made no attempt to improve the design, but merely reproduced the pictures as they appeared to my vision and the operation was always as I expected."[45]

A few weeks after Tesla started work in his new laboratory, a visitor stopped by to see what he was up to. Thomas Commerford Martin was much more than merely the editor of *Electrical World*. The same age as Tesla, he had come to the United States from England at twenty-one and worked briefly for Edison at Menlo Park before going off to Jamaica. He returned to America in 1883 and got a job at *Electrical World*, a once obscure journal that grew to prominence with the electrical industry itself. A year later Martin got himself elected vice president of the newly organized American Institute of Electrical Engineers (AIEE) and won the presidency in 1886. That same year he published his first book, *The Electric Motor and Its Applications*. With his British affectations, outsized ego, and charming manner, Martin became a formidable figure in the industry. A shaved head capped a round face dominated by large, soulful eyes and a flamboyant mustache. He could be an extremely helpful friend, and he took an immediate shine to Nikola Tesla.[46]

Perhaps it was Tesla's peculiar blend of charm, genius, and eccentricity that attracted Martin. Certainly he saw great promise in the lanky and often reclusive Serb, whom he persuaded to write an article for *Electrical World*. He also arranged for several distinguished Cornell professors to test Tesla's motors. One of them, William Anthony, was so pleased with the results that he urged Tesla to present a paper before the AIEE describing his motor. When Tesla balked, Martin added his powers of persuasion. The motor had, after all, performed well in Anthony's tests. The main obstacle for commercial development lay in adapting it to the high-frequency circuits then in use around the country. Although tired and overworked, Tesla finally yielded to Martin's and Anthony's urgings. Later Martin claimed that Tesla scribbled off his paper in pencil the night before the session.[47]

On May 15, 1888, a gaunt Tesla stood before the assembly of AIEE members and read the paper that changed his life. Entitled "A New System of Alternate Current Motors and Transformers," it described the rotary motor that Tesla hoped would revolutionize the power industry. During the lecture he paused to demonstrate the motor he had brought with him and to illustrate his points with diagrams. Already he had filed for fourteen of some forty patents basic to his AC system, and his backers were busy seeking advice from patent attorneys on prospective buyers for the rights. A gas manufacturer from San Francisco showed early interest, and so did George Westinghouse. One other interested party, Elihu Thomson, was actually in the audience and listened intently as Tesla described in mathematical terms how to construct single-, double-, and triple-phase motors, how to determine the number of poles and speed, and how to interlink the system with existing DC apparatus. When he had finished, Martin asked Professor Anthony to report on his independent tests of Tesla's motors. Anthony confirmed that they matched the efficiency of the best DC motors. Moreover, the reversal of current direction happened "so quickly that it was almost impossible to tell when the change took place."[48]

Thomson rose from his seat and, after praising Tesla's "new and admirable little motor," declared that he had for some time been working along similar lines toward the same goal. "The trials which I have made," he continued, "have been by the use of a single alternating circuit—not a double alternating circuit—a single circuit supplying a motor constructed to utilize the alternation and produce rotation." Tesla grasped the thrust of Thomson's point at once; he was trying to establish that his work on an AC motor preceded Tesla's, but he had blundered in describing the difference between the two versions. Thomson's single-circuit motor still required use of a commutator, which made it much less efficient. Smiling, Tesla bowed toward his rival and replied that "the testimony of such a man as Professor Thomson flatters me very much." He paused briefly, then added, "I had a motor identically the same as that of Professor Thomson, but I was anticipated by him . . . That peculiar motor represents the disadvantage that a pair of brushes [i.e., a commutator] must be employed."[49]

Thomson got the message and did not at all appreciate it. Shortly afterward, when Tesla's agents offered the patents to Thomson-Houston, Thomson dismissed them as not worth the cost of securing them. He may actually have believed this, or he may have seen them as a threat to his own system. Acquiring the patents would likely have brought Tesla to the company as well, and Thomson did not like other inventors mucking around in his domain. Whatever his reasons, Thomson followed Edison's lead in rejecting the doorway to the future proffered by Tesla. Westinghouse did not make the same mistake. Six days after Tesla's lecture he sent Henry M. Byllesby to Tesla's laboratory. Only a month earlier Westinghouse had learned of an article published by an Italian inventor, Galileo Ferraris, describing his earlier work on an AC motor. He promptly dispatched Guido Pantaleoni to Italy to buy the Ferraris patents.[50]

Pantaleoni managed to reach agreement with Ferraris for American rights on his AC patents, but closer inspection of Tesla's patents convinced Westinghouse that the Italian patents would be of little use. This discovery eliminated what little leverage Byllesby had in negotiating for Tesla's patents. After Tesla explained his motor, Byllesby reported that "his description was not of a nature which I was enabled, entirely, to comprehend. However, I saw several points which I think are of interest. In the first place . . . the underlying principle of this motor is the principle which Mr. Shallenberger is at work on at the present moment." Byllesby thought the motors operated very successfully. "They start from rest and the reversion of the direction of rotation can be suddenly accomplished without any short-circuiting." When he told Brown and Peck that he might be interested in buying Tesla's patents, they replied that a San Francisco capitalist had already offered them $200,000 plus a royalty of $2.50 per horsepower on all apparatus.[51]

"The terms, of course, are monstrous," Byllesby added, "and I so told them." But he left the door open for future negotiations. Westinghouse agreed that the terms were onerous, but he also saw the bigger picture. "If the Tesla patents are broad enough to control the alternating motor business," he told his lawyer, "then the figures . . . are not unreasonable; in fact the Westinghouse Electric Company cannot afford to have others own the patents that are necessary to enable it to make motors to work on the alternating current system . . . A successful alternating current motor . . . together with our meter will materially advance our interests." Shallenberger's chance discovery of the rotary principle thanks to the loose spring had occurred only three weeks before Tesla's lecture. From it might spring a workable AC motor as well as a meter, but it was too soon to tell. Shallenberger did not yet fully grasp the principles behind his discovery and had not yet filed any patents.[52]

Asked his views, William Stanley reacted much as Elihu Thomson had done. As early as September 1883 he had jotted down in a notebook that AC current could be used to excite an induction coil. "I have built an AC system on basically the same principle," he added, "which allows electromotive force to be transmitted from power stations to homes for the purpose of illuminating them." But Stanley's system had the same drawback as Thomson's: It used a commutator, and the object was to get rid of this bothersome device. Like Thomson, Stanley seemed unwilling to admit that someone else had beat him to the fundamental insight. Even Ferraris, who had come by Tesla's own admission "independently to the same theoretical results," had concluded that "an apparatus founded upon this principle cannot be of any commercial importance as a motor."[53]

Westinghouse intended to find out whether this was really the case. Never one to shrink from a challenge, he summoned Tesla to Pittsburgh early in July 1888 to close the deal. Once there, Tesla fell quickly under the spell of

Westinghouse's personal magnetism. They went to Westinghouse's home and then to Westinghouse Electric for a tour of the plant. Through it all Tesla found himself mesmerized by his host's boundless energy, enthusiasm, and utter sincerity. Nearly a quarter of a century later he paid tribute to the personality that he first saw that hot July day:

> Though past forty then, [Westinghouse] still had the enthusiasm of youth. Always smiling, affable and polite, he stood in marked contrast to the rough and ready men I met. Not one word which would have been objectionable, not a gesture which might have offended—one could imagine him as moving in the atmosphere of a court, so perfect was his bearing in manner and speech. And yet no fiercer adversary than Westinghouse could have been found when he was aroused. An athlete in ordinary life, he was transformed into a giant when confronted with difficulties which seemed unsurmountable [sic]. He enjoyed the struggle and never lost confidence. When others would give up in despair he triumphed.[54]

Even more impressive, Westinghouse went beyond mere show and tell. He offered Tesla $5,000 for a ninety-day option on his patents. If he picked up the option, Tesla would receive $20,000 in cash and $50,000 in notes payable in three installments at six-month intervals. He would also get a $2.50 royalty per horsepower for every motor with minimum royalties set at $5,000 the first year, $10,000 the second year, and $15,000 each succeeding year. Tesla agreed to come to Pittsburgh to help develop the motor for commercial use. His vision had found a home at last with a man who not only believed in the future of alternating-current power but had put his money on the table as well. Tesla moved into a Pittsburgh hotel—he would always live in hotels—and prepared to grapple with the difficult challenge of transforming his motor from an exhibit into a product.[55]

CHAPTER 12

GAINING TRACTION

*Some time elapsed after the development of the self-exciting ma-
chine before the marvelous characteristic of reversibility of function was discovered, with the necessary corollary, the electrical
transmission of energy by the use of two similar machines, one to
be driven by power and to generate electricity, and the other to receive electricity and to develop mechanical power.*

—FRANK J. SPRAGUE[1]

BY THE LATE 1880S ELECTRICITY had reached the brink of triggering
still another revolution in American life. Already it had established itself in the
fields of communication via the telegraph and telephone and in illumination
through arc and incandescent lamps. It had yet to penetrate two domains that
promised even more fabulous rewards and changes in the way Americans
lived: transportation and general power. The latter offered a bonanza of possibilities if efficient ways could be found to transmit and utilize electricity in
homes and factories. The key to developing both realms lay in one key component, the motor. It was the gateway through which electricity could be applied
to an almost infinite variety of applications.

A motor created mechanical motion by using the magnetism produced by
an electric current placed near a permanent magnet. Its origins traced back to
the epochal discoveries of Oersted and Faraday that electricity produced a
magnetic field and that mechanical action—moving a conductor through a
field—created electricity. Put another way, electrical and mechanical energy
were interchangeable, meaning that either one could be used to create the
other. A generator produced electricity through mechanical action; reversing
the process allowed electricity to create mechanical action. Curiously, scientists and inventors for decades treated these interrelated phenomena separately. Work on electric generators and electric motors proceeded separately

within narrow spheres. The generator, for example, was developed to serve one main function: the powering of lamps. Any other use had to await the emergence of other applications, all of which revolved around the creation of an electric motor.[2]

It is electricity's unique versatility that elevates it above all other power sources, but the full range of that versatility had scarcely been glimpsed by the 1880s. Most electricians knew three basic facts about the electric motor in 1880. It was basically a generator, or dynamo, running in reverse, which meant that knowledge about constructing dynamos could be applied to motors as well. Second, the motor was clearly the most vital component in the efficient transmission of electrical power. It was the device that actually performed whatever work needed to be done. Finally, the same current produced by dynamos to run lights could also serve motors. Building a reliable, efficient motor was but the first step to practical application; it would also require an efficient dynamo and a suitable network. There was plenty of incentive to create such a system, if only because the steam power used by most businesses to run machinery was so cumbersome and costly. However, the electric motor remained an idea in search of an application that would spur its development.[3]

During the 1880s that spur emerged in the form of urban transport. Growing cities groaned under the lack of adequate transportation that would enable their soaring populations to spread out and still get to work. Chicago, for example, doubled its size and tripled its population in a single decade. For decades urban public transportation consisted of the omnibus, a wagon drawn by horses or mules. Some improvement came by installing rails and running the wagons on them. By 1880 American cities boasted nineteen thousand such streetcars operating on about three thousand miles of track, but they still depended on animals for their power. Horses and mules were slow, costly, and vulnerable to disease, and they filled the streets with manure. They broke down under the best of handling and were expensive to maintain. By one estimate a team ate its original cost in feed every year and wore out in four years. Nevertheless, streetcar lines became a big business; investment in them reached $150 million by 1882.[4]

Attempts to create steam-powered wagons went back as far as John Fitch and Oliver Evans, but no one had succeeded in devising a commercially feasible version. One new form of transport, the cable system, appeared early in the 1870s. It used steam engines at a central station to move a revolving steel cable set in a slot between two rails. An arm or grip mounted on each streetcar engaged the cable, which moved the car along the track, but it took a pair of strong arms to force the arm onto the cable. Once engaged, the cars moved at twice the speed of horse-drawn cars, but the system was expensive to install and maintain. One expert estimated the installation cost at $111,000 per mile compared to $12,600 for a horsecar trolley. It was also fragile; when the cable broke, all movement had to be halted until it was repaired. The car could not move faster than the cable, and the system was inflexible in that it allowed no turnouts or sidings or any variation from the running track. Nevertheless, cable systems appeared in cities with populations dense enough to support them.[5]

Early efforts to develop an electric train in both England and the United States inevitably ran up against two major problems: The motors were crude, and batteries could not supply enough power. In 1860 Antonio Pacinotti of Italy devised the first commutator, but the significance of his discovery languished for ten years until Zénobe Gramme rediscovered it and ushered in a new age of dynamo development. In 1873 Gramme exhibited his generator in Vienna along with a motor driven by current from it. Since the two machines were identical, observers could readily grasp the relationship between them. As both motors and dynamos improved in design and power, interest renewed in applying them to transport as well as illumination. Only six years after Gramme's Vienna demonstration, Werner von Siemens titillated spectators at the Berlin Industrial Exhibition by displaying an electric railroad that carried eighteen passengers and a driver over a closed loop of track a thousand feet long. The little locomotive pulled three platform cars each fitted with two benches. *Scientific American* called it the Siemens "electrical merry-go round."[6]

The age of the electric railway had dawned, but the light was still faint. Major technical problems posed an obstacle to the creation of a commercially viable system. A suitable motor had to be devised, and then some way had to be found to supply it with current while on a moving vehicle. The car not only moved but stopped and started frequently and traveled at various speeds. It also needed an efficient means of transmitting power from the motor to the axles. The DC motor, like its sister dynamo, had brushes and a commutator to turn DC current into the alternating current that drove the motor. To do this, the brushes had to press firmly against the commutator to pass the current along. As a result both brushes and commutator endured heavy wear and required constant maintenance or replacement. Frequent stops and starts increased this wear, and the shift in voltages at start-up produced sparking and arcing. Moving power from the motor to the axles required some system of belts, chains, or gearing, which involved even more careful engineering.[7]

A host of inventors burrowed into these complex problems. That Edison was one of the first should not be surprising; what is amazing is that he did it while deep in the ordeal of developing his incandescent lamp. He managed this feat in 1880 by assigning some of his able lieutenants to oversee much of the work. His interest in what was then called "counter electric motive force" led him to create an electric railway at Menlo Park to study it. He had developed the low-resistance armature dynamo that served nicely as a motor. As Frank J. Sprague, who became the pioneer of electric streetcars, later observed, "Edison was perhaps nearer the verge of great electric railway possibilities than any other American. In the face of much adverse criticism he had developed the essentials of the low-internal-resistance dynamo with high-resistance field."[8]

During the spring of 1880 Edison's men laid some light-rail track in the shape of a *U* for about a third of a mile. Two "long-legged Mary Ann" dynamos supplied the power through the rails, one rail being positive and the other negative. Since these same generators also powered all the lights at Menlo Park, the electric railway could never operate at the same time as the lighting. For a locomotive John Kruesi built a third dynamo to serve as a motor and mounted it on a four-wheel truck. A friction pulley was used to transfer power from the motor to the wheels. On May 13 Charles Batchelor manned the throttle, and some of the boys climbed aboard the two cars, one roofed and the other canopied, for the first ride ever on an electric railway in America. Amid a flurry of cheers and waving hats the little train lurched forward and glided to the end of the line. On the return trip, Batchelor yanked the lever too violently and burst one of the friction pulleys. Somberly the boys pushed the train back to the waiting crowd of workers.[9]

During the next several months Edison and his men continued to tinker with the electric railway, making improvements and enjoying rides. Newspapers and scientific journals were quick to pick up the story. A few *Scientific American* men, invited for a ride, described it as a thrilling journey "at a breakneck rate up and down the grades, around sharp curves, over humps and bumps, at the rate of twenty five to thirty miles an hour." The little railway seemed almost like a hobby to Edison, yet it had great potential. His trip west in 1878 had given him the idea that small electric lines might serve as feeders from the vast grain fields, enabling farmers to move their crops to regular railroad stops. Although *Scientific American* enthused that "it is not entirely visionary to expect that our street and elevated railways may at no very distant day be successfully operated by electricity," most of the railroad men who flocked to Menlo Park to see the electric railway were unimpressed. Frank Thomson, head of the mighty Pennsylvania Railroad, examined the system, rode the train, and pronounced it impracticable. Electric systems, he declared, would never replace steam.[10]

Undaunted, Edison filed his first patent on the system in June. The following year he broached the idea to Henry Villard, who saw enough merit in it to underwrite construction of a three-mile road at Menlo Park to work out cost estimates and technical problems. The new test track and two new locomotives were ready for testing in 1882. By that time Edison was in New York immersed in getting his Pearl Street station ready. His assistants oversaw the tests, which went well. However, during 1883 Villard ran into deep financial trouble and had to abandon the project. Despite his own financial strains, a sympathetic Edison later refunded the $40,000 Villard had advanced for building the model system. Villard would not forget the kindness. Having lost his backer, and with the Electric Light Company showing no interest in the project, Edison let it go. In 1884 he said simply, "I could not go on with it. I had too many other things to attend to, especially in connection with electric lighting."[11]

But he did not let go entirely. Another inventor, Stephen D. Field, filed an interference suit against Edison's railway patents, which turned out to be poorly drawn. Busy with other projects, Edison agreed to join forces with Field in a new company. The Electric Railway Company of the United States, formed in May 1883 with Field as chief electrician, was capitalized at $2 million and held the patents of both inventors. Edison played no active role in the new firm, which shifted its focus from midwestern wheat fields to replacing steam on the New York elevated railways. Despite a number of demonstrations, Field never sold the electric system to elevated officials. His only installation consisted of a three-ton locomotive built for the Chicago Railway Exposition of June 1883. It did not help that Edison and Field despised each other and continued to wrangle over their patents. The new company limped along for seven years without doing any other business than the elevated railway experiments.[12]

Edison did not pursue the electric railway partly for lack of time but also because here, as in other projects, he envisioned the wrong future for it. He viewed it as a gradual replacement for the steam railroad rather than the urban horsecar. He also distrusted an overhead-trolley system with its high voltages as too dangerous. While he turned his attention elsewhere, other inventors pushed doggedly against the technical obstacles confronting them. In May 1881 Siemens unveiled an electrified line near Berlin. Running from the train station to a military academy a mile and a half away, it became the first commercial electric line in the world. That same year Belgian inventor E. Julien began work on an electric railway system. His efforts caught the attention of William Bracken, a New York lawyer, who in 1886 formed the Julien Electric Company to manufacture the Julien battery for electric streetcars. Julien also provided a thirty-passenger car for trials, but his system proved unreliable and gained few adoptions.[13]

Late in 1883 Edward M. Bentley and Walter H. Knight commenced work on an electric street railway in the Cleveland shops of Brush Electric. Their idea was to install a conduit between the rails, much like a cable car, and access power from it with a thin blade, called the plow, on the car. Their version used two Brush dynamos, one for power and another on the car as a motor. They formed the Bentley-Knight Electric Railway Company and managed to get a contract to install a short line in East Cleveland. Although the early reports chirped with optimism, the line was discontinued after a year because it proved so expensive and endured so many breakdowns. By then Bentley-Knight had severed ties with Brush and launched new experiments in the shops of the Rhode Island Locomotive Works in Providence.[14]

During 1886 the short test track attracted numerous executives from street railway companies, and the following year the company finally landed its first contracts: a three-mile line for Fulton Street in New York City, an electric railway for Woonsocket, Rhode Island, and another one for Allegheny City, Pennsylvania. This time Bentley-Knight went to Thomson-Houston for its generators, motors, and other equipment. The conduit system was installed on Fulton Street and on one mile of the Allegheny City system; the rest of the latter and the Woonsocket system used overhead conductors. Both approaches proved too costly and unreliable to maintain. Unable to develop a technically efficient and cost-effective system, Bentley-Knight struggled along only until 1889, when Thomson-Houston bought the company. The attempt to get around the perils of overhead wires proved a bust.[15]

A more promising system emerged from the work of another Belgian, Charles J. Van Depoele, who migrated to Detroit in 1869 at the age of twenty-three and began to manufacture furniture. As his firm prospered, Van Depoele turned his attention to electricity. He installed a laboratory in his factory, constructed some dynamos and arc lights, and in 1881 formed the Van Depoele Electric Light Company in Chicago with the financial backing of Aaron K. Stiles, a local capitalist. Once his arc-lighting system went into production, Van Depoele lost interest in it and turned to motors. By the winter of 1882–83 he was immersed in experiments on an electric railway, using a short track built near his lighting plant. He exhibited his creation at the Chicago Railway Exposition in 1883 and built a half-mile line at the Toronto Industrial Exposition the following summer to carry passengers to the exposition grounds. It worked well enough that in 1885 Van Depoele extended the track another mile, and his railway managed to haul between six thousand and ten thousand passengers daily to the exposition.[16]

That same fall Van Depoele got his first contract for a regular street railway system. His first car, in South Bend, Indiana, started operation on November 14, 1885. However, the company changed management and decided not to keep an electric system. Other orders came for temporary lines or for electric locomotives, but Van Depoele had yet to develop or install a complete system suitable for street railways. His first chance came in 1886 when he secured a contract for a two-car operation over two miles of track in Montgomery, Alabama. Using a single overhead trolley pole for a conductor, he soon won new contracts when the Montgomery line slashed operating costs by one third over the old mule-driven system. By the end of 1887 Van Depoele had installed a dozen street railways with a total of nearly one hundred cars and more than sixty miles of track.[17]

Van Depoele had seized the lead in an industry that had barely come into existence. The time had come to raise more capital and expand, but his partner, Aaron Stiles, balked. He did not believe the electric railway equipment business would amount to much, and in March 1888 he and Van Depoele sold the company to Thomson-Houston. Van Depoele went on the payroll of the new owners. Stiles at first accepted a royalty arrangement, but his pessimism over the industry's future prompted him to demand a cash settlement instead. He walked away with $300,000 in cash in lieu of an arrangement that ultimately would have earned him millions before the patents expired.[18]

Fate had almost but not quite crowned Van Depoele as the pioneer in electric street railway development. Although successful in reducing costs below those of animal power, his system contained several crippling drawbacks ranging from unwise placement of the motor to the wrong type of overhead trolley for a large urban system. Van Depoele might have perfected his system in time, but he became an employee of Thomson-Houston, which then took charge of its development. Instead the honor of pioneer went to yet another refugee from the Edison shops, Frank Sprague, whose remarkable achievement dwarfed that of Van Depoele.[19]

Born in 1857 in Milford, Connecticut, Sprague lost his mother at the age of eight and was sent along with his younger brother to live with an aunt in North Adams, Massachusetts. He revealed a gift for mathematics at an early age but lacked the means to attend college. His school principal advised him to take the entrance examination for West Point, which would give him a fine education in engineering. Sprague went dutifully to Springfield to take the competitive examination only to discover that it was for the Naval Academy, not West Point. He took the test anyway and entered the academy in 1874. There he excelled in mathematics, chemistry, and physics. It proved a wise move because Annapolis offered more advanced training in electrical engineering than

West Point or most universities. The navy collected, tested, and compared more dynamos than any other American organization because of its interest in the potential of electricity.[20]

After graduating seventh in a class of fifty in 1878, Sprague spent two years at sea with the Asiatic Squadron. During that time he filled a notebook with drawings and plans for fifty-seven inventions, including motors. While at Annapolis he had visited the Philadelphia Centennial Exhibition of 1876, which excited his interest in electricity. He followed the work of Edison and other electricians as closely as he could and was anxious to get ashore and pursue his own experiments. In the spring of 1880 he passed his ensign's examination and, during a short leave, performed experimental work at the Stevens Institute and the Brooklyn Navy Yard. He then landed an assignment to the Newport Torpedo Station, where he worked with Moses Farmer and in his own research developed an inverted dynamo.[21]

In the spring of 1882 Sprague managed to visit the Crystal Palace Electrical Exhibition in London and even served as secretary of an award jury. This duty provided an excellent opportunity to examine and test the devices on display. He overstayed his leave by six months but avoided court-martial by writing a comprehensive account of all the tests he had performed while in England along with his own recommendations. The navy was impressed enough to publish part of the report. While riding the London underground steam railroad, he gagged on its outpouring of smoke, dirt, and gas in the tunnel and wondered whether an electric motor might not be more suitable. He began working on a plan for an underground electric railroad. Part of his report also included descriptions of his experiments with a gas engine, which was then extremely new and mysterious. Altogether the London experience proved to be an education in itself for Sprague.[22]

It also became that most tantalizing of experiences, a doorway to his future. At the exhibition he met Edward H. Johnson, who was there to oversee the Edison exhibit. Impressed by the young man's knowledge of mathematics and electricity, Johnson suggested he leave the navy and go to work for Edison. By the time Sprague joined Edison's central station planning department in May 1883, he had piled up an impressive résumé of training and experience even though he was only twenty-five years old. Scarcely had he arrived in the planning department when he used his mathematical skills to show how the painstaking model method for planning central stations was not only tedious and expensive but also inaccurate. Hermann Claudius, the old German professor, was sent packing, and Sprague's method was adopted. Edison then ordered him to make a study of Hopkinson's three-wire system and sent him to help install the company's first three-wire power station in Sunbury, Pennsylvania.[23]

Ordered to start up the dynamo, an Armington & Sims that he had never seen before, Sprague burned out the babbitt metal in the bearings and spent the night fixing the machine. When Edison, Insull, and the chief engineer arrived the next day and demanded to know what happened, Sprague admitted, "there followed remarks which would not look well in print." But the plant started up, and Edison put Sprague in charge of the first underground three-wire installation at Brockton, Massachusetts. After completing that station, Sprague stayed on as operating engineer. During his free time he experimented with electric motors and devised a railway motor. So well did his experiments go that in the spring of 1884 he resigned his post with Edison and set out on his own. The parting was amicable; Sprague had developed a consuming passion for motors, but Edison was absorbed in lighting projects. All summer long Sprague worked at improving his motor, giving it new features that made it suitable for industrial use. In the fall he displayed several models at the Philadelphia Electrical Exhibition, where they received a warm reception. Edison called Sprague's inventions "the only true motor; the others are but dynamos turned into motors."[24]

Although Edison was too absorbed in his other projects to take an interest in the motor, Edward Johnson liked what he saw and helped Sprague form his own company, Sprague Electric Railway and Motor Company. Johnson agreed to pay Sprague's expenses in return for a share of the profits. At first it was clearly a shoestring operation. The manufacturing was contracted out to Edison Machine Works, and independent agents handled the marketing. For the first two years Sprague concentrated on industrial motors to serve Edison lighting plants and then clothing factories, mills, and printing presses. Where rival firms offered only small motors, Sprague designed them as large as 15 horsepower. Moreover, his motors worked smoothly with Edison power plants. By the end of 1886 he had sold 250 motors of varying sizes. In October he decided to lease a factory in New York and begin manufacturing motors for special purposes, leaving the production of standard motors to Edison Machine Works.[25]

Customers put their motors to a variety of uses. One Boston sugar refinery unwittingly provided a glimpse of the future by using its motor for a seven-hundred-foot railway carrying raw sugar from the waterfront to the refinery. Sprague himself devoted most of his time to working on a motor suitable for an electric railway; one of the models sent to the Philadelphia exhibition was intended for that use. He had in mind the New York elevated railways with their steam locomotives that spewed a constant stream of smoke, dirt, noise, and cinders on the populace below. Early in 1886 he arranged some trial tests for company officials. Jay Gould, who controlled Manhattan Elevated, was among those who came to witness the demonstration. After impressing the visitors with several ordinary car movements, Sprague abruptly reversed the controls. The rush of current "blew the safety-catch into a small volcano," prompting Gould to depart in haste and leaving him with an unshakable prejudice against electric engines.[26]

Despite this mishap, two Edison officials were impressed enough to invest more than $50,000 in Sprague's company. Another demonstration in April went off without a hitch before an audience of railway men and bankers. But none of Manhattan Elevated's directors or stockholders showed the slightest interest, even though the city's four lines groaned beneath the weight of their traffic and could not employ heavier locomotives without strengthening the track and superstructure. Their indifference led Sprague to turn his attention to street railways. The motor he had devised for the elevated cars could serve a street version with some modifications. His opportunity arrived in the form of an offer that common sense should have told him to refuse.[27]

IN THE SPRING of 1887 a group of New York investors approached Sprague with an offer to build a street railway system for Richmond, Virginia. They asked him to supply forty cars with two motors and all accessory equipment, a complete overhead current system for twelve miles of track, and an electric power plant capable of generating at least 375 horsepower. At any given time thirty of the forty cars were to be in service, sometimes on grades that reached 8 percent. No track had yet been laid, nor was the exact route completely settled, but it traversed steep grades, sharp curves, and clay streets. Sprague had never been to Richmond, yet he agreed to provide the equipment, bear all risk of failure, and complete the job in ninety days! Only after the system had run satisfactorily for sixty days would he receive $110,000. Sprague signed the contract in May 1887.[28]

To any sane observer, Sprague seemed to be embracing failure. "When the Richmond contract was signed," he recalled a decade later, "we had only a blueprint of a machine and some rough experimental apparatus. The hundred and one details that were essential to success were as yet undetermined. Fortunately for the future of electric railways, the difficulties ahead could not be foreseen or the contract would not have been signed." To make matters worse, Sprague fell ill with typhoid fever and had to leave the work in the hands of his two capable but inexperienced assistants for nine weeks. When he finally got to Richmond early in the fall, he found the newly laid track in wretched shape. It was light, crudely joined, unevenly laid, and insecurely attached, with twenty-nine curves, many of them sharp, and grades reaching 10 percent. "It was built for profit, not for permanence," said Sprague tersely. The car house was a joke—two rough-covered sheds squatting on an open lot.[29]

The task facing Sprague seemed impossible. All he had to do was gear motors to run independently on each axle; find a way to carry them underneath the car, control them from either end, and, despite their exposure to dirt and moisture, run them with fixed brushes in both directions and operate them without rheostats; utilize a 450-volt constant-potential circuit with track and ground return under conditions that electricians deemed impractical; operate a multiple-arc circuit on a large scale labeled insane by electricians; devise a small-sized overhead trolley wire supplied at intervals by a system of main conductors and feeders; somehow carry current via an underside contact onto the car; and mount grades beyond the contract's specification that would stymie a steam locomotive, let alone a self-propelled car. And do it all in the short space of a wet Richmond fall and winter. At the time the entire United States barely had sixty electric streetcars, no system had more than ten or twelve, and none had to climb the kind of grades presented in Richmond. As Harold I. Sharlin observed, "In 1887 there were no testing laboratories and no engineering experience . . . The electric street railway could not be tested anywhere but on the streets."[30]

Everything depended on whether the cars could climb those grades. During a tense conference Edward Johnson, who served as president of Sprague Electric Railway, ended discussion by saying, "Guess the best thing to do is to find out whether the car can get up the grade at all." While an enthusiastic crowd watched one evening, a self-propelled car went up and down several hills and navigated the curves smartly before approaching the highest point on the line. Aware that the motors were hot, Sprague paused to allow them to cool. When he started them up again, the car bucked and then stopped for good. Thinking quickly, Sprague told his assistant loudly that the circuits had a problem and ordered him to fetch the appropriate instruments to locate it. Then he switched off the light and lay down on the car seat to wait. As he hoped, the crowd finally grew tired of waiting for the instruments to arrive and gradually dispersed. After they had gone, the proper instruments arrived in the form of four powerful mules that hauled the car back to its stable.[31]

The problem lay with the armature and field coils, which required better gearing. From the Brown & Sharpe machine tool company in Rhode Island Sprague obtained new tools and jigs to revise the gearing, but new problems cropped up as fast as he solved existing ones. Sprague understood the pioneering nature of his work. All his predecessors had created cars with makeshift rigging and components that could not stand the rigors of constant use on a large system. His system had to be both reliable and cheap enough to earn a profit on daily running. By January it had become apparent that he could not meet the ninety-day deadline, and Sprague was forced to make some financial concessions. The payment was reduced from $110,000 to $90,000, with half of it to be in bonds of the streetcar company rather than cash.[32]

Early in February 1888 Sprague's line opened for regular service under a drizzling rain. To meet the contract he had to keep it running for sixty days, which meant that the real work had begun in earnest. "Cars would suddenly stop in the street, and refuse to move under any conditions, for the new gears had a freak of locking," Sprague recalled. An Irish mechanic who had worked with Sprague for years figured out that the gears lacked proper oiling. Still the problems kept coming. The constant stopping and starting of the cars put a strain on the motors that never affected stationary types. The brushes wore down rapidly and damaged the commutators. Cars had to be inspected after every trip, the brushes replaced frequently, and the commutators machined. A steady stream of motors and their parts flowed from Sprague's factory in New York and the Edison Machine Works in Schenectady to meet the demand.[33]

"We were using about nine dollars' worth of brass per day for brushes," lamented Sprague; "not a half trip was made without inspection, and generally a change of brushes: but the road must be kept in operation somehow while other experiments were being made." The armatures kept coming unwound; their replacements too were rushed by express from the factory. Parts kept having to be exchanged on cars that had not been designed "on a strictly interchangeable plan." To worsen matters, the rails often spread or sank into the mud, leading to derailments, especially on curves that lacked any rail guards. Every car had to be equipped with a scaling ladder and a large timber pole to jockey the car back onto the track. Thunderstorms, snow, and sleet bedeviled the operation; few jobs were more unpleasant than climbing atop the car with a broom to sweep ice or sleet off the overhead trolley line.[34]

Gradually, however, Sprague gained ground in the war of attrition. More than fifty different trolley poles were tested before finding one that worked suitably. Tracks were straightened, rails improved, and mudholes filled. Sections of rail along with a special electric connector enabled a derailed car to get back on track under its own power. Later carbon brushes replaced metallic ones and solved the brush and commutator problems. The armatures and field coils of the motors were redesigned and rewound. By May 4 Sprague finally reached the goal of thirty cars in operation; by the fall his line was providing reliable and economical service. This heroic effort rewarded Sprague's company with a financial loss of about $75,000, but the experience had been worth it. Sprague had done more than create the nation's first streetcar system; he had fashioned the model for what he himself called "the subsequent unparalleled growth of a great industry."[35]

WHEN SPRAGUE SIGNED his Richmond contract in 1887, the entire nation had only 8 electric street railways running on 35 miles of track. Five of those lines used Van Depoele's equipment. When the line opened for business, 21 cities had cars with electric motors. By the fall of 1889, only eighteen months after Richmond officials declared that Sprague had fulfilled his contract, the country had 180 electric railways either operating or under construction with 1,260 miles of track and 2,000 cars. A year later 200 electric railways were operating, and the New York Times predicted that "both horse power and cables are soon to be displaced on nearly all American street railways by electric force." By the new century electric street railways dominated urban transport; in 1902 they comprised 94 percent of all street railway mileage, with cable cars and horse trolleys reduced to a mere 1 percent each. In most major cities the traction companies became far and away the largest users of electric power. Sprague had set in motion a revolution in the electric industry, and all the major players scrambled to climb aboard.[36]

Thomson-Houston had already eased its way into the field. In the spring of 1887 it contracted to provide railway motors and generators for Bentley-Knight. The experience of building and installing this equipment convinced Charles Coffin that the company should get into the electric railway business. During the winter of 1888 he remarked that he thought the electric illumination business was about played out, and he wanted to find a new growth area. Uncertain whether the company's charter permitted such a move, he succeeded in getting it amended to include electric railway work. Here too debate arose as to whether Thomson-Houston should develop its own system or buy competing firms. Thomson favored the first approach; Coffin disagreed, arguing that too much time had been lost to start from scratch. In February 1888 Coffin, on Thomson's recommendation, persuaded Van Depoele to sell his patents to Thomson-Houston and join the company. Van Depoele was amenable because he saw in the transaction a chance to continue his work with strong financial backing and a support staff. Eighteen months later Coffin scooped up Bentley-Knight as well.[37]

These acquisitions thrust Thomson-Houston to the forefront as the major competitor of Sprague's company. Both firms landed important contracts with the West End Street Railway Company of Boston after its president, Henry M. Whitney, made the bold decision that only an electric system could solve Boston's chronic transportation muddle. Whitney's company used eight thousand horses to pull 1,700 cars over 212 miles of track throughout the city. He made a careful study of electric traction and visited Richmond three times to observe Sprague's progress. Aware that West End's general manager favored cable because he doubted the ability of electricity to move bunched cars, Sprague arranged a surprise demonstration for his visitors. He lined twenty-two cars up on the track, raised the pressure to 500 volts, and watched in relief as every one started forward in turn. The spectacle convinced Whitney that the answer to Boston's commuter problem lay with electricity.[38]

George Westinghouse did not sit idle while his competitors marched forward. As early as 1886 the invention of a Pittsburgh dentist had drawn his attention to the electric railway and its potential. The following year he formed a new company, saying it would "actively push this important industry with the same energy that has characterized its work in electric lighting." His Overhead Conductor Electric Railway Company experimented in the field but never did any business. From its work he concluded that DC traction systems harbored two major defects: The motors depended on brushes and commutators, and the voltage of the transmission lines had to remain as low as that of the conductors at the track. Tesla's new motor could answer both objections if it could be adapted to streetcar use. Unlike his rivals, Westinghouse started with idea of creating a traction system based on AC. During 1888 and 1889 his engineers toiled valiantly at the task of adapting Tesla's motor for streetcars.[39]

By the fall of 1889 it became painfully clear that Tesla's motor could not be made to serve street railway cars. It was a constant-speed motor, which meant it could run efficiently only at one speed, while streetcars required motors that could operate at widely varying speeds. Moreover, it could not exert enough force at start-up to power a vehicle that constantly stopped and started. Reluctantly Westinghouse conceded that he needed a DC traction system to compete and assigned a team of engineers to develop a DC motor for that purpose. To expedite their work, he lured some engineers from Sprague's firm to work for Westinghouse. Not until 1890 did Albert Schmid produce the first Westinghouse railway motor. It was a bipolar, double-reduction-gear type with carbon brushes similar to those used by Sprague, Thomson-Houston, and others.[40]

As the name suggests, the standard railway motor required two sets of gears that were expensive to maintain and extremely noisy. Its location underneath the car rendered the motor hard to get at, making repairs expensive. Yet the motor was completely open from below, exposing it to dirt, water, and debris thrown up from the street. Schmid's version enclosed the gears in grease to reduce the noise and wear; it also covered the motor on the underside and at one end to keep it cleaner. These alterations improved the motor but did not solve the problems posed by the two-gear system. Later Westinghouse tried to eliminate all obstacles with a gearless motor, but it had too many other drawbacks to be commercially viable.[41]

As usual Westinghouse kept his cards close to the vest. "Perhaps no electrical system ever was so thoroughly exploited with more of secrecy than the Westinghouse electric railway," wrote one observer. "Its managers have kept its details secret even from purchasers, and yet without models or even drawings have taken contract after contract." His first system, in Lansing, Michigan, revealed no radical innovations of design or machinery. Westinghouse knew full well that he had come late to the game and was playing catch-up. By the time he installed the Lansing system in 1890, events threatened once again to overtake him.[42]

THOMAS A. EDISON was not even in the game, at least directly. He had done no research in the electric railway field since 1883, and his relationship with Stephen Field deteriorated steadily. But a curious twist of fate brought the subject back to him. In his never-ending struggle to support the new laboratory and raise capital for his work, he had agreed to let his friend Henry Villard put together a plan to consolidate the Edison companies. As he later admitted to Villard, "When I sold out, one of the greatest inducements was the sum of cash received, which I thought I could always have on hand, so as to free my mind from financial stress, and thus enable me to go ahead in the technical field." His gloom over raising money to support the laboratory reached the point where he said to Insull one day, "This looks pretty bad. I think I could go back and earn my living as a telegraph operator. Do you think, Sammy, that you could earn your living as a stenographer?" A merger would also eliminate the haggling between Edison and the manufacturing companies over their financial relationships with the laboratory.[43]

The negotiations soon unmasked the strains that had developed in the ranks. Edward Johnson, who found himself in the awkward position of heading both Edison Electric Light and Sprague Electric Railway, wanted to bring the latter company into the merger as well. Edison balked, partly because he did not like Sprague's caustic personality. Neither did Samuel Insull, who could be pretty caustic himself, but Insull faced an interesting dilemma. By 1889 the Edison Machine Works, over which he presided, got fully two thirds of its orders for new equipment from Sprague Electric Railway. At one point Johnson accused Edison of "knifing" him during an interview with Villard, prompting the inventor to fire off a bitter letter suggesting that Johnson had not done justice to the Edison electric railway. "You know," he fumed, "that your bringing Mr. Sprague in the family has been a galling thorn in the side of all the boys but they kept still because you were their friend, and as I told you two years ago . . . it was only a matter of time when it would break up old associations which it has about done."[44]

Sigmund Bergmann, who with Edwin Johnson had formed Bergmann & Company to manufacture light fixtures for Edison, added his own long and testy letter to the growing literature of discontent. His complaints had mostly to do with Edison's handling of the new phonograph and certain accounting issues. In particular he resented the enlarged role played by Insull in the accounting done for the merger. An exasperated Edison complained that Insull's "plan of division don't seem to work with some of my friends." Through these troubled waters Villard cruised with tact and caution while pushing the merger forward. He enlisted the support of Siemens & Halske and the General Electric Company of Berlin for the project along with his longtime backer, Deutsche Bank. Drexel, Morgan signed on as well, and one of its brightest partners, Charles H. Coster, proved invaluable in helping to design a plan of consolidation. However, as Edison had insisted, Drexel, Morgan took only a small share of the stock in the new company even though it managed the $3.63 million offering. By far the largest share went to Deutsche Bank, which took 62.2 percent. However, J. P. Morgan himself had become the leading stockholder of the Edison Electric Light Company.[45]

On April 24, 1889, the new Edison General Electric Company incorporated in New Jersey, which had just passed its landmark law allowing corporations to own and manage companies in other states. Edison Electric Light, Edison Lamp, Edison Machine Works, Bergmann & Company, and Sprague Electric Railway were all folded into the new corporation. Villard, with the backing of Deutsche Bank, assumed the presidency, but Samuel Insull became vice president and took charge of operations. Johnson resigned as president of the Electric Light Company and remained in charge of Sprague Electric Railway. Bergmann stayed on to manage his firm until 1891, when he left to organize a new manufacturing company in Germany. Edison himself held no official position other than as member of the board and contract inventor, but the new company needed both his technical expertise and the prestige of his name. It would take care to defer to him on policy decisions. The company moved into handsome quarters on the eighth floor at the northeast corner of Wall and Williams streets.[46]

Acquisition of the Sprague company made Edison General Electric a major player in the traction industry. It could sell streetcar systems, manufacture the equipment needed, and provide power stations for operating them. Like Thomson-Houston, it had expanded the reach of electrical manufacturing beyond illumination to an entirely separate function, one that used enormous amounts of electric power. Westinghouse too was struggling to expand its business in the same direction. To meet this growing competition, Villard asked Edison in 1890 to develop an electric railway system suitable for cities that could not use overhead trolley lines. Edison revived the method used at Menlo Park of using the rails as conductors, but his low-voltage system failed to compete with trolleys and later subways.[47]

As for Sprague, he signed on as consulting engineer with the new company but soon came into conflict with Insull and other top managers over several technical issues. He resigned early in 1890 and turned his attention to designing electric elevators. Two years later he formed the Sprague Electric Elevator Company and developed a thriving business installing elevators in the tall buildings that were fast becoming the rage. In devising a control system for his elevators, Sprague realized that the same principles could be applied to electric railway cars. From this insight flowed the creation of the multiple-unit control scheme that enabled each car to be operated by itself or in combination with other cars and in any sequence. Each car, or an entire train of them, could be operated from either end, eliminating the need for turnaround tracks, and would have its own power, heat, lighting, and braking system. Such independent units needed no locomotives, which made them ideal for elevated railways.[48]

As with his streetcar, Sprague needed a venue to demonstrate his new system. His opportunity came in 1897 when he signed a contract with the South Side Elevated Railway Company of Chicago to install a six-car system. It was Richmond all over again. Sprague had no complete plan finalized, and his idea met with responses ranging from skepticism to ridicule. Against long odds and despite numerous setbacks, he succeeded in creating the system and pushing urban travel yet another giant step forward. In August 1897 *Electrical World* honored his achievement:

> It is now more than ten years ago that Mr. Sprague first proposed the operation of multiple car trains, each unit of which was separately propelled by its own motor, each forming in fact a distinct and separate self-contained motor car which could be operated as a single unit or any number of which could be combined in a multiple unit train, the various units of which were all synchronously controlled from one of their number.[49]

Twice Sprague had become a pioneer in urban transport, and in the process he shoved the commercial significance of electric motive power down new paths of development. Like every other important inventor and entrepreneur, he thrived on resistance and conflict in pursuing his goals. He insisted on following his own vision and his own path to it. That, at least, was a trait he shared with Edison, Westinghouse, and other of his peers.

CHAPTER 13

COMPETITION AND ELECTROCUTION

The struggle for control of the electric light and power business has never been exceeded in bitterness by any of the historical commercial controversies of a former day. Thousands of persons have large pecuniary interests at stake, and, as might be expected, many of them view this great subject solely from the stand-point of self-interest.

—GEORGE WESTINGHOUSE JR. (1889) [1]

COMPETITION WAS THE FUEL THAT drove American business during the formative years of the late nineteenth century. It was the heart and soul of the free-market system, yet most businessmen regarded it as a frustrating stew of blessing and curse. They relished the challenge of pitting their own wits and products against those of rivals while hating the damage it caused, the costs it incurred, and the maddening uncertainties it bred. The central paradox of business competition was that it drove the combatants to eliminate it as quickly as they could. The mythical free market was for them a maze of uncertainties that hindered their efforts to impose order on a given industry. Sooner or later the efforts of business leaders to replace these uncertainties with what business historian Alfred D. Chandler Jr. called the "visible hand" of managerial control ran headlong into the obstacles posed by competition. [2]

To meet the challenge of competition, businessmen could resort to several tactics. They could reduce their costs below those of rivals, sell their products more cheaply, create products of superior quality, offer a more varied line of goods, make their products more attractive to the buyer, or provide better terms for purchase. They could also seek to weaken rivals by waging a negative campaign against their products in hopes of driving buyers from them. They could enter into cooperative or collusive agreements with competitors or simply eliminate them through merger or acquisition. Over time rival firms in virtually every industry utilized most if not all of these tactics.

Competition in the electrical industry revolved chiefly around dueling technologies. Firms created superior products by developing or acquiring new technologies and improving them while trying to deny their opponents access to them. Thus arose the interminable legal battles over patents that consumed so much time, energy, and money. Three basic issues emerged as paramount in the competitive wars of these early years: cost, reliability, and safety. To most people electricity remained a mysterious and therefore dangerous technology. It promised great advances in everyday living, yet it could also kill without warning. Edison understood these mixed feelings and had always promoted direct current as the only safe form of electricity. Thomson too had always emphasized safety and railed against central stations that economized by not purchasing the safety devices he had devised for his systems. It outraged him that some operators resorted to such practices as putting two dynamos on the same line even though its insulation was not built to handle the higher voltage. "I do not believe in this kind of economy," he wrote Coffin. "Rather it would be better not to have the business than incur the risks which are thus involved."[3]

Edison had long been concerned about the safety issue. He boasted that "there is no danger to life, health, or person, in the current generated by any of the Edison dynamos" and claimed that "the wires at any part of the system, and even the poles of the generator itself, may be grasped by the naked hand without the slightest effect." Safety was a primary reason for his rejection of any AC system, with its higher voltages. He followed the development of Westinghouse's system and, in a long 1886 memo, assured Edward Johnson that it posed no business threat. The AC dynamos were not as efficient, the transformer wasted energy, Westinghouse had no meter, and he still lacked a motor that could compete with a DC motor. Moreover, it was a dangerous system. "Just as certain as death," he predicted, "Westinghouse will kill a customer within 6 months after he puts in a system of any size. He has got a new thing, and it will require a great deal of experimenting to get it working practically. It will never be free from danger."[4]

Like Thomson, Edison relished holding what he conceived to be the superior moral and ethical ground, but competitive forces were fast eroding that position. The harsh reality was that the Edison company continued to lose business to Thomson-Houston and Westinghouse, both of which had developed AC systems more suitable to the vast part of America that was not densely urban. By 1887 the Edison company had installed 121 central stations, but Westinghouse already had 68 after being in the business only a year. By 1891 Thomson-Houston could boast of three times as many central stations as

Edison and more than twice as many as Westinghouse. The latter's AC system not only transmitted over much longer distances but was cheaper as well. After losing contracts for Portland, Oregon, and for Tacoma, Washington, Edison's agent in the West let his displeasure be known by asking, "Are we going to sit still and be called old fashioned fossils, etc. and let the other fellows get a lot of the very best paying business?"[5]

Other Edison men voiced similar complaints. In New Orleans, where Westinghouse had installed a large AC plant after Edison had claimed the city with a DC station of his own, the Edison representative howled that "they are robbing our business pretty badly." The chief engineer complained that "we could get at least 2000 or 3000 more lights in at very profitable rates in a nice residential district from one to two miles away if only we had an alternating or continuous current system to support our regular three-wire system. It is idle to scoff at the Westinghouse people—they are hard and persistent workers." Edward Johnson, stung by Edison's charge that his business methods were flawed, pulled no punches in responding. "The three-wire system is the only thing you have given me to work with," he growled. "Give me a system that will *enable* me to compete [or] accept the fact that we will do no small town business, or even much headway in cities of minor size. A development of our system can only come from you. Is it fair to hold us accountable for the absence of this development?" Some Edison agents voted with their feet by leaving his employ and joining Westinghouse.[6]

To make matters worse for Edison, the price of copper began an ungodly rise, thanks to the maneuvers of a French businessman with the unlikely name of Hyacinthe Secrétan to corner the commodity. In March 1888 an engineering journal reported that "all the electrolitic [sic] copper in this country is now firmly in the grasp of the syndicate. There appears, in fact, nothing to prevent prices from being advanced to any figure the syndicate may wish." This threatened to undermine Edison's efforts to get new business. One bid for a Minneapolis power station reckoned the cost of the copper needed at $52,000. Every rise of a penny in the price of copper added $3,056 to that figure; a three-cent hike sent the total cost spiraling to $61,168. Since Westinghouse's AC system used only about a third as much copper as the Edison system, he gained a clear advantage as the price of copper continued to climb.[7]

While Edison grappled with these difficulties, opportunity knocked at his door in the guise of Dr. Alfred P. Southwick, a Buffalo dentist and member of a New York state commission charged with finding some more humane way to execute criminals. From his inquiry flowed a chain of events that turned an increasingly bitter but not untypical business competition into what amounted to theater of the macabre. The newspapers, always eager for a show, celebrated it as the "war of the currents." It was in fact three different battles that got entangled in each other through a bizarre sequence of events. The most obvious clash was the competitive fight between the major electrical companies. A second issue concerned the debate over the use of electricity for executions. The third involved the perils inherent in stringing electrical wires overhead in crowded cities, especially New York, and the growing public fear over the danger of accidental electrocution.

The American debate over capital punishment had roots deep into the nineteenth century, but opponents made little headway in eliminating it and turned instead to alleviating the suffering of the condemned. Hanging, the most common form of execution, was denounced as barbaric because improperly placed nooses resulted in slow, agonizing strangulation if the drop was too short or grisly decapitation if it was too long. In 1876 *Scientific American* decried "the disadvantages of hanging as a mode of capital punishment" and recommended "electric shock" as the "most certain and painless death." That same year Maine abolished capital punishment after a botched hanging outraged public opinion. As the century waned, public attitudes toward pain were also changing. For centuries it had been regarded as an inescapable part of life. By the late nineteenth century, however, compassion had begun to replace indifference and resignation. William James spoke of a "moral transformation" in which "we no longer think that we are called on to face physical pain with equanimity."[8]

The development of anesthesia and the unprecedented rapidity of its adoption by the medical profession did much to speed this growing aversion to pain and with it the belief that it could and should be alleviated. Not only people but animals benefited from this evolving attitude. The Society for the Prevention of Cruelty to Animals (SPCA) was founded in 1866 and won the allegiance of numerous prominent people in its mission not only to protect animals from abuse but to educate the public that it was wrong to mistreat animals. From these movements arose a growing desire to find more humane ways of killing both doomed criminals and unwanted animals. The Pennsylvania SPCA pioneered in this effort by building in 1874 a brick room to kill dogs with carbon monoxide instead of the usual shooting or drowning. A Philadelphia physician then suggested that a similar method be used on condemned criminals.[9]

In 1886 Governor David B. Hill of New York formed a three-man commission to seek a more humane form of execution. As chairman he appointed Elbridge T. Gerry, a distinguished philanthropist and descendant of a signer of the Declaration of Independence. Gerry served as counsel to the American SPCA and had in 1874 founded the first Society for the Prevention of Cruelty to Children. The other members were Alfred Southwick and Matthew Hale, an Albany lawyer. They plunged at once into diligent research on every method of execution ever used. For some time Southwick had been keenly interested in the use of electricity as a means for execution. In Buffalo he began collecting stray dogs and killing them with electric shocks; in 1887 the local SPCA engaged him to dispatch its overflow of stray dogs. This work convinced Southwick that electric shock was the most humane form of execution because it was fast and virtually painless.[10]

The commission uncovered thirty-four forms of execution in the historical record, of which only four seemed practical: the guillotine, the garrote, the firing squad, and the gallows. None of these appealed to them because they were cruel, mutilated the body, and sometimes produced uncertain results. In its search for more options the commission came up with two newer possibilities: death by electricity or lethal injection. The latter raised several objections. Use of a hypodermic needle was a rather new procedure, and physicians feared that employing it for execution would prejudice public opinion against any use of it. And since a doctor was required to deliver the fatal dose, it would undercut his oath to preserve life. That left electricity, for which Southwick had been the most persistent advocate. He convinced Hale, but Gerry held out for injection.[11]

Determined to buttress his position, Southwick wrote Edison on November 8, 1887, asking him to support the use of "electricity as an agent to supplant the gallows." He also asked Edison to recommend the best current strength for an execution and the most suitable equipment for the task. Edison declined, saying he would rather "join heartily in an effort to abolish capital punishment." Undaunted, Southwick wrote again on December 8 and added an appeal to rid civilization of the grisly practice of hanging. "The rope is a relic of barbarism and should be relegated to the past," he declared. He also stroked Edison's considerable ego by stressing that "the reputation you have as an electrician would be invaluable in persuading the legislature to abolish the gallows and substitute electricity." On this point he proved to be entirely correct.[12]

This appeal changed Edison's mind. On December 19 he wrote Southwick agreeing to endorse death by electrocution as more humane. With tongue firmly in cheek he also urged the use of alternating current generators "manufactured principally in this country by Geo. Westinghouse . . . The passage of the current from these machines through the human body even by the slightest contacts, produces instantaneous death." He referred Southwick to New Orleans, where in recent months "two men have been killed and others injured by this quality of current." This letter induced Gerry to join the other commissioners in recommending electricity. In January 1888 he delivered the commission's report to the legislature along with the draft of a new capital punishment bill. The report included an extract from Edison's letter.[13]

Here was an ideal marriage of humanitarianism and self-interest. Why Edison decided to aid Southwick and to use his request as a platform to attack Westinghouse remains an intriguing mystery. There is no doubt that he sincerely desired to find a more humane form of execution and that he ardently believed in the lethal nature of alternating current. These two convictions dovetailed conveniently in his mind, and they might have remained unconnected without the stimulus of Southwick's invitation, which arrived at a critical juncture in Edison's career. He was beset with immediate woes such as the loss of business to his major rivals, the rising price of copper, and the ongoing legal war over his lamp patent. As long as the validity of the patent remained in doubt, all of Edison's rivals clung to the hope that it might be struck down. To make matters worse, the internal disputes among the Edison companies over support of the laboratory continued to strain relations between Edison and his men.

And there was something more. Edison was not only losing business to Westinghouse; he was losing ground as an inventor. The Wizard's great claim to electrical fame was that he had created not merely a lamp but an entire system, and he could not help but take any challenge to it personally. He expected competitors to imitate his creation in one form or another, but Westinghouse had chosen a different route entirely. He had created a truly alternative system, one that still had problems to overcome but that promised to solve the one great defect in Edison's system: the ability to transmit power efficiently over long distances. Nor did it help that so many of Edison's own men nagged at him to find a solution to this problem. Their impatience could be seen early in 1888 when Edward Johnson and the Electric Light Company published an eighty-four-page tract entitled *A Warning*.[14]

This curious document amounted to a broadside against all Edison company rivals, especially Westinghouse. Organized around five "cautions" and twenty-one appendices, it took aim at several targets. The first part vigorously defended the Edison lamp patent and admonished those who hoped it would be overturned in court not to count on that outcome. Investors were warned that any guarantees issued by rival companies against patent infringement liability were worthless and that other systems were limited in their use. "Any 'system' which is available only for the single purpose of lighting," it stressed, "is necessarily so heavily handicapped as against a system which is capable of universal application, as to be practically out of the race." A direct comparison of the Edison and Westinghouse systems predictably found the latter to be more expensive as well as limited. Another section railed righteously against what it deemed the shady capitalization of Westinghouse's company compared with the honest financing of the Edison firm.[15]

The most inflammatory portions portrayed the Westinghouse AC system as not only inadequate but lethal. Appendix J cited a string of press accounts of deaths and injuries from electrical shock, while Appendix K spelled out "Some Dangers of the Alternating System." Yet another appendix highlighted the "Recognized Dangers of High Electrical Pressure," contrasting it with the safety of the Edison system. The tract claimed that no fatality had ever been caused by the Edison system. "It is a matter of fact," it emphasized, "that any system employing high pressure, i.e. 500 to 2,000 units [volts] jeopardizes life . . . Any *interruption* of the flow of the current adds to its destructive property, whilst its complete reversal, as in the Alternating (Westinghouse) system increases this destructiveness enormously." From this reasoning flowed the conclusion that "high pressure, particularly if accompanied by rapid alterations, is not destined to assume any permanent position. It would be legislated out of existence in a very brief period even if it did not previously die a natural death."[16]

Clearly for the Edison company this was a consummation devoutly to be wished. Whatever Edison's own role in this publication, and there is no evidence concerning it, the pamphlet revealed not only a growing animosity toward Westinghouse and his company but even more the desperation of the Edison company in the face of competitors that threatened to pass it by. Westinghouse was squeezing the Edison company on two fronts: the patent suits and his competing AC system. In its annual report for 1887 Edison Electric Light pointed out the drawbacks of an alternating-current system: dangerously high voltage, inefficient transformers, severe lamp breakage, and generators that could not be operated in parallel. *Warning* continued the attack on both fronts, paying as much attention to the patent issue as to rival systems. The very aggressiveness of its attack betrayed the defensive nature of its origins.[17]

The *Warning* pamphlet might simply have vanished into the maw of corporate warfare had not a sequence of events come together during the next few months. On the night of March 12, 1888, New York City was buried beneath a driving rainstorm that turned into sleet and then into twenty-two inches of snow. The Great Blizzard of 1888 piled up not only snow but thick tangles of telegraph, telephone, and electric wires as well. This massive mess prompted some newspapers to renew an old call to bury all the wires underground. On April 15 a boy grabbed hold of a broken telegraph wire dangling overhead and was electrocuted. Two weeks later a bookkeeper touched an overhead light and died instantly. Less than a month later a lineman neglected to wear his heavy gloves while working and suffered the same fate. Suddenly the press and public alike awoke to the danger electricity—especially high voltage—posed of "death by wire." As a Westinghouse biographer noted, "All overhead cables suddenly leaped into prominence not only as eyesores but as a public peril."[18]

The danger was real. Amid the thick jungle of overhead telegraph, telephone, and electric wires lay some arc light cables carrying as much as 6,000 volts delivered by Manhattan's numerous electric companies. Where Edison's wires were all buried in conduits under the streets, this mishmash of wires was strung on poles bearing a dozen or more cross-arms and several hundred wires of all kinds. Some wires were even looped across the facades and rooftops of buildings. Few if any of them had adequate insulation, which in any case wind and storms soon eroded. Nearly a third of all the overhead wires were dead—abandoned by defunct companies or owners and never removed as their insulation rotted away. Sagging raw wires sliced into the insulation of newer ones, putting metal on metal and sending currents along new pathways. Laws were passed ordering companies to bury their wires but were not enforced until just after the 1888 blizzard brought down so many of the lines. Even then the United States Illuminating Company, which operated arc lights with alternating current, challenged the law in court.[19]

A campaign had been under way for some time to force companies to bury their lines underground, but it floundered in the swamp of New York City politics. Mayor Abram S. Hewitt railed against the ineptitude of the Board of Electrical Control, which had been created in 1885 to oversee the burial of wires, and tried in vain to get the board abolished. In 1888 he was replaced by Hugh L. Grant, who renewed the campaign to get the wires underground. The mayor was by law a member of the board but lacked enough votes to dominate it. For months the issue hung fire amid clashes between the board and the electric companies and infighting among the board members.[20]

While public anxiety grew, the "execution" bill marched steadily through the legislature, and on June 4, 1888, Governor Hill signed it. The new law abolished hanging and substituted death by electrocution as of January 1, 1889. The *New York Evening Post* proclaimed that it would put an end to "revolting barbarism." That same paper also marked the entry of a newcomer to the fray in the form of an obscure engineer named Harold P. Brown. In a long, highly charged rant he blistered public officials and corporate officers alike for ignoring the perils of high-voltage wires and warned ominously of new casualties to come. In particular he directed his wrath at alternating current. "The only excuse for the use of the fatal 'alternating' current," he thundered, "is that it saves the company operating it from spending a larger sum of money for the heavier copper wires, which are required by the safe incandescent systems. That is, the public must submit to *constant danger from sudden death* in order that a corporation may pay a *little larger dividend.*" To alleviate this danger, Brown proposed a set of regulations that he wished to see enacted, including a prohibition on all AC current transmissions above 300 volts.[21]

What prompted this outburst from Brown remains a mystery. He had worked in the electrical industry and called himself an electrical consultant, but he had no known ties to Edison or grievance against Westinghouse. Certainly he believed what he preached in all sincerity. Whatever his reasons, Brown's entry into the fray changed the situation dramatically. That he had the zeal of a fanatic soon became evident. The "war of the currents" had amounted only to a series of skirmishes until Brown became the self-appointed catalyst that escalated it. Three days after his letter appeared, New York's Board of Electrical Control held its regular meeting. Brown came personally to demand that his letter be read verbatim into the record. His proposed rules for safety were then circulated among numerous electric companies and electricians for comment. Several of those who favored AC included personal attacks on Brown in their replies. The recipients included Westinghouse, who directed Henry Byllesby to reply. Byllesby argued that alternating current was in fact safer than its rival.[22]

Westinghouse himself sent out a peace feeler to Edison. Rumors were flying in Pittsburgh and winging their way east that the Edison and Westinghouse interests were going to merge. On June 7 Westinghouse wrote Edison to assure him that the stories had no substance. "I believe there has been a systematic attempt on the part of some people to do a great deal of mischief and create as great a difference as possible between the Edison Company and The Westinghouse Electric Co.," he observed. Recalling the time Edison had spent with him nearly a decade earlier when Westinghouse had visited Menlo Park in search of a system for his home, and a later meeting at Sigmund Bergmann's factory, Westinghouse extended an invitation for Edison to come visit his factory in Pittsburgh. Edison responded by dismissing the rumors, saying that his laboratory consumed all his time and he had no voice in the affairs of the Electric Light Company. As for the invitation to visit, he merely thanked Westinghouse for it. His reply revealed neither thaw nor freeze.[23]

Late that same month a *New York World* reporter approached Edison with questions about execution by electricity. The inventor obliged by having Arthur Kennelly, his chief electrician, rig up an informal experiment using a dog provided by the reporter. While the reporter, Edison, and some of the lab crew watched, the hapless dog was killed after "a quick contortion" and a "smothered yelp." The reporter asked how long it would take for electricity to kill a man. "In the ten-thousandth part of a second," replied Edison. He did a quick sketch of how the electrodes should be attached and added, "The current should come from an alternating machine." The *Herald* remained unconvinced. "The willful infliction on a human being of death by electricity," it reminded readers, "is an experiment which they have never tried and of which science possesses no record."[24]

Possibly inspired by these articles, Brown approached Edison early in July with a request to use some of his laboratory space for some experiments to prove the deadliness of alternating current. Edison obliged Brown with not only space but also the aid of Kennelly. A brilliant mathematician, Kennelly later helped crystallize the theories and applications of alternating current after leaving Edison in 1894. Working late at night when the dynamos were free, the two men carried out a number of grisly experiments electrocuting dogs and cats. After slaughtering a large number of luckless animals, Brown went public with his demonstrations. On July 30 at the Columbia School of Mines, before an audience of electricians, reporters, and other interested parties, he brought out a caged seventy-six-pound black retriever and proceeded to torture it with increasing jolts of direct current until he reached 1,000 volts. The helpless animal thrashed and yelped in pain—"heartrending in the extreme," one reporter noted—but remained alive as the horrified audience yelled for Brown to stop.[25]

Brown ignored the rising cries of outrage. He switched off the direct current, saying, "He will have less trouble when we try the alternating current." Using a Siemens generator, Brown zapped the terrified dog with 330 volts of alternating current. It fell dead almost at once. A representative of the SPCA then stepped forward and called a halt to any further executions. Brown grudgingly relented. As the disgusted spectators filed out, he remarked that "the only places where an alternating current ought to be used were the dog pound, the slaughter house, and the state prison." One spectator growled that a Spanish bullfight seemed a moral and innocent spectacle compared to Brown's experiment.[26]

Still Brown had not proved his case. The dog, after all, had been weakened by the DC shocks before succumbing to alternating current. More such displays would be needed to demonstrate the efficacy of electricity—and alternating current in particular—for executions. At another demonstration on August 3, presided over by a member of the Board of Health and a vivisectionist from Bellevue Hospital, three large dogs were killed with alternating-current voltages of less than 400. "We made a fine exhibit yesterday," Brown boasted to Kennelly the next day. "It is certain that yesterday's work will get a law passed by the legislature in the fall limiting the voltage of alternating currents to 300 volts." Although the legislature did not oblige Brown, it had designated electrocution as its official form of execution. But how to kill a man as painlessly as possible? No one knew the best method for such an execution or what apparatus to use.[27]

The legislature turned this task over to the Medico-Legal Society of New York, which created a committee to explore the subject. Its chairman, Dr. Frederick Peterson, had helped Brown in his Columbia demonstrations. Back they went to Edison's laboratory for more slaughtering of the innocents to determine how best to do the job. Aware that the Medico-Legal Society would render its decision in December, Brown worked zealously to shape the outcome. At the society's November meeting Peterson expressed a preference for alternating current as the type to be used, but a final decision was put off until the next meeting. On December 5, a week before the decisive meeting, Brown arranged another demonstration, this one at the Edison laboratory. The audience included several prominent physicians from the society, Dr. Southwick, Elbridge T. Gerry, and Edison himself, whose presence gave the event an aura of legitimacy and prestige.[28]

Larger animals were needed to prove that electricity could kill a man as easily as a beast. Brown opened his hideous show with a 124-pound calf, which lasted only thirty seconds at 700 volts before dying. A second calf weighing 145 pounds was brought out and killed in only five seconds. For his showstopper Brown brought out a healthy horse weighing 1,230 pounds. Here was an animal far larger than any person who might be electrocuted. It withstood two successive shocks at 600 volts; Brown then hit the poor animal with 700 volts of alternating current, and it fell dead. Kennelly's own notes revealed that all three animals suffered from the ordeal, but most public accounts glossed over their deaths as quick and painless. Next morning, under the headline "Surer Than the Rope," the *New York Times* declared that "The experiments proved the alternating current to be the most deadly force known to science" and predicted that after January 1 it would "drive the hangman out of business in this State." At its December 12 meeting the Medico-Legal Society unanimously approved a proposal for "death by alternating current." By this action it formally associated alternating current with death.[29]

The Edison people chortled in delight. During these months Edison himself had devoted nearly all his time to getting his new version of the phonograph finished by the year's end, but his staff was busy. When the *Electrician* and other publications attacked Brown's credibility, Kennelly defended him. Frank Hastings, treasurer of Edison Electric Light, had first steered Brown to Edison. He also arranged for the December 5 demonstration and pushed the campaign emphasizing safety. Brown developed a pamphlet on the comparative dangers of direct and alternating currents along with a circular letter on the society's decision. The latter went to mayors, city officials, insurance men, and important businessmen in every town with a population of five thousand or more. Although evidence is lacking, it is likely that the Edison company paid for it. Edison forbade the laboratory staff from commenting on the demonstration, preferring to let the results speak for themselves.[30]

A furious Westinghouse responded to the society's decision with a letter to New York newspapers refuting Brown's claims about alternating current and reminding readers that his firm had received orders for more central station lighting in October than the Edison company had for the entire year. "The business would not have had this enormous and rapidly increasing growth," he emphasized, "if there had been connected with it the dangerous features which Mr. Harold P. Brown and his associates of the Edison company so loudly proclaim." Brown wrote an insulting response capped by an absurd challenge to a duel by currents. Before an audience of "competent electrical experts" he offered to be wired to direct current if Westinghouse would tether himself to alternating current. Pressure of 100 volts would be inflicted on their bodies and increased 50 volts at a time until one of them relented and confessed his error. Westinghouse ignored the challenge.[31]

The whole episode reflected the extent to which the effect of electricity on the human body had become a topic of keen public interest. Already it was being touted as a potential cure for everything from insomnia to impotence. The public interest in electrocution was merely the more macabre end of this spectrum of curiosity, and the *Electrician* feared it might go too far. "If this sort of thing goes on, with the accidental killing of men and the experimental killing of dogs," it warned, "the public will soon become as familiar with the idea that electricity is death as with the old superstition that it is life." The most recent developments triggered a search for some new word to describe the pending form of execution. Edison had his own preference, which he hoped would become common usage: Let the criminal be "westinghoused."[32]

As 1889 opened, Edison was clearly winning the war of the currents. Thanks to Brown, he had seized the offensive and shifted the public's attention from the comparative efficiency of the two systems to their relative safety. Edward Dean Adams, a Boston financier who sat on Edison's board and was also friendly with Westinghouse, tried to end the madness by bringing the two men together. He urged Edison to visit Pittsburgh and see what was being done with AC equipment at the Westinghouse plant. Edison would have none of it. "Am very well aware of his resources and plant," he fired back in a telegram, "and his methods of doing business are lately such that the man has gone crazy over sudden accession of wealth or something unknown to me and is flying a kite that will land him in the mud sooner or later." Other Edison men pleaded with him to relent in his opposition to the AC system, but the inventor remained, in Edward Johnson's term, "granitic" in his belief that AC would destroy the reputation of anyone who touched it.[33]

The war went on. New York's state prisons hired Brown as their technical expert in devising their electrocution apparatus. Brown made certain that Westinghouse dynamos were used even though he had to acquire them in a roundabout way. He managed to persuade Charles Coffin to locate some utility companies that owned Westinghouse equipment and acquire their generators by replacing them with Thomson-Houston models. Brown then used funds given him by Edison Electric to buy the generators from Thomson-Houston.[34]

Elihu Thomson regarded the whole affair as distasteful. "I regret very much that our Company should take an active part in furthering Mr. Brown's personal fight with Westinghouse . . . ," he wrote Coffin. "I dislike to see steps taken to confer a reputation on Westinghouse apparatus which our own similar apparatus must share." He reminded Coffin that he had always "urged that high voltage of alternating current needs especial precautions to prevent accident." Thomson also thought the use of 1,000 volts as the standard for executions too low, and he harbored a low opinion of Westinghouse. "The methods of the Westinghouse people are, as we know, of the most unfair and undignified character," he declared, "and it would redound to our credit, I think, to let them keep to such methods."[35]

March brought more good news to Edison when the French corner in copper collapsed and the price stabilized. The following month witnessed the birth of Edison General Electric, which relieved the inventor of his immediate cash strains. In May the public learned that Westinghouse dynamos would be used by the state prisons for the first execution. The victim would be William Kemmler, who had been convicted on May 10 of brutally murdering his wife—who was not really his wife—with an axe. The sentence was appealed on the grounds that the new form of execution was unconstitutional.[36]

Hearings on the case began July 8 with W. Bourke Cockran, a prominent litigator, defending Kemmler. Although Cockran insisted otherwise, critics suspected Westinghouse of paying his fees. The *New York Times* charged that Cockran was actually defending "the objection of the Westinghouse Company to having its alternating current employed" for the execution. Although Cockran's agile questioning exposed Edison's basic ignorance of the physical factors involved in death by electricity, he could not overcome the inventor's reputation as *the* authority on things electrical. The Wizard had become the oracle, and his views prevailed. On August 3 a satisfied Edison and his young wife embarked on a leisurely trip to Europe that turned into a triumphal march. Feted and decorated everywhere he went, Edison met such fellow luminaries as Louis Pasteur and Werner von Siemens.[37]

While Edison was gone, another sensation broke on August 25 in the form of revelations in the *New York Sun* under the headline "FOR SHAME, BROWN!" Somehow the paper had obtained forty-five letters, later revealed to have been purloined from Brown's locked desk, that unmasked his relationship with the Edison interests. In his testimony Brown had denied any such connection. The letters also showed that in acquiring the Westinghouse dynamos by stealthy means, Brown had been aided not only by Edison officials but also by Charles Coffin of Thomson-Houston. Brown claimed lamely that the letters were fake but offered no evidence to that effect. Most of the criticism from the press landed on Brown rather than the Edison or Thomson-Houston interests. Arthur Kennelly later admitted that he had conducted most of the executions at the West Orange lab and that Brown had played only a minor role. From these revelations emerged what some critics had always suspected to be the truth: The Edison interests had sponsored a campaign to discredit Westinghouse, using Brown as their attack dog. However, the suspicion also lingered that Westinghouse had paid someone to steal the incriminating letters in the first place.[38]

The *Sun* took quite a different position. In an editorial entitled "An Electrical Quack Grounded," it concluded that "there is a conspiracy against the concern known as the Westinghouse Company. That HAROLD P. BROWN is the appointed agent of this conspiracy. That a great deal of the testimony elicited respecting the alternating current and the voltage required to insure immediate death was perjury or the equivalent of perjury."[39]

ON OCTOBER 6, after a two-month sojourn, Edison returned home a national hero, lionized abroad as well as at home. Samuel Insull, not usually noted for his sense of humor, met him with tape measure in hand to see whether "hob-nobbing with titled personages and great men in Europe had increased the girth of his intellectual temple any." Edison enjoyed the joke. "I come back decorated with the red ribbon of the Legion of Honor of France," he told a reporter. "I have attended more dinners since I left these shores than in my whole previous existence. I haven't had time to think hardly since I landed in Europe. My mind has been in a constant whirl."[40]

Good news on the patent wars also greeted him on his return. An earlier Canadian decision that had voided the Edison patent had already been reversed. Westinghouse brought his own retaliatory suit against Edison for violating the Sawyer-Man patent. Shrewdly he had the Consolidated Light

Company, which he controlled, bring the suit in a court just outside Pittsburgh, where he hoped to gain an advantage. The Westinghouse forces mounted a solid case with expert witnesses headed by Frank Pope, who presented a forceful argument in favor of Sawyer's lamp. However, the Edison lawyers trumped his appearance by putting on the stand Sawyer's brother George, who admitted that William had lied to just about everyone and deceived them into thinking that he had a working lamp. In October 1889 the U.S. Circuit Court rendered a decision upholding the priority of Edison's lamp over that of Sawyer, stressing that the "high resistance in the conductor with a small illuminating surface and a corresponding diminution in the strength of current . . . was really the grand discovery in the art of electric lighting."[41]

Edison hailed the decision and used it to denounce his foe. "Westinghouse simply grabbed fifty-four of my patents and started into business, saying that he could sell his manufactures cheaper because he did not have to pay out money experimenting," he told a reporter, adding that "Westinghouse used to be a pretty solid fellow, but he has lately taken to shystering."[42]

Westinghouse bought time by appealing the decision, but things grew worse. On October 9 the judge rejected Kemmler's appeal and sentenced him to die by electricity. More appeals ultimately carried the case to the U.S. Supreme Court and delayed the execution until August 1890. However, two days after the judge's decision another horrific accident in New York aroused public wrath as none had before. On October 11 lineman John Feeks climbed a telegraph pole with fifteen crossbars crowded with more than 250 wires to cut away the dead lines. Because he was working with low-pressure wires, he neglected to wear rubber boots or gloves. Working at noontime in a busy district of Manhattan, Feeks lost his balance, grabbed at a wire to regain it, and was electrocuted.[43]

Prior to this grisly spectacle New York had undergone a rash of electrical fatalities. Late in June a veteran Brush employee suffered a fatal shock. On September 2 a lineman died in Manhattan after making contact with a live wire. "The recent experiments with electricity as a means of executing criminals were fresh in the minds of the physicians," reported the New York Tribune, "and they made a careful record of what they did and of the effects of the electric current on the dead man's body." Three days later, in Buffalo, a young lineman employed by Brush Electric writhed in agony for forty minutes after receiving a shock while hundreds of spectators looked on helplessly. "Every person who saw the way poor Perry was tortured to death by electricity today," said the New York World, "is a firm foe of electric capital punishment." On September 13 an Italian fruit dealer was electrocuted in lower Manahattan. Four days later another lineman for the Brush company died while handling some wires. Then, on October 8, another lineman was electrocuted while working downtown.[44]

An outraged Mayor Grant summoned the Board of Electrical Control to a special meeting to consider ways for "preventing indiscriminate slaughter . . . by contact with electric wires." During an acrimonious discussion a representative of one electric company insisted that wires carrying alternating current could be made safe for handling. "You call them 'safe,'" snapped the mayor, "but we get the news of all who touch them through the coroner's office." After more heated rhetoric he offered a resolution that all electric companies be notified to cease using any overhead wires that were not properly insulated until the board's expert could certify them. The resolution was approved unanimously. Two days later came the Feeks tragedy.[45]

Normally Feeks would have suffered no ill effects from the low current in telegraph wires. However, some blocks away an alternating-current lighting wire had crossed the same wire Feeks held, and its insulation had been worn away. High-voltage current raced down the line and through Feeks's body into his right hand and out of his left foot. Like Perry, he fell into the tangle of wires and dangled as if in a spider's web while horrified onlookers below watched helplessly. More people crowded into the street and onto rooftops to observe the hideous scene. After forty-five minutes other linemen arrived to climb the pole and with difficulty cut their dead comrade down. The *World* prefaced its detailed account by saying, "A man was either instantly done to death or was slowly roasted . . . through this murderous neglect of one of the companies owning the wires which thread every street in the city." The reporter's *Sun* colleague described "a poor fellow roasted upon a gridiron of fire-spitting threads that had never before shown a sign of danger."[46]

"Stop the crime!" cried a *World* editorial. "This was not an accident. It was a crime committed for money . . . A few rich corporations, soulless, merciless and lawless, have secured a sort of license to girdle New York with deadly peril in order that they may grow rich quickly." The *Times*, like most other dailies, denounced the electric companies. "Five years ago the law required them to put their own wires under ground . . . but they defied the law, disregarded public interests, and treated public opinion with contempt." Nearly all the editors demanded that the lines be buried at once. Asked his opinion, Edison said he thought it a "disgrace to modern civilization that such crimes are permitted, for it is nothing more or less than a crime to use deadly currents in a city like New York."[47]

For the Westinghouse interests the gruesome Feeks spectacle became a worst-case scenario. No one knew who owned the wire that killed Feeks. Some of the companies that provided alternating current for electric lighting were

Westinghouse subsidiaries. Even more, the public associated Westinghouse with alternating current. It was easy for the public to assume, and for his adversaries to charge, that the fatal wires belonged to one of the Westinghouse companies. "How ignorant or how deceitful and how contemptible ... ," stormed the *New York Tribune*, "appear the assurances of safety with which the community has been mocked for years past."[48]

Next day, while the press thundered its outrage, the mayor rose from a sickbed to order all unsafe wires cut down at once. Westinghouse's lawyers hurried to court and got an injunction staying this order, but doing so cast him as the very heartless villain that Edison and Brown had labored so hard to portray. Thus began what one writer called the "electric wire panic" that succeeded in blowing the threat of electricity vastly out of proportion. "There is no safety, and death lurks all around us," warned one expert ominously. Another worried that "a man ringing a door-bell or leaning up against a lamp post might be struck dead any instant." The furor revived the fading star of Harold Brown, who signed on with the *World* as its electrical expert. He consulted several times with Edison, who greeted one reporter by asking, "Have they killed anyone there today?"[49]

In all, six electric companies got injunctions but shut down their machinery while the courts sorted out the legal tangle. "A Darker But a Safer City," declared the *Tribune* as large sections of New York went without streetlights. Efforts were made to reconnect at least some of the old gaslights to relieve the darkness. The Board of Electrical Control endured criticism heaped on it from all sides but could do nothing except argue until the injunctions were lifted, beyond preventing the electric companies from stringing any more wire. The *Herald* denounced the board as "Conceived in Sin, Born in Iniquity." Edison talked freely on the matter and offered advice on how to make electricity in the city safe. On October 29 a state supreme court judge prolonged the controversy by sustaining the injunction.[50]

Edison had gained the upper hand in both the war of the currents and the patent war. He was a public icon while Westinghouse, languishing in the temple of privacy he had built around himself, was pelted with opprobrium. Not until October 24 did he issue a statement saying that the furor in New York was exaggerated. With public concern aroused, orders began falling off for Westinghouse equipment. Brown fed the campaign by sending letters to New York papers condemning Westinghouse for its use of alternating current and itemizing the alleged casualties from it. Westinghouse ordered his sales agents to investigate the charges; they found that only one of the thirty deaths listed by Brown *might* have come from contact with alternating current. Another one was uncertain, twelve occurred in cities that had no Westinghouse plant at the time of the incident, and sixteen involved direct-current systems feeding arc lights. That autumn Westinghouse put out a booklet on the safety of alternating current.[51]

In November 1889 the war of words reached its peak when Brown and Edison published articles in the venerable *North American Review*. Identified as the "New York State Expert on Electrical Execution," Brown tried to explain the scientific and experimental basis for "the life-destroying power possessed by the alternating current," which made it the ideal medium for executions. Again he called for the legislature to "banish it from streets and buildings, thus ending the terrible, needless, slaughter of unoffending men." Brown was singing an old tune, but little had been heard directly from Edison until his decision to go public with an article entitled "The Dangers of Electric Lighting." With this move Edison entered the public debate directly as a foe of both alternating current and Westinghouse. He was moved to write, he said, by the Feeks accident.[52]

"The opportunities for repetitions of the accident . . . ," Edison warned, "will be practically unlimited. I can write upon this subject only as one convinced." He described four kinds of current used for electric lighting: 1) low-tension continuous [i.e., direct] current with pressure not exceeding 200 volts; 2) high-tension continuous current with voltages of 2,000 or more; 3) high-tension semi-continuous current with voltages of 2,000 or more; and 4) alternating current with voltages ranging from 1,000 to 3,000 or more. "The first is harmless," he noted. "The second is dangerous to life. Momentary contact with a conductor of the third results in paralysis or death . . . and the passage of the fourth, or alternating, current through any living body means instantaneous death." Putting the wires underground would not help, Edison stressed, because no known insulation could protect them for any length of time. "If a nitro-glycerine factory were being operated in the city . . . ," he reasoned, "and the people desired to remove the danger, no one would suggest putting it underground."[53]

Edison renewed Brown's charge that alternating current was being "employed solely to reduce investment in copper wire and real estate." The only solution, he insisted, was to "restrict electrical pressures." His own company had acquired patents for a complete alternating-current system, but he "succeeded in inducing them not to offer this system to the public, nor will they ever do so with my consent. My personal desire would be to prohibit entirely the use of alternating currents." Until such action was taken, he promised, "nothing better can be looked for than a multiplication of the casualties of the past few months."[54]

This diatribe was too much for Westinghouse. He responded in the December issue of *North America Review* with "A Reply to Mr. Edison" in which he used temperate language to shred Edison's arguments. Electricity had become a staple of everyday life, he declared, and if one were to prohibit everything that was dangerous in life, people "would be deprived of most of the necessaries and comforts of existence." Where Edison had conveniently sidestepped the bitter corporate struggle for business in his article, Westinghouse reminded readers that it lay behind much of the conflict. He provided a brief survey of that fight and concluded that "the energy and money since expended by each of these corporations in efforts to thwart the progress of the others has mutually embittered the interested parties to a degree that can with difficulty be comprehended by those not immediately concerned in the strife."[55]

Put more baldly, the clash was as much over money as over safety. Westinghouse walked readers through the fundamentals of the Edison system, which required low-voltage direct current to operate its lamps. He noted that it was "common practice of the Edison Company to use uninsulated overhead wires for its 220-volt current for the purpose of economizing," even though it was well known that "such a current is capable of burning a body." It was most likely a direct continuous current that killed the lineman alluded to by Edison. Westinghouse took issue with Edison's four categories of current, noting that he had witnessed "the roasting of a large piece of fresh beef by a direct continuous current of less than one hundred volts within two minutes."[56]

Westinghouse understood the need to educate his readers on the nature of alternating current. Using the published experiments of Edison's own laboratory, he underscored several erroneous conclusions while noting Edison's failure to mention that the current actually entering a consumer's home or business was not alternating but direct current. "It is one of the great advantages of this system," he emphasized, "that it admits of the use of high voltages for the street mains, and of wholly separate and independent currents, with absolutely safe voltages, for all wires within buildings." In many respects the AC system was actually safer, he insisted. As for Edison's concerns about burying wires underground, they were possibly "suggested by difficulties experienced in the working of his own system." Again Westinghouse spelled out the technical issues in laymen's language.[57]

So too with the ominous forebodings expressed by both Brown and Edison about the casualties connected with alternating current. Westinghouse pointed out that in 1888 only five people in New York died from contact with electricity compared with twenty-three from illuminating gas, fifty-five from wagons or omnibuses, and sixty-four from streetcars. Even those few deaths could be avoided by placing the wires underground. "There is not on record a solitary instance," he added, "of a person having been injured or shocked from the consumers' current of an alternating system." The real danger of electricity stemmed from fire, not shock. After suggesting a few regulations of his own, Westinghouse reminded his readers that "for three years past the purchasers of apparatus for electric lighting . . . have, for the most part, preferred to use the alternating system, so that to-day the extension of that system for central station incandescent lighting is at least five times as great as that of the direct current."[58]

Despite the forcefulness of his reply, Westinghouse continued to struggle against the tide. Within a month three more New Yorkers died in electrical accidents. Two of them worked for light companies, but the third, a dry-goods clerk, touched a low-hanging Brush arc lamp as he was moving a metal outdoor display back into the store and fell dead. Here was the nightmare come true: an ordinary citizen slain while going about his ordinary business. "One scarcely ventures to put a latch key into his own door," warned the *Evening Post*. On December 13 the state supreme court dismissed in harsh language the injunction against cutting down the wires. The Board of Electrical Control ordered the Westinghouse lighting interests to shut down their current by 8:00 A.M. Ninety minutes later, workmen swarmed to the lines as cheering crowds urged on their work. During that first day twenty-three poles and fifty thousand feet of wire came down; when the campaign tailed off at year's end, more than a million feet of wire—a quarter of the city's entire total—had vanished from the streets.[59]

New York had become a safer city, and a darker one, as New Year's Eve revelers soon discovered. The stripped lines belonged chiefly to companies operating outdoor arc lights; the underground Edison cables powered mostly indoor incandescent lamps. Underground cables could not be laid until the spring thaw, leaving the city in the dark for several months while the city and electric companies continued to bicker. "It is reasonably certain," declared the *Herald*, "that the people have become tired of this dangerous electric light business and are ready to go back to gas for awhile." That assertion proved to be wrong. As the *Tribune* observed, "It cannot be supposed that a great city which has once been well lighted by electricity will ever again be contented with ancient standards of illumination."[60]

Meanwhile, experiments went forward at the Auburn prison, where another calf and horse were electrocuted to test the Westinghouse generators. An apparatus for the execution had to be designed along with some plan for making electrical contact with the victim. After some debate it was decided to use a chair rather than a table, but controversy continued over how best to hook up the electrodes. The final version, designed by Dr. George Fell, applied the current through electrodes attached to the head and lower back.[61]

During these long months William Kemmler, freed at last from his perpetual drunkenness, found God and turned into a model prisoner willing to be the pioneer in this most modern form of execution. However, stays kept postponing the day of his demise. Some newspapers blamed Westinghouse for the continued delays. Then, in the spring of 1890, the lower house of the New York state legislature startled everyone by passing a bill to abolish capital punishment. The bill sailed through without debate by a 74–30 vote, leading one member to charge that "Westinghouse money is passing this bill." The *World* agreed, and the *Times* called it "probably the most disgraceful exhibition ever made of itself by a legislative body in a civilized country." Westinghouse hurried to publish a denial in the New York papers. "I make this denial without any reservation of any character," he wrote, and the editors mocked him for that as well. After reaching the state senate, the bill died a quiet death in committee.[62]

The appeals finally ran out that summer, and on the morning of August 6 Kemmler made his fateful trip to the new and untried electric chair. No fewer than fourteen of the twenty-five official witnesses were physicians, including Dr. Fell and Alfred Southwick; two wire service reporters had also been invited. The warden stopped to consult two physicians on a point no one had considered: how long to keep the juice flowing. They agreed that one of them, Dr. Edward C. Spitzka, president of the American Neurological Association, would decide when to turn the current on and off. Spitzka borrowed a stopwatch and made ready. Kemmler was brought forward and, after a brief ceremony, placed in the chair. He remained calm and helpful as the straps and electrodes were applied. After Kemmler said, "Good-bye," Spitzka gave the signal to send more than 1,000 volts of electricity over the line to the chair.[63]

Kemmler convulsed and strained against the leather straps. Spitzka waited seventeen seconds, then ordered the power stopped. The dynamos, located in an adjoining room, were shut down. The physicians stepped forward to examine the figure slumped in the chair. Southwick marveled that the execution had been as quick, clean, and painless as they had hoped. Before that thought could mature, however, one doctor noticed a pulse. "Great God!" he cried. "He is alive!" Kemmler's body shivered as he struggled for breath, saliva dripping from his mouth through a purple foam on his lips. An animal cry escaped his mouth. "Turn on the current! Turn on the current!" someone shouted as the electrode was hastily jammed back down on Kemmler's head. Two minutes passed before the current began flowing again. Kemmler's body heaved up again as his muscles contracted. To avoid further embarrassment, the current was allowed to flow for more than a minute. The sponge on the rear electrode dried out and burned away, giving the electrode direct contact with Kemmler's skin. The stench of burning hair and flesh wafted over the room.

One witness threw up, and a reporter fainted. The district attorney rushed out of the room. When Spitzka ordered the current stopped, the smell of urine and feces mingled with the acrid smoke. Dr. Fell hurried to put out a small fire that had started on Kemmler's coat near the rear electrode. The doctors certified that Kemmler was finally dead, and the witnesses gratefully filed out of the room; one of them, a sheriff, was in tears. The newspapers ignored the ban imposed on printing details about the execution and gave vivid accounts of the botched performance. Not content with the outcome, the *Herald* concluded "either that the dynamos were faulty or that the interested company had bribed some one to make them seem so." No one wanted to deal with the fact that the preparations had been sloppy and ill conceived. Later the man running the dynamo estimated that it had run at 700 volts, about half of what the physicians assumed they were using.[64]

Southwick and some other witnesses put on a brave front, insisting that all had gone well. "A party of ladies could have been in that room and not known what was going on, so silent was the process," he insisted. Others called it a gruesome spectacle. "I will see that bound figure and hear those sounds until my dying day," said one. Nevertheless, most of the physicians thought that Kemmler had suffered no pain, believing that the first shock had knocked him out. None explained how they came to this belief. Edison was sanguine over the outcome. Every new device, he observed, requires a few trials to get it right. Next time would be different. More people sided with reaction by George Westinghouse. "It has been a brutal affair," he told a reporter. "They could have done better with an axe."[65]

Civilization had not leaped quite as far forward as Southwick and others had hoped. Some newspapers echoed the *World* in saying, "The first experiment in electric execution should be the last." But it was not. As more bungled executions followed, attempts were made to refine the apparatus. Edison proposed dipping the victim's hands in liquid electrodes, but an execution utilizing this method produced even more grisly results, and his notion was discarded. Nevertheless, electrocution remained the way of execution in New York. Imperfect as it was, the alternatives seemed even worse. "Electricity seems on the whole," said the *Herald*, "to answer the purpose better than anything else." Harold Brown, having savored his fifteen minutes of fame, vanished from the front pages but continued to work for Edison's company and later became one of "Edison's Pioneers."[66]

By the time of Kemmler's execution the war of the currents had taken yet another turn. Both Edison and Westinghouse found themselves squeezed not only by the bitter competition in the industry but also by a deteriorating financial climate that shifted their attention from safety to the more basic problem of survival. The failure in November 1890 of the prestigious London banking firm Baring Brothers sent financial markets into turmoil and led to a sell-off of American securities by British holders. As gold flowed out of the United States, credit grew tighter and an already slow business climate grew gloomier. In this darkening atmosphere the fierce competition among the three leading electrical firms could not help but bruise them even more severely.[67]

CHAPTER 14

Money, Mergers, and Motors

With the advent of the dynamo electricity has taken a new and very much larger place in the commercial activities of the world. It runs and warms our cars, it furnishes our light, it plates our metals, it runs our elevators, it electrocutes our criminals; and a thousand other things it performs for us with secrecy and dispatch in its silent and forceful way.

—Edward W. Byrn (1900) [1]

By 1890 the war of the currents had cast the electrical industry in an entirely new light. Only a decade earlier it had been an infant struggling to be born, a flicker of development suddenly propelled by Edison's new lamp. First lighting and then the street railway, or traction, fed its growth, and with it came the fierce competitive wars so typical of fledgling industries grounded in new technological breakthroughs. Within a remarkably short time the list of eager young firms had been winnowed into three dominant ones: Edison General Electric, Thomson-Houston, and Westinghouse. Competition enabled, indeed forced, all three companies to grow, and size intensified the competition among them by raising the stakes of failure. The larger they became, the more costly grew failures, missteps, or the competitive wars that bled their balance sheets. [2]

To make matters worse, they had created a capital-intensive industry in a sector that was itself novel. Securities in industrial corporations were largely terra incognita to most investors. The most traditional source of investment for people with money had always been land or its upscale urban relative, real estate. Government securities constituted another outlet for surplus funds, along with public improvement projects like canals, docks, or roadways. The

advent of the railroad, the nation's first and most capital-intensive big business, transformed the capital market no less than the business and physical
landscape. The railroad mania literally created the modern American capital
market and with it the stock exchanges that came to be dominated by rail securities along with a smattering of government issues.

The term "industrial securities" did not even come into existence until 1889.
In that year no fewer than ten railroads had a net worth exceeding $100 mil-
lion, led by the mighty Pennsylvania Railroad at more than $200 million. Only
a handful of industrial firms boasted a capital of even $10 million. Except for
coal mines and a few New England textile mills, industrial stocks did not exist
on public exchanges. Most manufacturing firms were privately owned and resorted to private borrowing to raise capital. Railroads deviated from this pattern only because their capital needs were far too large to obtain from any
form of organization other than a corporation featuring limited liability. Even
giant firms like Standard Oil or Carnegie's iron and steel works did not sell
stock to the public but kept it in the hands of its own circle. These companies
had the ability to raise capital internally and so rarely needed bankers to do it
for them. Down to 1890 investment bankers had waxed fat first on government
securities, especially the plethora of offerings during the Civil War, and then
on railway securities. This profitable progression reached its peak during the
railroad expansion orgy of the 1880s and was beginning to lose its luster as the
rail industry built far more mileage than it could sustain.[3]

To alert eyes the industrial sector seemed promising as the next big thing,
but it also harbored dangers that made bankers cautious. The competitive
wars in any industry drove income lower even as they fostered a need for more
capital to gain a competitive edge through superior technology or growth or
merger. The electrical industry fit this pattern and embodied other risks as
well. The furor over safety clouded its future as nothing else had. Most obviously it blurred the issue of which technology would prevail, as well as the uses
to which it would be put. No one doubted that electricity had immense potential, but in what form and direction? To most people it remained a mystery,
and a scary one at that. Other industries had the advantage of dealing in products that were more palpable and easily understood. Moreover, the electrical
industry had already revealed a pattern that would later become commonplace
in American business: the tendency for technological innovation to shorten
and sometimes destroy the life span of a product. Who could predict what
"they" would think of next?

The ravenous hunger of the electrical firms for capital came to shape much
of what they did. Money was the fuel that drove Edison; he had, after all, proclaimed himself to be in the invention business. His need for funds became es-

pecially acute in his quest to support the new laboratory. It was bankers who underwrote his start in the electrical industry, and it was a banker in the person of Henry Villard who had eased his immediate needs by the formation of Edison General Electric. Thomson had floundered in his career as an inventor until the Lynn businessmen stepped forward to underwrite his work. They in turn found a prominent Boston banking house, Lee, Higginson & Company, willing to provide the capital needed for their expansion. In 1889 Higginson underwrote an issue of preferred stock by Thomson-Houston, one of the first such offerings by an industrial firm.[4]

George Westinghouse had no such banking alliance and did not want one. He borrowed constantly from bankers but kept them at arm's length from the business by meeting his obligations. Edison resented the toll exacted by the bankers; Coffin had a comfortable relationship with Henry L. Higginson, head of the banking firm, and freely shared information on the company with him. Westinghouse, however, did not like to share leadership with anyone. The companies were *his* companies, and he intended to keep them that way. For this and other reasons his Achilles' heel would always be the lack of adequate capital to underwrite his ambitious programs.

ON THE SURFACE Thomson-Houston looked to have the best of the war of the currents. It had stayed clear of both the patent and safety brawls and devoted its energy to selling as much product as possible. However, the controversy over safety exacerbated the growing split between Elihu Thomson and Charles Coffin over the direction of the company and Thomson's role in it. "I am a believer in the establishment of all safeguards which conduce to the good working of a system," Thomson observed in March 1888, "especially when they do not add greatly to the cost of making the installation." His disillusionment began when it became clear that customers tended to ignore even the most basic safety precautions in their zeal to get a system online and producing as much power as possible. Thomson urged Coffin to promote safety as an important part of the sales effort, but he soon realized that Coffin had no intention of doing so. The object, after all, was to sell products, not lecture the buyers on what to do with them. Thomson-Houston did not at first install the systems they sold and so could not control what was done with them.[5]

In many ways Coffin had left Thomson behind in his vision of the future. He understood the savage nature of competition in the industry, which threatened to get worse rather than better, and he saw clearly the strategy that must be followed to survive and prosper. The company had to diversify its products and manufacture them all as cheaply as possible. Given the pace of change in the field and Thomson's methodical, tightly controlled development program in the Model Room, Coffin could not wait for the inventor to invent or innovate. The logical course was to emulate Westinghouse and buy more outside patents, which not only expanded the product line quickly but also brought other inventive talent into the company. Although no one said as much, it had become clear that Thomson could no longer serve as the epicenter of creativity in the firm. Accordingly Coffin began in 1888 to buy more outside patents and looked for companies to acquire as well. Between 1888 and 1891 Thomson-Houston spent about $4 million to scoop up seven smaller firms.[6]

Thomson bitterly resented this strategy. An increasing amount of his time was consumed in patent litigation. By the end of 1889 he found himself involved in more than sixty interference cases, many of them requiring extensive work or testimony on his part. Nor did he care for the bevy of inventors who invaded his space after their patent or company had been purchased. Evaluating the patents of others also took up time he needed for his own work. As his control over innovation in Thomson-Houston slipped away, his anger mounted. When the company bought a dubious patent for "electric water" in 1889, he protested that "if our Company should go seriously to work to expend any money in making tests ... I should feel like resigning on the spot."[7]

The safety controversy brought his resentment to a head. In October 1889, amid the furor over the death of John Feeks, Coffin asked Thomson to write an article on "How to Make Electricity Safe" in defense of alternating current. Thomson flatly refused. The accident occurred, he insisted, because of poor installation, which he found unacceptable. If Coffin had listened to him earlier on the safety issue, the present crisis might not have arisen. Within a few weeks Brown, Edison, and Westinghouse published their crossfire articles, but Thomson remained silent and seething. Finally, on Christmas Eve 1889, his anger exploded in a letter to Coffin insisting that the company accept him as an inventor, not a publicist. He again refused to give alternating current an unequivocal endorsement:

> I have no panacea—for all the ills which may follow the use of high potential currents under conditions usually found in large cities. I can no more say how to make electricity safe in such cases than I can say how to make a railroad safe, or how to make steamship travel safe, or how to make the use of illuminating gas safe, nor the use of steam boilers safe. No improvement of our modern civilization has ever been introduced but that involved considerable risk.[8]

The last sentence echoed the message delivered by Westinghouse in his article, but Thomson kept his views in house. Gradually Coffin realized that the old relationship could never be restored. During the summer of 1890 Thomson agreed to replace his existing contract with an informal agreement that allowed him to concentrate more on his own research. He began devoting more time to professional organizations and was elected president of the American Institute of Electrical Engineers in 1889 and 1890. As his role in Thomson-Houston ebbed, other inventors began assuming larger roles. Thomson did not stop inventing; in 1890 he produced a gem of a mechanical meter that attained widespread use. He also played a role in bringing to Thomson-Houston a frail hunchback dwarf who proved to be a genius in electrical design. Charles Steinmetz, or Carl, as they called him, began his illustrious career with Thomson-Houston.[9]

Although Coffin did what he could to placate Thomson, he could hardly ignore the reality of the industry's fiercely competitive nature. What Thomson-Houston needed was either an alliance or a merger with one of the other top firms. But which one? That was the question Coffin set out to explore.

THE KEMMLER EXECUTION fiasco lost for Edison much of the ground gained in the war of the currents and created a situation in which, as Harold C. Passer so aptly put it, "the outcome was a victory for Westinghouse and yet not a defeat for Edison." As the furor over safety gradually diminished, the old competitive issues reasserted themselves. The suit over Edison's lamp patent wound its tortuous way through the court system, its outcome threatening to inflict severe damage on whichever side lost. The stakes were large and growing. Incandescent lamp sales had mushroomed from about 250,000 in 1885 to 3 million in 1890; of that latter figure, nearly 1.3 million were Edison lights, 600,000 Thomson-Houston, 500,000 Westinghouse, and the rest smaller companies. At the same time the surging traction market had intensified the struggle among the three leaders to capture as much of it as possible.[10]

Much depended on which system would ultimately prevail, and here Edison's stubborn refusal to consider an AC system greatly handicapped Edison General Electric. Samuel Insull, who ran the operation under the title of second vice president, chafed at Edison's hesitation. In July 1890, unable to contain his impatience any longer, he reminded Edison that a year earlier "the Edison General Electric Company promised to have an Alternating system, at the latest within six months. A year has expired since that meeting was held, and we are no further ahead to-day than we were at that time. The matter of

Alternating apparatus is a very serious one indeed to our business." Insull had arranged to have one of Edison's dynamos wound for alternating current, but he needed more information on a converter. "It is of the utmost possible importance that we should be able to go ahead on alternating apparatus, so as to be ready for our next season's business," he emphasized. Insull had to tell the district managers something, and he didn't want to promise anything he couldn't deliver.[11]

Late in August 1890 Edison resigned himself to the distasteful necessity of developing some kind of practical AC system to compete with his rivals. He put Arthur Kennelly to work designing one, but it would take months of hard work to come up with a working version. Meanwhile, Edison continued to struggle with the financial woes he thought had vanished with the creation of Edison General Electric. "I have been under a desperate strain for money for 22 years," he told Villard in February 1890, "and when I sold out, one of the greatest inducements was the sum of cash received." The consolidation had actually reduced his income and forced him in April to lay off most of his chemical staff. Not until October, after a year of negotiations, did he finally get a contract from the company to support his light and power experiments. The new contract eased his financial burden and turned his attention at least partly back to the electrical field.[12]

While Edison fretted, Henry Villard looked for a way to rationalize the industry itself. In his view the best way to eliminate cutthroat competition was to eliminate competitors, either through cooperation or consolidation. No sooner had Villard put together Edison General Electric than he began to explore both options. He established relationships with both Coffin and Westinghouse by offering to exchange information on production, sales, and earnings. That done, he sounded out Westinghouse on a patent agreement and Coffin on an agreement to fix bids on contracts. Coffin was more amenable than Villard knew; as early as March 1889 he had suggested a possible merger between the two companies. Villard ran the suggestion by Edison and got a predictable response. He dismissed Thomson-Houston as "amateurs" who had "boldly appropriated and infringed every patent we use," and he warned Villard that "if you make the coalition, my usefulness as an inventor is gone. My services wouldn't be worth a penny. I can invent only under powerful incentive. No competition means no invention."[13]

The idea of joining with Westinghouse appalled Edison even more. "No one can ever convince me," he told Villard, "that a competitor (whose system as shown by tests made at Johns Hopkins University by scientific men absolutely above reproach, gives an average efficiency of only 47 per cent) can ever prove a permanent competitor for large installations in cities against a system giving

79 to 80 per cent efficiency." Ignoring the crucial issue of transmission over long distances, Edison added that if he was wrong, "it is very clear that my usefulness is gone—that working night and day under strain to advance efficiency from 80 to 85 per cent and in other ways cheapen electric lighting is an absurdity. Viewing it in this light you will see how impossible it is for me to spur on my mind under the shadow of probable affiliations with competitors, to be entered into for financial reasons . . . I would now ask you not to oppose my gradual retirement from the lighting business, which will enable me to enter fresh and more congenial fields of work."[14]

Although Villard tried to soothe his friend's ruffled feathers, he persisted in his quest. Edison had admitted that Villard viewed the Westinghouse alliance "through a telescope, while I see the subject through a microscope." What Villard saw was not only that the rival companies needed to stop the bloodletting but also that they possessed complementary strengths that could fit well together. The right merger could stabilize the industry, end price wars, and improve profits. Both Thomson-Houston and Westinghouse had AC systems that could fill the most glaring gap in Edison General Electric's product line. Henry Higginson agreed with Villard. For the bankers it meant protecting their investment and improving their return on it, along with giving them some profitable business in underwriting a merger. Thus began the delicate process of deciding who should dance with whom.[15]

WESTINGHOUSE'S BUSINESS HAD grown fast—too fast, Edison observed sourly—and in the process outrun his resources. Annual sales jumped from $800,000 in 1887 to $4.7 million in 1890. The war of the currents had cost him money as well as bruises to his reputation; he could endure the latter much more easily than the former. Expansion into both lighting and traction, as well as developmental work on Tesla's motor, required more factory space and a larger engineering staff. The number of employees mushroomed from 200 in 1887 to 1,200 in 1890. By June of the latter year Westinghouse had 301 central stations built or under contract. More money went into buying out smaller firms and the bottomless pit of patent litigation. Nor was electricity his sole concern. By the spring of 1890 he was manufacturing air brakes, switches, and signals for railroads, providing natural gas to a large region near Pittsburgh, and running some plants that utilized natural gas along with his lighting, central station, and electric railway business. Westinghouse Electric alone operated four factories—two in Pittsburgh, one in New York, and another in Newark.[16]

Edison was America's Wizard, who loved the headlines even as he pretended to be indifferent to them. He had mastered this role. In his own way Westinghouse was no less charming and charismatic, but he was also an intensely private man who loathed publicity and submitted reluctantly to it only because he knew the company needed it. "There is in America no man more difficult to interview than George Westinghouse, Jr.," admitted a *New York Times* reporter. "He has issued positive instructions to all the local newspapers never to send to his house or office to get a personal interview, as he will not talk on business to a reporter." Granted a rare exception to the rule, the reporter hurried to Westinghouse's home and was favored with three terse paragraphs before being shown out.[17]

In the electrical field Westinghouse was the brash newcomer who not only refused to play by the book but was trying to rewrite the book on his own terms. It became a game among his closest lieutenants trying to guess "where the Old Man was going to break out next." Once he handed a foreman at the Machine Works a bundle of drawings for an elaborate rotary motor and asked him to make a model of it. The foreman protested that it would cost too much and could not possibly work but was overruled. He built the model, which failed miserably as he had predicted. "Mr. Westinghouse," he said, "I hated to see you throw your money into the ditch like that."[18]

"Oh, it wasn't thrown away," Westinghouse replied cheerfully. "Think how many men it kept employed; and besides it is one more step toward ultimate success."[19]

For all its growth, Westinghouse remained a homegrown product. The founder and his associates held most of the stock, which had little market outside Pittsburgh. To finance expansion, Westinghouse had put $1.2 million of his own money into the company and also borrowed heavily from local banks. By midsummer of 1890 the company had more than $2 million in short-term liabilities. In July Westinghouse got his board to double the capital stock from $5 million to $10 million and offered the $50 par value shares to the stockholders for $40 a share. To do this, he utilized a charter he had procured to reorganize Westinghouse Electric as the Westinghouse Electric and Manufacturing Company. Westinghouse was said to have taken $1.25 million of the new stock himself, but the sale flopped; stockholders took less than $1.8 million of the offering.[20]

Short-term obligations could be a ticking bomb if a crisis arose, and in November 1890 the bomb exploded in the form of the sharp financial contraction following the Baring Brothers failure as banks began calling in their loans. Westinghouse had gone to his beautiful new country place in Lenox, Massachusetts, when word reached him of the failure. Recognizing the danger at once, he hurried back to Pittsburgh and summoned a meeting of his board.

Convinced that the company only needed immediate cash to get through the crisis, he invited Pittsburgh's leading bankers to an informal meeting. Several came; a few stayed away, fearing that Westinghouse's well-known charm would lull them into making a loan they would later regret. After a tour of the facilities and a reminder of what he had done for the city of Pittsburgh, Westinghouse asked the bankers for a $500,000 loan to meet his immediate obligations.[21]

The financial pinch was temporary, he assured them. Westinghouse Electric, after all, had just completed its best year. Altogether his eight companies were worth $23 million; they had earned $16.5 million with profits of $4.2 million, or more than 25 percent. Against that they had only about $1 million in bonds outstanding but more than $6 million in accounts receivable and $2.5 million in material on hand. Duly impressed, the bankers formed a committee to look over the Westinghouse books and determine the company's actual worth. The committee reported favorably and recommended that the loan be made. On December 24 *Electrical Engineer* declared that "only minor details are to be looked after before a deal is closed. This will put the electric company—certainly a money-making concern when properly managed—on a sound footing."[22]

The phrase "when properly managed" rang a sour note. The bankers demanded a voice in the management of the company until the loan was repaid. "Mr. Westinghouse wastes so much on experimentation, and pays so liberally for whatever he wishes in the way of service and patent rights," said one, "that we are taking a pretty large risk if we give him a free hand with the fund . . . We ought at least to know what he is doing with our money." A second committee drew up a contract with stipulations that the bankers should be kept informed of all activities and also have the right to name a general manager for the company. Westinghouse countered genially but firmly that he would brook no interference with his management. After some more sparring, Westinghouse demanded an answer. The bankers responded with a flat refusal. If the bankers expected Westinghouse to relent, they were disappointed. He stood up and said, smiling, "Well, thank God I know the worst at last!" He told them a humorous story, bade them good-bye, and left the room.[23]

It was neither the first nor the last time Westinghouse had come to an impasse with bankers. They looked at the world so very differently than he did, viewing money as an end in itself rather than as a tool for getting things done. So too with any merger. Both Higginson and Villard wanted him to merge with their companies, which amounted to transferring the leadership somewhere else so that all parties could profit handsomely. Westinghouse wanted nothing to do with Edison's company, and he couldn't abide Coffin. "Mr. Coffin has a very swelled head," he confided to Clarence Barron. "He talks about making [his] . . . Company bigger than the Standard Oil Company . . . Coffin will make a man about ten different propositions in ten minutes. I have had many interviews with him. I suppose you know how he takes people into his inside room and then locks the door on them, so as not to be interrupted."[24]

Barron listened intently to what amounted to a rare display of pique by Westinghouse. "At one of our interviews," he continued, "I think it was in New York, he asked me if I would be willing to go into any electrical combination of which I was not the head. I said most emphatically that I would not go into any electrical combination of which I was to be head . . . Furthermore, I told him that I would not go into an electrical combination of which he was to be head." Coffin then confided how he had run the stock of Thomson-Houston down and deprived both Thomson and Houston of any benefit from an increased stock issue. Westinghouse fixed him with a look. "You tell me how you treated Thomson and Houston, why should I trust you after what you tell me?"

On another occasion Coffin and Henry Pevear met Westinghouse and Samuel Insull in a New York hotel room and proposed a merger plan for the three companies "so that whatever occurred, whether the companies were successful or unsuccessful, we should make money under any circumstances. I said to Coffin . . . that I was not in the habit of robbing my stockholders. There was no reply." Nor was that all. On another occasion Coffin came to Westinghouse with the request that he raise the price of lighting from six dollars per lamp to eight dollars. He admitted that six dollars gave a fair profit but said he had to pay boodle to aldermen and other officials, which the extra two dollars could cover. Westinghouse added proudly that some officials from Siemens & Halske had gone through both his factory and that of Thomson-Houston and afterward complimented Westinghouse on having superior arrangement and workmanship. He also heard that Coffin had taken them into his office, locked the door, and offered them $1.5 million to join with him in the United States or else leave the field clear. The offer was declined.[25]

To Westinghouse's mind, bankers and hustlers had the same trait. They followed the money rather than the work itself and lacked the vision to see what it might do if put in the right hands. However, the harsh truth was that, like it or not, Westinghouse needed money desperately, and the only place to get it was from the bankers. By March 1891 the floating debt had mounted to $3.3 million, and rumors swirled of an impending receivership. After leaving the Pittsburgh bankers, he climbed aboard a train to New York and hurried to the firm of August Belmont & Company, which had taken the lead in the relatively new field of recapitalization. Belmont was receptive and, together with Henry Higginson, formed a committee to underwrite a plan of voluntary reorganization. On April 7 the Westinghouse board approved the plan.[26]

Westinghouse then sent the stockholders a circular urging them to accept its provisions. "Your officers," he admitted, "have for some time been confronted with a condition of affairs which has kept the Company upon the brink of a receivership." That condition was the perpetual lack of sufficient capital. The proposed plan would pay off the floating debt, leave some cash in reserve, and provide the means to acquire outright the two companies that Westinghouse already leased, United States Electric Lighting and Consolidated Electric Light. They would be folded into Westinghouse Electric and Manufacturing.[27]

The plan went public in May. It asked holders of common to sacrifice nearly 40 percent of their stock, about $2.7 million, by returning it to the treasury. In return their remaining 60 percent, which would be called "assenting" stock, would be entitled to 7 percent dividends ahead of the remaining common, or "unassenting," stock. Another $3.2 million of unsold common remained in the treasury, making a total of $5.9 million. Of this amount, $4 million would be converted into preferred stock, of which $3 million would be issued immediately. Belmont's syndicate agreed to take this $3 million in preferred if the stockholders approved the reorganization plan; in fact, they underwrote only $1 million of it, but creditors of the company agreed to accept part of the other $2 million as payment for their obligations. The funds would be used to pay off the floating debt and acquire the subsidiary companies. When the smoke cleared, the total capitalization would remain at $10 million while interest and lease obligations were reduced by $400,000 annually.[28]

On July 15, 1891, the plan was approved and went into effect. It satisfied virtually everyone. The floating debt had been converted into a stock liability with a corresponding reduction of interest charges. Altogether the company had outstanding only about $8.8 million in stock and no bonds or notes. Westinghouse Electric and Manufacturing had not only survived the crisis but emerged in much sounder financial shape. The only question that remained was whether George Westinghouse remained in full command of the company.[29]

The newly elected board included Westinghouse; Belmont; Charles Fairchild of Lee, Higginson; Henry B. Hyde of Equitable Life; and Charles Francis Adams Jr., the descendant of two presidents who had chosen a business career for himself. The delicate question of management still loomed, for here too the bankers wanted a voice. Fairchild regarded Westinghouse as "a bright & fertile mechanic" but one lacking in tact and a firm grasp of high finance. He and other members of the committee looked for some way to curtail Westinghouse's control over the company. Their solution was to have Thomson-Houston placed in control of the new company, but Westinghouse's low opinion of Coffin put an end to that hope. He knew Adams well and got him to persuade the bankers to leave Westinghouse in charge. Adams liked Westinghouse but regarded him as a "curious man," one of a kind. "He reminds me in a way of John Adams," he confided to his diary, "in essence a really great man but cursed with a vanity which limits him in every direction."[30]

There was truth in this observation. Westinghouse suffered from hubris in at least one respect. He wanted always to command, to lead and not to follow or even share leadership. He consulted freely with others, but then, as Henry Prout observed, "he made up his own mind, and nothing milder than an earthquake could budge him. We have seen him sitting like a rock, serene, gentle, and unmoved when every member of the board of directors was against him." This attitude made him a poor prospect for any merger that did not leave him in charge. He would buy or absorb rivals, make them part of his own company, but he would not become a piece of some larger entity that he did not direct. The new arrangement kept the financial wolf from the door, but it also left Westinghouse beholden to the bankers. His need for capital would only grow if the business continued to expand, which meant that he could never shake free from their influence unless he did well enough to finance his operations out of earnings. That seemed unlikely for some time to come because so much of his work remained in the developmental stage.[31]

By far the most promising project remained Tesla's motor, but it lay in limbo. Tesla himself had given up in frustration after a year in Pittsburgh and gone back to his laboratory in New York, leaving work on the motor to Westinghouse's own engineers. For a time young Charles F. Scott, who had assisted Tesla, worked on the project. However, the financial crunch of 1890 forced Westinghouse to suspend work on it. When Tesla protested, Westinghouse replied that he had little choice in the matter. No one knows exactly what passed between them, but Tesla apparently agreed to drop the onerous $2.50-per-horsepower royalty clause in his contract if Westinghouse would resume work on the motor. It was a generous gesture that later cost him a considerable sum. A relieved Westinghouse wanted to pursue the motor's development anyway and in 1891 put another of his bright young men, Benjamin Carver Lamme, to work on the project.[32]

Lamme had already shown his talent with an electric railway motor. Having come late to the business, Westinghouse needed a better motor than those used by Sprague or Thomson-Houston. In 1889 Albert Schmid came up with the No. 1 double-reduction-gear motor that, unlike rival versions, enclosed the gear. Although the new motor sold well in 1890, Lamme was put to work designing a superior version even before he received the Tesla motor assignment. Lamme understood that "the double-reduction motor . . . with its exposed ar-

MONEY, MERGERS, AND MOTORS   291

mature and field windings, and its two sets of gears, could not persist very long." In August, when a strike shut down the Westinghouse plant, he used the time to pursue a new approach. Within a remarkably short time he produced sketches that turned into drawings and then into working models. The single-reduction No. 3 motor progressed so rapidly that it went on sale in the spring of 1891 and immediately rendered its predecessors obsolete. It remained the standard motor design for the lifetime of streetcars.[33]

The method behind Lamme's motor proved as important as the motor it-self. At an early age he showed a gift for mathematics; where other boys learned the multiplication tables to twelve, Lamme memorized them to twenty-five and then to thirty-six. In 1888 he graduated from Ohio State with a degree in mechanical engineering and an undying love for calculation. It was this talent that impressed Schmid and other Westinghouse officials, who saw to it that Lamme got work that challenged him. The early No. 1 motor became the coming-out party that led to his being assigned the Tesla motor conun-drum. Unlike Edison and his generation of inventors, Lamme had not arrived at his creation through endless trial-and-error experiments. "A radically new type of machine had been produced by calculation," he recalled proudly, "and had showed a tendency to revolutionize the practice in an important section of the electrical field. This gave me the idea that other types of electrical ma-chines could be improved, or even revolutionized, by calculation."[34]

So superior was Lamme's motor that every company, including Westing-house, found itself saddled with an expensive inventory of double-reduction motors that had to be scrapped. "This was a situation which Westinghouse rather enjoyed," said an associate, "for progress was always a good deal more important to him than profit." The episode also illustrated certain key differ-ences between Edison and Westinghouse. The former was at heart an inven-tor, the latter an engineer. Edison was most at home in his chemistry laboratory, Westinghouse at the drafting table or in the foundry and ma-chine shop. Both men had the core values of persistence and determination but in different forms. Edison relished his reputation as the wizard capable of inventing wonders; Westinghouse cared less for the glory of invention than for the practical business of turning possibilities into products. Where Edison insisted on solving a problem in-house, Westinghouse constantly sought out the ideas and inventions of others. He cheerfully gave credit where it was due, which is what led Tesla late in life to call him "one of the world's noblemen."[35]

For all of Edison's emphasis on invention as a business, Westinghouse was even more focused on converting ideas to workable products. "He was ex-tremely quick to see a situation and to judge of the possible merits of any device," said Herbert T. Herr, who became a vice president in the company. Another old associate called Westinghouse "a thirty-day man. The profits of the new idea or the new enterprise would begin to appear in about thirty days." Historian Steven W. Usselman noted that Westinghouse "seldom exhibited either the inventive playfulness that enabled Edison to turn out one clever amusement after another or the compulsive pursuit of technical perfection that sustained the Wizard of Menlo Park's efforts to improve the phonograph." Instead he looked always to what return the solution of a problem would produce. "Not one of [his] patents is a flash out of the blue sky or a vision on the horizon," wrote Prout. "Every one is calculated to meet a situation that he has seen in his own practice . . . and not one of them was invented to sell or as a speculation."[36]

Beneath the bleak financial skies of the early 1890s Westinghouse remained steadfast in his faith that his men would develop the products he needed to keep the bankers at bay. Lamme had made a superb start with the No. 3 motor, which gave the company a solid lead in the electric railway business. Could he perform the same feat with Tesla's motor?

THE SITUATION LOOKED pretty much the same to both Henry Villard and Charles Coffin. The ideal merger would bring together all three major rivals in one giant electrical concern, pooling their resources, patents, and expertise. However, the personalities involved rendered that goal unrealistic. Both men recognized that Westinghouse would not join in any merger unless he managed the new company. Coffin had the same ambition, which made it all but impossible to get an agreement between them; he also resented Westinghouse's "attitude of bitter and hostile competition." Villard had no desire to manage a new company, and Westinghouse seemed small enough to acquire outright. Although some negotiations took place, Villard realized that Westinghouse could not be made a willing partner to any merger except on his own terms. The logical, indeed the obvious, course was to merge Edison General Electric and Thomson-Houston, preferably on Villard's terms.[37]

In February 1891 Villard paid a secret visit to the Thomson-Houston plant in Lynn. "Mr. Villard is to be shown through our plant . . . ," Coffin instructed the superintendent. "Please see to it that *his identity does not become generally known.*" That same month, at a meeting of his board, Westinghouse made it clear that "the only understanding that exists between us and the Thomson-Houston people is as far as relates to certain patents in which we are mutually interested. There will not be any consolidation of interests of the two companies." Rumors had been flying in every direction for months; if nothing else,

Westinghouse pointed the gossips away from his company, which was absorbed in working out a reorganization plan.[38]

From Villard's perspective Edison General Electric, as the larger firm, should be the parent and dominate the management of the new enterprise. To that end he formed in June 1890 the North American Company, a holding company capitalized at $50 million to sweep together the Edison central stations and also serve as parent for the merger. His plan was to assign Thomson-Houston a lower valuation than Edison General Electric, thereby giving the latter control over the new entity. The Baring failure in November lent fresh urgency to Villard's efforts as his German financial supporters began selling their American holdings and recalling their loans. In both his electric and much larger railroad interests Villard found the ground dissolving beneath him as it had in 1883. By the end of 1890 the North American Company was insolvent and Villard's influence in financial circles was waning. Rumors circulated that he had been asked to resign his presidency of Edison General Electric.[39]

Nor was he the only banker eager to promote a merger. Henry Higginson and his investors had come to believe that consolidation offered the best solution to the industry's intense competition and need for capital. "What we all want," declared Fairchild, "is the union of the large Electrical Companies." J. P. Morgan did not at first share that view. "I do not see myself how the two things can be brought together," he wrote Higginson in February 1891, "certainly not on any such basis as was talked about a year ago." However, his attitude began to shift as both Thomson-Houston and Westinghouse pulled ahead of Edison General Electric in the marketplace. During 1891 Edison General Electric earned only $1.4 million in profits on $11 million worth of business, compared to Thomson-Houston's $2.4 million on $10 million in sales. Coffin boasted to Fairchild that he was "knocking the stuffing out of them all along the line."[40]

Thwarted in his attempt to put Westinghouse under Thomson-Houston's wing, Higginson encouraged Villard and Coffin to pursue a merger of their companies. However, Villard had little leverage left. To meet the onrush of financial demands he had sold his Edison General Electric stock. Some of it had gone to Hamilton McKown Twombly of the Vanderbilts, the rest to Morgan. Although still president of Edison General Electric, he lost most of his support when his German friends sold out their stock in the company for a tidy profit. Edison himself still dawdled over an AC system and informed Villard in February 1891 that "the use of alternating current instead of direct current is unworthy of practical men." A merger might salvage something for Villard, or at least provide a graceful exit from the electrical field.[41]

In December 1891 Villard began a round of secret talks with Coffin and his lawyer, Frederick P. Fish. Meanwhile, J. P. Morgan had changed his views on the virtues of an electrical merger. He represented large holdings in Edison General Electric, which had a sizable floating debt and was losing business to its competitors. Knowing little about Thomson-Houston, Morgan asked Twombly to look over the company's books and properties. To Twombly's surprise, Coffin and his team seemed reluctant to join a merger. One Thomson-Houston executive reportedly said that "we don't think much of the way the Edison company has been managed." Apprised of this attitude by Twombly, Morgan asked Coffin to come down to 23 Wall Street and make his case in person.[42]

Villard had assumed that Edison General Electric would control the merger and receive the higher valuation. Charles Fairchild of Lee, Higginson even urged Coffin to give the Edison shares a higher value on the grounds that "for the sake of union T-H can afford to give them a good trade." However, Coffin convinced Morgan that Thomson-Houston was the better-managed company of the two; it had, after all, earned 50 percent more per share in 1891 than Edison General Electric. He proposed converting three shares of Thomson-Houston to five of the new company, with Edison General Electric being exchanged one for one. The terms were accepted, and in February 1892 a committee was formed to execute them. By then Morgan had come full circle in his views, telling Boston financier T. Jefferson Coolidge that "I entirely agree with you that it is desirable to bring about closer management between these two companies."[43]

The committee consisted of Morgan, Twombly, D. O. Mills, and three Boston financiers: Higginson, Coolidge, and Frederick L. Ames. Villard was nowhere to be found; Morgan had taken charge. Official announcement of the proposed merger went public the second week in February. Villard gained reelection as president of Edison General Electric on February 12, but only temporarily. Morgan had come to see that Thomson-Houston had the best and most aggressive managers as well as the best balance sheet. Villard was not present at the meeting and sent a letter saying he would remain president only until a suitable successor could be found. Morgan partner J. Hood Wright offered a resolution expressing regret that Villard chose not to stay on; in fact, Morgan had already arranged for his resignation. Later Villard wrote that he disapproved of the merger. Rumors of bad feelings swirled around the transaction. One version had Edison deeply upset with Villard for "working contrary to his interests in bringing about the consolidation." According to this version, Villard turned to Insull and said, "Were it not for your mismanagement no consolidation would have been necessary."[44]

Insull later conceded that the negotiations were "not particularly pleasant."

Edison didn't want it even though Insull thought it was to his advantage that the consolidation take place. As for his own role, he admitted that "I was quite young . . . and doubtless indiscreet in arousing the hostility of people in the conduct of business. Consequently, I made a good many enemies and . . . these enemies took advantage of the situation and prejudiced Mr. Edison against me."[45]

Edison denied the rumors, saying that "the stories in which some of the newspapers have called this a game of 'freeze-out' are founded on a misapprehension. I expect the consolidation will result financially to my advantage." He also denied the allegations of friction between Samuel Insull and himself and professed to know nothing of the strain between Morgan and Villard. The consolidation was nearly arranged, Coffin would head the new company, and both concerns would be the stronger for it. The whole affair was one of "pure business" and had been negotiated on a "friendly and satisfactory basis." As for himself, he admitted that "the business grew and I could not attend to it," so he sold out and held less than 10 percent of Edison General Electric stock. "I cannot waste my time over electric lighting matters," he added, "for they are old. I ceased to worry over those things ten years ago, and I have a lot more new material on which to work . . . I am not business man enough to spend my time at that end of the concern. I think I was the first to urge the consolidation."[46]

Incorporated on April 15, 1892, the new General Electric Company boasted a capitalization of $50 million (five times that of Westinghouse), $15 million of which went to Edison General Electric shareholders, $18 million to Thomson-Houston stockholders, and the remainder into the company treasury. Coffin became the president and led the company until his retirement in 1922. Thomson-Houston men dominated the management at the executive level; five of the six top officers came from that company. The name of the new company reflected the most obvious change. Gone were the names of the inventors who had done so much to create the electrical revolution, subsumed beneath a moniker that could not have been more generic. Gone too were most of Edison's men, most notably Edward Johnson. Samuel Insull was the only one offered a top post in the new company, but he chose to go elsewhere.[47]

After the merger Elihu Thomson stayed on as a senior scientist, adviser, and researcher. For Edison, however, the merger was the unkindest cut of all. Morgan, who had helped finance his early lamp experiments, did not bother to inform him about the pending merger or the dropping of his name from the new company. Alfred Tate, Edison's secretary, said later that he first heard the story from a Wall Street reporter, Herbert Sinclair. Tate left at once to catch a train to West Orange, where he broke the news to Edison. "I never before had seen him change color," he recalled. "His complexion naturally was pale . . . but following my announcement it turned as white as his collar." Edison said only, "Send for Insull." Although he had been elected to the board, Edison attended only one meeting. He had once said, "I will not go on the board of a company that I don't control," and he had no voice in this new entity. After learning that Insull had favored the consolidation, Edison thought he had been sold out. Insull claimed the misunderstanding lasted only a few weeks, but he admitted that "the spell was broken."[48]

However upset he was over the merger, Edison did not cut his ties to the new company. He could not afford to lose the business, given the costs of his laboratory, and he got along well enough with Coffin and his officers. Publicly he supported the merger, telling a reporter, "I simply want to get as large dividends as possible from such stocks as I hold." Privately he voiced a very different tone. Tate once asked him a technical question and was startled by the response, which came with more vehemence than he had ever heard from Edison. "If you want to know anything about electricity," he snapped, "go out to the galvanometer room and ask Kennelly. He knows far more about it than I do. In fact, I've come to the conclusion that I never did know anything about it."[49]

The problem, though Edison would not admit it, was not that electricity had grown old for him but rather that the field had passed him by. "I'm going to do something now so different and so much bigger than anything I've ever done before," Edison insisted, "that people will forget that my name ever was connected with anything electrical." But Tate suspected otherwise. "These were bitter words spoken by a tongue in the mouth of a wound that never healed . . . ," he noted. "I knew that something had died in Edison's heart and that it had not been replaced by the different and bigger thing to which he had referred. His pride had been wounded . . . He had a deep-seated, enduring pride in his name. And this name had been violated, torn from the title of the great industry created by his genius." But Edison was dead wrong in one respect. The world most certainly did not forget that he was connected to things electrical.[50]

THE PROBLEM WAS easy to describe but fiendishly difficult to solve. Tesla's motor was polyphase, meaning that it used two or more currents with periods of alternation that had to be out of phase with each other. However, the existing alternating-current system was single phase and used as a standard frequency (the rate at which the current reversed direction) 16,000 alterations per minute, or 133 cycles per second. This high frequency had been adopted to prevent flickering in the carbon-filament lamps. To their dismay, Tesla and the other Westinghouse engineers discovered after considerable research that a frequency that high simply wouldn't work for any kind of alternating-current motor. Nor could Tesla's motor be adapted to work at lower frequencies. By contrast the speed of a DC motor could be easily adjusted and therefore made suitable for a variety of uses. But DC motors required a commutator, with all the maintenance and other problems that entailed.[51]

Westinghouse understood the importance of developing an AC motor. It held the key to extending the use of electricity beyond illumination to a wide variety of applications. Only gradually did the magnitude of the challenge become clear to the engineers. They had to design a motor from the ground up and create for it some system of polyphase generation and transmission. Having done that, they had to choose some frequency well below 133 cycles per second that would be acceptable to manufacturers and customers as well. A decision also had to be made on how many phases or voltages the motor should use. A two-phase motor could be readily adapted to single-phase lighting circuits, and lighting was still the primary market for electricity, along with street railways. A three-phase system provided a steadier flow of power to the motor and used 15 percent less copper in transmission lines. It was more versatile but also required a more complicated network. Put another way, what served lighting well served motors poorly or not at all.[52]

Tesla had given the engineers the prototype but not the finished version, which remained tantalizingly out of reach. His patents would do Westinghouse little good unless the engineers could unlock the riddles baffling them. "The first three years of work on the induction motor were mostly valuable in showing what could not be done," observed Henry Prout. "The engineers played a losing game against high frequency and single phase, and they simply developed something that could not be used until standards of phase and frequency were changed." After Tesla returned to New York, Charles Scott took up the task and produced a motor with many good features but was still unable to solve the frequency disconnect. For a time the work was dropped. Once Westinghouse eased past his financial crisis, attention turned again to the induction motor dilemma. Lamme had observed Tesla and Scott working on the project. He noticed a problem and brought it to Scott's attention. Scott suggested that he tackle it as soon as he could find the time.[53]

Lamme's star continued to rise in the company. He had done impressive work on both AC and DC generators, working from calculations he made that often ran against conventional wisdom. During 1891 he began work in earnest on two projects that proved vital to the company: the induction motor and the rotary converter. While he busied himself with calculations, the Westinghouse engineers grappled with the issues of phase and frequency. Electrical designers and engineers tended to choose the frequency that best suited the immediate needs of whatever equipment they were creating. At least nine different frequencies, ranging from 25 to 133, were already in use to match the coupling of various incandescent lamps, transformers, arc lights, induction motors, and other apparatus. As new products and fresh designs of old products grew in number, the need for standardizing frequency grew more urgent.[54]

So too with phase. If the engineers could decide between two and three phases for a motor, and settle on one frequency, progress toward a design could move ahead more rapidly. In an industry where the technology evolved quickly and equipment design mutated constantly, the ideal became some form of universal system that could accommodate a broad range of usages. Nothing even close to such a system existed in 1890, but efforts to rationalize phase and frequency marked an important step in that direction. Lewis B. Stillwell had studied the phase and frequency issues while on business in Europe and reported his findings. In 1890 he joined Shallenberger, Scott, Lamme, and Albert Schmid as an informal committee to come up with standards. They decided to adopt 60 cycles and two-phase as their standards.[55]

During 1891 Lamme started work on a rotary converter, a device on which another inventor, Charles S. Bradley, had already received a patent and a version of which Siemens & Halske exhibited at the Frankfurt International Electrical Exposition in 1891. As the name indicated, this device converted AC to DC power for those applications that required the latter. Since all electric railways used DC current, they needed either numerous power stations as the system expanded or the ability to transmit alternating current over distance and change it to DC. Work on the rotary converter proved to be far more crucial than anyone suspected. Bradley established a factory in Yonkers, New York, to manufacture his version but soon sold out to General Electric, giving that company the lead. However, Lamme had done some calculations in 1891 and even built a 150-horsepower converter to try out. Work began in earnest the following year and helped the engineers decide on 30 cycles as a standard for the prototype machines.[56]

Once developed, the rotary converter became not only a versatile device but one that eased the transition from older DC to newer AC systems. "The transformer was not enough," observed Henry Prout. "To transmit electric energy over considerable distances, high-potential alternating current must be used; but there are many important uses for direct current which alternating current does not meet. Moreover, . . . there were great investments in direct-current motors which would be sacrificed reluctantly, if at all." The rotary converter enabled both currents to coexist in those areas like electric railways and chemical processes that relied heavily on direct current. Westinghouse seized the lead in developing the converter, but it would take time to bring to market.[57]

The rise of men like Lamme and Scott to prominence reflected the change occurring in electrical research. Old-school electricians, including Edison, Thomson, and Westinghouse, learned the art in home workshops and private tinkering. They were self-taught men who became expert in the practicum of problem solving rather than theory. The newer breed got their education from the proliferating number of schools offering courses and programs in engineering. When Edison opened the Pearl Street station in 1882, only about seventy American institutions offered courses in engineering, none of them in electrical engineering. By 1899 the number of colleges and universities had jumped to eighty-nine, nearly all of which offered full four-year programs in engineering and forty-nine of which had courses in electrical engineering. Scott graduated from Ohio State and did advanced work in mathematics, physics, and chemistry at Johns Hopkins before coming to Westinghouse in 1888, the same year Lamme graduated from Ohio State. They were the vanguard of a corps of bright young men rising to prominence in the most technologically advanced field around.[58]

In grappling with the development of induction motors, rotary converters, and AC transmission systems, they sensed that they were on the verge of a major breakthrough in the field. So much about alternating current remained a blend of myth and mystery even to those who worked with it. The challenge of working at the frontier of knowledge excited men like Scott and Lamme, who had known each other at Ohio State. "He developed methods of calculation for replacing the 'cut-and-try' method," Scott said of his friend. "His inventive ingenuity proposed new types to which his intelligent calculation gave definite size and proportion. Contrary to the experience and advice and warning of others, he made radical changes which have now become universally accepted practice." They would need all the ingenuity they could muster, for the future of Westinghouse hinged on finding practical solutions to the application of alternating current. Two projects in particular offered immense potential for profit and publicity if an alternating current system could be developed: the Chicago Columbian Exposition and the proposed power plant utilizing the waters of Niagara Falls.[59]

CHAPTER 15

A Show of Lights: Chicago 1893

The Columbian Exposition is a magnificent triumph of the age of Electricity . . . With the exception of some of the exhibits in Machinery Hall all the exhibits in all the buildings are operated by electrical transmission. The Intramural Elevated Railway, the launches that ply the Lagoons, the Sliding Railway on the thousand foot pier, the great Ferris Wheel, the machinery of the Libby Glass Company on the Midway, are all operated by electrically transmitted energy . . . Beginning with the pressure of President Cleveland's thumb upon a "Victor" telegraph key that set the machinery of the Fair in motion . . . everything pulsates with quickening influence of the subtle and vivifying current.

—JOHN P. BARRETT (1893) [1]

Paris was lighted with 1,100 arc lamps and 9,000 incandescents, and the display was justly regarded as the wonder of the world . . . The Columbian Exposition contains the far greater number of 5,000 arcs and 110,000 incandescent lamps, and the power is applied in ways not dreamed of only four years ago.

—CHICAGO TRIBUNE, JUNE 2, 1893

HAVING ALREADY SEEN ONE WORLD'S FAIR, Ned could not resist the temptation to visit another, especially when it was being held so much closer to home. As a nine-year-old he had been awed by almost everything he saw in Philadelphia. He still remembered how small he felt standing in front of the mighty Corliss engine that seemed to loom above him almost as far as he could see. Since then he had finished school, had become a schoolteacher, and was engaged to be married. The trip to Chicago might well be his last fling as a single man. Ned would have liked to come with his father, as he had in 1876, but he had died a year earlier. Still, his memory and their experience together in Philadelphia accompanied Ned to Chicago and resonated in everything he saw.

Even though he was older and presumably wiser, Ned did not know what to expect. He had read about the fair in the papers, where it was described as a fairyland rising out of what had been a swamp alongside Lake Michigan, but he had no idea what a fairyland looked like. The city itself hit him in the face with its loud, bustling, chaotic energy. As the train wheezed toward its heart, Ned passed through "an industrial amphitheater bigger and blacker than Pittsburgh—endless reaches of factories, marshaling yards, slaughterhouses, grain elevators, and iron mills, and slag heaps and coal piles that looked like small mountains . . . everywhere, covering everything, were wind-driven clouds of black and gray smoke." After reaching Chicago on a pleasant if sooty June day and finding a boardinghouse he could afford, Ned took a horsecar to the fair's entrance. Admission cost fifty cents—the same as at Philadelphia—but fifty cents bought more in 1893 than it had in 1876. Once he stepped inside the grounds, Ned saw at once what was meant by a fairyland.[2]

Before him shimmered the beautiful Grand Basin with a gilded sixty-five-foot Statue of the Republic at the far end and, at the end nearest him, the magnificent snow-white Columbian Fountain. It was, he learned, an allegory of the great explorer's discovery of America, the commemoration being celebrated a year late by the exposition. Two smaller fountains flanked the Columbian, and two canals flowed from the Basin; one of them passed the Manufactures and Liberal Arts Building before emptying into a lovely lagoon. Around the Basin stood a cluster of buildings with roofs of a uniform height, all of them designed in the neoclassical style and painted a dazzling white. This was the Court of Honor, the centerpiece of the fair that earned it the name "White City." Behind the Columbian Fountain stood the Administration Building with its imposing black and gold dome. Two buildings flanked the Basin itself: the gigantic Manufactures and Liberal Arts edifice and Agriculture. To the left of the Administration Building stood Electricity, and Mines and Mining; to the right, Machinery and its annex. Manufactures claimed the title of the world's largest roofed building; 787 feet wide and 1,687 feet long, it housed 44 acres of exhibits and featured a clock tower 125 feet high.[3]

For a time Ned simply stared, savoring the spectacle before him. Nothing in his memories of Philadelphia prepared him for this dazzling display of buildings, water, esplanades, and sculpture. He walked north between the Electricity and Mines buildings only to confront another beautiful body of water, the lagoon, which surrounded an island reached by four walkways, one of them directly in front of him. To his left loomed the Transportation Building. He stared in wonder at its gleaming Romanesque entryway with five concentric semicircular arches painted in vivid reds, oranges, and yellows and brimming with ornamentation. How different it looked from the chaste neoclassical designs he had just left! Beyond it stood the Horticulture and Women's buildings, and across the lagoon from them the U. S. Government and Fisheries buildings. Ned strolled across the walkway onto Wooded Island and headed toward its far end, where he found to his delight the delicate Japanese pavilion modeled after an ancient temple. Gondolas cruised in the waters around him.[4]

North of the lagoon, just beyond the North Pond, lay the Palace of Fine Arts, but Ned saved that visit for another day. Instead he headed west between the Horticulture and Women's buildings to the Midway Plaisance with its rows of exotic and entertaining exhibits ranging from a Moorish palace to a Samoan village to the North African women performing the *danse du ventre*, or "belly dance," that Americans later called the "hootchy-kootchy." He passed the Blarney Castle and the Irish, German, Turkish, Javanese, Austrian, Chinese, Lapland, Tunisian, Algerian, and Dahomean villages—most of them inhabited by genuine natives.[5]

What grabbed Ned's attention, however, was the most stupendous sight of all, the enormous Ferris wheel. Billed as the "World's Greatest Ride," it carried riders 265 feet into the air, higher than the crown of the Statue of Liberty, giving them a fabulous view of the exposition and the city at the top. Eagerly Ned plunked down half a dollar and climbed aboard one of the thirty-six wood-paneled cars, each larger than a Pullman Palace Car. Every car had thirty-eight plush seats and room for twenty-two standees. Altogether the giant wheel could accommodate 2,160 passengers. Two 1,000-horsepower reversible steam engines supplied its power. For his fifty cents Ned enjoyed two complete revolutions with six stops on each turn, including one at the very top.[6]

Using his own funds and initiative, the thirty-four-year-old Ferris had responded to a challenge from Daniel H. Burnham, the prominent architect who oversaw the design and construction of the fair. Burnham had urged the architects and builders to give the fair something spectacular to rival the Eiffel Tower that was the centerpiece of the 1889 Paris Exposition. Ferris built it in only five months and then had to confront skeptics who feared disaster when it went into operation. To convince them of its safety, Ferris, his wife, and a reporter climbed aboard the wheel for a ride one day during a gale that buffeted them with hundred-mile-per-hour winds. No mishap occurred. The Ferris wheel became the signature emblem of the exposition as the Corliss engine had in 1876.[7]

Once back on solid ground, having caught his breath, Ned finished the mile-long circuit of the Midway and then headed back to the Court of Honor. During the next few days he would return often to the Midway; that was where the fun was. For now, however, he wanted to see what the fair had—besides the Ferris wheel—that could compare to the fabulous Corliss engine. He strolled over to the Machinery Building, which, with its annex, was the second only to Manufactures and Liberal Arts in size. Inside could be found an endless array of machines for a broad range of uses from putting out fires to making books. "In 79 groups and nearly 200 classes of exhibits is here represented almost every mechanical device fashioned by the ingenuity of man," noted one observer. This array of technology reminded Ned greatly of Philadelphia, yet the machines belonged to a different age that often made Philadelphia seem primitive. "The Centennial exposition created an epoch in Machinery exhibits at International Fairs," proclaimed the *Official Guide*. "Compared with the Centennial the Columbian exhibit is full of surprises."[8]

Impressive as the machinery exhibits were, it was the power plant in the south nave of the building that took Ned's breath away. There he found the mightiest source of power ever assembled for one purpose. No fewer than forty-four steam engines churned out power to other machines. The largest of them, an Allis, did not dominate the room as the Corliss had at Philadelphia; it stood at the end of the main aisle but produced 2,000 horsepower—a third more than the Corliss. Six other engines cranked out 1,000 horsepower each, with the other engines ranging from 150 to 650 horsepower. A huge white marble switchboard controlled nearly all the incandescent and many of the arc lights throughout the fairgrounds, including the Midway. The boiler plant featured a continuous line of giant steel boilers eight hundred feet long. Ned saw no smoke, dust, or dirt. The furnaces operated on fuel oil fed by automatic pressure gauges. At the center of the building, amid displays of pumps and water elevators, stood a waterfall; fountains adorned each end of the structure.[9]

Ned noticed something more that had not been at Philadelphia. Electricity was everywhere. The machines in the annex used electricity as their power source. The enormous Allis steam engine drove two Westinghouse dynamos that powered ten thousand lights each. Ned had read about electricity, but it had not yet come to the small Iowa town where he lived, except in the form of the telegraph at the train depot. Where Philadelphia had celebrated the age of steam, Chicago was the showcase for electricity. As Ned discovered to his delight that evening, the exposition was awash in the wonders of electricity. "Illuminated," marveled an English visitor, "the grounds and buildings become an enchanted world." They glowed with ninety thousand incandescent and five thousand arc lights. Giant searchlights splashed the grounds in color, including one six-thousand-pound General Electric monster that was the largest of its kind in the world. The fountains sparkled with colored lights and became a show in themselves. The Ferris wheel alone gleamed with 2,440 lights.[10]

But it was far more than a light show. The elevated "intramural" railway inside the grounds ran on electric power, using an innovative third rail instead of an overhead trolley line. Fifteen three-car trains traveled a 2.8-mile loop around the perimeter of the fair with ten station stops. Even the gondolas ran on electric batteries. Electricity powered the elevators in the buildings and a moving walkway from the pier on the lake to the esplanade. Three enormous electric cranes traveled the length of Machinery Hall, allowing visitors a bird's-eye view of the exhibits. The two fountains flanking the Columbian each shot twenty-two thousand gallons of water into the air every minute and had 250,000 candlepower worth of lights along with rotating filters and different nozzles that created an ever-changing light spectacle.[11]

At Philadelphia the Corliss had provided power directly to all the machines. At Chicago the steam engines created power to generate the electric power that powered everything else. "The Fair, considered as an electrical exposition only," rhapsodized veteran reporter Murat Halstead, "would be well worthy of the attention of the world. Look from a distance at night, upon the broad spaces it fills, and the majestic sweep of the searching lights, and it is as if the earth and sky were transformed by the immeasurable wands of colossal magicians . . . It is electricity! When the whole casket is illuminated, the cornices of the palaces of the White City are defined with celestial fire. The waters that are at play leap and flash with it . . . It is all an electrical exhibit."[12]

Eagerly Ned hurried over to the Electricity Building to get a first glimpse of its wonders. At its south entrance stood a statue of Benjamin Franklin clutching his kite and looking skyward for lightning. Above the various entries were carved the names of great electricians—Franklin, Galvani, Ampère, Faraday, Ohm, Morse, Siemens, Davy, Volta, and others—but, tactfully, none who still lived. The building itself contained forty thousand panes of glass, more than any other structure in the fair. Just inside the south entrance Ned paused at the American Bell Telephone exhibit with its row of "Hello girls" manning a giant switchboard and making actual calls. He marveled at the photophone, a strange device that used a ray of light to send messages and hold conversations without using any wires. All of the building's exhibits ran as motors, drawing current from the dynamos in Machinery Hall. The wires were all strung along the sides of underground passageways tall enough for workers to stand up in, making them both safe and easy to repair.[13]

To the right Ned found a huge display by the Western Electric Company, which filled three pavilions. One display showed Moses Farmer lighting his home with an incandescent lamp in 1847. The Brush company filled the space across the building from Western Electric with its line of central station and railway equipment. In the center of the building loomed the General Electric Tower of Light, a graceful cylindrical shaft rising eighty-two feet through the center of a circular colonnade and adorned with thousands of red, white, and blue lamps capped by the giant Edison bulb. Ned gaped at the bulb, ten feet high and four feet wide at its bulge, a prism composed of forty thousand small pieces of cut glass that flashed light in every direction. The lamps on the shaft were wired to glow in shifting patterns of light and color. At its regular exhibit General Electric featured a 1,200-horsepower dynamo and two 800-kilowatt generators powering a waterwheel and other apparatus. The exhibit included Edison's first direct-connected engine and dynamo, which had created such a sensation at the 1889 Paris show.[14]

After taking in the General Electric spectacle, Ned wandered over to the nearby Westinghouse exhibit, which occupied fifteen thousand square feet and featured a forty-five-foot monument proclaiming in bold letters WESTINGHOUSE ELECTRIC & MANUFACTURING CO. TESLA POLYPHASE SYSTEM. On the south wall Westinghouse had created a beautiful mural honoring Columbus with all the lettering illuminated by 1,988 incandescent lamps in frosted, white, and colored bulbs. The major product lines were all on display, and Tesla had his own section of the exhibit that included motors along with armatures, generators, and even phosphorescent signs that glowed the names of famous electricians. His "Egg of Columbus" delighted Ned as several balls, pivoted discs, and other devices flew about at a distance from the rotating field when current was applied to the iron ring beneath them. In one corner stood a special dark room for simulating a spectacular flash of lightning complete with a thunderous noise that could be heard throughout the building.[15]

And the other applications! Ned stared wide-eyed at machines washing, wringing, drying, and ironing clothes, baking bread, washing dishes, heating rooms and even whole houses, lighting cigars, curling hair, running elevators. Ned understood little of what he saw, but it left a deep impression on him.[16]

Ned left the building and went in search of some other major exhibits listed in the *Official Guide* he had purchased. At the Agricultural Building he smiled at the enormous eleven-ton cheese from Canada and at the thirty-eight-foot-high temple made from thirty thousand pounds of chocolate. The Mines and Mining Building offered the largest gold nugget in existence (344.78 ounces) and Forestry the largest plank, as well as a cross-section of a redwood tree with rings that showed it to have been six and a half feet thick when Columbus discovered America. He walked over to the Krupp gun works to view the celebrated largest gun in the world, a 120-ton rifled cannon that could supposedly hurl a one-ton projectile twenty miles. In the huge Manufactures Building he admired the giant Yerkes telescope, a gift to the University of Chicago, and the breathtaking chapel assembled by jeweler Louis C. Tiffany from hundreds of thousands of pieces of colored glass.[17]

As twilight loomed, Ned dragged his weary legs out of the building and chose a restaurant for dinner. The sheer scale of the exposition overwhelmed him even more than it had in Philadelphia. This was hardly surprising, for the World's Columbian Exposition dwarfed all previous world's fairs. Tomorrow he would tackle another round of buildings; another day would be devoted to exploring the Palace of Fine Arts, which harbored a mind-boggling 10,040 individual exhibits including no fewer than 1,075 American oil paintings alone. He could take in one of the many band concerts, some of them by Sousa's New Marine Band, and one of the symphony concerts arranged by the fair's musical director, Theodore Thomas. However, the classical concerts charged a dollar for admission, a steep price for Ned, while most of the bands played for free. And amid all of these demands he could succumb happily to the lure of the Midway with all its exotica.[18]

Most of all, Ned liked to pause and simply admire the scene before him—the sheer scale of it all, the beauty of the buildings themselves, the harmony of the entire layout. It was a magical city unto itself. "The grandest thing about the exhibition," said Andrew Carnegie, who came there from Pittsburgh—a city under a perpetual cloud of soot and dirt from his steel mills—"was the scene without. The frame was finer than the picture, and more valuable." Like all world's fairs but better than any other, it showed life not as it was but as it should be.[19]

PARIS 1889 WAS the inspiration, the model, the mountain to be climbed. Chicago wanted not merely to emulate but to outdo that fabulous exposition in every way. It might not create any one structure as spectacular as the Eiffel Tower, but it could concoct something entirely new: an exposition with exquisite composition and consistency of design. Something that would advertise to the nation and to the world this city that in a mere two decades had not only rebuilt but reinvented itself after the disastrous fire of 1871. It had fought and won a spirited battle with New York City for the right to host the Columbian Exposition, and now it had to deliver a finished product in a little over two years. The managing board made two wise decisions early in choosing the firm of Burnham & Root as consulting architects and Frederick Law Olmsted as consulting landscape architect. Burnham and Olmsted, along with their partners, worked smoothly together at this monumental undertaking.[20]

Olmsted chose a lakeside site to the north but acceded to one farther south because it centered on an undeveloped park he had designed twenty years earlier. Only part of his original plan, called Washington Park, had been realized; the rest of the area remained a sandy, desolate swamp. The task of transforming it into a site of incomparable beauty would be Herculean, the more so for its sheer size. The grounds encompassed 1,037 acres, nearly triple the size of any previous exposition. Somehow Olmsted completed the huge dredging and filling operation in four months, enabling the erection of buildings to begin in May 1891. Burnham oversaw the construction effort with masterful skill. Through sheer force of personality he put together a tag team of renowned architects to design the buildings, created enthusiasm for the project where none had existed, and promoted the themes of harmony and unity. Through bitter winter cold and drenching rain he pushed and prodded the work gangs forward. In a distinguished career it may well have been his finest performance.[21]

Not since Pierre L'Enfant laid out his ambitious plan for the nation's capital had there been such an attempt to integrate architecture and landscape planning on so grand a scale. Burnham presided over a workforce of thousands, eighteen of whom died and more than seven hundred suffered injuries in 1891 alone. None of the buildings were to be permanent structures except the Palace of Fine Arts, which alone used brick and cast iron in its construction. The other buildings used a composite material called staff, made of plaster, cement, and fiber spread over lathe. The major buildings alone swallowed up thirty thousand tons of staff, twenty thousand tons of iron, untold thousands of tons of glass, and seventy million feet of lumber. Mindful that the exposition was a business proposition that needed constant publicity, Burnham allowed spectators to come watch the work for a charge of twenty-five cents (later raised to fifty cents) even though they often got in the way of the workers.[22]

From this concerted effort arose the magical White City, created in only two years and doomed to exist for only six months at a cost of $18.3 million for construction alone. Like all cities, it needed sources of power and light. Steam engines would furnish the primary source of power as they always had, but not directly to other machines as the Corliss had done. Instead they would power a phalanx of dynamos that would flood the fairgrounds with electricity. Among its many accomplishments, the Columbian Exposition served as the coming-out party for electricity in all its varied roles. Hardly any of its planners suspected in the beginning what role it would play by the end or how strong an impact it would make on visitors. The man most responsible for changing the public perception of electricity at Chicago was George Westinghouse.[23]

The call for bids to light the fair came at an awkward time for the major electrical companies. Edison Electric and Thomson-Houston were still perfecting their merger and for the moment retained their separate identities. Westing-house was easing out of the financial crisis that had all but paralyzed his company for a time. To make matters worse, the country was sliding into a depression that arrived in 1893 along with the fair. Moreover, the interminable suit over Edison's lamp patent finally looked to be settled in the fall of 1892. Westinghouse expected to lose his appeal of the July 1891 decision against him and, preoccupied with other problems, did not even submit a bid for the exposition.

Not so the energetic Charles Coffin. Sitting atop the newly formed General Electric Company, he saw a golden opportunity to profit handsomely from the fair. Expecting no competition, Coffin submitted a bid in mid-March 1892 to provide six thousand arc lamps at $38.50 each, noting demurely that the figure "meant no profit to them as the bid only contemplated giving them an advertisement and helping out the World's Fair corporation." The exposition committee blanched at the price; during the past fall they had paid Chicago Edison $11 per lamp to light the grounds during construction. Rather than submit to what they considered extortion, they jury-rigged an arrangement with several local companies for lamps at $20 each. Undeterred, Coffin then offered to furnish dynamos at the steep price of $15.78 per horsepower. The committee again sought out local contractors and got a price of $2.50 per horsepower.[24]

Next came the biggest and most coveted contract of all, covering some 92,622 incandescent lights to illuminate the fairgrounds. Confident that he had no competitors for this prize, Coffin put in separate bids from the Edison, Thomson-Houston, Brush, and Fort Wayne electric companies—all General Electric subsidiaries—ranging from $13.98 to $18.51 per lamp. Altogether the bids totaled $1,750,000. At that figure the company stood to make a profit of more than $1 million. When the bid box was opened on April 5, however, Coffin learned to his chagrin that an obscure local firm called the Electrical Construction Company had offered a bid of $5.49 per lamp. No one knew much about the firm or its head, Charles F. Locksteadt, who was president of the South Side Machine and Metal Works.[25]

This unexpected bid threw the exposition committee into a quandary. The local company had no factory; how could it deliver so large an order? Nevertheless, the gap between the bids was large enough to arouse suspicion. The committee decided to ask the "Electrical Trust," as the General Electric consortium of companies was being called, to submit a revised bid. Coffin unblushingly responded with a bid of $5.95 per lamp, less than a third of his original figure. Although this price still exceeded that of the local firm, the committee agreed to award the contract to Coffin's consortium without asking for any other competitive bids.[26]

Outraged at this slight, the Electrical Construction Company wrote a vigorous letter of protest, arguing that the committee had not played fair in privately allowing the General Electric firms to submit another bid. At least one fair director was known to be a stockholder in General Electric, and others were suspected as well. The protest letter said pointedly that "the committee called publicly for something which in the light of later developments they did not want, and later and privately arranged to get what they did want from a particular source . . . The fact that several of the members thereof are stockholders in a company which profits by their action would seem to make their position peculiarly delicate."[27]

The source of this letter remains an intriguing mystery. Did it come from Locksteadt or was it orchestrated by George Westinghouse, who involved himself in the situation immediately after its publication? One version claims that Locksteadt approached Westinghouse in hopes of getting him to assume a contract that Electrical Construction could not possibly fulfill. However, it is equally plausible that Westinghouse interjected himself into the dispute, or even that Locksteadt had been his stalking horse. Whatever the case, Locksteadt assigned the bid to Westinghouse, who then sent his agents to protest the "sandbagging process" and to demand that the bidding be reopened. Doing so, he insisted, could save the fair another $50,000 at a time when it was looking to cut corners wherever possible. The committee hesitated, fearing that reopening the bidding might instead produce higher prices instead of savings. Most of the electrical firms had just combined into General Electric; what if Westinghouse joined their ranks as well?[28]

The press immediately sided with Westinghouse against the Electrical Trust. "It remains now," complained the *Chicago Tribune*, "for some enterprising syndicate to obtain control of all the air and drinking water in the United States to put all its citizens completely in the corporate grasp . . . Many people fear that the general electric company intends to stamp out all rivals in the manner generally adopted by such trusts." And what if Westinghouse did join? "None of the men in the trust and few of those in the Westinghouse company are in business for their health, and it may dawn on them to effect a reconciliation."[29]

Clearly they did not know George Westinghouse. To sway the committee, he offered to post a bond of $500,000 and a cash deposit of $50,000 to back his guarantee that he could do the job for $50,000 less than the Electrical Trust. Given the publicity that had been generated, this move all but forced the committee to reopen the bidding process on May 3. The papers applauded Westinghouse for saving the fair from being hijacked by the trust and eagerly awaited the next round.[30]

Westinghouse had not bid originally because his finances had just been stabilized and so large a contract seemed unduly risky that soon after placating
the bankers. Certainly more money could be made through the regular product line than by tying up the shops and staff with so huge a project. Why then
did he change his mind? Most likely the challenge appealed to him. The fair offered fringe benefits that were hard to resist. Little if any profit would come
from what would be a gigantic effort, but the exposition promised to be the
grandest showcase ever seen. It would introduce millions of people to electricity and in the process to the Westinghouse product line. For six months the
public would admire the wonders of electricity displayed in a myriad of forms
in the most ideal of settings. Amid a city of wonders electricity could become
the most dazzling wonder of all. It might in a single stroke send the stigma of
"killer electricity" packing once and for all. As a bonus, scooping the contract
would allow Westinghouse to stick it to Coffin and General Electric. It was, in
short, an unprecedented opportunity that might never come again.

Never one to shirk risk, Westinghouse hurried to Chicago, where his genial
personality quickly won over the local reporters. Already his public relations
man had smoothed the path by casting him as the hero who would save the
fair from the monopolistic greed of that giant trust General Electric. Coffin
now realized he was in a fight and sent his representatives to convince the committee that only General Electric had the resources to make good on a bid. Eugene Griffin, GE's vice president of sales, warned that his company would sue
Westinghouse for infringement of the Edison lamp patent. This threat injected
a dose of uncertainty into the committee's deliberations, but Westinghouse
brushed it aside. "They claim the earth," he said briskly, "but they don't own
it."[31]

On May 16 Westinghouse and Charles Terry, his counsel, went to the fair's
offices in the Rookery, a fine building designed by Daniel Burnham, to await
the opening of the bids. Griffin was there as well. GE's bid, opened first, consisted of three parts. It quoted $577,485 to install a direct-current system,
$480,694 to install an alternating-current system, or $525,384 for a mixed system utilizing both currents. Westinghouse submitted a bid of $399,000 for an
AC system, a figure that amounted to $4.25 per lamp. When the committee
asked him to bid also on a mixed AC-DC system, he came up with $499,319, or
$5.38 per lamp. Whichever system the committee chose, Westinghouse had the
lower bid. However, he did something more that escaped notice. Along with
his bid Westinghouse included a detailed argument for the use of alternating
current.[32]

His argument was a compelling one. An AC system would be cheaper and
more reliable and would require less space for installation. "It is desirable," he

emphasized, "that there should be only one form of generators and one switchboard used in the station." This would ensure a simpler and more reliable system. Thanks to the rotary converter, the danger from fire would be reduced because the potential in any building would never exceed that needed to operate 100-volt lamps, and all the circuits within the building would be separated from the wires leading to the dynamos. To reduce the potential of direct current within safe limits would require twenty to fifty times more feeders and mains for the entire project, which would be both expensive and ugly. It would also increase the danger of leakage.[33]

These points convinced both Daniel Burnham and his electrical engineer, who recommended that the Westinghouse bid be accepted. Once again, however, the committee hesitated. Still perturbed by the threat of an injunction over the lamps, they retreated behind a locked door to deliberate for several hours, then agreed to meet again the next morning. Through the long wait Westinghouse maintained his unflappable good humor as well as his bulldog tenacity. As he left the Rookery he told a reporter candidly, "I feel as though I am entitled to the contract, and I think it will be awarded me . . . There is not much money in the work at the figures I have made, but the advertisement will be a valuable one and I want it."[34]

At the meeting next morning Griffin was joined by Bernard E. Sunny of Thomson-Houston and Samuel Insull and John Beggs of Edison Electric. Griffin renewed his pressure on the directors by reminding them again that the patent litigation cast doubt on Westinghouse's ability even to deliver lamps. Westinghouse laughed the charge away, though he well knew it was a serious issue. More hours of indecisive discussion behind closed doors ensued. After deliberating until 3:30 P.M., the committee wearily decided to defer a decision until they consulted their lawyers on the lamp question. Afterward a fair official assured one reporter that "all this talk about a decision from the courts that will affect Westinghouse is the merest folly." On May 20 the fair's lawyers listened patiently to their brethren from Thomson-Houston outline their case against Westinghouse. That afternoon they met with the committee, made their report, and recommended acceptance of the bid from Westinghouse.[35]

Despite their advice, the debate dragged on. Amid a thickening cloud of cigar smoke the committee wrangled for another six hours before reaching a decision of sorts: They invited the two companies to share the prize. Griffin and his cronies then huddled beneath their own canopy of cigar smoke debating what to do. At first they refused the offer; then, realizing that they could not avoid the unpleasant fact that their bid was higher, they reluctantly agreed. The relieved committee appointed a subcommittee to notify Westinghouse of their decision and request his presence on May 23. Westinghouse and Terry dutifully returned to Chicago for yet another round of talks. The committee went into session at 4:00 P.M. and listened first to Griffin and Beggs from the GE consortium. They expressed their displeasure with dividing the contract and again raised the specter of an injunction on the lamp question. Pressed on the issue, they promised that any legal action would be aimed at Westinghouse but would not interfere with his work for the fair should he be awarded the contract.[36]

Thus reassured, the committee summoned Westinghouse and asked if he would be willing to post a $1 million bond guaranteeing the contract against any such injunction. Of course, he replied without hesitation. The directors then huddled again behind closed doors and continued to deliberate until 7:30 that evening. Finally they voted unanimously to award the prize to Westinghouse on condition of the $1 million bond. Later one member explained that "there has been no other matter ever brought before the committee that occasioned us so much trouble as this. We desired to do the right thing, but at first we were afraid that Mr. Westinghouse would be unable to carry out the terms of the contract if we gave it to him." Informed of the decision, Griffin stalked out of the office and told a reporter angrily that once the court ruled on the patent appeal, Westinghouse would be "entirely in our power. He will not be able to make his own lamps and he can only buy from us. We will not injure the fair, but we will not let him continue his contract." As Westinghouse left, he told a reporter genially, "That is an easy task. There will be no difficulty in furnishing the entire plant by the time of the opening of the Exposition."[37]

FOR THE ENGINEERS in Pittsburgh the task would be anything but easy. Westinghouse had startled them by announcing that he was going to Chicago to secure the exposition contract, but few expected him to prevail. When he returned to Pittsburgh, the impact of his success was immediate. He summoned one of his top draftsmen, E. S. McClelland, to his office and informed him of the contract. "I want an engine," Westinghouse declared. "1,200 brake horse power . . . 200 revolutions per minute (Engines of that size usually run about 75 R. P. M.) . . . 150 pounds per square inch boiler pressure, non condensing . . . Splash lubrication . . . Must go in such and such space." At each specification McClelland gulped and said, "Yes sir." When Westinghouse had finished his list, he said, "I will be in again at 2:00 o'clock to see what you have."[38]

Westinghouse swept out of the room, leaving McClelland flabbergasted. Impossible, the draftsmen protested. They were building 250-horsepower engines at the time; how on earth could they come up with a 1,200-horsepower model operating at 200 revolutions? But the boss had ordered it, so they had to attempt the impossible. At 2:00 P.M. Westinghouse called to say he would not be in until the next morning. McClelland and another man toiled until 2:00 A.M. in the drafting room, struggling to come up with a design. A eureka moment gave them the notion of a vertical engine that could fit the required space. Eagerly they set to work drafting a version that might be acceptable. Westinghouse arrived and examined their drawings. He nodded approval and said, "How soon may I have four of them?"[39]

The lamp posed a thornier problem. The court ruling in July 1891 had upheld Edison's patent, and, as Westinghouse anticipated, the appeals court sustained it on October 4, 1892. Although the patent would soon expire, it blocked Westinghouse from using any form of lamp covered by Edison's patent for the fair. General Electric had no intention of selling him bulbs and had suits pending against other manufacturers; the only alternative was to come up with one of his own. The Edison patent covered two key elements of the lamp: the high-resistance filament and the one-piece glass bulb. The 50-volt circuit used by his AC system enabled Westinghouse to employ a low-resistance filament. After revisiting the Sawyer-Man patents he had acquired, Westinghouse devised a solution in the form of a "stopper" lamp. It consisted of two pieces; the bottom held the filament and fitted into the bulb-shaped top like a cork in a bottle. The plug was then sealed with cement to make it airtight. The stopper lamp was neither as efficient nor as long-lasting as the hermetically sealed Edison bulb, but it was cheaper to manufacture and enabled Westinghouse to skirt the patent.[40]

To manufacture his improvised lamp, Westinghouse set up a factory in Allegheny, Pennsylvania. He designed and produced both an apparatus to grind the stoppers and an air pump to exhaust the bulbs. Two days after the appeals court handed down its decision, Westinghouse issued a circular announcing that his new bulb, which did not infringe on the Edison patent, would be available in quantity by December 1. General Electric countered by trying to obtain a restraining order in Pittsburgh to prevent Westinghouse from selling its new lamp, but a chance encounter on the elevated railway in New York with Grosvenor P. Lowrey, GE's chief patent counsel, alerted Westinghouse to the danger. Learning through a casual remark that Frederick P. Fish, another GE lawyer, had gone to Pittsburgh, Westinghouse warned his own lawyers there to watch out for trouble. One of them showed up unexpectedly in court and managed to get the request for a restraining order denied. With his new factory up and running, Westinghouse was able to supply lamps for the fair and for his other customers.[41]

Despite its loss of the lighting contract, General Electric hardly went home empty-handed. Thomson-Houston obtained the contract to build the intramural railway through its interest in another company aptly titled the Chicago & Western Dummy Railway Company. Hailed as "the first road in America to operate heavy and high speed trains by electricity," the elevated consisted of four passenger-carrying cars with the lead one powered by a 500-horsepower engine and drawing its power from a third rail. The controlling switches for the motors were so massive as to require compressed air to turn them. Tucked into the southeast corner of the fairgrounds, the power plant featured a giant DC 1,500-kilowatt dynamo of a type never before built. Throughout the fair the elevated ran smoothly and silently, giving riders a taste of a future that never quite arrived. General Electric also supplied the giant searchlights, the 4,300-foot moving sidewalk complete with seats, and the batteries for the gondolas.[42]

For the fair Westinghouse wanted to create an efficient AC system that would showcase the virtues of that current and rid it forever of the image as a killer. Of necessity it would have to be an enormous system, much of which had to be developed from scratch. By November 1890 the company already had 350 central power stations online, but the installation for the exposition would dwarf any past effort. The final installation included a dozen 1,000-horsepower generators. Each unit, standing ten feet high and weighing about seventy-five tons, consisted of two 500-horsepower alternators placed side by side with separate fields. Their armature windings were staggered ninety degrees so that each one provided two single-phase circuits having a quarter phase relation to the other one. Westinghouse himself had suggested this clever arrangement to secure two-phase voltage from each pair of generators. "The idea," recalled Benjamin Lamme, "was that this would be a step toward a coming polyphase supply system." Lamme designed the alternators, which had a higher potential than any ever built in the nation.[43]

To run the power station the engineers put together the largest switchboard ever made, and the first made of marble. Divided into three sections, it covered about a thousand square feet and was mounted so that visitors could reach it by galleries with spiral iron stairways. It operated forty circuits arranged in such a way that if a break occurred in any one, another could immediately be substituted for it. Altogether it presided over 250,000 incandescent lamps and most of the arc lights illuminating the fair. Only 180,000 of the incandescent bulbs were used at any given time; the rest were held in reserve. This mammoth installation required only one operator, who received information by telephone or messenger from all over the fairgrounds and responded by throwing or cutting a switch. As Westinghouse had suspected from the beginning, the exposition's needs for illumination far outran the early contract figure of 92,622 lights.[44]

Impressive as the power plant in the Machinery Building was, Westinghouse hoped to outdo it in his exhibit. Like the central station, it displayed all the components of a complete polyphase transmission system, not as a mere exhibit but in actual operation. It covered about thirty feet and had both a receiving and a distribution station. A large two-phase induction motor, driven by current from the central generators, drove most of the machines. The system had transformers for raising and lowering the voltage, a short transmission line between them, a smaller marble switchboard, both induction and synchronous motors, a rotary converter that operated a DC-railway motor, meters, and other auxiliary devices. Here was the first showing of the progress that had been made on Tesla's motor, but only knowledgeable eyes fully grasped its significance. Visitors could appreciate the lights that illuminated the fair and thrill to the gadgets on the display, but the impact of power transmission and induction motors escaped most of them entirely.[45]

Part of the space was reserved for a display of Tesla's work. The small dark room in the southwest corner housed some early Tesla inventions: motors, armatures, generators, and the captivating phosphorescent signs bearing the name of noted electricians along with two neon signs inscribed WESTINGHOUSE and WELCOME ELECTRICIANS. The rotating Egg of Columbus excited audiences, as did the simulated lightning created by high-frequency charges between two insulated plates. A deafening noise that could be heard throughout Electricity Hall accompanied the ersatz lightning discharges. In August Tesla himself came to the fair for a week to give a demonstration as well as attend the International Electrical Congress, one of many gatherings of all kinds held in conjunction with the exposition. To demonstrate the absurdity of AC as a killing current, Tesla shocked crowds by passing 100,000 volts of current through his body without harm being done. Another popular exhibit rivaled the Egg of Columbus. It used a ring to show a rotating magnetic field in which brass balls were sent spinning and the smaller ones revolved around their larger cousins like moons around planets. Tesla also showed vacuum bulbs "in which small light metal discs were pivotally arranged on jewels and these would spin anywhere in the hall when the iron ring was energized."[46]

Crowds flocked eagerly to the Westinghouse exhibit to see these marvels. Other exhibits had their attractions. Along with the Tower of Light, the General Electric display offered a tribute to Edison with his phonograph, lighting systems, dynamos, telegraph devices, and his new "kinetograph," described as a combination of his phonograph and stereopticon that promised to "not only record and deliver a speech but shows the speaker on a screen, faithfully reproducing his every movement and facial expression." GE also had polyphase equipment on display but not in operation. Elihu Thomson contributed a high-frequency coil that shot out sparks five feet long. Alexander Graham Bell presented a telephone that could transmit voices via light beams, while Elisha Gray offered what he called his "telautograph." A precursor of the modern fax machine, it could reproduce a message or drawing electrically at a distance. The two company exhibits made an intriguing comparison. General Electric was a paean to Edison the Wizard; the other revealed the hand but not the presence of George Westinghouse.[47]

By opening day, May 1, 1893, the exposition was still a work in progress, but the power system was in place and ready to perform. After days of soaking rain and leaden skies, the weather obliged with a sunny morning for the ceremony. An overflow crowd estimated at two hundred thousand gathered before the enormous platform erected in front of the Administration Building in the Court of Honor. The platform seated three thousand and groaned beneath the weight of dignitaries ranging from President Grover Cleveland to governors, Supreme Court justices, congressmen, foreign notables, and the fair's executives. Theodore Thomas, who had also done the honors at Philadelphia in 1876, opened the ceremony by leading the fair orchestra in a snappy rendition of the "Columbian March." After a prayer, a poem by Harriet Monroe, the playing of Wagner's Rienzi Overture, and a speech by Director General George R. Davis, President Cleveland stepped forward and gave a brief speech that could actually be heard by much of the crowd. As the chorus and orchestra burst into the "Hallelujah Chorus," Cleveland pressed a telegraph key that, much to the relief of all the Westinghouse people, stirred the giant Allis steam engine to life. The generators hummed, and water leaped skyward from every fountain as the crowd cheered.[48]

"The electric age was ushered into being in the last decade of the nineteenth century," rhapsodized one reporter, "when President Cleveland, by pressing of a button, started the mighty machinery and the rushing waters and the revolving wheels in the World's Columbian exposition."[49]

One editor marveled at the "miracle of transformation" that in little more than three years had changed "a region largely of sand dunes and swamps into the White City of beautiful palaces, stately bridges, elegant lawns and flower beds, winding lagoons, vistas of incomparable beauty, profusion of statuary." He predicted that the fair would be "an epoch-maker in the development of taste, industrial activity, and greater comfort in our lives." Three days later the temporary telegraph poles holding some of the lights were finally removed and about half of the 1,200 lamps illuminating the walks and promenades were switched on. The electric launches made their debut on May 10, and three days later the Hale elevator made its first thrilling climb of 185 feet to the top of the

Manufactures Building, where a promenade offered spectacular views of the lake and grounds.[50]

Not until May 13 did the first "grand illumination" take place. The spectacle of the fully lit grounds, buildings, and fountains overwhelmed the large crowd that flocked to see it. The fountains dazzled the eye with their giant leaps of water that danced in the changing colors of light. Even more than in the daylight, the White City at night seemed a fairyland aglow in the sea of arc lights, giant spotlights that could be seen miles away, and buildings awash in their own displays of light. "It was a sight such as no man alive today ever saw," wrote one reporter, "nor ever again will see when the fair has closed its gates." The fair's board decided to open the grounds every Tuesday, Thursday, and Saturday night until 10:30, something that would not even have been conceivable at Philadelphia only seventeen years earlier. Nighttime at the fair became a show unto itself, and the fear that the cost of staying open would exceed the receipts quickly proved groundless.[51]

Despite a ragged start because of poor weather and incomplete exhibits, the exposition did spectacularly well. Altogether it drew more than 27.4 million visitors, nearly three times that of Philadelphia, half of them foreign. When the books closed, the fair even turned a modest profit, although the investors still lost money. The visitors who plunked down their cash for the spectacle got full value. Writer Hamlin Garland was so excited by what he saw that he wrote his parents, "Sell the cookstove if necessary and come. You *must* see this fair." The midwestern location made it easier for people from all over the nation to reach the grounds. While the summer wore on, the contrast between the fantasyland of the fair and the gritty, grimy city outside its gates became even more glaring as a deepening national depression threw more people out of work and into severe hardship.[52]

During the worst of the financial panic, the fair's officials made sure that George Westinghouse got his pay by arranging for him to receive a stipulated amount of the current receipts. At one point, when the banks balked at cashing checks because of a severe currency shortage, the officials handed over to the local Westinghouse representative a large quantity of dollars, half dollars, and quarters. He promptly shipped them to Pittsburgh, where Westinghouse used the loose cash to pay his workmen at a time when currency was commanding a premium of 5 percent.[53]

When the fair closed its doors on October 30, the ceremonies were austere and funereal not only because of the worsening depression but also because of a local tragedy. Two days earlier a disappointed office seeker had shot Mayor Carter Harrison to death and plunged the city into gloom. Harrison had been hard at work seeking ways to preserve the buildings after the fair closed. On the very day of his death he told a crowd at the exposition that it would be a crime "to think that it would be allowed to crumble into dust." Like all fairylands, however, it could not last and was not made to. Amid the worsening turmoil of the depression, arsonists destroyed most of the main buildings with two spectacular fires in January and July of 1894. The rest were reluctantly consigned to the wrecking ball except for the Palace of Fine Arts, which had been built to endure. It became first the Field Museum and much later the Museum of Science and Industry.[54]

However, its legacy lingered in many forms. For all its many charms, electricity had been the big show and alternating current its star. Westinghouse had given the fair three times more electricity than was being generated in the entire city of Chicago. His elaborate exhibit gave the nation its first glimpse of a working polyphase system. As Lamme described it, "The 250 hp. Motor received power from the 2-phase lighting machine . . . and drove, by belt, the alternating current-direct current generator, giving 3600 alternations. From the alternating side of this alternative-current-direct current generator, current was supplied to the two rotaries, so that the whole exhibit could be in motion at one time. In addition, one of these rotaries furnished direct current for the operation of some apparatus in another exhibit." Another rotary converter, rated at 500-horsepower capacity at 550 volts, served an electric railway display.[55]

Nor was this all. Waiting in the wings was an understudy ready to burst into stardom: the induction motor, which after seven years had finally reached a state not of perfection but of payoff. Henry Prout called its development "one of the great and splendid chapters in electrical history . . . The induction motor and the rotary converter made possible the prodigious development of the alternating-current system." Like the converter and so many other innovations, the induction motor evolved from the hands of several inventors working independently of each other. Tesla gained the most acclaim and had his key patents sustained. Galileo Ferraris conceived of the rotating magnetic field about the same time as Tesla; Westinghouse secured his patent but found it useless for his needs. Michael Dolivo-Dobrowolsky of Germany took out patents for a motor in 1889 and credited Ferraris for the principle. Several versions of induction motors were displayed at the Frankfurt Exposition in 1891, but no one had yet come up with a practical version for commercial use.[56]

The induction motor was rugged, durable, and watertight. It turned at nearly constant speeds regardless of the load and could be built in all sizes, speeds, and voltages. Its one shortcoming was that it ran only at the speed for which it was built and was difficult to adjust, while the speed of a DC motor could easily be adjusted. At the fair Westinghouse integrated the Tesla induction motor into an entire polyphase system that revealed for the first time the contours of a long-sought universal system. Historian Thomas Hughes described it this way:

> The motor in the display was intended to represent the prime mover of the universal system. The motor powered a two-phase generator, which in turn sent current to step-up transformers for transmission and to step-down transformers at distribution points. At these points a motor drove an arc-light generator, a rotary converter supplied direct current for streetcar motors, incandescent lamps were supplied, and various motors were driven.

Here, on a miniature scale, was a system capable of providing power for lights, traction, and machines of all kinds. The issue was no longer a technical one of how to create such a system. Rather it was fast becoming a practical question of how to implement a universal system without having to scrap the enormous investment in earlier systems and apparatus. The rotary converter had made AC and DC power compatible to a degree but could not solve the entire problem.[57]

Technological change occurs in two broad forms. The first type improves or upgrades existing technologies in various ways but maintains the basic format of the machine or system. The second and more radical type simply replaces the existing technology with one based on altogether different principles. Examples of this range from the diesel locomotive over the steam locomotive to the compact disc over the long-playing phonograph record to the calculator over the slide rule, as well as the automobile over the horse and buggy. Although the universal system seemed to embody elements of both types, knowing observers saw it as the doorway to the future of an electric industry geared to sell and distribute light and power. It would render the war of the currents a relic of the past. Chicago had demonstrated that electricity—and especially alternating current—was the wave of the future and that its potential had only begun to be tapped.[58]

"Very few of those who looked at this machinery," wrote Henry Prout, "who gazed with admiration at the great switchboard, so ingenious and complete, and who saw the beautiful lighting effects could have realized that they were living in a historical moment, that they were looking at the beginnings of a revolution."[59]

LATER ELIHU THOMSON paused in his busy schedule to ponder what he had seen at the fair and what it meant. He could not help comparing it to his experience at the Philadelphia exposition in 1876, which he had taken in as a wide-eyed youth. How much had changed since then! It was almost as if he lived in a different world. "No similar period in the world's history," he concluded, "has in any art shown so rapid development, so extensive and refined scientific study and experiment, so active invention, so varied application, such care and perfection in manufacture, as has taken place within the electrical field."[60]

In 1876 no dynamos could be found powering lights. Philadelphia displayed the crude Gramme and Wallace machines along with Weston's version, which was used for electroplating. One Gramme dynamo, connected by belt with another, enabled it to run as a motor operating a pump that produced a little cascade of water. The Wallace machines provided current for a single arc lamp, powered plating baths, and were reversed to run as motors. Apart from these examples, the electric motor was virtually nonexistent and relied on batteries for its power. At Philadelphia the great Corliss engine became the main attraction, the centerpiece from which all other machines drew their power. The Chicago fair used motors to deliver power to many places and confined the large generators to one building that fed a host of others.

In Philadelphia arc lamps were a mere curiosity; in Chicago they splashed light everywhere. "The whole of Jackson Park was dotted over with these miniature suns at night," Thomson marveled, "and the interiors lighted." Huge searchlights with sixty-inch mirrors threw beams of light into the heavens and across the fairgrounds with an intensity never before seen. Incandescent lamps, which did not exist in 1876, could be found everywhere in Chicago, not only as lighting but arranged in artistic configurations employing many colors and designs. The transformer was unknown in 1876 and the storage battery nearly so, yet in Chicago such batteries powered the fleet of launches that cruised the lagoon. And the intramural railway! Here was a device completely unknown or even imagined in Philadelphia. It drew power from a direct-coupled continuous current dynamo rated at 2,000 horsepower and built especially for the fair without a prototype or previous experiment.

Then there were the exhibits—the Edison tower, the telephone exchange, a variety of motors to operate pumps, hoists, elevators, locomotives, and mining machinery, a wide variety of lamps, and lighthouse equipment. Perhaps the most exciting exhibits for their promise were those utilizing alternating current, especially the motors employing single-, double-, and triple-phase power. "This class of electric machinery," Thomson noted, "is of so recent production that we must look to the future for its application." That future was arriving faster than anyone dared hope, and as Thomson himself admitted, "in dealing with the subject of electrical applications, . . . a new discovery might at any time change the aspect of every prophecy based on present knowledge and conditions."

THOMSON JOURNEYED TO Chicago on business as well as pleasure, though it proved to be the most pleasurable of business. Like so many other organizations, the International Electrical Congress arranged to hold its meeting at the fair in the Hall of Columbus. An impressive roster of luminaries in the field streamed into the hall for the opening session on the afternoon of August 21. The elderly Elisha Gray, his handsome face wreathed in a snowy beard, drew a wild round of applause from the audience when he escorted to the podium a squat, balding figure with unimpressive features. Hermann von Helmholtz, who had become the most famous physicist in the world for his work in mechanics, acoustics, optics, and numerous other fields, had come to attend the congress and view the wonders of the Chicago fair. The audience applauded every mention of his name, and he was promptly made honorary president of the meeting. He responded with a few gracious remarks, as did William Preece of England and Eleuthère Elie Nicolas Mascart of France, two other distinguished electricians.[61]

Elihu Thomson, the temporary chairman, surrendered his post to Gray. Silvanus Thompson of England and Galileo Ferraris of Italy, two other founding fathers of the electrical profession, were there as well. Except for Thomson, who looked surprisingly young with his dark hair and drooping mustache, they were the old hands of the trade, ready to anoint a new generation of inventive minds. One of the most brilliant of those young minds was scheduled to deliver a lecture on Friday the twenty-fifth. Nikola Tesla, exhausted and overworked as usual, was the headline attraction of the congress. An overflow crowd crammed into the Agricultural Assembly Hall to witness his performance. Thunderous applause greeted Elisha Gray as he escorted the gaunt figure with dark, sunken eyes to the platform. Dressed in a stylish four-button cutaway suit, eyes cast to the floor, he heard Gray introduce him by saying, "I give you the Wizard of Physics, Nikola Tesla."[62]

"I have with great reluctance accepted these compliments," Tesla began when the applause died down, "because I had no right to interrupt the flow of speech of our chairman." With that he began a lecture/demonstration using a jumble of apparatus that few in the audience recognized. He showed new steam generators and mechanical oscillators, some so small they could fit "in the crown of one's hat." He fetched up a continuous-wave radio transmitter that could turn lamps on and off without wires. Radio was to be another great frontier Tesla would explore, but few if any in his audience grasped the implications of his device. Another exhibit, similar to his Egg of Columbus, illustrated both a rotating magnetic field and Tesla's theory of planetary motion. Other displays gave some viewers the sense that Tesla was working toward the goal of creating light without heat. When he had finished, the audience showered him with applause.

No one put on a better show than Nikola Tesla. Where others dazzled the uninformed and gullible observer, he had the ability to enchant the most knowledgeable spectators with performances that spurred their imaginations. The next afternoon he went to the Westinghouse exhibit in Electricity Hall to work his wonders on the Tesla display there. One reporter thought he was on retainer for the company at a fee of $5,000 a month. "He is looked upon by electricians," he wrote, "as one of the most remarkable discoverers of the age."

ON THE TRAIN ride home Ned had ample time to sort out all the things he had seen and experienced at the exposition. It had been the most exciting four days of his life, and he was determined to store as much of what he had seen in his mind as possible and to make sense of as much of it as he could. The sheer scale and variety of all he had seen and done overwhelmed him. Some things he would never forget, like his ride on the Ferris wheel or the fabulous spray of the fountains in the Grand Basin with the water dancing in their kaleidoscopes of colored light. So too with his long hike through the Manufactures and Liberal Arts building, the sheer size of which left one numb. The ground floor enclosed more than thirty acres of exhibits, while the fifty-foot-wide galleries around the entire structure added another fourteen acres of floor space. The U.S. Capitol, the Great Pyramid, Madison Square Garden, St. Paul's Cathedral, and Winchester Cathedral could all fit onto its floor with room to spare. To walk every aisle and cross-aisle on the main floor and galleries amounted to a journey of fifty miles. To see everything on display in the building required more than a week of roaming the aisles nonstop for eight hours a day.[63]

Ned got only a modest taste of the building's treasures, but even they were enough to boggle his mind. Never in his life had he seen so many different kinds of things. Goods, many of them actually for sale and bearing price tags, stretched out for miles before him—everything from jewelry to furniture, tableware to bathroom fixtures, baby carriages to wallpaper. It was, as a historian observed, "the largest department store the world had ever seen," and Ned had not seen any department store except in Philadelphia, where his father had taken him to Wanamaker's new and different emporium. He had never imagined there could be so many things to buy or own. He wandered in a daze

among the wares until he felt utterly lost, then went outside to clear his head
before attempting another foray into the endless aisles and galleries.[64]

Some of the exhibits he saw throughout the fair stuck in his mind because
they seemed so bizarre to him. He smiled at a Venus de Milo made of chocolate, a map of the United States fashioned from pickles, and a knight on horseback composed entirely of prunes from California. Santa Barbara contributed
a fifty-foot obelisk of olive oil bottles, Denver a tower of beer bottles. On his
second day Ned finally bought another guidebook. Entitled *The Time Saver*, it
listed and located "5,000 Things . . . That Visitors Should Not Fail to See" and
ranked them as "Interesting," "Very Interesting," or "Remarkably Interesting."
Some exhibits in the latter category included a twenty-six-foot section of California redwood pierced by a stairway, a Japanese iron eagle adorned with three
thousand individually crafted feathers, guns that fired the final shots of the
Civil War, the Ferris wheel, and the original manuscript copy of the Declaration of Independence. Ned eagerly sought out all the top-rated displays. He
even browsed dutifully through the 10,040 exhibits in the Palace of Fine Arts
even though not one of the 1,075 American oil paintings rated a "Remarkably
Interesting" in the guide.[65]

For all the strange and wonderful things he had seen, however, two impressions lingered most vividly in Ned's memory. One was the sheer beauty and
unity of the fair itself, the blend of magnificent buildings, glistening waterways, gorgeous plantings, broad promenades, beautiful sculptures, and pleasant walkways to every corner of the grounds. From time to time Ned paused
simply to admire the overall picture that filled his eye from almost any place
he stood or sat. The second impression revolved around this new (to him) phenomenon of electricity, which seemed to be everywhere at the fair. The presence of bright, clear, and brilliantly colored displays of light excited him, as did
the relatively clean, quiet power that ran everything from the giant pumps to
the lights to an endless array of machines, including the Ferris wheel. He wondered how long it would be before electricity finally came to rural Iowa.

Peering out the window of his gently rocking coach, lulled to drowsiness by
the clacking of the rails, Ned thought about his future. He was going home
soon to be married and, he hoped, to raise a family. Maybe he would have children of his own that someday he could take to a world's fair if there was another one. What fabulous things, he wondered, would be on display by then?

CHAPTER 16

THE NIAGARA FALLOUT

The work on the American side of Niagara ... is really a colossal piece of engineering; and although it had been conducted during the past two or three years so quietly that but few reports of it have been made public, a great deal has been accomplished toward utilizing that gigantic source of energy. We are to-day on the very eve of electrical developments in the line of transmission of power that are destined, perhaps, to change profoundly our industrial methods.

—ELECTRICAL WORLD, FEBRUARY 6, 1892

The twentieth century is about to dawn on the world and still Niagara has not been "utilized" ... The conclusion was hastily jumped at that a great center like Niagara was also to become the center of industrial power. This is a great mistake ... It is safe to say that Niagara and all other greater water powers of the world will continue to waste their strength as they have done in the past. The hope of a wide diffusion of mechanical power by means of electricity lies in a fundamental misconception of the laws of electricity.

—CHICAGO TIMES, JUNE 8, 1892

THE THREE BATTLES THAT BECAME intertwined during the late 1880s disentangled themselves during the early 1890s and went their separate ways. The debate over electrocution as the preferred form of execution remained heated after the Kemmler debacle. Five more unsatisfactory executions led authorities to adopt Edison's recommendation of using liquid electrodes. In February 1892 convicted murderer Charles McElvaine was strapped into a chair with his hands placed in attached containers of saline solution. The result was another horror story. Like Kemmler, McElvaine had to be jolted twice with current before he died, leaving his corpse riddled with burns and blisters. Although Edison stood by his method, it was never tried again. Gradually the method of execution improved, the rhetoric cooled, and other states began adopting electrocution—but not quickly. By 1900 only two states, Ohio and Massachusetts, had joined New York; in 1949 West Virginia became the twenty-sixth and final state to adopt death by electricity.[1]

The war of the wires in New York City lingered on for several years at a lower level of intensity. Throughout 1890 the fight continued to rage over what company should provide the underground, or "subway," conduits for the lines and what rent they should charge the electric companies. Political infighting spiced with charges of corruption kept the level of public indignation high even as it slowed the progress of burying the wires. A new company prepared to take charge of the subway work amid cries that it was controlled by Westinghouse only to be thwarted by another injunction. As the original, much maligned company resumed work, the basic dilemma remained unresolved. The wires were being pulled down faster than replacements could be buried. Another brutal storm in January 1891 reminded all sides of the stakes as power had to be shut off again. By the spring of 1892 the city had nearly 1,600 miles of buried electrical wires. However, more than two years later the *Tribune* still referred to the Board of Electrical Control as "rotten with jobbery from the start" and "filled with political pets and parasites."[2]

The battle of the currents did not so much end as simply fade from the scene for several reasons. Edison's staunch defense of direct current became irrelevant after the creation of General Electric in 1892. His influence in the company dwindled to that of stockholder and scientist providing services. At the same time, the merger brought to power the Thomson-Houston managers, and with them the alternating-current system devised by Thomson. Under Coffin, General Electric quickly expanded its product line to compete with Westinghouse across the board. Technical innovations, improved equipment, and greater awareness of electricity's dangers reduced the number of accidents, driving them from the front page of metropolitan papers. Perhaps most important, the Chicago World's Fair demonstrated in spectacular fashion the efficiency and versatility of a polyphase system. Westinghouse's gamble had succeeded; he showed the nation and the world that the future belonged to alternating current.[3]

By 1892 two major firms dominated the electrical industry, and one dwarfed the other. General Electric's capitalization of $50 million was five times that of Westinghouse, its annual sales more than four times that of its rival. Competition between them grew increasingly fierce and extended across the entire gamut of products—lighting, central stations, electric railways, motors, transformers, and many others. However, the companies shared one major problem: the constant need for money in a capital-intensive industry. Even as they had expanded, both companies resorted to the same practice of financing their growth through short-term borrowing. For Thomson-Houston, Coffin had cushioned this policy through his relationship with Henry Higginson and other Boston bankers. Westinghouse chose in typical fashion to go it alone and got caught in the financial panic during the fall of 1890, which forced a needed reorganization of his company and nearly drove him from its management.[4]

Westinghouse's restructuring included a $1 million reserve in its treasury, which eased the cash crunch. General Electric began life with a whopping $17 million of its new stock in the treasury. The disparity in size and resources created by the merger of Edison General Electric and Thomson-Houston left Westinghouse no choice but to innovate and create new products as efficiently as possible. Always there lingered rumors that General Electric would absorb its last remaining rival and make the industry a virtual monopoly. General Electric had become a formidable firm. It had the financial backing of Morgan and Higginson among others, the organizational and managerial skills of Coffin, and the inventive genius of men like Edison, Thomson, and Charles Steinmetz. Once in charge, Coffin reorganized the firm around the Thomson-Houston model.[5]

The bitterness of the competition between Westinghouse and General Electric far transcended the personal animosity between Westinghouse and Edison and Westinghouse and Coffin. It was fueled in large part by the interminable patent wars over lamp design, which had taken more twists and turns than a Victorian melodrama. In February 1889 the Canadian patent commissioner abruptly voided the Edison patent in that country for not complying with certain local regulations. Hope among other manufacturers soared only to be dashed when the Canadian minister of agriculture reversed the decision. Then a suit brought by a Westinghouse subsidiary against the McKeesport Light Company, an Edison subsidiary, was decided in favor of the latter. The suit claimed an infringement of the 1885 Sawyer-Man lamp patent; instead the court ruled it invalid, thereby restoring the validity of the Edison patent.[6]

Buoyed by these events, the Edison lawyers launched the suit against United States Electric Light, which Westinghouse controlled, for infringing on the Edison patent. It was the decision in this suit, rendered July 14, 1891, and upheld in October 1892, that finally sustained the Edison patent and forced Westinghouse to devise the stopper lamp to supply his customers and the Chicago exposition. The Edison Electric Light Company had spent large sums developing the original lighting system only to watch helplessly as others, who had spent little or nothing, imitated the system and grabbed market share from it. With the patent's life fast dwindling, General Electric determined to maximize its return by using its monopoly power and vigorously suing other infringing manufacturers. However, this policy only reinforced the image of the company as a ruthless, monopolistic electric trust, and it failed to eliminate all competitors, most notably Westinghouse. The stopper lamp lacked the life span and efficiency of Edison's lamp, but it was cheaper and gave decent service.[7]

General Electric also had to deal with a quirk in the patent law. Patents usually protected an invention for seventeen years; however, a peculiar situation caused the Edison patent to expire not in 1897 but in November 1894, leveling the competitive field. General Electric controlled about half the lamp market and Westinghouse about 15 percent; the remainder belonged to thirty-five different smaller companies, some of which also became targets of GE suits. For two years bitter competitive wars raged until the contestants reached an agreement fixing prices and allotting market shares. The same savage struggle extended to dynamos (which could not be patented), distribution systems (which could be patented), electric railway systems, and the rest of the product line. In December 1886, for example, the Edison company sued a Westinghouse firm for infringing on its three-wire feeder-and-main system. In March 1893 it won in the circuit court and awaited the inevitable appeal.[8]

Although Westinghouse came late to the traction wars, he soon made up for lost time thanks to Albert Schmid and Benjamin Lamme. Their Motor No. 1 had been sold to 233 railways by May 1893, and a Motor No. 2 was created to serve narrow-gauge roads. When Motor No. 1 hit the market in 1890, Schmid and Lamme were already at work on something far superior, a design requiring an entirely new type of armature winding worked out by Lamme. When Motor No. 3 went on sale in the spring of 1891, it swept the market so rapidly that Westinghouse had to scrap its entire remaining stock of double-reduction motors. The first modern railway motor, its features became the standard for future models. Both Thomson-Houston and Sprague also produced new motors, as did smaller firms, but none matched Motor No. 3's performance. Not until February 1893 did GE's engineers produce a competitive new motor, the GE-800, which forced Westinghouse to counter with its own improved version called Motor No. 12.[9]

For all the attention given to lighting and railway systems, shrewd observers understood that a potentially greater payoff lay in the delivery of power for a variety of other uses as well. Two things were needed to realize this dream: an efficient transmission system and a universal power system versatile enough to perform an entire spectrum of tasks. A universal system could employ one or more large generators drawing power from a steam engine or waterwheel and transmit current to a number of substations, each of which could feed a different service ranging from arc lamps to incandescent lighting to streetcars to motors of all kinds. Existing power stations utilized separate systems for each type of service. The Buffalo General Electric Company, for example, had seven steam engines of several makes and fifteen generators of seven different types that provided power to five different types of customers on separate circuits. Under this arrangement economies of scale could never be realized as they might be with a universal system.[10]

As the sale of central power stations assumed greater importance, the competitive wars intensified. Moreover, as the number of central stations multiplied, their problems became more glaring. The success of a lighting system inevitably created a demand for expansion, but direct-current stations could not respond. Given the prohibitive cost of transmitting beyond a distance of a mile or less, they required more stations to serve outlying areas. But central stations in densely populated urban areas, which was their primary market, had to pay high real estate costs as well as haul all the coal and water they used to the station from outlying storage areas. These expenses, along with the heavy investment in copper, kept the price of electricity high, which favored the competing market for isolated power plants.[11]

The onslaught of new electric railway systems compounded the problem. Every new railway required electric power from either its own or a nearby central station. Within only a few years the demand for power by streetcars dwarfed that for lighting. Manufacturers and utility companies alike awoke to the realization that their primary market had mutated almost overnight into one hungry for both power and lighting as well as current for running motors of every kind. Industrial users had been slow to adopt motors, if only because suitable versions had not yet been devised for a wide variety of production and other uses. But Westinghouse and General Electric, as well as smaller firms, were hard at work on new motor designs that might capture that market just as Lamme's Motor No. 3 had done in the railway field. Here too General Electric won a major victory when in May 1893 the Patent Office ended years of contention by granting Charles Van Depoele two patents that gave him complete control of overhead trolley systems in the country.[12]

A scenario already familiar in other areas of technology was unfolding in the field of electricity. Every successful new invention begat hosts of innovations and improvements as well as imitators and alternative devices. One need only scan the pages of *Electrical World* from these years to marvel at how many people were hard at work trying to solve all the problems and how many novel inventions, upgrades, or imitations clamored for attention. The interplay of these factors created a domino effect that rippled through every firm. More lights or power required bigger engines, which demanded more power from bigger dynamos as well as improved transmission systems and better safeguards, meters, switches, and other devices. The elimination of a weak link in the system did not so much solve the problem as transfer it to some other link. For electric manufacturers and their engineers, problem solving became a way of life. Like it or not, they had to live by the maxim "Innovate or perish."

The demand for more power, then, involved a complex chain of issues. At its heart lay the problem of transmission. Already large cities had a plethora of central stations, isolated power plants in individual buildings, and power plants for electric railways. As cities of all sizes grew—and they were mushrooming during these years—how could utility companies possibly keep up with the demand for more lighting and power? They could hardly saturate the city with new plants because of the costs involved. The logical solution was to transmit large quantities of power from a facility outside or at the margin of the city to its various users. Land costs would be much cheaper, and coal, water, and other supplies could be stored in quantity on site. All these factors pointed to alternating current as the solution. The advantages of AC over DC had long been known, but their significance had grown with the passing years. More important, the technical obstacles that had prevented widespread adoption of alternating current were fast being overcome.

Westinghouse had seen this coming since his first commitment to alternating current. His foresight and resolution gave him the lead, and his talented engineers helped him keep it when the competitive wars heated up. As Steven W. Usselman has shown, Westinghouse brought to his work a different mindset from that of Edison. He had grown up around a workshop and made himself into a talented designer of machinery. Long before turning his attention to electricity he had honed this and other skills in all his innovative railroad work. "It was always a treat to see him at work to solve a problem in the workshop, drawing office, at home, or anywhere else," said an associate in the air brake business. "One could almost see the wheels go round in his brain. His resourcefulness was something marvellous. When one solution did not satisfy him, he had instantly several others ready at hand, as if his brain was a storehouse of original ideas and as if he had only to take them out as wanted."[13]

These talents along with his background as a mechanic proved invaluable to Westinghouse's work in electricity. "In a matter of days or even hours," noted Usselman, "Westinghouse could have a crude sketch transformed into a working prototype, without leaving his desk." He had but to draw it and leave it to his engineers with the resources at their disposal to produce a working model. Two of his most successful products, the air brake and the interlocking signaling system, shared two elements that revealed Westinghouse's basic interest: They involved transmission over distance and crucial linking mechanisms that connected the transmission lines to the rest of the system. So too did his system for sending and distributing natural gas. The challenge of transmitting alternating current over distance was merely an extension of this interest along with the transformer and the rotary converter, which connected the power lines to the rest of the system. The triple valve performed a comparable role in the air brake system along with another important valve in the cab of the locomotive.[14]

"All of these crucial regulating devices were mechanical marvels," wrote Usselman, "complex arrangements of parts, machined to fine tolerances, sensitive enough to detect subtle variations in electrical characteristics or pressure yet rugged enough to operate without fail when mounted under a railroad carriage or placed in other similarly demanding environments."[15]

Between 1886 and 1889, when the battle of the currents flared up, the Westinghouse AC system had moved beyond its crude early phase. Generators became more efficient (although they still could not yet operate in parallel as their DC counterparts did), a meter had been devised, the transformer upgraded, and progress made on the ever elusive motor. During the fall of 1889 tests at Johns Hopkins rated the efficiency of the Westinghouse system at full load at between 73 and 74 percent, well below that of the Edison DC system. However, advances and improvements kept coming, and the demand for high-voltage systems kept rising despite the furor over safety. During that same year a grandiose project that had long been the stuff of dreams and frustrations received for the first time strong financial backing and emerged as the largest, most ambitious electrical scheme yet conceived. For Westinghouse it became both opportunity and catalyst for even more rapid innovation.[16]

THE SIGHT WAS breathtaking, the potential even more spectacular. To those who gaped in awe at the spectacle over the years, Niagara Falls seemed the perfect embodiment of nature's majesty and power. Some visitors wanted to do much more than simply admire the falls. Since the 1750s efforts had been made to harness their mighty power, but neither the capital, the technology, nor the local market existed for so grandiose a project. The coming of the electrical age gave fresh impetus to such schemes by raising the possibility of not only capturing such immense waterpower but transmitting it over distance. By 1885 a canal had been built to divert water from the Niagara River above the falls to an outlet eight or ten miles below them. A few flour and paper mills, using waterwheels, lined the canal but used only a scant fraction of the potential power.[17]

The potential was so enormous that it defied measurement. The Niagara River had a total fall of 326 feet from its origins at Lake Erie to its outlet into Lake Ontario. Of that amount 164 feet came at the falls, 51 feet at the rapids just above the falls, and 94 feet from five miles of rapids below the falls between Lewiston and the suspension bridge. At the falls, which lay 23 miles below Lake Erie and 14 miles above Lake Ontario, the river made a right-angle turn into a gorge with walls rising 210 feet above the water. An estimate made in 1911 gave the discharge of water as 210,000 cubic feet per second, representing a theoretical energy of nearly 8 million horsepower. The flow of water was remarkably constant; the Niagara River served as an outlet for a huge drainage area that used four of the Great Lakes as reservoirs. And the location could hardly have been better. By one estimate, 20 percent of Americans lived within four hundred miles of the falls in 1890; Buffalo, then a city of 250,000 people, was only twenty miles away.[18]

Any major power project required commercial development on a large scale, which conflicted with the role of the falls as a scenic wonder and tourist attraction. During the 1850s a canal was constructed to provide power for some mills built alongside it, with the discharged water flowing over a cliff back into the river. The usable power created was very limited, and the mills became an eyesore to what had long been regarded as the nation's foremost scenic wonder. A movement to protect the falls led to the passage of a bill in 1883 to create a state reservation around the site. The reservation embraced a strip of land from about a mile above the falls to the suspension bridge below along with all of Goat Island, which lay between Niagara's two principal falls, the International, or Canadian, and American Falls. Within that area about 150 buildings were gradually demolished to create a scenic vista that opened to the public in 1885. The new arrangement meant that any power project would have to be longer, cost more, and serve fewer customers unless the electricity could be transmitted.[19]

Serious efforts to utilize the falls on a larger scale began with a plan published by Thomas Evershed, an engineer on the Erie Canal, in February 1886. Evershed proposed building not a canal but a tunnel two and a half miles long between the upper and lower rivers to carry the discharge flow after it had served as many as 238 waterwheels, each producing about 500 horsepower. A system of twelve canals would carry the water to a series of vertical shafts leading to the tunnel. The water would leave the river via the canals, flow into the shafts, turn the waterwheels as it dropped, and flow from the tunnel into the river. The Evershed plan placed the power project well above the reservation and projected "any number of mills and factories of any size, from the making of toothpicks to a Krupp's factory" along the widely spaced canals. Evershed estimated the cost at $9 million or $10 million.[20]

The Niagara River Hydraulic Tunnel, Power, and Sewer Company was formed to carry out the plan, but its efforts to raise capital floundered. Although intrigued by the project, neither American nor English investors could be induced to put money into it. Earlier Sir William Siemens had calculated that if all the coal produced by all the coal mines in the world were applied to steam engines in the most economical manner, the power output would not equal that of Niagara Falls. He was also struck by the possibility of using dynamos to transmit that power throughout the state of New York. But no funds were forthcoming. "Inviting as the scheme looks," concluded *Electrical World*, "it cannot be denied that it presents serious difficulties." In their quest the promoters came to Francis Lynde Stetson, one of the brightest, toughest, and best-paid corporation lawyers of the era and the legal right hand of J. P. Morgan. In this instance Stetson did not follow Morgan's bidding; evidently his interest was aroused by one of his legal associates, William B. Rankine, who hailed from the town of Niagara Falls and was involved in the project.[21]

In June 1888 Stetson took an option to purchase the company but let it expire without acting. Then, in February 1889, he suggested broadening the charter with an amendment, which was done. On July 5 Stetson, Rankine, and Edward A. Wickes, a Vanderbilt associate, appeared at a meeting of the tunnel company and unveiled their plan to acquire it, change its name to the Niagara Falls Power Company, and create a new company, Cataract Construction, to act as contractor. The heavy hitters had arrived, backed by the resources of several major banking houses as well as Morgan and Vanderbilt. By April 1890 these plans had been carried out, and Cataract Construction also acquired control of Niagara Falls Power, making it both owner and agent of the latter. Thus began one of the most audacious ventures in American business history.[22]

The project was not only huge but unprecedented. It soon became clear that Evershed's tunnel-and-canal scheme made no economic sense. The village of Niagara Falls had only about five thousand people; who would use all the power produced by the falls unless it could be transmitted elsewhere? The logical target for transmission was Buffalo, then a thriving industrial city and port, but how to get power to that city? The scale of the project was such that its managers dared not make a false step. Strong, sound leadership was required, and Morgan thought he knew who could provide it. He had hesitated to get involved in the scheme until Rankine asked him who should head it. "Well," replied Morgan, "there is Adams. If you can get him, I'll join you." Rankine recruited Adams, and Morgan came on board.[23]

Edward Dean Adams looked the part of a facilitator. Born in 1846, a descendant of *the* Adams family, he graduated from Norwich University in 1864, trav-

eled in Europe for a year, and spent another year at MIT before joining a banking firm. In 1878 he became a partner in the prestigious banking firm of
Winslow, Lanier. Through numerous railroad reorganizations he earned a reputation as an effective manager who got things done without ruffling too
many feathers. Nothing about his appearance craved attention. A thick, over-
grown mustache masked the lower part of his pleasant but unimpressive face
with its hooded eyes that revealed little. As president of Cataract Construction
Adams proceeded with utmost caution and deliberation. The Evershed plan
provided for as many as 238 mills, each with its own waterwheel, but the geol-
ogy and topology of the region made any such arrangement unfeasible. Moreover, mills would not produce income for the company at once; transmitted
power would.[24]

Gradually Adams and his fellow executives came around to the view that
Cataract should concentrate on building not mills but power stations to supply a wide variety of users in Buffalo and hopefully elsewhere. To explore that
possibility he sought out the best specialists on electric power. Predictably, he
began with Edison, who told him that it was feasible to transmit continuous
current to Buffalo. In November 1889 Edison came up with some figures:
about $5.2 million to produce (but not distribute) power that would net about
$880,000 annually. Adams next consulted Dr. Henry Morton of the Stevens
Institute of Technology, who concluded that "something new in the dimensions and proportions of electrical machinery must be developed in order to
meet requirements of such a problem as you propose." Frank Sprague was a
cheerful advocate of alternating current but declared that "I do not think the
problem to transmit power by electricity from Niagara Falls to several points
at varied distances up to 20 miles, a sound one, commercially . . . I can transmit and distribute this power, but I think it a problem so uncommercial, in
view of the attendant difficulties and risks, that it is better to keep out of it."[25]

Other experts chimed in with reservations of their own. Only Coleman
Sellers of Philadelphia offered a plan he thought would work. "The problem of
transmission of power to any considerable distance," he declared, "with the
object in view of renting that power, at a low rate, is only worth considering in
conjunction with a *first power* remarkable for its quantity, reliability, and its
cheapness of production." Although such a system did not yet exist, Sellers believed in the "practicability, and economy" of the project, and felt certain that
"if carried out judiciously" it would "be the means of presenting manufacturers with an abundant and cheap power." Going down the list of possible transmission forces, Sellers endorsed electricity as the only suitable one for long
distances. However, he admitted, "I do not think there can be any hope of
cheaply sending larger quantities of power through one plant[;] to increase the

amount transmitted, the whole plant will be duplicated or multiplied to the extent called for."[26]

Despite the parade of gloomy reports, the bankers put up $2.63 million in January 1890 to start work on the tunnel and other preparations. The following month Adams went to Europe to research what was being done there, especially in France and Switzerland, which had taken the lead in the development of hydraulic systems. He amassed a portfolio full of information and concluded "after careful investigation," as he cabled Stetson: "Practise here far ahead of ours. Recommend defer execution construction contracts." Instead Adams set up a distinguished five-man international commission with no less than William Thomson at its head. Sellers joined Adams in Europe, and they agreed on a new plan to "abandon all tunnel beyond 8000 feet, also all canals except short surface inlet; adopt one central station for entire power capacity."[27]

"While prosecuting the investigations," Adams explained in more detail, "I received the impression that the method we had planned (direct driving of mills by individual wheels) and were on the point of carrying out was a mistake . . . I came to the conclusion that our true way possibly might be to build this tunnel and develop the whole power in this one central station, transmitting the power to different places."[28]

In June Adams chaired a meeting at Brown's Hotel in London to present his new plan. He also cabled a summary to Morgan, who was then taking his leisure at Aix-les-Bains. "As well as can judge," replied Morgan, "your whole plan meets my entire approval." The International Niagara Commission promptly invited firms and individuals to compete for prizes by submitting "projects for the development, transmission and distribution of about 125,000 effective horse-power on the shafts of water motors at the Falls of Niagara." It received seventeen proposals by the January 1, 1891, deadline, three of which were discarded as irregular. Eight prizes were awarded, four each for electric and pneumatic proposals, but none struck the commission as worthy of adoption. Lewis Stillwell, who had met Sellers in London that summer, was eager for Westinghouse to submit a proposal, but the boss showed no interest in doing so. "These people are trying to secure $100,000 worth of information by offering prizes, the largest of which is $3000," Westinghouse explained later. "When they are ready to do business, we will show them how to do it."[29]

The competition led the commission to only three tentative conclusions. It favored using electricity as the chief, but perhaps not only, means of distributing power, and it agreed that the generators, which required constant service, should not be placed underground near the water wheels but rather on the surface for easier access. The commission also thought that, contrary to any of the proposals, the electric wires should be buried in subways rather than strung overhead. During 1891 and 1892 Adams tackled several major problems. No one had ever built a waterwheel or turbine anywhere near the size planned for Niagara. After careful study Adams found two Swiss firms with satisfactory designs for a 5,000-horsepower water turbine and let the contracts. By 1892 Cataract officials had finally dismissed compressed air and decided on electricity as the power system, but they disagreed on which current to use. While they deliberated, events did much to shape their final decision.[30]

ALTHOUGH FEW PEOPLE yet realized it, the battle of the currents had already been decided. During 1891 both German and American engineers succeeded in creating model systems that transmitted AC power over distance. The German project had a much higher profile as a component of the International Electrical Exposition in Frankfurt. In February Charles E. L. Brown, a brilliant engineer who worked for the Swiss firm Maschinenfabrik Oerlikon, presented a paper describing the virtues of transmitting alternating current in remarkable detail and his success at sending voltages as high as 40,000 over short lines with effective oil insulators and negligible line losses. His work led to the construction of an experimental 108-mile transmission system from Lauffen to the Frankfurt exhibition, where it powered a DC motor driving an artificial waterfall. Brown designed a three-phase generator that worked well enough to convince many engineers of its superiority over the Westinghouse two-phase versions.[31]

The Lauffen experiment got worldwide publicity and showed for the first time the possibilities of a long-distance AC power system. "The sensation in the engineer world was enormous," observed Georg von Siemens. "Professional men from all countries, particularly the United States, flocked to Frankfurt to see the almost incredible with their own eyes." No detail of this achievement escaped the alert eyes of Westinghouse and his engineers. Through the long dreary ordeal of scrounging for money and reorganizing the company in 1891, he had struggled to keep everyone's focus on the research and business at hand. Two projects in particular provided an impressive leap forward even if they got little attention. Both were firsts in the United States, and both used alternating current to transmit power over distance.[32]

The first project centered in Portland, Oregon, which lacked a nearby coal supply to fuel an electric plant. However, the falls of the Willamette River, thirteen miles away, could provide cheap hydroelectric power if it could be transmitted to Portland. Westinghouse responded with a system both efficient and elegant in its simplicity. From a power station located on an island in the middle of the river, 4,000 volts of current passed along lines to a substation in the city, where large transformers, arranged in banks of ten, lowered the 3,300 volts received to 1,100 volts for distribution by numerous circuits throughout the city to ordinary transformers, which stepped the power down to 50 or 100 volts. "The secondary current from the reducing transformers," explained Charles F. Scott, who played a large role in designing the system, "is treated as if it were received directly from a dynamo."[33]

The Portland system transmitted alternating current over distance, but only for lighting. Westinghouse's work for the Gold King Milling and Mining Company carried its pioneering work a step further. The company's low-grade mines near Telluride, Colorado, sat twelve thousand feet high, above the timberline and, having exhausted the supply of trees below, had to pack coal in by burro to operate. Unless the company could find some way to reduce its monthly power cost of $2,500, it could not continue. The only feasible solution lay in using waterpower from the San Miguel River, three miles away and two thousand feet below the mine. The obvious, indeed only, way to exploit the river was transmission via an AC system. Although no such system existed, the mine operators decided to invest in one. Unlike Portland, the challenge was to create a system that powered not only lights but motors as well.[34]

During the summer of 1890 the operators obtained from Westinghouse a 100-horsepower generator that operated at 3,000 volts, along with an identical motor for the mill near the mine. Housed in a rough cabin near the river, the generator was belt-connected to a waterwheel, and transmission lines were run three miles to the mine. The copper wire cost seven hundred dollars, only about 1 percent of what it would have been for a direct-current line. The Tesla motor at the mill, a daring new "split-phase" version, was self-excited and reached synchronous speed through a connection by belt with an induction starting motor. As the system neared completion in the spring of 1891, *Electrical World* hailed it as "unique in the electrical transmission of power among motor plants." Despite some technical glitches, it started in June and operated steadily for a month, although it required fifteen to twenty attendants. So successful was the system that in the fall of 1892 a new 600-horsepower generator was installed along with a 250-horsepower motor for a mill ten miles from the central station. Lighting was also extended to Telluride, eight miles from the station.[35]

In summarizing these pioneering projects, Scott drew a pointed moral. "The alternating current motor has been one of the most complicated and difficult electrical problems," he conceded, "and yet the first large alternating current motor installed in this country excels direct current apparatus in simplicity of construction and operation, in perfection of regulation . . . running for a week without adjustment, a pressure which with commutators is impracticable." The Tesla multiphase system promised an even greater advance in combining the distribution of power with transmission. It was self-starting even with load and had no commutator or brushes. It was, he emphasized, "the ideal motor in point of mechanical construction," simple yet flexible, perfect for a wide variety of uses. "The records of the plants at both Portland and Telluride demonstrate that these possibilities are being realized, and that work in this field is fast passing from experimental investigation into practical engineering."[36]

The Tesla motor had yet to be perfected, but improvements came steadily to every element in the AC system. By January 1892 an electrical writer recalled that only recently "tales of the deadly alternating current glared at us from every newspaper. It was portrayed as the god of destruction." Although some people still clung to this image, "absurd prejudices against it have spent themselves and are relegated to the scrap heap." Its potential seemed unlimited. "Of all electric manifestations of energy," he enthused, "none gives greater promise of wonderful future development than those produced by alternating current."[37]

During 1892 Westinghouse completed a thirty-mile transmission system in southern California that carried 10,000 volts. Borrowing from the Lauffen experience, it used oil-insulated transformers for raising and lowering the voltage, but the equipment was still relatively small. The largest generators produced only 150 horsepower, and the transformers did not reach 10 horsepower. Already the company had developed a reputation for fine workmanship. "The machines of the Westinghouse company," observed one writer, "are singularly beautiful in mechanical design and finish." And the machines, especially the dynamos, were growing larger. In November 1891 Westinghouse drew praise for its "colossal" new 500-horsepower generator for railway service. A year later *Electrical World* called attention to the "growing tendency of the times in the construction of larger units for central station use for both lighting and power purposes."[38]

Thanks to growing demand and especially the needs of the Chicago fair, the escalation of size had only begun. That same year the Westinghouse engineers toiled diligently at putting together the components of the Chicago system. Working under severe pressure against tight deadlines, they outdid themselves in both production and innovation. The indefatigable Lamme developed the crucial rotary converter as well as the large kodak generator. He also perfected the new slotted-type armature for alternators, which proved much more efficient for polyphase work than the old toothed type. Most of the machines for the fair exhibit were built at the Allegheny Works, where the engineers had to do all their testing on Sundays because construction work occupied the factory the rest of the week. Long, animated discussions about the potential and technical problems of induction motors, rotary converters, and alternating current itself punctuated their efforts.[39]

The pressure to produce excited and energized the engineers. Working on the frontier of alternating-current technology , they felt themselves constantly at the edge of dramatic new breakthroughs. "The latter part of 1892, and the beginning of 1893, and the next two or three years, as far as I was concerned," recalled Lamme, "might be called a period of experimentation rather than new development." The bitter competition with General Electric and the unexpected demands of fulfilling the Chicago fair contract drove them hard, and at the end of 1892 George Westinghouse added yet another huge challenge to the mix. He notified Cataract Construction that his company was now ready to present a plan for the Niagara facility based on a polyphase system operating at 30 cycles.[40]

The evolution of the Westinghouse plant from a workshop into a major electrical factory reflected the founder's own background as a mechanic. William Thomson once remarked that "an electrical engineer is nine-tenths mechanical." Even at the height of his career Westinghouse never ceased being a mechanic at heart. "I have a good deal on my mind," he told one of his officers, "but I like to talk to you about these mechanical things. They relieve me." He took pride in the way his companies always manufactured their own products and in the innovations in production his men pioneered. Now, in the midst of creating all the components needed for the gigantic installation at Chicago, he was about to take up the challenge for another enormous project, one that could not match the fair as a showcase but that would prove far more enduring in its consequences.[41]

TECHNOLOGICAL INNOVATION CONTINUED to parade before Adams like an unfolding panorama as he and his fellow officers wrestled with the key decisions that would determine the fate of the Niagara project. In April 1892 Adams hired Professor George Forbes, a Scottish authority on electricity, as a consultant along with Henry Rowland and Sellers. "His familiarity with electrical machine construction in many countries," noted Adams, "and his facility of mathematical determination of mechanical and electrical design, gave him special qualifications." A prickly, imperious snob, Forbes seldom agreed with Rowland on anything. However, he had been one of the early champions of alternating current. While the Niagara officials agonized over the question of which power system to choose, Forbes in his very first report dismissed two designs submitted by then Edison General Electric based on DC power.[42]

In December 1891 the Cataract board had sent letters to three major American and three major European manufacturers inviting them to submit proposals for an elaborate power installation at Niagara. The choice of current was left to the companies. By September 1892 only the European firms had responded, and two of the American invitees had merged into General Electric. Shortly afterward General Electric submitted a proposal and later followed with two more. The foreign firms labored under the handicap of having a 40 percent duty and 10 percent freight cost added to their bids, which made the American equipment much cheaper. The foreigners would get the work only if nothing suitable came from the American companies. Forbes favored the plan from the Swiss firm Oerlikon, but he also knew that Westinghouse was the leader in the field of AC systems. In December 1892 Westinghouse submitted his proposal for a polyphase system.[43]

This willingness revealed Westinghouse's faith that the motor, long the weak link in the AC system, was at last ready to perform. In 1890 the Tesla motor was, as Edward Dean Adams himself admitted, "a prophecy rather than a completely demonstrated reality. The immediate demand was for motors to operate from the existing light circuits, which were single phase and had a frequency of about 130 cycles per second." As late as February 1892 William Stanley declared "a commercial AC motor" to be "a thing unknown to the practical engineer." But Stanley failed to notice what the Tesla motor had done at Telluride, and he could hardly even guess what it would demonstrate in Chicago. Throughout 1892, despite the harshest of weather, the Telluride motor had performed steadily. Lamme recalled that it was not until 1893 that the induction motor "began to receive more serious consideration and attention as a commercial possibility."[44]

To bolster his brain trust, Westinghouse had visited Tesla in New York in May 1892 and persuaded him to help with the work for both the Chicago and Niagara projects. Tesla began making trips to Pittsburgh; when he was in New York Lamme, Scott, and Albert Schmid sought him out there for advice and helped prepare his exhibit for the fair. Tesla also arranged for Westinghouse to get some business in Germany and helped with other connections in Europe. Some of the Westinghouse engineers resented Tesla for his pomposity and the large sum he received for an invention they thought worthless until Scott and Oliver Shallenberger improved it. However, no one could deny the importance of Tesla's multiphase system for the company's future. Scott had suggested sticking with the old tried-and-true single-phase system to run the stopper lamps at the fair, but Westinghouse said, "No, they have been telling me that the two-phase is the right system and I want to find out whether it is or not."[45]

The Tesla induction motor required a new breed of alternator as well, one that could run smoothly at lower frequencies. Westinghouse had decided to use 60 cycles for lighting and 30 for motors and rotary converters. Confusion still reigned among engineers over the worth and use of lower frequencies, but the Westinghouse men understood their importance to a multiphase system even though the concept spelled doom to all the units already installed in central power stations. "Commercial circuits were single-phase at a frequency of 133 cycles," Scott explained many years later. "Strenuous efforts made to adapt the Tesla motor to this circuit were in vain. The little motor insisted on getting what it wanted, and the mountain came to Mahomet. Lower-frequency polyphase generators inflicted obsolescence on their predecessors in a thousand central stations—such was the potency of the Tesla motor."[46]

In January 1893 Coleman Sellers and Henry Rowland journeyed to Pittsburgh to inspect the plant and observe a string of demonstrations. After four days both men came away deeply impressed by the facilities and the engineers. Sellers marveled at the "excellent workmanship and correct engineering design in all the machinery examined." Rowland concluded that Westinghouse had "the greatest experience in the practical use of the alternating system and they seem to control the most important patents." They also praised a staff that included men like Schmid, Stillwell, Shallenberger, Lamme, Scott, and Westinghouse himself. The following month they visited the GE plant in Lynn, Massachusetts, and found the equipment similar but not necessarily as good. Sellers was wary of GE's three-phase system. "I should incline to the biphase," he recommended, "on account of its greater simplicity and its adaptability to a broader field of uses."[47]

Sellers opposed the foreign bids that Forbes favored, not only arguing that Americans should have the honor of building the system but also expressing doubt that a foreign firm could maintain and repair the facility in a timely manner. He also reminded Adams that Westinghouse controlled the key AC patents and offered his belief that "no foreign company can secure the Cataract Construction Co. against all losses from patent litigation." Adams was concerned enough to ask Tesla his opinion on the patent question along with some technical matters. During their exchange of letters that winter, Tesla gave Adams a short course in motors, dismissed much of the writing on the subject as ill-informed, and assured him that Thomson's earlier patent had "absolutely nothing to do with my discovery of the rotating magnetic field and the radically novel features of my system of transmission of power disclosed in my foundation patents of 1888."[48]

In March 1893, less than two months before the opening of the Chicago fair, both Westinghouse and General Electric submitted their complete proposals for three 5,000-horsepower alternators, transformers, copper wires for transmission, safety and switching appliances, rotary converters, and motors. The Westinghouse proposal featured a two-phase system, the General Electric a three-phase system. Adams and his advisers had to ponder not only the question of which phase to choose but, before that, which current to use. He sounded Tesla on this point as well and was told vigorously that any DC system would be "disadvantageous, if not fatal, to your enterprise . . . but I do not think it possible that your engineers could consider seriously such a proposition."[49]

Tesla also stressed the advantages of a two-phase over a three-phase system. "The probability of a break-down is much smaller in the two phase than in the three phase," he noted. "If there is a trouble in one circuit the two phase motor will still operate and perform 60% of the work, and if liberally designed will probably carry all load until the damage is repaired. In the three phase it would, as a rule, mean a complete break down." The two-phase motor had less slip and worked more efficiently as a synchronizing machine. It was also better adapted to the relatively high frequencies needed to run lamps, and it was easier to construct and repair. For these reasons, Tesla advised Adams to "accept the two phase plan, though it would be much more for my personal interest if you would decide for the three phase."[50]

Forbes, Rowland, and Sellers all agreed that an AC system would be superior. Sellers and Rowland liked the mechanical design of the Westinghouse proposal, and Forbes preferred the two-phase over the three-phase system. But he was not happy with either generator design because neither fulfilled the conditions imposed by the designers of the water turbine that had already been built. Months earlier Adams had asked him to develop a design of his own in the event that none of the submissions met the needs of the project. Forbes had done so, and the working drawings of his version were being prepared even as the deliberations took place. For reasons that are not clear, the process of evaluation dragged on through April. On May 1 Adams received a cable from William Thomson saying, "Trust you avoid gigantic mistake of adoption alternating current." However, on that very same day the Chicago fair opened and revealed its marvels of alternating current to every knowing observer.[51]

Apart from its more spectacular effects, the fair showed the Tesla induction motor operating within an AC system. For Adams and his experts, the display evidently satisfied them that alternating current held the key to future electrical development. Five days later they formally specified alternating current for the Niagara system. Later Adams called it "one of the red letter days in the history of The Niagara Falls Power Company and of the electrical industry as well." It was in fact a bold commitment made at a time when many if not most electrical engineers still did not grasp the nature and ramifications of alternating current. Well might George Westinghouse have smiled broadly at the news, but his good humor was short-circuited almost at once by two shocking developments.[52]

On May 11 Adams wrote all four remaining contestants—two American and two European—seeking the Niagara contract and informed them that the Cataract board had carefully considered their proposals along with the extensive reports from their technical advisors and come to five conclusions: 1) The electrical system had to be uniform for both local and long-distance transmission; 2) it had to use alternating current; 3) it had to use two rather than three phases; 4) none of the two-phase generators submitted met all requirements of the proposed system; and 5) therefore, the board's own technical advisers were designing a special generator for that purpose. "The drawings of the generator specially designed by our advisors . . . are well advanced," Adams added, "and we expect to submit the same to you, as well as to others, for proposals for construction within a brief period."[53]

This stunning letter, especially points four and five, appalled Westinghouse and his men, as it no doubt did the other recipients. Throughout the long bidding process Forbes had been secretly at work on the "special generator" and had the benefit of information and solutions laboriously arrived at by the other engineers struggling to design their systems. Adams told the applicants nothing about Forbes's work until this letter, which said, in effect: Thanks for your help; we'll design our own generator and let you bid on the privilege of building it for us. Later the well-known electrician Silvanus Thompson blasted this "ungenerous picking of the brains of others" and called the "contemptible collaring of rival plans . . . the one discreditable episode the savour of which will ever cling about the undertaking."[54]

This display of dubious ethics, which seemed so out of character for Adams, was influenced by a bizarre and wholly unexpected event that burst into the newspapers a few days before he wrote this letter.

ON MAY 5 Morris M. Mead, the city electrician of Pittsburgh, was arrested and charged by the Westinghouse company with conspiracy in the theft of numerous Westinghouse plans and patents. According to the Westinghouse spokesman, the company had for several weeks been puzzling over how plans submitted by General Electric for competing projects turned out to be nearly identical to its own. One customer told Westinghouse, "These plans are so similar that one of two things must have happened: Either you have stolen the plans of the General Electric Company or they have stolen yours." When Westinghouse showed another customer designs for a particular engine he had ordered, the customer said he had seen virtually the same plans at GE's Lynn plant. A similar experience occurred with Westinghouse's specifications for the Niagara project. After Westinghouse submitted its proposal, GE withdrew its original version and countered with one strikingly similar to that of Westinghouse, which had been worked on for months in secret.[55]

Late in March Westinghouse hired detectives to investigate the matter. They reported that in July 1892 Charles Coffin, William Clarke, and a few other GE executives arranged with Mead to find someone to get work at the Westinghouse plant. A man named H. F. Ashton took the assignment and got hired as a draftsman. He received $150 a month from General Electric along with a slush fund to bribe other draftsmen into giving him copies of blueprint drawings, confidential papers, patents, and other useful material. Clarke provided a code for Ashton and his cohorts to use in telegraphing information to him. So well did Ashton do his work that the plans and blueprints reached GE quickly enough for them to "utilize the designs of the Westinghouse engineers almost as promptly as the Westinghouse Company itself."[56]

With this report in hand, Westinghouse learned that a number of the company's plans, including those for Niagara, were kept at the Lynn plant. He obtained a writ to search the plant but the superintendent never appeared. However, the sheriff did get Walter Knight, the chief engineer, to admit to having some blueprints, and these were surrendered. On Friday, May 5, Westinghouse lawyers filed a bill in equity against General Electric and several of its officers, including Coffin and Clarke, seeking return of their blueprints and papers. The following day the complaint was taken to Pittsburgh's district attorney, who issued warrants for the arrest of four Westinghouse employees alleged to be in the pay of General Electric on the charge of larceny. Warrants on the charge of felony were also issued against Mead, Coffin, Clarke, Knight, and some other General Electric officials. "The action which has just been taken," said Westinghouse, "was an absolute necessity in view of the great injury being worked to the company by the conspiracy."

Westinghouse also released excerpts from a letter received from one of his engineers bidding on a project in San Francisco. "You have, of course, heard the rumor that the General Electric company has obtained the drawings of our Niagara generator," he wrote, "and I think there is no doubt that they had all of the preliminary information which was sent to the Livermore as to the Sacramento plant . . . because of the remarkable similarity in our plans."[57]

Clarke dismissed the charges as "ridiculous" and "ill advised," and threatened to file charges against Westinghouse for bribing Connecticut legislators to block an effort to modify the Thomson-Houston charter four years earlier. He admitted that General Electric might have some Westinghouse plans but contended they were obtained only through the usual industry and customer channels. The charges were brought, he insisted, simply to influence the price of stock and "by implication to assert superior engineering ability." Coffin denounced the charges as "false and malicious" and agreed they had been brought to influence the stock market. Asked his opinion, Francis Lynde Stetson declared that he "did not for a moment believe that either competitor has stolen the plan of any other."[58]

"We did not take this matter up without having in our possession ample evidence of our charges," replied a Westinghouse spokesman, "not only against the defendants who have been served with orders of arrest, but also those who are still outside the State. We propose to push this matter and to exhaust every lawful means to bring the conspirators to punishment." There the matter rested through the summer as the wheels of justice ground slowly through their labyrinth of procedure.[59]

Evidently the episode influenced Adams's decision to stiff both companies on the contract. Tesla admitted to him that "it has caused me some regret that you have not adopted the design of Mr. Schmid for I am convinced that no living practical engineer could have furnished a more satisfactory one. If the design does not meet your requirements it is certainly not his fault." However, he also understood how unsavory the charges brought by Westinghouse seemed to Adams. "That one of the competing companies should have had the misfortune of having some inferior employes capable of contemptible acts," he noted, "and that the other should have made the grave mistake of calling public attention to these wrongs and thus has cast a shadow upon the whole profession,—such a trifle, to be sure, was not worthy of any consideration."[60]

Meanwhile, Forbes continued to work on his dynamo while the electric giants pondered their next step in the confused Niagara competition. The battle of the currents had for all practical purposes ended, but the war for business went on unchecked. The fight had turned ugly, and the broader economy had begun to turn even uglier.

CHAPTER 17

HARD TIMES

In certain electrochemical processes, direct current is essential; where storage batteries are used, it is requisite; for the operation of street cars it is desirable; in certain isolated plants and in industries where motors require a particular range of speed control it is preferable. But aside from such cases, it is used but little today, except that the early direct-current lighting systems in the centers of a number of our larger cities are still in operation. In practically all these cases, however, the power is initially produced in large central stations as alternating current and is transmitted as such to a number of sub-stations where it is converted into direct current in amounts as required for local use. In general, the direct-current distribution in cities is not being extended, but is being replaced by the alternating service, and in most cases, it has disappeared entirely.

—EDWARD DEAN ADAMS (1927)[1]

THROUGH THE SUMMER OF 1893 George Forbes labored on the design of his generator. While working he lived on the Canadian side of the falls, dismissing the town as a dirty tourist trap with its "cheap restaurants, merry-go-rounds, itinerant photographers, and museums of Indian and other curiosities." An egocentric Scotsman, he had few kind words for things American or for his engineering colleagues. The plans submitted by GE and Westinghouse he scorned as "the crudest work conceivable." Aware that everything hinged on the perfection of a suitable AC generator, Forbes basked in the limelight while carping about the "intriguing, underhand dealing, and jobbery" that surrounded his work. By early August he had completed the preliminary design that Cataract wanted. On August 10 the unflappable Adams informed General Electric and Westinghouse that Forbes had finished his work and invited both companies to send representatives to Niagara to study the results for the purpose of bidding on construction of the system.[2]

The two companies did not receive complete drawings, Adams added, because Cataract wanted them to feel free to "*modify the design to adapt it to your machine tool and your mode of manufacture. Any alterations that you may propose in the design will be carefully considered, and if acceptable, will be appreciated in awarding the contract.*" Westinghouse hesitated, then wrote Coleman Sellers a week later that he had been put in an unfair position. Having spent an enormous amount of time and money working out plans for power systems, he declared, he was now being asked to suggest modifications or improvements that might be incorporated and manufactured by someone else. "We do not feel," he stressed, "that your company can ask us to put that knowledge at your disposal so that you may in any manner use it to our disadvantage." Despite his protest, Westinghouse still wanted the contract badly. He dispatched Lewis Stillwell and another engineer to Niagara.[3]

The engineers arrived on August 21 and pored over the drawings of the Forbes generator. It did not take them long to find numerous flaws. Forbes chose a frequency of $16\frac{2}{3}$ cycles per second, which Stillwell knew would not work at all. It would cause lights to flicker, would provide a limited range of speed for induction motors, and could not operate most polyphase power equipment such as the rotary converter. Forbes was thinking in terms of a motor with brushes and a commutator rather than an induction type. He also wanted the high voltage of 20,000, which would require a form of insulation that did not yet exist. In specifying this combination of high voltage and low frequency, Forbes had virtually nullified the chief advantages of alternating current. It could be generated at a low current because increasing the voltage for transmission was easy to do. Moreover, AC allowed the use of the rugged induction motor, which had no brushes or commutators.[4]

"Until I went to America," Forbes recounted later, "the manufacturers of electrical machinery never had a consulting engineer to reckon with, but dealt directly with the financiers, who knew nothing about cost or efficiency of machinery. When they knew that I was to advise the Niagara company, they tried by every means in their power to revert to the old plan. Every game of bluff, bounce, and threats was used, but without the least effect. One of our engineers . . . , who knew nothing of electricity, was completely taken in by one of these firms. This and other intrigues hindered me a good deal in carrying out my plans. Then again, I had great difficulty in keeping the president and vice-presidents in hand."[5]

Or so Forbes viewed events through the narrow lens of his interpretation, which portrayed himself as the Hamlet of the play. *Electrical World* ripped

Forbes for his "lordly tone" and "many uncalled-for disparagements . . . not only of individuals, but of Americans in general." The London *Engineer* blasted him even more severely for his conceit in an editorial that began: "Who has harnessed Niagara? Prof. George Forbes. Who says so? Prof. George Forbes." Apart from the rancor, the facts told a strikingly different version of the story.[6]

Stillwell and his colleague submitted their report to Westinghouse. On September 15 they returned to Niagara to confer with the company consultants. They expressed their reservations about the design of the dynamo first to Sellers and then to Forbes, who discussed some of the points but refused even to consider others, saying he had weighed the matter fully and was certain he was right. On the twenty-seventh Westinghouse informed Cataract that he advised a frequency of 33⅓ cycles and would build a generator at 16⅔ cycles but would not guarantee its performance. General Electric recommended 41.6 cycles for its system. Cataract replied that it would not award any contract unless the manufacturer guaranteed the efficiency of the dynamo's operation. Later Adams explained that Cataract's board had "instructed its officers to bring the details of the specifications as close as possible to the desires of its technical advisors, combined with sufficiency of guarantees from the manufacturers."[7]

A stalemate loomed. Early in October Westinghouse and General Electric submitted their bids for building three 5,000-horsepower generators. However, Westinghouse refused to guarantee performance if the Forbes specifications were required. Adams and his experts showed surprisingly little interest in the General Electric proposal; later that company's sales committee blamed their failure on Forbes's pro-Westinghouse bias. This was hardly the case; Forbes didn't care for either American company. Moreover, General Electric had historic ties to Morgan and other financiers who controlled Cataract. For his part Adams had in 1889 sold his shares in the Edison companies and resigned from the board of Edison Electric Light to avoid any hint of partiality on his part in the crucial decisions facing Cataract. Adams also realized that Westinghouse had long been the leader in the use of alternating current and that his company held most of the key patents. In one way or another he had to deal with them.[8]

During that same September the Westinghouse charges of industrial espionage went to trial in Pittsburgh. On the stand H. F. Ashton told his story of being hired by Morris Mead and William Clarke and procuring Westinghouse plans for them. The accused General Electric officials admitted to having taken some materials but insisted they were only looking for violations by Westinghouse of the Sawyer-Man injunction pertaining to incandescent lamps. The defense tried to show that Ashton was actually in league with Westinghouse to frame General Electric. Curiously, the defense did not summon Mead to testify. Despite his confession, Ashton made a poor showing. The judge informed the jury that his testimony "was of no value if not corroborated." The jury pondered the case for one day, then reported that they could not agree. They were said to be evenly divided between conviction and acquittal.[9]

With that unpleasantness laid to rest, the diplomatic Adams arranged a dinner at New York's Union Club with the Westinghouse people to discuss the disputed issues. Amid the sumptuous parade of courses the engineers thrashed out their differences. They settled on a voltage of 2,200, but Cataract would not let go of Forbes's low frequency. As the dinner broke up, Adams sought out Stillwell and asked him whether Westinghouse could build a satisfactory alternator that would deliver current at 25 cycles and guarantee its efficiency. Stillwell said he would discuss it with Westinghouse and the other engineers.[10]

Lamme was asked to study the question. After poring over the ramifications with his usual thoroughness, he concluded that 25 cycles served quite nicely as a compromise. It worked well with both an induction motor and a rotary converter. Westinghouse informed Adams that his company would guarantee an alternator with that specification. On October 6 the Cataract consultants formally recommended acceptance of the Westinghouse bid. By the twenty-seventh the details had been hammered out, and the news went public that Westinghouse had won the contract for the three generators, the switchgear, and auxiliary equipment for the power station. General Electric was awarded a contract for the transformers, the transmission line to Buffalo, and the equipment for the substation in that city. The long ordeal of the bidding competition had at last ended, and the real work could begin. Schmid was placed in charge of the mechanical and Lamme of the electrical design and construction.[11]

To win the contract Westinghouse offered an impressive package of guarantees. The generators were to be interchangeable with the loss in each one not to exceed 2.5 percent. The efficiency of the transformers at full load was to exceed 97.5 percent, while that of the motors was to be "approximately equal to the efficiency of the direct-current motors of corresponding sizes." The project generated several technical breakthroughs that the Westinghouse engineers incorporated into later work. Once it became known that the Niagara project had adopted the 25-cycle frequency, that figure gradually became the standard for engine work of many kinds. The chief obstacle to the work turned out to be George Forbes. The Westinghouse engineers resented his condescending attitude and thought him incompetent. Coleman Sellers tried to mediate the mounting friction but to no avail. Forbes inflamed matters by delivering a lecture that led Westinghouse to believe he intended to become "a possible rival in dynamo design." Ever suspicious, Westinghouse did not care to impart any design or technical information to Forbes. In December he announced in his forthright way that he and his men would no longer work with Forbes.[12]

On February 6, 1894, Edward A. Wickes, one of Cataract's early financial backers, met with Westinghouse and two of his engineers in Manhattan. To his dismay, Westinghouse would not budge in his stand on Forbes. Admitting that the standoff created "considerable difficulty," Wickes told Adams that "we must get along the best way we can." Coleman Sellers, who did have Westinghouse's respect and friendship, became the indispensable man by simply going around Forbes on all crucial matters. Predictably, the arrogant Forbes resented his dwindling influence in the project; later he penned a nasty article on the project for an English magazine and was roundly chastised for his churlishness from several quarters. In September 1895, when the Westinghouse engineers had installed and tested the first two generators, they informed Adams that several "minor modifications" had been made. "These modifications," they added, "have in all cases received the approval of Dr. Coleman Sellers."[13]

The developmental work for the project stretched the ingenuity and stamina of the engineers to their limits. Only the recent advent of nickel steel made possible the construction of a new type of alternator five times the size of any yet built. "The switching devices, indicating and measuring instruments, busbars and other auxiliary apparatus," admitted the engineers, "have been designed and constructed on lines departing radically from our usual practise. The conditions of the problem presented especially as regards the magnitude of the amount of power to be dealt with, have been so far beyond all precedent that it has been necessary to design a considerable amount of new apparatus. Nearly every device used differs from what has hitherto been our standard practise." Since 1890 Lamme had spent three hours a night, five days a week, developing his methods of calculation and checking them with whatever apparatus he could get. The Niagara project accelerated this overtime effort, which in turn enhanced his contribution to the developmental work.[14]

The results were more spectacular than even Adams had dared hope. The handsome Power Station No. 1 opened for business in August 1895 with the first two alternators. Adams wanted the plant to be "*attractive*, artistic in grandeur, dignified, impressive, enduring and monumental." He envisioned the building as part of the souvenir pictures taken away by tourists. It should also be "*protective*, invite confidence and afford shelter." The celebrated architectural firm of McKim, Mead & White undertook this tall order and to a large extent fulfilled it. The power station became an attraction in itself even as it began to churn out power. More important, it proved to be almost instantly profitable.[15]

To the astonishment of almost everyone, the demand for power from the facility grew so rapidly that the company had to add an average of two new alternators and turbines every year until the station housed ten units by 1900. That same year work began on Power Station No. 2 directly across the canal from No. 1. The latter's ten alternators, all of them supplied by Westinghouse, generated 50,000 horsepower. For Station No. 2 General Electric won the contract to provide eleven alternators for another 55,000 horsepower. By 1904 the long-cherished dream of harnessing the falls for power had become a spectacular reality as the last of these units went online. To professionals in the field it was a spectacle almost as awesome as that of the falls themselves.[16]

Not only the size but also the source of the demand for power surprised Cataract officials. The very first company to receive power from the Niagara facility, in August 1895, was the Pittsburgh Reduction Company, a firm that made a relatively scarce product called aluminum. Its founder, Charles M. Hall, had produced the first globules of aluminum in February 1886 and started his company two years later. His output remained small because its manufacture involved an electrochemical process that required large amounts of electricity for both heating and the chemical process. At Niagara the company could obtain all the power it needed, and it proved to be the ideal customer for Cataract as well. Given the cramped space limitations for locating industrial firms at Niagara, Pittsburgh Reduction used more electricity per square foot of floor space than any other type of mill or factory. A year later Hall doubled the size of his factory and built a second plant. A third soon followed. In 1907 the firm changed its name to the Aluminum Company of America, better known as Alcoa.[17]

Nor was Pittsburgh Reduction alone in grasping the opportunity offered by Niagara's power supply. Silicon carbide, or carborundum, was the hardest abrasive known next to diamond dust. The Carborundum Company built a plant near Power Station No. 1 and became a major customer. Its founder, Edward Goodrich Acheson, a self-educated inventor, made his first batch of carborundum in a small furnace fashioned from a small iron bowl and sold it for $440 a pound, less than half the price of diamond dust. His little plant at Monongahela City, Pennsylvania, used 135 horsepower of electricity to produce forty-five tons a year of a product used primarily for polishing gems. He lost money steadily until the operation moved to Niagara, where the ability to produce in quantity lowered costs and prices as well as opening new markets. While experimenting with carborundum, Acheson also discovered by accident how to produce graphite with electricity. Another new industry was born.[18]

The electrochemical industry was new and had never been able to produce at scale. "The development of Niagara power in 1895 marked the beginning of the electric furnace art," declared Frank J. Tone, Carborundum's president. "It stands today as one of the big factors in our industrial life. Up to 1895, when Charles M. Hall came to Niagara, the aluminum industry depending on steam power had given little promise of commercial success. Its almost incredible development has been due to the impetus of Niagara power. Dr. E. G. Acheson with a 150 horse-power furnace operated by electric power generated from steam had made a commercial failure of carborundum. Coming to Niagara in 1895 he was at once enabled to found the artificial abrasive industry." Two years later Acheson doubled the plant and boasted that he could sell carborundum for half of what it had originally cost him to produce it before coming to Niagara. [19]

When Cataract first began its work, artificial abrasives, calcium chloride, artificial graphite, ferro-alloys, and even aluminum amounted to little more than laboratory curiosities. Niagara provided all of them with the power they needed for large-scale production. The inventor of calcium carbide had made little progress with his 200-horsepower furnace in Spray, North Carolina. Once located at Niagara, his furnace produced more carbide in a day than he had earlier in a year. Two kinds of electrochemical plants moved into Niagara. Some employed electrothermic processes that used high temperatures in an electric furnace to combine two or more natural elements into a new product such as carborundum. Others relied on the electrolytic process, which used electricity to break down a natural element to produce new materials such as chlorine. [20]

What they shared was the need to run electric furnaces at high temperatures twenty-four hours a day, which delighted the power company. Within two years these and other industries, which the promoters had not even considered in their original plans, were consuming more power than all the lighting and motor needs combined. Niagara emerged as the site of the largest electrochemical complex in the world, and it developed as an unexpected by-product.

During the six long years of planning, design, and construction, the focus of Cataract officials had been on the transmission of electric power to other locales, most notably Buffalo. This vision finally became a reality on November 15, 1896, when Niagara delivered its first commercial power to the city. A decade earlier Buffalo had opened the first AC power station in the nation; by 1892 it had four central stations delivering light to different parts of the city. In 1887 leading businessmen of Buffalo had joined forces to offer a $100,000 prize to the "inventors of the World" who could design a system for transmitting power from Niagara to the city. The offer drew worldwide publicity and a heavy response, but no project was deemed worthy of the award. Finally the city got its wish, and without having to pay up on the award. Having inaugurated the

AC central station, Buffalo could claim the honor of the first major commercial AC power transmission system in the United States.[21]

However, Buffalo did not at once fulfill the destiny envisioned by Cataract's officers. The first transmission line to Buffalo delivered only 1,000 horsepower, which proved ample for the demand. In 1897 Power Station No. 1 had four alternators in operation and produced 18,500 horsepower. Of that amount only 5,000 horsepower went to Buffalo; the rest was consumed at Niagara Falls, 90 percent of it by the growing cluster of electrochemical plants that proved to be the real bonanza customers for power. The Niagara project created not only a new source of power but a new industrial base for the country as well as pioneering the transmission system that would become the standard for power stations. The future had arrived in the form of the central station delivering AC power to substations located at ever-increasing distances.[22]

"Who could have visualized or pictured even in dreams the magical growth and wonderful attainments of the electrical industry at the time the first Niagara development was under contemplation?" marveled Adams in retrospect. "The art was then nebulous and its future dubious." Yet they had accomplished one of the greatest engineering feats ever with ample resources furnished by hard-nosed financiers who might normally have fled from so risky an enterprise. And they had done something even greater. "Upon the broad foundations laid by the pioneers at Niagara Falls," wrote Adams, "has been erected the great electrical structure we see today."[23]

GEORGE WESTINGHOUSE HAD won a major battle, but the war was far from over. The fierce competition between the two electrical giants, already embittered by the charges of espionage, was exacerbated by the deepening depression that had fastened itself on the country in 1893 and maintained its stranglehold for four years. In the desperate scramble for business, buyers forced the rival companies to bid well below list prices. As the downward spiral of prices continued, both General Electric and Westinghouse found themselves bidding for business that fetched ever-shrinking profit margins. In the true competitive spirit, they were bleeding each other to death with help from a broader deflationary trend in the economy as money grew steadily tighter.

Prior to the Chicago fair, General Electric had dominated the arc light market and charged sixty dollars per light for larger installations. During the winter of 1894 Westinghouse advertised that it would install arc-lighting plants of at least sixty lamps for thirty-five dollars a light. Incandescent lamps that sold for sixty cents apiece in 1890 retailed for twenty cents in 1896. In 1888 a complete trolley car cost $4,500; by 1893 the price had dropped to $2,000 for a much better car. A year later seven firms submitted bids for a single car that ranged from $1,500 to $640. "We do not believe," said one observer, "that any company can make and sell satisfactory car equipment at a profit for $640." Some analysts argued that falling prices would speed widespread development of electrical facilities; others warned that buyers would hesitate to purchase supplies until they were convinced that prices had touched bottom. Not until the spring of 1896 did an equipment buyer complain that he had to pay 30 percent more for the same goods he had bought three months earlier. Even then, the price was only half of what he had paid in 1892.[24]

One telling example of the rivalry occurred in the bitterly contested trolley arena. In June 1893 Westinghouse got its first order from the Brooklyn City Railroad, which had previously bought its equipment from Thomson-Houston. Despite the worsening economy, the boss exuded confidence in the future of electric power, especially for street railways. Brushing aside yet another rumor of a merger with General Electric, Westinghouse predicted that both Boston and Philadelphia would abandon cable cars in favor of electric trolleys. With his usual confidence he declared that his new electric railway system would "revolutionize present methods of street car propulsion." As orders developed, he happily put men on the night shift to build motors. Late in 1894 Boston's West End Street Railway Company, soon to be the largest single street rail system in the world, divided an order for 110 motors between Westinghouse and General Electric, saying it would compare them before placing a larger order in the spring. Six weeks later West End awarded Westinghouse a contract for 269 motors at a price said to be less than $250 each. General Electric was offered half the contract but declined, saying the price was too low and the specifications too rigid. One observer thought the real problem was GE's unwillingness to settle for half a loaf.[25]

Competition unmasked a striking difference between the rivals. Both companies had risen from small firms dominated by a single personality to become large corporations, although only one was still ruled by its founder. Size exposed them to the fickle vicissitudes not only of the economy but of Wall Street as well. After the merger General Electric was immediately branded as the "Electrical Trust." Once past its financial ordeal, Westinghouse Electric managed to dwell in the shadows of the financial world so long as it kept its balance sheet stable. The sheer size of GE, however, made it an inviting target for speculators and market players ready to pounce on any morsel of news that might push its stock up or down. If the news wasn't there, it could be fabricated in Wall Street's unsleeping rumor mill. Like it or not, GE's stock had become a plaything of the street, subject to the machinations of bear operators and anyone willing to inflict damage if he could see a profit in it. Westinghouse stock was far less in play because it was more tightly held, there was less of it, and the company maintained a much lower profile.[26]

Ironically, General Electric suffered the same kind of cash-flow squeeze that had plagued Westinghouse earlier. A large portion of its assets consisted of the securities taken from utilities by both Thomson-Houston and Edison General Electric in exchange for equipment. Their value dropped steadily in a declining market, and few customers were prompt in making cash payments on their equipment in a deteriorating business climate. This left GE with a treasury full of securities on which it could not realize any cash. As its floating debt mounted, Wall Street buzzed with rumors of financial weakness. During the winter of 1893, even before the depression hit in earnest, bears began to hammer GE stock, claiming that its "policy of crushing out opposition is leading to complications which must eventually do the company more harm than good, because of animosities and hostile legislation which it is arousing."[27]

That April, in the company's first annual report, Charles Coffin tried to put the best possible spin on the financial situation. The treasury had nearly $3.9 million in cash along with $12.2 million in notes and accounts receivable to cover its floating debt of $5.2 million. However, the securities of local companies comprised 21 percent of GE's total assets, and the notes and accounts receivable another 24 percent. Moreover, the victory in the incandescent lamp patent case did not, as many observers had predicted, result in a sharp increase in the price of lamps. Instead General Electric lowered the price in order to stimulate business and avoid charges of monopoly. The board also created a committee on accounts to grapple with the difficulties of meshing the figures from the original companies.[28]

The bear interests responded by issuing a scarlet-colored pamphlet, reminiscent of the Edison Electric *Warning* brochure, filled with lurid accusations of the company's weaknesses. Coffin insisted that the company had no floating debt and that collections on accounts were coming in, but to no avail. He also had to deny what proved to be the first wave of rumors that the company was about to be reorganized. In May 1893 General Electric sold its holdings in both the New York and Boston Edison Electric Illuminating companies, thereby feeding the stories of a cash crunch. As the bears continued to pound the stock, rumors circulated of a split between the Boston and New York interests over what to do with the remaining securities of local companies. Despite repeated assurances that the company was sound financially, the stock continued to decline. In its heyday it had stood above 120; during the depression it sank into the twenties.[29]

In June, to Coffin's dismay, business fell off sharply, and General Electric was forced to pay unheard-of interest rates to get cash for its payrolls. By July

1893 the company's short-term debt had mushroomed to $10 million with only $1.3 million cash on hand. Apart from its own debts, GE had also to deal with the notes of local companies that it had endorsed and the companies had then turned into cash at discount. Taking care of these notes was said to have "driven the General Electric managers almost wild and has made some serious financial sacrifices necessary." Fresh stories hinted at a growing schism between the Boston and New York directors over how to meet the crisis. Coffin was said to be on the verge of resigning his post in favor of Twombly, who was chairman of the board. Versions of the rumor persisted for more than two years despite repeated denials by Coffin and other officials. Through the weary months that followed, Coffin toiled under constant criticism. Years later he admitted that "there were months that seemed like scalding centuries."[30]

The rumors kept coming. One had the company on the brink of receivership; another claimed that Drexel, Morgan had secured control. "The bears," reported *Electrical World*, "spread any number of discrediting stories." Company officials did little to help analysts get a straight story. "It cannot be denied," observed one, "that the company has been hard hit by the frightful slaughter of its securities, not so much by the stock market decline as by the distrust created thereby."[31]

To weather the storm, Coffin resorted to a desperate tactic. He put $12.2 million of the best utility securities in a special trust and offered them to the stockholders at a third of their face value. Finding value in the securities became a major challenge. Two companies in which General Electric had large holdings, the Fort Wayne Electric Company and the Northwest Thomson-Houston Company, were carried on the books as assets totaling $5.5 million. So drastic was their fall that General Electric marked their value down to $1 each, and Northwest was liquidated. To raise immediate cash, Henry L. Higginson's banking firm agreed to help underwrite the new issue, but Coffin also needed Morgan's approval and collaboration in the syndicate. He was at his summer place in Maine when Coffin sent him the plan to look over. The wait for a reply was agonizing. Asked by a subordinate how the plan was going, Coffin replied, "Everything is either going through or falling through by tomorrow morning."[32]

It went through. Although Morgan grumbled at the terms of what he called a "Boston idea," he joined the underwriting syndicate. The bankers advanced General Electric $4.5 million for the $12.2 million in securities and held them in trust until the crisis eased and their value increased enough to be sold. For their pains they extracted a fee of 10 percent, or $450,000, from the proceeds. The funds enabled Coffin to skirt the threat of bankruptcy but did not ease the company's broader problems. Nor did it squash the persistent rumors of his removal and dissension within the board. On August 13 the *Chicago Tribune* asserted that Coffin would retire, that he had sold most of his GE stock, and that the Edison or New York people would bring Edward Johnson back as president. The article moved Samuel Insull, then president of Chicago Edison, to deny the stories even though he was no longer associated with General Electric. Meanwhile, General Electric began laying off workers and shrank its workweek to three days to save money.[33]

The Coffin plan created a new company called Street Railway and Illuminating Properties. The bankers bought the local company securities from General Electric, deposited them in the new company as assets, and on October 17 offered its shares to General Electric stockholders. The following month the board decided to pass the November dividend; by then the workforces at Schenectady and Lynn had dwindled to about half their usual size. Bad luck also dogged General Electric. Frederick L. Ames owned 7,500 shares of GE stock and was considered the richest man in Boston. A member of the board, he had been a stalwart of the Boston group supporting the company. On September 4 he predicted that the stock would soon rise and that no more than one dividend would be passed. Eight days later, while traveling from Fall River to New York by boat, he died of apoplexy much as his mother and sister had before him. His untimely death left the disposition of his stock in doubt, giving the bears another morsel of uncertainty to chew on: what would happen if the shares were thrown onto the market?[34]

There was no mystery behind the company's woes, insisted one observer. "The company undertook to furnish the entire country with electrical machinery and supplies and to do it on credit. It received stocks and bonds for goods . . . but the moment these securities could not be sold . . . General Electric reached the end of its borrowing capacity." To remedy that problem, Coffin instituted a new policy of making sales on a strictly cash basis even though it meant a severe drop in business. Still the pressure did not ease. A Philadelphia writer called General Electric the "mystery of Wall Street" and offered a fanciful tale that Morgan had tried in vain to lift the stock's price, that the stress of its fall killed Frederick Ames, and that another influential director, D. O. Mills, had broken with Morgan by refusing to furnish his share of funds to support the stock. Detractors pointed out that Morgan and Mills were also heavily involved with the Cataract Construction Company, which had awarded major contracts to Westinghouse. For weeks Mills was obliged to deny rumors that he was leaving the board.[35]

In November 1893 another round of rumors flared up over the split in the company between the former Edison and Thomson-Houston factions. A story circulated that at the Lynn plant the old Thomson-Houston lettering was scraped off all the machines and replaced with G. E. CO. Elihu Thomson was said to have invented a new incandescent light and assigned it to former Thomson-Houston friends outside of General Electric. "There can be no doubt, however," said one pundit, "that the old Thomson-Houston crowd would be glad to get their company back." The rumors were denounced as absurd as fast as they appeared; one even had GE merging with Westinghouse until George Westinghouse quashed it by saying he "would not take the General Electric Company as a gift." Still the stories kept coming. Coffin and his board had to beat down tales that the company was in financial trouble. One set of rumors he did not bother to contradict because they contained some truth: The company had decided to centralize many of its operations in Schenectady.[36]

Before the merger Thomson-Houston employed about 3,600 people at the Lynn plant. By December 1893 their number had dwindled to about 1,800 compared to 1,000 to 1,200 in Schenectady. Coffin wanted to centralize and streamline the organization even though he realized that the Schenectady plant would be inadequate once business picked up again. As a first step he folded the outlying district offices, which had been largely independent of central control, into the general office in order to exert tighter control over accounts, securities, and inventories. He also moved the corporate headquarters from the Edison Building in New York City, where it occupied more than three floors, to Schenectady, leaving behind only a few offices. "This will result in a much more compact organization," he declared, and it would also save money. The Lynn workers feared that all their facilities would also be transferred to Schenectady, but only the engineering and design departments were moved. When Thomson himself decided not to go to Schenectady, rumors flared that he was leaving the company. By July 1895 Schenectady had 4,400 workers compared to 2,000 at Lynn.[37]

Rumors continued to push Coffin out the door, one version insisting that "the Drexel-Morgan-Vanderbilt-Twombly interests now hold control of the stock to such an extent as to practically corner it, and already talk is heard of the election to the presidency next April of Mr. Twombly." Early in April, however, the same analyst reported a "remarkable reconciliation of inside quarrels." At the annual meeting Coffin stayed put in the presidency and all the directors kept their seats. Morgan partner Charles H. Coster moved to address the lingering problem of overcapitalization and provide for its reduction. Retrenchment became the order of the day. "A summary stop has been put to the practice of getting business at prices that do not admit of a profit."[38]

In his annual report for the year ending January 31, 1894, Coffin admitted candidly that the financial and industrial collapse, especially in the South and

West, had hit the company hard. "Very many accounts, notes and investments," he stressed, "assumed as good a year ago must now be recognized as of
diminished value." So too with the Northwest company, the collapse of which
had shrunk its assets to virtually nothing. Everywhere, he added, "even old and
strong customers were obliged to ask for leniency in paying their accounts and
notes." Inventories too suffered from the fall in prices that reduced their value,
and especially from innovations that rendered large amounts of old apparatus
obsolete and worth little more than scrap value. However, the floating debt
had been reduced, the organization tightened and centralized, sales made only
on the basis of cash or short-term credit, and costs reduced. General Electric
had survived all these harsh blows and was in sound shape for the future.[39]

Coffin's candor impressed most observers, but at a steep price. The volume
of business had shriveled nearly 46 percent from the previous year, and the
stock continued to sink. One bear operator declared that "the company has
lost all its prestige by the exposé it had to make in the annual report. It is no
longer an industrial company of very large earning power . . . It is feeling very
sharply the pinch of competition at the hands of powerful rivals." The sudden
shrinkage of assets posed other thorny problems as well. General Electric had
outstanding some $44.7 million in securities that at current prices were worth
just under $26 million. An analyst used this fact to emphasize the amount of
water in the stock. Moreover, the impaired capital ran afoul of a New York law
that barred any payment of dividends until that discrepancy in value was recti-
fied. This could be done in three ways: reduce the amount of outstanding stock
to reflect the current value of assets, subscribe fresh capital to make up the difference, or simply wait until the assets appreciated enough to close the gap.[40]

For the moment the inability to pay dividends posed no problem. Between
1894 and 1897 a chastened Coffin ignored the clamor of certain stockholders
and paid no dividends while he continued to write down assets and use earnings to buy back the company's debenture bonds. Not until August 1898 did he
finally gain approval to reduce outstanding GE stock from $34.7 million to
$20.8 million. "This reduction of 30 per cent . . . ," said one observer, "at last
corrects the error made when it [GE] was originally formed." The following
year Coffin further adjusted the capital account by slicing the book value of
GE's patents in half, from $8 million to $4 million. By 1900 the company's balance sheet would list assets of $27 million compared to $50 million in 1893.
During those four dismal years sales remained static at around $12.6 million.
The floundering Fort Wayne company was liquidated and replaced by a new
company with a greatly reduced capitalization. After the Northwest company
folded, its business reverted back to General Electric.[41]

Credibility came hard to the company. The bear attacks, fueled by the rumor mill, persisted through the summer of 1894, as did the Fort Wayne company saga. In November the *Wall Street Journal*, after looking into General Electric's affairs, reported that "financially, the company has never been stronger." Still the stock languished despite repeated assurances that earnings were solid and business improving. In February 1895 a reporter stunned one of the GE directors by asking if there was any reason to suspect a collapse in the company. "No!" he replied indignantly. "There is nothing in connection with its affairs to bring about any collapse." Another reporter noted that with prices of only three years earlier General Electric would be earning 30 percent on its stock. New contracts were coming in, the factories had all increased their workforces, and large orders began to surface. "A day at the Schenectady plant . . . ," wrote one reporter, "would convince the most skeptical bear that the company has valuable assets, and that its business is very large."[42]

Despite such assurances, the company and the stock had lost their luster. Rumors of disagreement within the management cropped up yet again. The board was expanded from eleven to thirteen members, and two Bostonians, Gordon Abbott and Robert Treat Paine, took the new places. Both also went on the executive committee with Coolidge, Coster, Higginson, and Twombly. Outside publications resumed their attack on the company. By the spring of 1895 a new question had arisen to divide opinion among the directors: whether to negotiate some sort of patent agreement with their archrival, Westinghouse.[43]

WHILE GENERAL ELECTRIC stumbled and struggled, Westinghouse seemed to be moving briskly along despite the hard times. His strong board of directors included Charles Francis Adams, August Belmont, Henry B. Hyde of Equitable Life, and Brayton Ives, president of the Western National Bank, but no one doubted that Westinghouse was in command. During the first year after the reorganization the electric company earned $1.6 million in profits, or about 18 percent on the $9 million of outstanding stock. Like Coffin, Westinghouse dispensed with many of his sales agents and centralized the operation to save money. The Chicago fair had given him both publicity and marketable products. Buoyed by his success, Westinghouse decided in the spring of 1893 to expand his facilities by acquiring twenty-three acres in East Pittsburgh and building a new manufacturing complex there.[44]

The move was as bold as the timing was unfortunate. The onset of the depression delayed the start of construction until January 1894 and prompted Westinghouse to do new business on a cash basis. Although orders declined, Westinghouse kept its workforce of 2,000 fully occupied. The company turned out new stopper lamps at a rate of five thousand a day when the Chicago fair opened and hoped to reach twenty thousand within a month. Westinghouse cheerfully advertised it as "the best and cheapest Incandescent Lamp offered to the public." The company also issued a letter noting that the Edison system had a year earlier been officially condemned by the Board of Fire Underwriters for "greatly increasing fire dangers." The fault lay in the feeder-and-main system, which involved "the interconnection of all of their underground mains" and the leakage from them. "With the alternating system," the letter stressed, "there is no electrical connection between the mains in the streets and the wires in the buildings, so that this system is absolutely free from the great fire danger due to the practice of connecting house wires directly to the street mains."[45]

Andrew Carnegie once described Westinghouse as "a genius who can't be downed." Boldness had gained him the Chicago fair and Niagara contracts, and it moved him to construct the new cluster of buildings at East Pittsburgh. He sensed that he had seized the initiative from General Electric, and he did not intend to let go of it. "Westinghouse attacked with energy and audacity," said one of his engineers, "and he held on with tenacity and fortitude." Having scooped General Electric on the two prizes of Chicago and Niagara, he proceeded to underbid them for some work on the Erie Canal. When GE reduced the price of its lamps despite its victory in the patent case, Westinghouse saw it as a sign of weakness in competing with his stopper lamp. Wall Street viewed his efforts as favorably as it distrusted General Electric. One broker declared that Westinghouse was "making prices so low as to afford General Electric a very small profit on the cash business obtained."[46]

As work proceeded on the new plant, Westinghouse declared that his capacity would soon equal that of General Electric. He added fourteen acres to the original parcel for two more buildings as well as a glass factory to increase lamp production to thirty thousand a day. More than one broker compared the capitalization of the two companies and conceded a clear advantage to Westinghouse. It had only $10 million in stock, half of it common, and less than $1 million in bonds. General Electric had nearly $30.5 million in common stock, $4.3 million in preferred, and $10 million in bonds. Put another way, GE needed a lot more income to support its heavier capitalization, which meant that lower prices caused it more pain than they did Westinghouse. "The difference in capitalization," said one broker, "gives Westinghouse an advantage in making prices when competition is close and it is close enough to make every advantage count."[47]

The new plant at East Pittsburgh extended that advantage when it finally opened in February 1895. A complex of seven buildings, it could swallow four thousand workers once fully completed. A visitor hailed it as "the most complete and extensive manufactory in the world ... The machine tools are among the finest ever made in this country. All shafting and machinery are driven by Tesla alternating current multiphase motors, and the whole plant in full operation furnish one of the best illustrations of electric power distribution to be found in the country." That was exactly what Westinghouse wanted: not only a superbly efficient factory but a showcase as well. Located on the main line of the Pennsylvania Railroad, it was in full operation by June 1895 and had absorbed the work formerly done at the Pittsburgh, Allegheny, and Newark plants. Everything was state-of-the-art, especially the power system that featured direct-connected alternators coupled in multiple. Electricity drove everything in the facilities, including seventeen freight and passenger elevators and the two thirty-ton Shaw cranes that could travel the full length of the shop.[48]

East Pittsburgh was a thing of beauty; it was also expensive. The floating debt in 1895 jumped above $3 million compared to $1.9 million the previous year, and earnings suffered from the time lost in making the move. To boost earnings Westinghouse remained as aggressive as ever in stalking new business. He folded the two New York City companies, United States Illuminating and Brush Illuminating, into United Electric Light and Power Company of New York and proceeded to erect a new AC central station in Manhattan that promised to be the largest in the nation. Equipped partly with apparatus from the Chicago fair, it posed a direct challenge to the dominance of the Edison company in the city where electric lighting made its formal debut. It was, said one observer, "expected to lead to the remodeling of all important electric stations in the country, as the old style continuous current apparatus is comparatively expensive and complicated." The Baltimore Brush Electric Company also boasted a fine new plant with equipment that included three of the engines and generators from the Chicago fair.[49]

In October 1894 Westinghouse officials claimed they were "now doing a gross annual business nearly equal to that done by the General Electric Company, and assert just as confidently that their margin of profit is much greater." By the fall of 1895 some analysts declared that Westinghouse had moved past General Electric in the competitive wars. "The street is rapidly growing to believe," said the *Wall Street Journal*, "that the Westinghouse Company has won the battle in the electric field ... The Tesla patents and ability to make goods at a lower cost are said to be the secrets of the Westinghouse success." The selling of GE stock, observed another writer, "is supposed to have been induced by certain trade conditions known to insiders which are reported turning against General Electric in competitive fields in favor of Westinghouse." With the momentum in his favor and business apparently picking up, why would

George Westinghouse even consider a patent-sharing agreement with his bitter rival?[50]

FOR ALL ITS woes, General Electric was hardly a fallen giant. Nor was Coffin on his way out. Turning fifty in 1894, he had lost none of his drive or determination. Like his nemesis George Westinghouse, Coffin shunned publicity and granted interviews only when they served his own immediate purpose. One of his oldest friends liked to say that Coffin was a man for whom nothing seemed too difficult, but the depression had brutally challenged that premise. One trade journal had hurled sharp accusations against the company's bookkeeping and impugned Coffin's personal integrity in the bargain. More harsh criticism was heaped on him for his plan to sacrifice so many securities to raise cash even though the securities had little immediate value. A Quaker, Coffin practiced the patience and consideration he urged on others.[51]

Years later Owen D. Young, who became chairman of General Electric's board, recalled the words Coffin imparted to his officers. "Every one has had mornings when he hates to hear the telephone ring, or see the office door open," he said. "I beg of you, gentlemen, when next you meet such a morning, take a stick of dynamite and blow up one of our plants. But do not take it out on a customer of the General Electric. We can replace the plant you have destroyed; we know its value; we have a reserve from which we can rebuild. But we cannot measure the good-will you have destroyed, and we can never know if we have replaced it."[52]

Once past the immediate crisis, Coffin succeeded in streamlining GE's organization and revamping its strategy. The depression had dealt the central station market a crippling blow and flattened the street railway market. Arc-lighting contracts sagged because local utility companies could not secure capital for their work. Moreover, the arc light, unlike other equipment, had undergone no significant improvements since its introduction. To survive the crunch, GE needed new approaches and new products. In the past both Thomson-Houston and General Electric had gained success by encouraging businessmen to set up separate companies and power stations for arc lighting, incandescent lamps, and traction. Westinghouse had developed a radically different model by providing a single central station with such interconnecting devices as rotary converters that enabled them to expand and diversify their load to serve all these uses.[53]

With Edison's influence no longer a factor, General Electric hastened to develop not only alternating-current equipment but systems that emulated the Westinghouse approach. Elihu Thomson developed a constant-current transformer that allowed DC arc lights to operate on an AC network. He and Charles Steinmetz also worked on an AC arc light. In 1893 the company began encasing the carbons of arc lights in glass globes, which increased their life tenfold. Its engineers developed a line of equipment for mining and motors for textile mills. At Lynn, Thomson worked to develop a three-phase AC transmission system. New motors came along in the continuing race with Westinghouse to create a superior version. In one area GE managed to scoop Westinghouse, thanks to an enterprising salesman named Sidney B. Paine.[54]

The enterprising Paine began his career at nineteen in a Fall River, Massachusetts, textile mill before becoming a sales agent for the Edison Electric Light Company in 1881. During the next few years, he traveled the circuit of mills in New England and convinced many of their owners to install Edison lights. By the 1890s the textile industry had begun its historic shift to the South, and the electric industry had moved beyond lighting to its early power systems. In 1893 Paine managed to persuade a Columbia, South Carolina, mill owner to buy a daring AC power system for his plant. The mill wanted to draw power from the rapids of the Congaree River, only eight hundred feet from it, but a canal lay between the river and the mill. Electricity seemed the only way to convey the power, and three firms bid for the job.[55]

GE's bid proved the highest and also the only one to employ alternating current. The mill owner wanted a cheaper DC system until Paine convinced him that it would not work. Both the lights and the motors, he explained, would be powered by the same direct-current generators. Turning on the lights would automatically raise the voltage of the generators and the feeders supplying the motors. This would in turn increase the speed of the motors, but textile machinery required a constant speed to run efficiently. Paine's argument won him the contract. Then he had to sell it to the company as well. To make matters worse, the motors—which did not yet exist—had to be mounted on the ceiling because the mill had no room on the floor for any more machinery.[56]

The innovative package included seventeen 65-horsepower induction motors. General Electric's executive committee blanched at the proposal. "We have never manufactured an AC motor larger than ten horsepower," protested Coffin, "and have never sold any larger than five horsepower . . . You must use DC motors for such a job as that." But Paine's explanation finally won Coffin over, and GE developed the system. When the motors began operation in April 1894, the engineers discovered that they needed an 85- rather than a 65-horsepower engine. However, once installed, the motors performed beautifully and were still working thirty years later. Paine's bold move gave General Electric a lead in textile mill installations that it maintained for many years. It also sent the company down a different road from that taken by Westinghouse in its cultivation of the polyphase power business.[57]

Through the 1890s both companies developed polyphase power systems and induction motors. Westinghouse concentrated on polyphase central stations, GE on supplying isolated industrial plants. Not surprisingly, the induction motors they created were designed for different purposes, and both firms expanded their equipment line to supply their particular approach. While the two companies struggled to top one another, one of the many small-fry firms in the industry elbowed its way into the competition for AC generators. Its leader was none other than William Stanley, the intense young inventor who had developed the first transformer for Westinghouse. After leaving that company in 1890 he started his own firm, building a factory in Pittsfield, Massachusetts, to manufacture transformers. In 1893 he added AC motors to the product line. A year later, when Westinghouse played hardball by refusing to sell AC generators to customers that bought Stanley's transformers, he developed his own model.[58]

All the Stanley equipment, including his improved transformer, differed from any rival versions on the market. Originality of design enabled Stanley to skirt the Tesla patents. His motor did not equal the Westinghouse induction version, but it performed well enough; his generator, called an "inductor alternator," was simple, rugged, and inexpensive. It appealed to inexperienced buyers because it lacked the brushes and collector rings used by a synchronous alternator. A Westinghouse report dismissed this last feature as one "of very little significance from a practical engineering standpoint . . . It is, of course, folly to sacrifice good engineering for such considerations." However, when the Stanley products cut sharply into Westinghouse's market share, the sales people demanded an inductor alternator of their own to push. Reluctantly the engineers provided one in 1896. That same year General Electric countered with a new design that combined the best features of the synchronous and inductor alternators. Two years later Westinghouse followed suit.[59]

Stanley's modest enterprise swelled his workforce from fifteen to three hundred and required a larger factory. In six years the firm turned a $400,000 profit. Although still a small player, he had forced the two giant companies into changing their product design to meet his competition. He also designed and patented a "transformer motor" to install in a mill at Housatonic, Massachusetts, and a "compensated alternating motor" for use on electric railways. Westinghouse did not appreciate his presence in these markets. Years later Stanley recalled that a Westinghouse officer visited his primary bank in Pittsfield and warned its manager that "they were going to 'drive me out of business no matter how much it would cost' and advised them 'not to loan the company or myself any money as they would certainly wreck us.' "[60]

But Stanley persisted. Together with his two partners, Cummings C. Chesney and John F. Kelly, he began experimenting with transmissions of much higher voltages than yet used commercially. In 1897 he installed a 40,000-volt line in California; later he increased it to 60,000 volts at a time when neither GE nor Westinghouse would install lines above 40,000 volts. "We take it from this," reported a GE sales manager, "that the Stanley Company are willing to be quite courageous in their exploiting power transmission work." Charles Steinmetz, one of GE's resident geniuses, was not impressed. "This voltage has never been used or approached on any transmission line . . . ," he argued. "I do not believe it can be operated successfully at the present." He was wrong. Within another few years more than twice that amount of voltage was being sent over lines. The higher voltages enabled utilities to exploit distant water power and, even more important, encouraged the development of larger, more economical steam-generating stations to supply power to broader areas at lower cost.[61]

Nor was Stanley the only small firm nipping at the heels of the two giants. The Walker Manufacturing Company of Cleveland began life as a machine shop in 1883, turned to electrical equipment, and soon had a complete line of railway, lighting, and power apparatus. Late in 1897 it passed into the hands of four financiers, including Anthony N. Brady, who was known as the "trolley king" because of his large investments in street railways. Siemens & Halske, the formidable German company, entered the American market in Chicago only to suffer the misfortune in 1894 of losing its plant to a fire. The Lorain Steel Company of Cleveland also made electrical goods. A number of smaller firms manufactured lamps as well as other equipment. Many of them boldly infringed on the patents held by General Electric and Westinghouse, figuring that the two giants were too busy fighting each other to worry about smaller game. For a time this strategy worked, but it also aggravated the whole patent question, which had become the biggest stumbling block to progress in the electrical industry.[62]

Hard times increased the pressure on all the companies in the industry regardless of size. Cutthroat competition aggravated the financial drain by bleeding profits from everyone. The patent wars injected yet another element of uncertainty in that no one knew what devastating effect might result from a decision once reached by the seemingly interminable judicial process. Even such doggedly optimistic men as Coffin and Westinghouse could not help wondering how long this war of attrition could go on. Hard times, it seemed, had a way of trumping hard feelings.

CHAPTER 18

THE FUTURE ARRIVES

How vastly must the character of our civilization be changed by further discoveries in this field of effort, and the complete application of principles already known to practical use? . . . What a vast source of power will be at the service of man . . . When that force is made not only to drive our machines in the factories and workshops and vehicles on the railroads and streets, and as a universal illuminant, but also to warm our buildings and cook our food . . . surely there must have been effected a revolution from existing conditions.

—CHICAGO TRIBUNE, JUNE 2, 1893

THE COMPETITION BETWEEN General Electric and Westinghouse extended even to matters of phase and frequency. As new technologies would demonstrate repeatedly in the twentieth century, competition often became a barrier to progress by hindering the process of standardization. Each major producer insisted on its own standards as the basis for any universal usage. In developing its first polyphase system, Westinghouse had chosen two instead of three phases because it eased the transition from existing single-phase systems. In the realm of frequency it had selected 30 cycles per second for power circuits only to change to 25 cycles when the Niagara project required it. Neither frequency proved satisfactory for lighting purposes. At 25 cycles incandescent lights tended to flicker, while arc lights did poorly at any figure below 60 cycles. [1]

General Electric took a different tack. Thomson-Houston had used 125 cycles, but GE viewed that as undesirable and chose $66\frac{2}{3}$ cycles for general power and light installations. For power circuits alone it decided first on 36 and then on 40 cycles; the latter figure went into its Niagara bid. As for phase, General Electric in 1894 introduced a "monocyclic" system devised by Steinmetz in an effort to skirt the Tesla patents. Despite the name, it was a three-phase system in which one had much more capacity than the other two. It worked decently enough and, like the Westinghouse two-phase system, bridged the gap between single and polyphase systems. However, it was theoretically complex and harder to explain to customers than a two- or three-phase system. Since it was an unbalanced system, induction motors had trouble running on it unless they were carefully designed and built. In the fight for customers, both Westinghouse and Stanley used these drawbacks against General Electric in their publicity.[2]

A French critic chided Steinmetz for what amounted to artful deception. "The monocyclic system," he sneered, "is, in fact nothing else than a *polyphased system* by means of which two phases are transformed (for what reason we do not know, unless it is to avoid the patents of another company), into three phases or one, according to whether the motors are starting up or working under load." He could see no advantage of Steinmetz's creation over a well-designed two-phase system, "while I do see very well that it is inferior to such a system, both in power and efficiency of its machinery."[3]

Like Westinghouse, Stanley decided on two-phase because it was "well suited for increasing existing single phase plants." But Stanley used the higher frequencies of 132, 125, and 66⅔ cycles for both lighting and power installations. The Stanley motors required a condenser to operate properly, and the higher frequencies were necessary to manufacture a small and cheap condenser. As long as every company clung to its own approach, no standardization was possible. Gradually, however, engineers became more familiar with polyphase power and appreciated the advantages offered by a three-phase system, especially those of steadier power and lower copper costs. One of Westinghouse's brightest engineers, Charles F. Scott, eased the transition by inventing a transformer that changed phases. With it engineers could generate in two phases but transmit in three, which enabled the company to move smoothly from one to the other. In 1895 Westinghouse began to sell three-phase installations.[4]

By that year the standardization of phase and frequency seemed far over the horizon, as did resolution of the long and bitter patent wars. But appearances proved deceiving. The string of mergers had eased matters by gathering whole clusters of patents in the hands of the parent companies. Court decisions had given General Electric control of key patents for the incandescent lamp and the Van Depoele overhead trolley. Westinghouse held sway over polyphase power, thanks largely to the Tesla patents. This domination made it difficult for one company to sell products in the area controlled by the other because customers demanded guarantees against patent infringement. Both General Electric and

Westinghouse recognized the threat to their product lines posed by the resolution of these cases. They were also weary of the financial bleeding caused by endless litigation over patents. As 1895 drew to a close, the two giants had over three hundred patent suits pending against each other.[5]

Nor was that all. Along with the costly legal battles against each other, the two companies had to fight smaller firms that infringed on their patents. Having already won suits in New York, Boston, and St. Louis against companies that challenged the Edison lamp patent, General Electric successfully enjoined another manufacturer in Milwaukee. But the challenges kept coming, and the clock was ticking on the lamp patent. In September 1894 GE lost a decision that gave Westinghouse the right to use the feeder-and-main system pioneered by Edison. When merger rumors surfaced yet again in May 1894, one analyst conceded that "insurmountable obstacles have arisen, largely in the nature of the suits which General Electric has against the Westinghouse company for alleged patent infringements." Despite their public belligerence, both sides had grown weary of supporting battalions of lawyers with little to show for the expense.[6]

Although the deep personal antagonism between Westinghouse and Coffin made a negotiated agreement unlikely, other factors pushed them toward such a settlement. Neither man wanted to abandon any sector of his business or produce inferior products to get around the other's patents. Both wanted to stop enriching generations of lawyers in the endless patent suits. Gradually the inexorable logic of business needs prevailed over animosity, but not easily. Sometime during the winter of 1895 General Electric and Westinghouse began discussing the possibility of some sort of patent arrangement. General Electric controlled more than a thousand patents and supported an entire department to oversee them. In March the *Wall Street Journal* declared that an agreement had been reached to pool patents, which was hailed as "the most important thing that has happened in the history of electrical competition."[7]

For nearly two weeks more details of the agreement leaked out, leading to speculation that "the agreement proposed is much broader than has been definitely stated, and covers not only patents but prices of electrical apparatus." *Electrical World* stressed that such a pact "will not only cut off enormous expenses for litigation, but give a great impetus to electrical development by removing from the minds of purchasers the fear of trouble from infringement suits." Then rumors began to surface of a snag. T. Jefferson Coolidge was said to have been unhappy with certain details of the deal and blocked approval. Westinghouse and Charles Francis Adams tried to smooth over the difficulty but refused to accept the changes demanded by Coolidge. Westinghouse left New York and went back to Pittsburgh without a signed agreement. The next day a notice appeared in a trade journal that his company intended to pursue suits against General Electric for infringement of the Tesla patents.[8]

By late April the agreement was dead. Puzzled financial writers tried to decipher what had happened. Some blamed a "prominent financier" who had been left out when the stock began climbing on rumors of the pact. One blamed Westinghouse and another Coolidge, who was said to have vetoed the deal. Still others thought that the proposal originated with the bankers on both boards and was scuttled by the officers of the company who opposed it. Or did the two sides, as one account had it, simply disagree over the valuation of their patents? The rumor mill gained fuel in May when both Hamilton Twombly and D. O. Mills left the General Electric board. One pundit insisted that Twombly resigned "due to the efforts of that particular interest in the General Electric board, which was responsible directly for the breaking off of negotiations with the Westinghouse Company." By early May the situation was referred to as "the Westinghouse and General Electric fiasco."[9]

The failure to reach agreement left all elements of the industry unsettled. The resignations of Twombly and Mills fueled new rumors that the Boston interests on the board, who supported Coffin, planned to take a more aggressive approach to the company's affairs. They were credited with summarily breaking off the negotiations with Westinghouse. Morgan was said to be unhappy with the outcome and contemplating "something like the organization of a gigantic electric power company to absorb both the General Electric and Westinghouse Companies." In May Twombly made another effort to rekindle negotiations by approaching Westinghouse. A memorandum of agreement was drawn up, and both companies appointed committees to negotiate terms, but they never met. Other committees were formed in their place and met for three straight days. Just when it appeared that an agreement had been reached, the General Electric committee backed away.[10]

This was too much for Westinghouse, who granted a rare interview to give his version of events. "Every suggestion made by Mr. Coffin," he said, "was of such a nature that . . . harmonious relations under the patents would have been impossible if the suggestions were carried out." Although rumors of a revival flickered anew in June, the negotiations were again pronounced dead in mid-August with little chance of their being renewed. For three months they remained moribund while General Electric's managers debated the question of whether to keep their operation in Schenectady or move it to one of several cities wooing them, and perhaps close the Lynn factory and move it as well.[11]

The GE complex at Schenectady consisted of forty-six buildings spread across a twelve-acre site. It could support a workforce of about 4,000 but fluctuated during 1894 between 2,000 and 3,300. In 1895, however, the payroll jumped to 4,400, and the directors, painfully aware of Westinghouse's leap in productivity thanks to its new facilities, wanted to increase output by 25 percent. The need for more space was imperative, partly because newer electrical machinery kept growing in size and required larger housing. However, GE could not enlarge its existing buildings economically, and it lacked space to erect new ones. Its triangular-shaped grounds, bordered by the Erie Canal on one side and a New York Central Railroad embankment on another, were cut in two obliquely by Kruesi Avenue. A patch of land across Kruesi Avenue offered the only possibility for expansion.[12]

Several cities, including Buffalo, offered General Electric attractive packages of inducements to move their facilities. The company coyly entertained these propositions while the city fathers of Schenectady scrambled to top them. In August 1895 the city agreed to close Kruesi Avenue and turn the parcel of land beyond it over to General Electric. The company in turn announced that it would stay in Schenectady and began construction of a large building on the newly acquired land. Although rumors cropped up in 1896 that GE was about to concentrate its Schenectady, Lynn, and East Newark factories in Elizabeth, New Jersey, nothing came of them. The company was in Schenectady to stay.[13]

November 1895 brought fresh rumors that secret negotiations for a patent agreement had been revived. The word on Wall Street was that Morgan, who loathed corporate strife, had brought pressure through the New York directors to get an agreement done. Some said that Coffin had resigned and that the Morgan interests would run the company, but that proved to be untrue. That same month the Supreme Court upheld the decision upholding the Edison incandescent light patent. In February 1896 the Court of Appeals sustained GE's monopoly of the overhead trolley system with the Van Depoele patent. These two victories increased the pressure on Westinghouse to reach an agreement. Compromise made sense; as one analyst observed, "Both companies have spent several million dollars in patent contests, none of which money has come back, while prices have been cut in the most desperate fashion in order that one company could get the better of the other."[14]

On March 12 the two companies finally announced a cross-licensing agreement that pooled patents and allowed both companies to manufacture most of each other's patented equipment for the next fifteen years. The patents would be managed by a board of five people, two appointed by each company and the fifth chosen by those four. General Electric contributed 62.5 and Westinghouse 37.5 percent in value of the total patents. Each company would pay a royalty to the other for any use of patents in excess of its value, with some exclusions. Coffin declared it to be "practically a working agreement for life." Another observer welcomed the end of the patent wars. "Whenever a patent has been broken down," he noted, "it has opened the field to countless small manufacturers with the result of flooding the market with apparatus . . . The field for electrical appliances has grown with inconceivable rapidity. But these two great companies have thrown away great profits by the war."[15]

No longer. Although some disputes and problems arose, the agreement enabled General Electric and Westinghouse to confine their warfare to the time-honored spheres of price and quality competition. It also strengthened their hands against the smaller companies they accused of infringing on their patents. Stanley had been sued for violating the Tesla patents in 1895; shortly after the agreement the two giants launched twenty suits against Walker. Rumors flared that the smaller rivals might be brought into the pact; instead they were swallowed up. Walker remained a tough competitor until September 1898, when Westinghouse acquired the company. A few years later Westinghouse also bought the electrical business of Lorain Steel. In 1900 Siemens & Halske, plagued with a strike and financial troubles, sold out to General Electric. Three years later Stanley followed suit. He then formed a meter company, which was quickly sued by Westinghouse and forced out of business.[16]

The historic agreement of 1896 proved a milestone in the evolution of the electrical industry. General Electric and Westinghouse solidified their dominance of the industry even while they continued to compete. During the next few years they also moved closer together on phase and frequency. The agreement enabled General Electric to develop polyphase equipment without worrying about the Tesla patents. By 1900 both companies, thanks to the invention of new devices, settled on two standard frequencies: 60 cycles for lighting and small power and 25 cycles for heavier power. Within another twenty years the engineers put nearly all AC lights and equipment on 60 cycles, which became the standard frequency. Since 1885 frequencies had come full circle from the original 133 cycles to multiple cycles for different purposes to a single cycle for nearly all usages. A once fledgling industry was fast approaching maturity even as innovations continued to characterize its development.[17]

ONCE AGAIN THE domino effect of development reared its persistent head. During the development of an AC power system, the emphasis had for years been on its weakest link, the motor. The success of the induction motor did much to galvanize the development of AC systems to the point where the weak link shifted to the generator and its prime mover, the steam engine. By the mid-1890s it had become painfully apparent that the AC generator was fast approaching its upper limit in size. The amount of power generated depended on the size of the alternator, which was limited by the size of the slow-moving steam engine that drove it. To obtain the economies of scale made possible by the growing demand for electric power, a central station required the largest possible generators. However, it was still cheaper to buy, install, and operate two 8,000-kilowatt alternators than one 16,000-kilowatt unit.[18]

The reciprocal steam engine, that reliable old warhorse of power supply, had undergone numerous upgrades over the years, but it could never overcome certain inherent limitations. Although its efficiency improved, it still devoured enormous amounts of coal and water, a constant supply of which had to be brought to the site. Nor could steam engines or the generators they fed keep growing in size indefinitely. Dynamos could not produce more power without some newer, more efficient, and preferably smaller primary source. This need led both Westinghouse and General Electric to explore an intriguing alternative device that first gained attention during the 1880s: the steam turbine.

The turbine is basically a device in which a moving fluid, usually water or steam, moves the blades on a rotor to create a rotating motion that can be used to produce electrical or mechanical power. It moves the blades by shooting them with steam or water forced through nozzles. Instead of relying on pressure, the turbine utilizes kinetic energy in two ways. A *reaction* water turbine uses the sheer weight of the water falling on the wheel to generate energy; an *impulse* turbine uses the force of the water to drive the wheel. The reciprocal steam engine converts steam into energy through direct pressure of the steam on a piston. By contrast, the steam in a turbine acts as a mass that is set in motion by its own power to expand. In a reaction steam turbine, the steam's expansion occurs in both the stationary nozzles and the rotating blades, while in the impulse turbine it takes place only in the nozzles.[19]

Early versions traced back to ancient times, but nobody invented one that harnessed steam for practical purposes until Charles A. Parsons took up the challenge. Born in 1854, the son of a well-known British astronomer, Parsons became an engineer and developed an interest in both dynamos and turbines. While still a young man, he realized that the familiar water turbine could serve as a model for a steam version. As he observed much later:

> It seemed to me to be reasonable to suppose that if the total drop of pressure in a steam turbine were to be divided up into a large number of small stages, and an elemental turbine like a water turbine were placed at each stage . . . then each individual turbine of the series ought to give an efficiency similar to that of the water turbine, and that a high efficiency for the whole aggregate turbine would result; further, that only a moderate speed of revolution would be necessary to reach the maximum efficiency.

Drawing on this insight, Parsons worked through the mechanical and lubricating obstacles to build his first reaction-type, high-speed steam turbine, which he patented in April 1884. It proved to be a breakthrough rivaling in importance the Watt patent for a condenser.[20]

Two years earlier Gustav de Laval, a Swedish engineer, had devised an impulse-type steam turbine that operated at high speeds. Both Parsons and Laval tried coupling their turbines to generators, but the latter's high speed required reducing gears, which kept their turbogenerators at small capacities. They also guzzled steam hungrily, which meant high fuel costs. Parsons designed and patented his own generator, but both men struggled for years to refine their turbines into commercial products that could compete in size and efficiency with reciprocal steam engines. In 1889 Parsons lost control of his original patents and had to come up with a new design for a turbine. Five years later he regained possession of the patents and could develop his original design again. By 1895 more than four hundred Parsons units had been installed in central power stations and isolated plants.[21]

Turbines were hardly known in the United States. As late as 1898 *Electrical World* thought it "strange that more attention has not been paid to the steam turbine in this country, and that the excellent results that are attending its use abroad have not been appreciated here." However, news of the Parsons unit did reach the two rival electrical giants in the United States. General Electric had the first crack at securing the rights to the machine but declined on the advice of Charles Steinmetz. "I think this field very promising," Steinmetz wrote, "and believe we should push experimental work in this regard with the greatest possible speed . . . I do not think very much of Parsons' steam turbine but think that a simpler design, even if not quite as efficient in steam consumption, would be desirable." He urged delaying any agreement with Parsons. Westinghouse had been following Parsons's work in his usual way and saw the potential for a new and more efficient prime mover. Rotary engines had long intrigued him; his very first patent in 1865 had been for a crude rotary steam engine. In 1895 he sent one of his men to England to report on the Parsons turbine. Upon receiving a favorable recommendation, he bought the American rights that same fall.[22]

Once again Westinghouse secured rights to an invaluable patent and set out to improve it for commercial production. The Machine Company under his brother Herman built the turbines while the Electrical and Manufacturing Company produced the generators that would be linked to them. The first unit, completed in 1896, was a DC machine that proved unsatisfactory because of commutation difficulties. The next experimental unit used an AC generator that posed problems of its own. By 1898 Westinghouse was ready to install the first commercial 400-kilowatt units at his own Air Brake Company plant. Two years later he made the first central station installation in Hartford, Connecticut. This unit delivered 2,000 kilowatts. General Electric did not develop its first turbine until 1901, and at one point seemed on the verge of abandoning the project altogether until an engineer named William L. Emmet took hold of it. By 1903 General Electric finally had its first unit ready for sale. The buyer was none other than Samuel Insull, who had declined a position with General Electric and become president of the Chicago Edison Company.[23]

In this modest way the future of primary power arrived. Three factors determine the power output of a generator: the number of conductors, the strength of the magnetic field, and the speed. Gramme's winding had increased the number of the conductors, and the drum winding improved the effectiveness of each conductor. The landmark 1886 paper by the Hopkinson brothers had advanced theoretical understanding of the magnetic field and with it the design of generators. At that point physical size alone limited the output of generators, and the invention of the turbine removed this barrier. As one well-known engineer put it in 1905, "In no other piece of electrical machinery is so large an output concentrated within so small a volume."[24]

The turbine took little time to demonstrate its enormous advantages over the reciprocal steam engine. It ran at speeds of 1,000 to 2,000 rpms compared to the 100 rpms of a large reciprocal engine, which meant that a smaller unit could produce even more power than a larger reciprocal version. Smaller units meant lower costs; they also offered the potential of building much larger versions, producing unheard-of amounts of power. A 5,000-kilowatt steam turbine, for example, weighed only 20 percent as much as a reciprocal steam engine putting out the same amount of power. In 1901 Westinghouse built the largest reciprocal engines ever made for the Manhattan Elevated railway. These seventeen beasts each stood forty-two feet high, weighed nearly a million pounds, and produced 6,000 kilowatts. They were awesome dinosaurs. "This obsolescence of the engine-type alternator was almost pitiful," wrote Henry Prout. "Here was a branch of heavy engineering built up at great cost and backed by years of experience. In the coming of the turbo-generator this experience was mostly thrown away."[25]

Compared to reciprocal engines, the turbines were as economical as they were compact. They cost less to run and cut maintenance costs by about 80 percent. They ran for weeks and months with little need for attention, while reciprocal engines had to be inspected, adjusted, and lubricated regularly. As the demand for electricity soared, the pressure grew on central power stations to deliver more energy. Direct-power generators could not produce more than 3,000 kilowatts because of commutation problems. Alternate-current generators eliminated that limitation. So too with the turbine units. The low speed and efficiency of a reciprocating steam engine meant that alternators could not be built larger than about 10,000 kilowatts. The turbine removed that ceiling. By 1907 turboalternators of 10,000-kilowatt size were built; ten years later units three times that size were produced, and by 1935 units producing 200,000 kilowatts were built. The turbine proved to be another crucial breakthrough that enabled the spectacular growth of electric power in the twentieth century. Small wonder that Prout called the turbogenerator "the greatest contrivance for the manufacture of power yet produced by man."[26]

BY 1900 ELECTRICITY had become an integral part of American life, especially in the cities. Between 1890 and 1905 the output of electric power in the United States increased a hundredfold. By revolutionizing production and manufacturing, electricity made possible the rise of the consumer economy that was to dominate the twentieth century and transform every corner of American life. Already factories consumed more than half of the electricity generated, even though less than 5 percent of factory power came from electricity. Arc lights illuminated the streets even of small towns and flooded with light the avenues of large cities. In 1902 some 51,000 electric streetcars whisked urban passengers along 22,000 miles of track. For the folks at home the Sears catalog listed electric arc lamps, battery plasters, bells, fans, insoles, liniment, rings for rheumatism, and the cure-all belt. For four dollars any sufferer could buy the "Primary Heidelberg Electric Belt," which also came in a twelve-dollar version and the majestic eighteen-dollar "Giant Power" model. It promised to relieve "disorders of the nerves, stomach, liver and kidneys, for weakness, diseased or debilitated condition of the sexual organs from any cause whatsoever," and was "worth all the drugs and chemicals, pills, tablets, washes, injections and other remedies put together."[27]

Only twenty years had passed since Thomas A. Edison opened the Pearl Street station, yet the electrical manufacturers had proliferated, expanded, clashed, and been largely whittled down to two giant firms. The day of the pioneers had passed, and to a large extent the technology had passed the pioneers by. Most of them lived long lives but never again played a pivotal role in the field. William Stanley, after selling out to General Electric, stayed on as a consultant and continued to obtain patents for electrical inventions until his death in 1916. Charles A. Coffin defied the rumors of his dismissal and remained head of General Electric until 1913, when he surrendered the presidency to Edwin W. Rice, the man who had begun his career as Elihu Thomson's pupil and assistant. Coffin stayed on as chairman until 1922 and lived until 1926. He and Rice gave way to the next generation of outstanding GE leaders, Owen D. Young and Gerard Swope.[28]

Elihu Thomson continued his dual role as scholar and research scientist for General Electric until 1900, when the company finally established its industrial research laboratory. At that moment he terminated his career as an inventor without a murmur of regret. During the 1890s Thomson did extensive research and development work on X-rays, automobiles, and incandescent lamps, but he grew weary of product development. "My ambition," he admitted in 1897, "has been to be able to devote more and more of my time to scientific work." After 1900 he did exactly that, turning his keen mind to astronomy and lens improvement for telescopes along with his activities in scholarly and professional associations. He had close ties to MIT, where he never taught but served as acting president twice between 1920 and 1923. Like Edison, he basked in the honors heaped on him at home and abroad.[29]

In the end his reputation clung as closely to General Electric as did that of Edison. Charles Coffin liked to tell a story about the young boy who visited the plant years earlier and asked to be introduced to "General" Electric. Coffin sent for Thomson, who happened to be there that day. When Thomson walked in, Coffin said, "This, my boy, is 'General' Electric." The "General" lived to be eighty-four, dying in 1937. Of him MIT President Karl T. Compton wrote, "Perhaps no inventor save Edison has brought so much renown to our country or contributed so much to its recent progress."[30]

Save Edison. Always save Edison. Although the Wizard never returned to the electrical field, his name was indelibly associated with its pivotal moment. For three more decades Edison solidified his reputation as America's inventor. Despite the costly fiasco of his ore separator business and other missteps, he developed the phonograph, motion pictures, an improved storage battery, a prefabricated concrete house, and numerous other inventions. He created the first motion picture studio and made the first silent movies. Like Thomson, he became interested in X-rays and worked on developing fluorescent lights and fluoroscopes. His discovery of the "Edison effect" led to the development by others of the vacuum tube, radio, and television.

As an old man Edison became close friends with Henry Ford, who idolized him to the point of reconstructing the Menlo Park laboratory at Greenfield Village in Dearborn, Michigan. The dedication in October 1929 featured Edison and an elderly Francis Jehl re-creating the 1879 birth of the carbon filament lamp for an adoring crowd of dignitaries, including President Herbert Hoover, and a nationwide radio audience. Twenty years earlier he had been honored by the creation of the Edison Pioneers, a social organization of his former associates and fellow inventors who met every year at a luncheon to honor the master inventor.

By then Edison had long since become a national icon, his name invoked in advertisements by the company that had stripped that same name from its own. He died in October 1931, having outlived all his old friends and rivals except Thomson and that most eccentric of geniuses, Nikola Tesla.[31]

Life did not treat Tesla as kindly as it had Edison. Although he remained in the public eye throughout his long life and was honored by several awards, his career flamed out in a succession of near misses and unfinished projects, each more grandiose than the last. His downhill run began in 1895 when a fire destroyed his laboratory in New York. *Electrical World* called the loss "one of the most severe that could have befallen electrical science." Edison gave him temporary lab space at West Orange, and Westinghouse donated some of the early Tesla models left over from the Chicago fair, but the fire accelerated what became Tesla's lifelong quest for funds to support both his lifestyle and his increasingly ambitious visions. He still owned a reputation as an electrical genius and a brilliant scientist, but he struggled at parlaying these assets into enduring tangible results.[32]

"One is naturally disappointed that nothing practical has as yet proceeded from the magnificent experimental investigations with which Tesla has dazzled the world," observed *Electrical World* in 1894, "and what is still more disappointing, that they hold out no definite hope of the realization of the great promise they were supposed to contain."[33]

Money became the root of all frustration for him. Over time Tesla grew increasingly bitter at the ability of others to profit from what he deemed to be his creations. Both Westinghouse and General Electric produced AC systems that had come to dominate the electrical world, but Tesla collected nothing more from the patents he had sold to George Westinghouse. According to one estimate, if he had retained the $2.50 royalty per horsepower contained in his original contract, he would have earned between $12 million and $17 million by the time his patents expired in 1905. Late in life the Westinghouse company agreed to cover his rent at the New Yorker Hotel, where Tesla spent his final years, but he got little else from his old affiliation.[34]

Tall, gaunt, impeccably dressed, an improbable blend of charm and eccentricity, Tesla gained access to the highest moneyed and social circles. He induced both John Jacob Astor IV and J. P. Morgan along with other wealthy men to put money behind his projects but never delivered on his promises. Like Edison and Thomson, he explored the intriguing realm of the Roentgen ray, or X-ray, which he called "shadowgraphs." Tesla experimented with them on himself as well as on small animals in the belief that "it would take centuries to accumulate enough of such matter to interfere seriously with the process of life," but his work produced nothing of commercial use. His attention turned to the study of high frequencies in hopes of developing a wireless communication system. During 1896 he received eight patents, mostly for different types of oscillators, which were devices capable of generating AC power to one specific frequency. By 1901 he had acquired thirty-three patents for "transmitting electrical energy through the natural medium."[35]

From this work evolved an audacious scheme to create a global wireless telegraph system using giant oscillators to harness the earth's own electrical currents as carrier waves. Tesla unveiled this vision in an 1897 lecture in New York. Even more, he imagined a system that could also transmit power, change weather patterns, and even communicate with other planets. A year later he astounded the public with another invention, which he called the "teleautomaton," based on wireless motors he had displayed in 1892. It took the form of a boat four feet long and three feet high, exhibited for a special audience of potential investors that included Westinghouse, Morgan, and Cornelius Vanderbilt. Using several transmitters, Tesla started, stopped, and steered the boat by remote control; he could even turn its lights on and off. From this remarkable model emerged the dawn of radio and remote control, but the investors did not bite. Neither could he interest the government in his vision of a wireless torpedo delivered by a crewless boat even though the nation was at war with Spain at the time.[36]

In 1898 Tesla moved into the Waldorf-Astoria Hotel, the social center of the moneyed elite. For twenty years he lived there while pursuing funds for his projects, and commenced what became a lifelong pattern of running up bills he could not afford to pay. After showing Astor a laundry list of his projects, he extracted an investment of $100,000 from him for development of a fluorescent lighting system. Other investors put up smaller amounts for what seemed a sound, practical project. Instead Tesla used the money to build a new laboratory in Colorado Springs, Colorado, where he intended to build a resonant transformer "which would be capable of disturbing the electrical condition [of part] if not the entire globe . . . thus enabling me to transmit intelligence to great distance without wire." At one point in 1899 he grew convinced that he had received impulses that could only have come from extraterrestrial sources. "I felt as though I were present at the birth of a new knowledge or the revelation of a great truth," he wrote. The source, he concluded, was most likely Mars.[37]

Tesla returned to New York in 1900 oblivious to the fact that he had not developed the practical system of fluorescent lights or oscillators anticipated by his investors. Neither had he shaped his wireless system into a practical one. While he dawdled, Guglielmo Marconi had been hard at work trying to perfect a more modest but workable wireless system. He happened to be in Manhattan in 1900 and met Tesla at the New York Science Club. After asking for and receiving an explanation of a technical point from Tesla, Marconi pronounced it impractical. He left before the debate turned nasty and continued to focus on his system. By contrast Tesla kept trying to sell too many potential inventions to different investors and refused to focus on pushing one of them to market. Increasingly he became convinced that Marconi was stealing his ideas and designs. Tesla obtained a loan from Westinghouse to construct a receiving station in Great Britain but could not interest him in the wireless system.[38]

His next move was to solicit funds from Morgan in 1901, the same year that the great financier was putting together what became the nation's largest company, United States Steel. Morgan showed enough interest to advance Tesla $150,000 in return for a 51 percent share of the inventor's patents and the lighting concern originally contracted with Astor. An overjoyed Tesla promptly set to work creating what he called his World Telegraphy Center. He bought a two-hundred-acre site at Wardenclyffe on Long Island and hired his friend Stanford White to design a massive tower for sending wireless messages around the globe. Tesla projected heights as great as 600 feet before reluctantly settling on what became 187 feet. Predictably, he soon ran through the funds advanced by Morgan and went back to 23 Wall Street for more. Morgan reiterated what he had said before: no more money. Distraught, Tesla decided to continue work on the tower while continuing to plead with Morgan and others for more support.[39]

Tesla had gained access to Morgan through friendship with his daughter Anne, who enjoyed the company of eccentrics. However, his timing turned out to be as poor as his approach was impractical. During 1901 a clash between E. H. Harriman and James J. Hill rocked the stock market in the spring and a crazed anarchist shot President William McKinley in the fall, elevating the unpredictable Theodore Roosevelt to the White House. Even worse for Tesla, Marconi carried off the first successful transmission from England to Newfoundland. Radio had been born, although as yet it remained only a form of wireless communication by which signals could be sent in Morse code. Tesla took the news hard, convinced that Marconi had utilized the Tesla coil, oscillators, and even design to pull off his triumph. His bitterness mounted when a glittering roster of electricians, including Elihu Thomson, Charles Steinmetz, Michael Pupin, Carl Hering, William Stanley, Frank Sprague, T. C. Martin, Alexander Graham Bell, and Mrs. Thomas Edison, representing her husband, gathered at the Waldorf to honor Marconi's feat.[40]

Tesla was not there, having slipped out of the hotel and gone to Wardenclyffe to ponder his tower, on which construction had finally begun. He had much earlier promised to render Edison's incandescent bulb, which he deemed wasteful and expensive, obsolete with his own fluorescent, or cool, light, but he had failed to produce it. He aimed to astonish the world with wireless transmission, but Marconi had scooped him, working diligently while Tesla concocted ever grander schemes and made wild boasts. He had wheedled financial backing from two of the richest men in the nation and given them nothing to show for it. For all his genius, Tesla remained the brilliant but erratic visionary while others turned ideas into practical products. Without realizing it, he had burned his best bridges. For years he continued to implore Morgan to advance him more money, but Morgan would not budge. Nor would he return Tesla's patents to him so that he could try to entice other investors.

Years earlier, in 1896, Tesla had explained that "it is my honest conviction that I am far in advance in certain lines of scientific investigation, which I consider of greater importance for mankind, and I see before me clearly near success. I am impelled by my sense of duty to follow the dictation of my own reason." For the rest of his life this explanation became his mantra for a career that seemed always on the verge of a great discovery.[41]

During 1903 Tesla began to manufacture oscillators and work on his fluorescent lamps. He scraped up enough money to push construction of the tower that he hoped would transform his fortunes. Once again he scrounged everywhere for funds to keep the work going, but in most cases his reputation preceded him and the deals never materialized. By 1907 he had reached the end of his rope. Creditors had been hauling away equipment from Wardenclyffe, and Tesla could no longer support even the modest operation there. For over a decade the giant tower sat empty until it was finally demolished in 1917. With it went Tesla's dream of a global wireless communication system. Disconsolate and broke, Tesla conjured up new visions. He began work on a flying machine and in a speech prophesied both a jet-powered airplane and a curious aircraft that would take off and land vertically. He also designed a hovercraft that he hoped could fly on wind currents without fuel. These were bold, original, and influential notions, but Tesla realized nothing practical from them.[42]

It galled Tesla to find other, more focused inventors turning his concepts into actual products. Radio, aviation, and fluorescent lighting all continued to advance without him. Still hoping to revive Wardenclyffe, he turned to yet another bold vision, a bladeless turbine simple and light enough to drive an airplane or automobile or to serve as a pump. "The Tesla turbine is the apotheosis of simplicity," wrote one journalist. "It is so violently opposed to all precedent that it seems unbelievable." Once again Tesla conjured up great possibilities for his creation. "I am now at work on new ideas of an automobile, locomotive and lathe in which these inventions of mine are embodied and which cannot help [but] prove a colossal success," he wrote in 1909. "The only trouble is to

get the cash, but it cannot last very long before my money will come in a torrent." However, it continued to come in a trickle.[43]

As other frustrations followed in the old pattern, Tesla turned increasingly
inward. Legal entanglements and credit for inventions falsely given others (or
so he thought) drained and embittered him. A celibate of dubious sexuality, he
had no family and only a few close friends, most notably Robert Underwood
Johnson and his wife, Katharine. In his loneliness Tesla developed a fondness
for pigeons. He loved to feed them and smuggled sick or wounded ones into
his hotel room to nurse back to health. Of his favorite white pigeon, a beautiful specimen, he said, "I loved her as a man loves a woman, and she loved me.
When she was ill I knew, and understood; she came to my room and I stayed
beside her for days . . . That pigeon was the joy of my life. If she needed me,
nothing else mattered. As long as I had her, there was a purpose in my life . . .
When that pigeon died, something went out of my life."[44]

As Tesla aged, his eccentricities grew even more marked, especially his fear
of germs. Perpetually broke and forced to move from hotel to hotel until the
Westinghouse company relieved his distress, he became a prophet with some
honor. The electrical profession still prized his genius, and occasionally the
larger world acknowledged it. In 1917 he was awarded the Edison Medal, an
irony he swallowed with great difficulty. At the ceremony Arthur Kennelly, the
former Edison assistant then teaching at Harvard and the man who had assisted in the execution of animals during the war of the currents, gave the
opening speech. Shortly following his seventy-fifth birthday on July 10, 1931,
Time magazine honored Tesla by putting him on its cover above the caption
"All the world's his power house." Although he continued to come up with
concepts for mysterious new inventions such as a death ray, Tesla retreated
ever more into embittered loneliness. He died on January 7, 1943, at the age of
eighty-six, having outlived all the men who had done what he never managed
to do: turn ideas into practical inventions. The *New York Sun* rendered him
this tribute:

> Granting that he was a difficult man to deal with, and that sometimes
> his predictions would affront the ordinary human's intelligence, here,
> still, was an extraordinary man of genius . . . His guesses were right so
> often that he would be frightening. Probably we shall appreciate him
> better a few million years from now.[45]

LIKE TESLA, GEORGE WESTINGHOUSE deserved a better deal than fate
handed him. However, like Tesla, too, he did much to engineer the downfall

that came so swiftly and unexpectedly to him. Of all the electric pioneers, he alone still kept at the business even while veering into other things. Of course, he had been engaged in other products long before venturing into electricity, and his attention to these areas did not falter. Ironically, his downfall with the electrical company occurred not because he failed but because he succeeded too well at what he had set out to do. He realized his grand vision on a scale that induced, indeed compelled, him to repeat the very mistake that had almost done him in fifteen years earlier. The crisis of 1890–91 did not imprint its lesson on him. Ever the optimist, always eager for the next challenge, confident of his ability to meet any situation, he looked always to build and to expand on what he had done.

Westinghouse had been present at the creation. Since the mid-1880s he had believed in the potential of alternating current and boldly thrown his resources behind its development without hesitation. He must have beamed with satisfaction when in September 1893 *Electrical World* proclaimed that "the alternating current may now be said to have the lead over its continuous current rival, and it appears that the latter will cut but little figure hereafter except for comparatively minor or special applications." Having survived his financial crisis, he spent the next decade building Westinghouse Electric into an industrial giant only to find himself caught in the same vicious cycle. He had placed his faith in the belief that the demand for electricity would continue to expand, that AC power provided the best solution to that demand, and that therefore his business would continue to grow. This is exactly what happened. By the 1900s his problem—and that of General Electric—was not a lack of orders but a rush of business that overwhelmed their ability to meet it without constant expansion of facilities.[46]

The heart of the matter lay in a familiar problem: the voracious need for capital. Even after the depression lifted, competition kept profit margins low, leaving little reserve for expansion. The cost of both labor and goods rose steadily after 1897. Electrical manufacturers needed not only expanded facilities but more workers and large inventories of supplies. Inventories required a delicate balancing act between having enough material on hand to fill orders and not overstocking on products that might become obsolete. Westinghouse had learned this lesson many times, most notably when Motor No. 3 transformed a large inventory of double-reduction motors into scrap metal. Coffin had suffered similar losses that accounted in part for General Electric's writedown of assets during its crisis. Copper remained a major expense; by one estimate the electrical manufacturers used between 140 million and 160 million pounds a year and had to stockpile it. When the price dropped, they had to write off the difference between the purchase and current prices as a loss.[47]

As prosperity returned, George Westinghouse took up the challenge of expansion with his usual gusto. He turned out an impressive list of new products, including a gas engine. In 1901 Westinghouse Electric employed 7,971 men; by 1907 its ranks had swelled to 17,500. During those same years its business jumped from $12.5 million to $40 million. Westinghouse also invested in subsidiary companies in Canada, Great Britain, France, Germany, and Russia. Altogether his business empire, including the nonelectrical firms, operated twenty-four factories with 38,000 employees in the United States and Europe. The total capital of his companies was $120 million, of which the American group exceeded $71 million. By February 1907 Westinghouse Electric was turning out three thousand motors a month in anticipation of a growing market. To keep up with demand, the company started work on a new eight-story factory at East Pittsburgh. The other Westinghouse companies were expanding as well; Union Switch and Signal purchased more land to double its output.[48]

All these activities required more money. Generating stations and transmission lines, as well as electrical equipment, swallowed large sums. Like General Electric, Westinghouse found it necessary to take notes from local companies in payment, but these notes, secured by deposits of the utility company's securities, were hard to discount in tough times. The foreign ventures broadened Westinghouse's market but getting them up and running also drained cash. Earnings remained strong but went largely to dividends. Since his earlier crisis Westinghouse had rewarded his stockholders with generous dividends. Between 1891 and 1902 the company paid 7 percent on its preferred stock; in 1900 it began paying dividends on the assenting common as well. In 1903 it raised the dividend on both classes to 9 percent, and then to 10 percent after 1904. By 1907 Westinghouse Air Brake was paying 20 percent, Union Switch and Signal 12 percent, and Westinghouse Machine Company 10 percent.[49]

To make matters worse, profit margins dwindled after 1905 when General Electric decided to drop prices 10 to 15 percent even though the cost of labor and material was increasing. Westinghouse objected but had to follow suit. Business was booming. In the spring of 1907 General Electric claimed to have booked 40 percent more orders than the previous year, while Westinghouse said its plants were "taxed to capacity limits." The problem was that orders did not produce immediate cash. A Westinghouse manager estimated that it took six months to turn an order into cash; meanwhile, the company needed capital for the men and material to fill that order. However, the large dividends precluded the accumulation of a surplus for such needs. In 1907, for example, the company earned $2.7 million, of which $2.5 million went to dividends alone.[50]

Once again Westinghouse found himself trapped on the old treadmill of constantly seeking funds to keep up with his expanding business. Loans grew steadily more expensive and harder to procure. Faced with another floating debt of more than $16 million as 1906 opened, Westinghouse got Kuhn, Loeb & Company to underwrite a $15 million issue of convertible debentures, which brought a net return of about $13.7 million. The money provided little more than a reprieve; by March 1907 the floating debt had risen past $10 million. The dilemma puzzled and frustrated Westinghouse. Was it possible for a company to suffer from too much prosperity?[51]

During the first quarter of 1907 the stock market tanked, sending railroad stocks plummeting an average of 24.05 and industrials 14.10, by far the largest declines in the history of the New York Stock Exchange. Yet business seemed sound, especially in the electrical field. "Salesmen are undoubtedly the barometers of business," observed the *Wall Street Journal*, "and if any recession had set in they would be the first to give warning." But the electrical salesmen reported excellent business and no negatives in sight. For the fiscal year ending in March 1907 Westinghouse Electric received orders totaling $38.3 million, $10 million more than the previous year. At the month's end it still had $14.6 million in unfilled orders. "This great increase in your Company's business," admitted Westinghouse, "coupled with higher costs of labor and material, has necessitated a large increase in working capital."[52]

To raise capital, Westinghouse decided on a bold move. In 1905 the stockholders had authorized an increase in capital stock from $21 million to $50 million. None of the new stock had yet been issued. Westinghouse decided to offer stockholders $5 million worth at a price of $75 for each $50-par-value share, a figure higher than the market price. Most corporations offered new issues at par if the market value was above that figure in order to induce sales and enable stockholders who could not afford new shares to sell their rights. The previous year General Electric had sold $17 million in new stock at par at a time when its market price stood at 165. By offering the stock at a premium, Westinghouse hoped to raise $7.5 million to pay off the floating debt incurred by the new building, rising production costs, and other expenses. Once again he invited the stockholders—including himself as the largest holder—to invest in the company's future. It was a gamble born of his unflagging faith that the stockholders would see that they were underwriting a sound expansion.[53]

Here was an eerie example of history repeating itself. Westinghouse had done the same thing during the financial crisis of 1890–91 except that he had offered those shares at 20 percent below par. Since then the number of small stockholders had increased, and small stockholders had never been a good market for new stock offerings. For Westinghouse this gamble marked yet another triumph of faith over experience, and it was not rewarded. To his chagrin, the stockholders took only about a third of the offering, nearly all of it going to Westinghouse himself and one or two other larger holders.[54]

Business remained good into the summer, but the capital crunch continued. "While the earnings of the Westinghouse Company have shown a large increase," noted an analyst, "they have not been sufficient to cover the costs of new construction required to keep pace with the enormous and growing demand for electrical goods. Consequently, it has been necessary to raise funds through the sale of new securities. The General Electric Co. had had the same experience." One company officer, asked why the firms did not pursue the electrification of steam railway suburban lines more vigorously, admitted that "the capacity of the electric plants is not large enough for the railroads to generally adopt electricity as a motive power." In July Westinghouse issued $6 million in three-year notes to pay off a like amount of notes maturing August 1. The new notes paid 6 percent compared to 5 percent on the older ones, which raised the interest costs $60,000 a year. Moreover, the notes were offered at 97½, which hiked the real rate to 8 percent.[55]

Feeling the financial pinch, Westinghouse startled the financial world in July by issuing a detailed annual report for the Electric and Manufacturing Company. Since 1897 he had issued only vague generalities in his reports and held no stockholders' meetings except in 1906. Always one to keep his cards close to his vest, he had avoided revealing much to prevent competitors from gleaning any information that might be of use to them. By summer, however, the need to show how well the company was doing outweighed all other considerations. In the report Westinghouse emphasized the company's growth, its conservative policy of selling for cash rather than securities, the splendid prospects of the foreign affiliates, the policy of "selling new shares at the highest price obtainable and to limit the issue of capital stock . . . to actual necessities," and the great value of the company's patents. "The striking feature of the report," wrote one analyst, "is found in the enormous growth in the business of the company," which had tripled in seven years. Although the surplus after paying dividends was small, the company appeared to be "in a stronger position physically and financially than ever before in its history."[56]

However, manufacturing profit margins remained slim, totaling only 12.6 percent of the gross income. General Electric managed a margin of only 11.6 percent that same year compared to 21.3 percent in 1903. Moreover, Westinghouse had invested $6.3 million in the securities of the Lackawanna & Wyoming Valley Rapid Transit Company and held nearly $16 million in the stocks and bonds of other companies, foreign and domestic. These securities comprised nearly 31 percent of Westinghouse's total assets and, as General Electric had learned the hard way, could rarely be converted into cash in a crisis.

Westinghouse defended the purchase of the Lackawanna securities as benefit-
ing the company; he also conceded that in a tight market the company had
paid higher interest than usual for the money it needed. Although the future
seemed rosy, the report revealed that Westinghouse had again expanded too
far and too fast if the markets and the economy took a downturn.[57]

Although the report was praised for its candor, it did not mention the complications posed by two holding companies. Interests "friendly to General
Electric and Westinghouse" had formed holding companies to provide capital
for development by holding the securities received by the manufacturing companies. The General Electric people created Electrical Securities Corporation
in 1904 and Electric Bond and Share Company in 1905; Westinghouse investors followed suit in 1906 by incorporating the Electrical Properties Company. To complicate matters, George Westinghouse had much earlier created
the Security Investment Company to help carry the financial burden. When
the original issue of $6 million in notes due August 1 was made, Security In-
vestment underwrote it by agreeing to pay thirty dollars a share for the thirty
thousand shares of Lackawanna given Westinghouse Electric as a bonus in that
purchase. To raise the money for such transactions, Security Investment borrowed from more than two hundred banks ranging from New York to Denver
to New Orleans. Its affairs posed another source of weakness to Westinghouse
if times turned tough, which they proceeded to do.[58]

Seldom had the economy been more at odds with the financial markets.
The very fact that business was robust in many sectors strained the ability of
the money market to provide credit. The problem was not merely local or national but global. For the quarter ending June 30 United States Steel had
earned a whopping $45.5 million and was spending $10 million a month for
expansion. At the same time the city of San Francisco failed to float a loan in
New York and the Bank of England had to rush $3 million in gold to Alexan-
dria to ease the crash of the Egyptian stock market. The Tokyo exchange also
crashed, sending banks all over that country into ruin. French investors sold
off their American holdings and bought gold to ship home, which tightened
money in the United States. The cities of Boston and New York failed to sell
bond issues. New York's Metropolitan Traction Company went bust along
with a major iron manufacturer. "There is no getting around the fact," declared the flamboyant capitalist John W. Gates, "that a slowing up in business
has set in . . . The truth of the matter is there is not sufficient money for legiti-
mate business purposes. Business has grown too fast in comparison with the
growth of our money supply."[59]

A suddenly communicative George Westinghouse published a letter to his
stockholders praising the efficiency of his officers, pointing with pride to the

company's expansion, and urging them to get more involved. "In its infancy your business occupied a space having less than 100 square feet of surface," he reminded them. "To-day it is carried on in the most complete workshops ever constructed, having a floor area of over seventy-five acres and employing 23,000 operatives." Earnings dipped somewhat in July but not seriously, and the price of copper was dropping. However, in August money to pay for crops began to flow out of New York to rural banks in the annual ritual that always tightened the money supply. The stock market fell sharply again, and on the twenty-eighth the *Wall Street Journal* reported that that fifty leading stocks had suffered a depreciation of more than $1.2 billion since January. During 1907 trading on the exchange declined by more than 100 million shares or 31 percent less than the previous year, and bond sales were the smallest in a decade.[60]

By October 1 the new eight-story Westinghouse factory was ready for occupancy, just in time to witness the financial markets convulse into the worst panic since the bleak days of 1893. The nation in 1907 had 6,422 national and 14,939 state banks with total assets of $8.4 billion. They lacked even a semblance of coordination, cooperation, or any common reserve pool. Most of them deposited their reserves in correspondent New York banks, which in turn lent the money to the New York Stock Exchange, businesses, and individuals. The national banks were required to keep 25 percent of their deposits in cash, which made them vulnerable to any wholesale withdrawal by outlying banks or a run by individual depositors. The banking system of the world's largest industrial nation was anything but a system. It was rather, in the words of Jean Strouse, "like an immense tangle of dry brush and timber waiting for a spark."[61]

That spark came in October when an audacious attempt by speculators F. August Heinze and Charles F. Morse to corner the stock of a copper company collapsed and brought down a mining company, two brokerage houses, and the Mercantile National Bank. Rumors that Charles T. Barney, head of the Knickerbocker Trust Company, had invested large sums in the venture prompted a run on that bank and triggered a broader panic that threatened to spread through Wall Street like an epidemic. The trust companies were the weakest link in the feeble banking chain. Chartered by the state, they operated like regular banks but had no reserves requirement and went unsupervised. As frightened depositors swarmed the financial district seeking to pull their money out of banks, J. P. Morgan, then seventy years old, summoned the city's leading bankers into conference. During the critical weeks that followed, Morgan virtually assumed the role of a central bank that the nation lacked. It was his finest performance.[62]

For Westinghouse the panic could not have come at a worse time. The floating debt had climbed above $14 million, much of it maturing soon. His stock
offering had fizzled, and the banks in Pittsburgh and elsewhere, which had advanced some $9 million, refused to extend the loans in the face of the crisis en-
gulfing Wall Street. Merchandise creditors, who had claims amounting to
about $5 million, pressed hard for their money. Somehow Westinghouse had
to raise $4 million in cash to stave off immediate demands. The New York
banks, wallowing in panic, could not help; in desperation Westinghouse hurried to Pittsburgh to make the same appeal he had made in 1890: The company
was sound and thriving but temporarily embarrassed and needed help. Before
anything could be done, however, word came on October 22 that the Knickerbocker Trust had suspended. The banks could offer no help; they were too
busy trying to save themselves from the spreading panic.[63]

At 5:30 that same afternoon Westinghouse vice president Herbert T. Herr
was sitting in the boss's office discussing a routine matter. As their conversation finished Westinghouse said almost casually, "Herr, I shall have a new job
for you to-morrow."

"What's that?" Herr replied.

"Receiver of the Electric Company."[64]

This time Westinghouse could find no way out. The panic had undermined
any chance of getting through the immediate cash crunch, but Westinghouse
had put himself in position to fail by not heeding the lessons of his first crisis.
As always the chief put on a good face. As the news swirled through the plants
and the city he insisted that "the Company is not insolvent—only hampered
for the moment. It is doing more business than ever before. It will come out all
right." While preparing the court papers next morning, a subordinate asked
him about some current piece of business. After disposing of it, Westinghouse
said cheerfully, "By the way, Macfarland, I've got an idea now for our turbine
that will make a sensation when we bring it out!"[65]

Was it a stiff upper lip or simply denial or that unflappable Westinghouse
optimism? He had been in trouble before and come out all right, but this time
was different. In 1891 the bankers had been able to sell company stock; in 1907,
however, no market existed for junior securities. Stock prices continued to
shrivel; Westinghouse preferred fell from a high of 80 in 1907 to a low of 30,
common from 77 to 16. The fall prompted the Pittsburgh Stock Exchange to
close. Somehow the financial basis of the company had to be adjusted to reflect the reality that it would always need liquid assets in reserve to provide for
capital needs. Any reorganization plan had to square the claims of four
groups: the merchandise creditors, the banks holding short-term notes, hold-

ers of the funded debt, and the stockholders. Each group formed a committee to represent it in the negotiations that followed.[66]

On November 13, barely three weeks after the receivership, the first plan of readjustment was made public. It asked holders of the floating debt to accept a mix of stock, bonds, and notes for their claims, in effect making them partners in the company. The merchandise creditors and Pittsburgh banks quickly approved it, but other banks rejected it outright. Westinghouse hoped for a quick fix to the cash flow problem and a return to things as they were, but it was not to be. Early in December the creditor groups created a Readjustment Committee with banker James N. Jarvie at its head. On January 20, 1908, it put forward a plan of its own featuring the creation of a $35 million first mortgage on the company along with issuing $7 million in new assenting stock. The bankers loved the plan and urged creditors to support it. However, the continued fall of Westinghouse securities on the market made it painfully clear that the new stock could not be sold. By mid-February the plan was all but dead.[67]

A stalemate had arisen. The stockholders seemed the best source for raising new capital, but they would not step forward if the bankers saddled the company with a heavy mortgage. The bankers and other creditors had priority claims on the property but did not want to put more money into the company. Meanwhile, the value of all their holdings continued to decline. Behind this conflict lay another more fundamental clash of interests. The collapse had not tempered Westinghouse's belief in the future of the company and the industry. He worked tirelessly for a settlement that would restore the *status quo ante bellum* and let him get back to work building the business. The creditors, even those sympathetic to Westinghouse, clung to the more immediate goal of getting paid. Beyond that, they had grown skittish about Westinghouse's lack of conservative instincts and wanted the company to pursue a more cautious course.[68]

The foreign subsidiaries became one flash point in this conflict. Westinghouse insisted that they be sustained at all costs. The Jarvie Committee was adamant in not wanting a cent diverted to them. Even more, it demanded a management that promised a more conservative financial policy. Westinghouse saw no need for a change in direction. All the company needed, he argued, was to extricate itself from the current difficulties and let the growth of business bring it relief. Through the dreary winter the committees wrangled, first negotiating and then discarding one plan after another without getting at the heart of their basic differences. The one point on which all sides could agree was that the company urgently needed more working capital, but how to get it? Gradually it dawned on the bankers that, like it or not, the only solution lay in some modified version of the plan originally put forward by the merchandise creditors and Westinghouse.[69]

Near the end of March all sides came grudgingly to agreement on what became known as the Merchandise Creditors' Modified Plan, largely because that group agreed to make the largest sacrifice. Given the ordeal that produced it, the plan was remarkably simple; it also bore a striking resemblance to the original plan of November 1907. The merchandise creditors agreed to take $4 million in new assenting stock for their claims; the bankers would take a mix of new stock, convertible long-term bonds, and notes averaging five years in length for their debt. To make this happen, the existing stockholders had to buy $6 million in new stock at par. Existing bonds and short-term notes were left undisturbed. The beauty of the plan was that it transformed debt into stock, thereby reducing fixed charges. It also created no new mortgage and left untouched the security behind older obligations.[70]

The weakness of the plan lay in getting the 3,800 stockholders to buy at par ($50) shares that in May were selling around 27 in the market. Small holders were notoriously lax in subscribing to any new issue, let alone one that involved a sacrifice on their part. The average holder owned eighty-three shares, or about $4,150 worth at par value. Westinghouse was eager for the plan to work but had a problem of his own. He owned 14 percent of the company's stock outright and held another 42 percent of it in the Security Investment Company. However, much of the latter holding had been pledged as collateral for loans, which meant he had either to redeem the notes or get the lenders to cooperate in the subscription. Nor were the banks thrilled with the prospect of accepting stock. The large New York banks remained aloof from the plan through the summer, as did many of the smaller banks scattered across the country.[71]

Stockholder resistance proved even more fierce. Some who had bought Westinghouse stock at high figures declared themselves "angry and disgusted with the past management, [and] professed to have no confidence that the management would be different in the future." The Stockholders' Committee appealed to them in a series of letters, reminding them of their duty to support the company and warning of the consequences if they did not. Westinghouse himself sent out a circular that said bluntly, "If you do not subscribe you in effect vote for a sale of the property by the creditors and thus the elimination of all stock interests." The threat of a forced foreclosure sale was a potent one, but still the subscriptions dribbled in slowly; by late May only 1,200 of the 3,800 stockholders had responded. The June 1 deadline was extended to June 23 amid another flurry of appeals, warnings, and upbeat news about the company's prospects once the reorganization was finished. The foreign subsidiaries reported much improved performance and less need for cash from the parent company.[72]

Still the subscriptions lagged. On June 18 the plan was reported to be hanging in the balance. The deadline was extended to September 1 and again to October 1. One group of stockholders demonstrated their loyalty to the company: Some five thousand Westinghouse employees stepped forward with subscriptions totaling $611,250. Improved business in July nudged the stock price upward, a trend that continued into September. As the stock climbed, the bankers softened their opposition to the plan; on September 25 the Readjustment Committee finally agreed formally to accept the plan in lieu of its earlier one. The deadline was extended past October 1 as Westinghouse and the plan's supporters mounted a furious campaign to induce the 1,800 holdouts to subscribe. The large New York banks and New York Life Insurance Company finally signed on as business continued to improve and the stock continued its rise in price. Jacob Schiff of Kuhn, Loeb, who had earlier dismissed the plan as too chimerical, changed his views and worked on its behalf.[73]

Not until November 20, 1908, nearly thirteen months after the receivers were appointed, was the Merchandise Creditors' Modified Plan declared operable. Praise went to the stockholders, the merchants, the bankers, and to Westinghouse himself for being "indefatigable in his efforts to get the company back on its feet." After tending to some formalities Westinghouse Electric and Manufacturing finally emerged from receivership on December 5. Financially it appeared to be in excellent shape. Some $12 million in current liabilities had been transformed into stock and long-term obligations, and $6 million in cash had been raised for working capital. The net debt of the company was reduced from $44.5 million to $30.9 million, and interest charges from $2.7 million to $1.6 million. The company had been reborn, but under changed management.[74]

In broad terms the reorganization followed the scenario of 1890 except for its denouement. George Westinghouse remained president but under a new structure of management. As early as June 1908 it was reported that he had a plan for relieving himself "to a large extent from many duties connected with the financial end of the corporation and that this work will be under the supervision of a committee made up of representative banking interests. Mr. Westinghouse will devote most of his time to the operating and productive branch of the business." This was the price the bankers extracted for their help, and at first Westinghouse seemed unaware of how steep it would be. He remained as president but agreed to place his 56 percent of the stock in the hands of five trustees who had absolute authority to choose the new board of directors. The sixteen directors chosen represented a blend of members from the reorganization committees. They in turn handed financial policy over to a committee of six headed by Edward F. Atkins of Boston. The other members included three bankers and two of the merchandise creditors.[75]

The board of directors then chose Robert Mather as their chairman. A conservative lawyer with no experience in the electrical industry, Mather had gained his reputation among the bankers as counsel for the Rock Island Railway. He was a cold fish with an unblinking eye on finance, and he seemed immune to the legendary Westinghouse charm. Late one Friday afternoon, at the end of a meeting in Pittsburgh, Westinghouse told Mather pleasantly, "If you are going to New York this evening, I shall be glad to take you in my car. I am going East myself." There was an awkward pause as Mather shuffled papers into his briefcase. "I prefer to go to New York by myself," he said finally, "and pay my own fare."[76]

That moment proved a harbinger of things to come. Westinghouse still ran the company, but he was no longer the boss. He chafed under a financial leash that restricted his efforts to expand and develop the business. He knew this was coming; it was, after all, part of the deal that brought the bankers on board. What he could not have anticipated was how poorly he would get along with Mather, a man who could not have been more different from him. Westinghouse believed in progress as well as profits and took enormous pride in the welfare programs he had created for the benefit of his employees. When the new plant at East Pittsburgh was built, Westinghouse noticed that it lacked overhead protection and sidewalks. Although cost estimates had already been exceeded, he insisted on adding an enclosed walkway to shield the women workers from rain. "We employ a great many women," he explained, "and when it storms they will be exposed to the rain in their thin dresses, or walk in unprotected shoes . . . They will catch cold, and if any harm comes to them it will be our fault."[77]

Mather and the bankers never let their eye stray from the bottom line. They regarded Westinghouse as a profligate spendthrift and took a dim view of the money lavished on the employees. During the campaign to get the smaller stockholders to subscribe, Westinghouse had assured them that the dividends would be paid on the new assenting stock. When business turned sour during the first year after the reorganization, Mather and the board did not hesitate to renege on that pledge in the name of fiscal prudence. To Westinghouse it was a question of honor; to the board it was merely sound financial policy. Nor did Mather hesitate to charge off nearly $2 million from the patent accounts, foreign investments, and other accounts. He had reason to be conservative; the year ending March 31, 1909 proved to be the worst in the company's history, resulting in a deficit of $3 million.[78]

Westinghouse found this approach intolerable, but he could do nothing to oppose it. All his life he had insisted on doing things his own way, and for the first time he could not dictate what went on in his own company. "They say I'm no financier," he growled on one occasion. "So I suppose all those great works built themselves!" Yet he had always counted "prospective earnings" as a part of capitalization, and it was true that "no manufacturing plant of his was ever built big enough to suit him; he never inspected an installation in one of his shops without beginning to calculate how soon it would be outgrown." And he believed it his duty to maintain dividends even in hard times. His financial balance sheet was rooted in faith, that of the bankers in prudence and caution.[79]

It did not take Westinghouse long to realize that he could not work under these restraints. In January 1910 the board passed a resolution overflowing with compliments and granting Westinghouse six months' leave of absence. He did not appear at the annual meeting in July, and the directors chose Atkins to be the new president. The following summer Westinghouse gathered proxies in an effort to unseat the bankers but fell far short. Although he still had his other companies, he had lost the electric firm, and with it the love of his business life. "The loss of the Electric Company was to Mr. Westinghouse a disappointment from which he never recovered," recalled his public relations man, Ernest Heinrichs. "There is no doubt that it broke his spirit." Shortly afterward his health began to falter; he died in 1914. Two years earlier he too had been awarded the Edison Medal for "meritorious achievements in the development of the alternating current system." Unlike Tesla, he savored the delicious irony of it all.[80]

In November 1909, just before Westinghouse became president of the American Society of Mechanical Engineers, a writer praised him as "one of the greatest engineers, inventors, manufacturers, and financiers of this mechanical and electrical age. With him to live is to be young . . . This enormous business is sensible of the personal touch; it responds to the influence of the man who insists that quality in men and output stands first in the requirements of business, that faith be kept, that the spoken word is as firm a pledge as a signed bond."[81]

Like Edison and Thomson before him, Westinghouse built an enterprise to the point where the bankers and businessmen nudged him out, largely because times had changed and the company needed direction of a kind he could not provide. He too was a victim of his own success in realizing his vision so well that the enterprise simply outgrew him. Back in the early days he took pride in handing out turkeys to each worker on Thanksgiving; before long that proved impossible, and in time the turkeys gave way to elaborate benefit programs.

The instinct was still there, but the personal touch had been lost. So too with the business itself. Once Westinghouse left, the boss was no longer the "old man" but a faceless corporation stripped of sentiment.

Nor was this pattern confined to the electrical industry. Everywhere it seemed that the bankers were coming in and taking over enterprises once ruled by their creators. The very success of the founders had turned their firms into something much larger than themselves, something that required a different mind-set and approach than most entrepreneurs possessed.

In the electrical industry the pioneers were gone, or so it seemed. In fact one more stood waiting in the wings, his presence so inconspicuous at first that few if any observers grasped the significance of what he was doing.

CHAPTER 19

MASTERING THE MYSTERIES OF DISTRIBUTION

It is of prime importance to central-station managers that they should sell their product, electricity, to the greatest number of consumers at the lowest possible price, and yet obtain a reasonable profit.

—SAMUEL INSULL (1898)[1]

DURING THE 1890s, A DECADE scarred by a long and deep depression, the use of electric power increased thirtyfold. Yet it had barely scratched the surface of industrial and domestic use. The traction companies readily embraced electricity when they could afford it, but most manufacturers still clung to their old ways. In 1900 only 5 percent of industrial power came from electricity, two thirds of it from isolated power plants rather than central stations. Despite its rapid growth, only 8 percent of American homes had electricity in 1907. Although still the wondrous technology of the age, it remained a spectacle to most people, a luxury experienced elsewhere rather than a convenience enjoyed at home.[2]

By the 1920s these conditions had been reversed in striking fashion. Electricity had become a staple necessity of everyday life for vast numbers of Americans, especially in the cities and fast-growing suburbs. More and more manufacturers relied on it to power their production lines, which turned out an astonishing array of products for the emerging consumer economy. Isolated power plants gave way to ever larger central power stations that funneled electricity in huge quantities across entire regions. Electric utilities became one of the largest industries in the nation while scrambling to keep up with the demand for more power.

In the pioneering days the manufacturers, most notably General Electric and Westinghouse, dominated the world of electricity. During the 1920s the utility companies moved to the forefront. Central station companies were relatively small and diffuse, varying widely in efficiency and profitability. By World War I they had changed the relationship between them and manufacturers. One man above all others was responsible for this remarkable transformation. He was not one of the original pioneers, but he had been present at the creation, and all of them knew him well.

SAMMY INSULL, AS Edison always called him, had come a long way from the scrawny boy who arrived in 1881 to become the great man's secretary. Unlike his mentor, he invented nothing; his great talent lay in figuring out how best to use what others created. In true Horatio Alger fashion he sprang from humble origins and worked his way to the top through hard work, intelligence, and a generous helping of good luck. He had come to America to seek his destiny, never dreaming how far it would carry him. His parents were solid lower-middle-class people who did not want him to give up a perfectly good job in London for an unknown position in a strange land, but they could not dissuade him. Insull read everything he could find on Edison, who had become his hero. Although his position as private secretary to an American banker in London earned him about four hundred pounds a year, a handsome income for one so young, he did not hesitate to dump it for the chance to work with Edison.[3]

At twenty-one Sam Insull embodied the best traits of his parents. His father was a bright, idealistic enthusiast who embraced good causes but utterly lacked the practicality and energy to get things done. All his life he remained the well-meaning visionary dedicated to promoting religion and crusading against the demon rum that had destroyed his great-grandfather. His wife, Sam's mother, could not have been a better match for such a husband. Strong, realistic, tough-minded, she kept the family together while Samuel senior tilted against his favorite windmills. She had the backbone and the shrewdness he lacked, and passed them along to her son. Through a stroke of good luck Samuel senior landed a job as secretary in the Oxfordshire district that enabled him to give young Sam six years in a good private school at Oxford. There he revealed a mind strongly attuned to the practical and indifferent to abstractions whether of religion, philosophy, or mathematics.[4]

The one trait Insull inherited from both his parents was a belief in temperance. When his mother, who was the family boss, finally agreed to let him go to America, she made him promise that he would never touch liquor. He kept the pledge all his life. Upon his arrival in New York on February 28, 1881, he was met by the only person he knew in America, Edward Johnson, who stunned him by announcing that he was sailing for England in twelve hours. They went at once to Edison's headquarters at 65 Fifth Avenue, where Sam met his hero for the first time. "With my strict English idea as to the class of clothes to be worn by a prominent man," he recalled, "there was nothing in Mr. Edison's dress to impress me." What did strike him at once was the brightness of Edison's eyes and an expression that seemed to exude intelligence. There was also a magnetism about his presence that Insull found irresistible.[5]

Johnson took Insull off to meet his family and have a quick supper before returning to the office, where Edison got down to business right away. He showed Insull his checkbook and said he needed to raise $150,000 in Europe. Insull went through Edison's books and, drawing on his experience at the bank of his former employer in London, suggested how money could be raised by modifying some of the contracts for patent rights. They worked straight through the night until Johnson had to leave for his ship. Johnson took with him the program worked out by Insull and managed to get all the funds he needed while abroad. This first experience convinced Edison that he had found the right man. Much later, when Edison was near death, Insull asked him about his impression at their first meeting. "I thought I had made a hell of a mistake," Edison admitted, "but the next day I knew that there was no mistake about it."[6]

Insull was given a bed on the third floor of the building and a desk in Edison's office. Within a short time he came to appreciate Edison's "wonderful resourcefulness and grasp, and his immediate appreciation of any suggestion of consequence bearing on the subject under discussion." For his part Edison quickly saw the amazing talents and dedication of his new secretary, who handled the inventor's erratic hours and schedule with ease. "Edison's whole method of work," Insull said later, "would upset the system of any business office. He cared not for the hours of the day or the days of the week. I would get at him whenever it suited his convenience . . . I used more often to get at him at night than in the day time, as it left my days free to transact his affairs." It helped that during the long hours they spent together both men liked to smoke cigars.[7]

During these years, as Edison piled more responsibilities on his secretary, Insull not only grew and flourished but learned a wide variety of new skills as well. "Whatever technical knowledge I have of the electric light and power business," he declared, "I owe primarily to Mr. Edison and his painstaking effort to make me familiar with its fundamental, technical, and economical principles." Insull absorbed everything around him. He had always been a voracious reader, especially of history and biography, and in business he proved a shrewd observer as well. Although the other Edison men respected his ability, they did not always like him personally. Insull retained his English sense of propriety in dress and manners. He was invariably polite and proper but possessed one unfortunate habit: When surprised or puzzled, he did not raise an eyebrow but instead curled his lip, which resembled a sneer. Coupled with his rather bulging eyes and aloof, often brusque manner, it gave many people an entirely wrong impression. Told of the habit, he addressed it by shaving off his muttonchops and growing a thick mustache.[8]

The education of Sam Insull escalated sharply when Edison ordered him to take charge of moving the Machine Works to Schenectady. He had located and purchased the property; when Edison told him to oversee the move as well, Insull pushed the work hard even though it left bad feelings among the men on whose toes he stepped. "It was my habit," he said, to "ride roughshod over opposition to accomplish what my chief desired should be done." Within a month he reported that all the machinery was either installed or on its way to Schenectady. Edison then told him to move to Schenectady and manage the plant there, exhorting him to make it a big success or a big failure. Thus, in December 1886, having just turned twenty-seven, Insull took up his first independent command. Alfred O. Tate replaced him as Edison's secretary, although Insull remained in charge of the inventor's financial affairs.[9]

The new post entailed huge responsibilities. That summer the Electric Tube Company had moved its operation to the new ten-acre site. Edison reassigned his trusted lieutenant Charles Batchelor, who had managed the Machine Works, to another position and sent the ever loyal John Kruesi, who oversaw the Electric Tube Company, to be the assistant general manager to Insull. Kruesi's help was sorely needed, for Insull knew hardly anything about manufacturing. His strength was finance, and the Machine Works needed a constant flow of money. Insull turned into a perpetual motion machine, running back and forth between New York and Schenectady to consult with Edison or visit the banks in search of funds. He often spent six nights a week on the train between the two cities. On a typical day he arrived at the Machine Works at seven in the morning, worked all day, handled his correspondence in the early evening, and then caught the night train for New York.[10]

The money chase became nerve-racking for Insull, who had to raise funds not only for the Machine Works but also for other Edison projects. Insull liked to do things in a proper manner; Edison was careless about his finances and piled on bills for the new West Orange laboratory. "He of course wants a great deal more money than he at first anticipated," Insull wrote after a visit with his mentor, "but this is simply a repetition of what has occurred so frequently before . . . Exactly how I am going to carry out his wishes and give him what he requires, I don't know." Edison also had a habit of informing Insull about a note only a day or so before it was due even though he had promised always to give at least a week's notice. "I am, as you well know," Insull complained, "perfectly willing to run all round the country to hunt up money for you, and I can by hook or by crook, supply your wants, but I cannot do it unless you advise me ahead of your requirements." [11]

Exasperating as the money chase was, it gave Insull a thorough grounding in practical finance along with a deep-seated dislike of banks and bankers. Apart from Edison's insatiable demands for money, the very success of the Machine Works demanded more capital. "We moved here into two buildings which were practically only shells," he reminded Edison when the bills piled up. "Our business suddenly jumped up to about three times its former dimensions; our plant had to be increased; more buildings had to be built; more material kept on stock; and a larger amount carried in Accounts receivable . . . I do not like liabilities any more than you do, but the owners of a concern must invest more money if they want to accept the business, or else they must borrow money on which to conduct it." For Insull the borrowing was a frantic and never-ending experience. [12]

Edison understood the need for more capital and hoped that Villard's plan to merge his firms would solve it. After the creation of Edison General Electric in 1889, Insull became second vice president in charge of the manufacturing and selling departments. In that position he reorganized and integrated the scattered manufacturing operations to make them more efficient. From this work flowed the basic principles that governed his later career. He expanded operations whenever possible by plowing back profits and borrowing freely. He avoided labor strife by treating his workers well and tried to spread his fixed costs by diversifying operations to keep the plants working at all times. He also believed in cutting prices in order to increase sales volume and thereby lower unit costs. Lamps that had sold for a dollar in 1886 went for only forty-four cents in 1891. Insull also revamped the sales organization, or lack thereof, into a system of seven regional divisions, each headed by a district manager. [13]

So well did Insull do his job that when the merger with Thomson-Houston occurred, he was the only Edison man offered a high executive position: second vice president at $36,000 a year. However, the merger had poisoned his relationship with Edison, if only for a time, and even though he was only thirty-two years old Insull's considerable ego chafed at not being made president instead of Coffin. Convinced that the Thomson-Houston men did not really understand the business, he told Coffin he would take the vice presidency but only temporarily. Henry Villard offered him the vice presidency of the North American Company, but that did not suit Insull either. He wanted his own command even though it meant separation from Edison. "I felt a devotion to his interests . . . ," he explained later. "It was the parting from the creator of that business—from the man himself—that stirred my innermost feelings."[14]

Insull's devotion to Edison never wavered despite their differences at the time of the merger. "He was my hero when I was a boy," he recalled late in life, "and after fifty-five years engaged in various businesses of which his inventions were the fundamental basis, my admiration of him and his accomplishments is greater than ever." He remained for Insull "the greatest inventor, to my mind, that America has produced, and one of the great world figures of his time."[15]

A career crisis loomed for Insull. Although manufacturing still dominated the electrical industry, Insull decided to leave the field. Instead he offered himself as a candidate for president of the Chicago Edison Company, a central station that had asked him to recommend a suitable person. They were quick to offer him the job but could not come close to his current salary. On July 1, 1892, Insull left a job with a $50 million company that paid him $36,000 a year for one with a $900,000 company and a salary of $12,000 a year. But for the first time he was in charge and had the opportunity to learn the central station business and develop it as far as he could. Later Edison reminded him wryly that "Central Station was coined I think either by myself or some of the boys at . . . Menlo Park in the days when guessing was a substitute for mathematics."[16]

The General Electric gang bade him farewell with a lovely dinner at Delmonico's attended by Edison, Coffin, Villard, and forty-six other colleagues. After the speeches they presented Insull a beautiful silver punch bowl with all their names inscribed in the bottom. Coffin and Frederick Fish also sent him warm notes wishing him well. On that wave of good feelings Insull arrived in Chicago, a stranger in a strange land. Previous visits to the city had mainly impressed him with its filth. He had occasionally idled away an evening on the porch of the Sherman House wagering other guests on how many rats would scurry out of the street drains. "Since my arrival in America I had been used to living in New York . . . ," he admitted. "I had a New Yorker's point of view that anything outside of New York was not much worth while. I felt that I was coming to a frontier town." As insurance against his early impressions, Insull had insisted on a three-year contract to keep him from bolting at the first opportunity.[17]

He need not have worried. Chicago proved to be far more congenial to him than even his outsized ambitions could have imagined. At the farewell dinner Insull had predicted grandly that Chicago Edison, modest as it was, would someday equal or exceed the size of General Electric. Some guests found this amusing, believing naturally that Insull was going to a dead-end job. In taking the position Insull had made two demands that were accepted: He wanted nothing to do with raising capital for the company; the harrowing experience with Edison had cured him of that. He also wanted the company to begin construction at once on a new plant to increase its generating capacity and to pay for it by issuing $250,000 in new stock, which Insull would take himself. To purchase it he managed to borrow the entire amount from Marshall Field, the king of Chicago's merchandisers. It was a bold bet on his future and that of the company.[18]

PRIOR TO INSULL'S arrival the utilities business in Chicago had been a constant struggle between the gas and electric interests and the politicos who ran the city. In 1887 a group of Philadelphia speculators managed to organize a monopoly of the gas supply and also consolidated several arc-lighting companies into one firm, the Chicago Arc Light and Power Company. The Gas Trust soon emerged as an arrogant symbol of corporate power, but new small electrical companies continued to defy efforts at achieving a comparable monopoly in arc lighting. The gas interests, brought together by Charles T. Yerkes, aroused widespread public opposition that overshadowed the more modest attempts at consolidation in electric lighting. Chicago Edison remained the major central station company, but no fewer than twenty-four new central stations appeared between 1887 and 1893. As DC motors came into more widespread use, Chicago Edison and Chicago Arc Light found themselves competing with each other for customers as well.[19]

All the Chicago central stations used DC power, which limited their ability to extend service into new areas without building new stations. They also had to contend with the large number of isolated systems in service, especially within the Loop, as the central business district was called. Chicago Edison opened the city's first central station on Adams Street in 1888. By 1892 the city contained eighteen central stations and 498 isolated systems, many of the latter sold by Chicago Edison and its competitors. Demand for electricity continued to grow even though it remained too costly for the vast majority of people. Affluent neighborhoods and upscale new suburbs formed small companies to provide power for their homes. The result was a crazy quilt of distribution grids created by small central stations in a chaotic pattern of growth. It seemed unlikely that any one utility company could impose order on this mishmash of lines, let alone produce power at a cost low enough to challenge the isolated units.[20]

Yet Insull set out to do exactly that. He wanted to consolidate the central stations under his command. At his very first meeting in July 1892 Insull persuaded his executive committee to buy their second largest competitor, the

Fort Wayne Electric Company. Then he went after bigger game. Three months later the head of Chicago Arc Light invited Insull to lunch and asked him to join a syndicate for taking some of the company's securities. Insull declined, saying it was improper because they were competitors and, more important, he intended to acquire Chicago Arc Light even though it was much larger than Chicago Edison. His bold statement drew an amused laugh. "Relations at the end of the lunch were not quite as cordial as they were at the start," noted Insull wryly, but he proceeded to carry out his prediction. In February 1893 Chicago Edison bought control of the rival company for $2.195 million, paying $120 a share for stock then selling at around $105 and funding it with 6 percent debentures that did not require using cash or stock. During the next five years Insull managed to acquire all fourteen remaining central stations in the Loop, giving him a monopoly in the central business district.[21]

However, control of the central station business meant little unless he could deliver power cheaply and efficiently. The existing plants, scattered about the business district and adjoining neighborhoods, were small, noisy, and inefficient. His own company's station on Adams Street groaned beneath the strains placed on it and could not begin to keep up with growing demand. "One of the first things that impressed me," he recalled, "was that, if we ever expected to offer energy at low prices to our customers, we must produce that energy at the lowest possible price." That was the point to the new station Insull insisted on constructing. Located on the Chicago River, the Harrison Street station could use condensing marine generators, which required large supplies of water. Insull relied on two bright engineers, Louis A. Ferguson and Frederick Sargent, to design a new type of plant that was not only huge but included a capacity for growth as well. They came through in splendid fashion.[22]

When completed in August 1894 the Harrison Street station had a total capacity of 16,400 kilowatts, four times that of the Adams Street station. So efficient were the high-pressure condenser engines that they reduced the amount of coal needed to generate one kilowatt hour from between ten and fourteen pounds to between three and five. As part of its equipment Insull pulled off a coup by purchasing the 1,200-horsepower engine and two 800-kilowatt generators General Electric had used at the World's Fair. He got them for $50,000, a fraction of their cost, and for an interest-free note for one year with a replacement guarantee included. The machines helped make Harrison Street the largest electric power station in the world. When the station opened in August 1894, Adams Street was shut down and a new era of power generation opened. Harrison Street began life as a giant among central stations; it became in a surprisingly short time a dwarf.[23]

Insull also moved to dominate the supply of electrical equipment in

Chicago. The purchase of Chicago Arc Light and Power included that company's exclusive right to sell Thomson-Houston equipment. With Coffin's help, Insull gained the rights to sell every kind of American electrical equipment in Chicago except that manufactured by Westinghouse. The latter company tried to capitalize on its World's Fair popularity by starting a lighting company in Evanston, giving it an exclusive license to sell Westinghouse equipment in all of Cook County. Later Insull acquired this company as well, giving him a complete monopoly. He even bought a company that held the Illinois rights to sell equipment manufactured by the leading German firm, Siemens & Halske.[24]

Unlike the usual practice, however, Insull created a monopoly not to raise prices but to lower them by providing more efficient service. He knew little about central station management, confessing freely that "my knowledge of the manufacture of electric energy and the conditions governing its distribution were, at this time, largely theoretical." But Insull had above all else an impressive learning curve if not yet a clear sense of what he wanted to do. The central station business was still so new that no one had a handle on its economics, which seemed to defy logic. Edison had discovered his error in using the gas industry as a model. Gas could be stored, which made it possible to produce on an orderly, rational basis like other manufactured products. It could maintain reserves to meet peak requirements and level out demand over a twenty-four-hour period. Not so electricity. It had to be produced, sold, delivered, and used all at once, which meant that the plant supplying it needed the capacity to deliver the total maximum load demanded by customers at any given moment.[25]

The critical factor in electricity usage, Insull soon realized, was load. If a plant had to stand ready to produce its maximum load at any time, then the obvious policy was to obtain as much load as possible and spread it across the entire twenty-four-hour cycle. To complicate matters, the load factor was tied to another seemingly unrelated mystery: The expenses of a central station remained relatively constant regardless of how much or how little electricity it sold. The reason, Insull thought, was that plants capable of meeting maximum demand were expensive to build, and the cost of laying the cables and wires for transmission was four or five times that of the powerhouse equipment. Fixed costs such as interest, taxes, insurance, maintenance, and depreciation, which did not vary with the amount of electricity sold, exceeded total operating costs, which did depend on volume of sales.[26]

Put another way, the large initial investment in building a central station saddled the balance sheet with large interest obligations on the money borrowed for construction. That monthly cost had to be paid from the proceeds of electricity sales. Selling more electricity produced more income; spreading the load across the hours of every day kept the generators busier and created more income. The variable costs of generating electricity could be reduced by improving the operation of the plant in every way possible. When this occurred, however, it created an unexpected financial effect. The more efficient a plant's operation grew, the more imbalanced became the ratio between fixed and operating costs.

The lesson this impressed on most men in the field was that adding new customers only increased investment and lowered profits; the proper policy, therefore, was to maintain a small pool of customers and induce them to pay more or increase their usage. On this basis most managers priced their product high in the belief that electricity was doomed to remain a luxury product. Chicago in 1892 had a million people, of whom fewer than five thousand used electricity. Optimists thought that figure might someday reach as high as twenty-five thousand. Moreover, the introduction of the Welsbach gas mantle during the early 1890s rendered gas lighting much more efficient, and gas manufacturers pushed hard to conquer an ever larger share of the cooking and heating markets as well. In all these areas they offered a product that was cheaper and would remain so as long as the price of electricity stayed high. This competition did much to discourage expansion of central station power.[27]

During the 1890s Insull rethought this equation and came to radically different conclusions. Gaining control of all the central stations enabled him to produce electricity more efficiently, but he still could not sell cheaply enough to compete with either gas lighting or the isolated power stations favored by most customers in the central business district. The gas model for production was flawed, but manufacturing offered him a crucial insight into the concept of producing in quantity to reduce unit costs. Lower rates would bring more customers, which would reduce unit costs, thereby permitting rates to go still lower and perhaps launch a self-perpetuating cycle. The key lay not just in selling at cheaper rates but also in finding new customers that would spread the load as well. Thus was born the approach that, once refined, became known as the "gospel of consumption."

Years later Insull explained some of its key elements. "By far the most serious problem of central-station management," he argued, "and by far the greatest item of cost of the product, is interest on the investment." As the amount of business increased, the interest cost per unit fell because it was spread across a growing number of units. The key lay in the load factor, which Insull called "the most important one in central-station economy." Load factor measured the percentage of time the plant (and therefore its investment) was in actual use. Plants were built to meet maximum demand, which happened only on rare occasion. Most of the time the generating capacity was barely used. An office building, for example, had a load factor of only 3.7, meaning that it used only 3.7 percent of the maximum demand at any one time during an entire year. In effect it used the equivalent of only seventy-five to one hundred hours of electricity a year. A large dry-goods store might have a load factor of 25 percent, which meant it paid seven times the return on investment of the office building.[28]

The ideal load factor was one with a low maximum use but a long average use. "If your maximum is very high," Insull said, "and your average consumption very low, heavy interest charges will necessarily follow. The nearer you can bring your average to your maximum load the closer you approximate to the most economical conditions of production, and the lower you can afford to sell your current." Insull did not arrive at these insights quickly or all at once, but his experience kept reinforcing one central theme: The way to reduce costs was by increasing output, which meant going after more customers and finding a mix of usages that spread the load across the clock. And the way to get more customers was to offer them lower rates.[29]

Other operators tried to entice customers by offering them better rates if they used more electricity. Insull tried this approach at first but soon discarded it in favor of one that told salesmen to get new business at whatever rate they could. An early test came from Ferguson, who found an opportunity to wire a prestigious new hotel and furnish its power at a price other operators considered ridiculously low. Insull agreed to do it providing the hotel took a five-year contract; he brushed aside the criticisms of men lacking "the courage to cut so deeply and try so hard to get business. They were unwilling to take risks in trying to develop a real knowledge of the economic conditions governing the business." Thereafter he relied on the combination of low rates and long-term contracts to secure large lighting customers.[30]

This results of this approach stunned Insull's fellow central station operators. Despite the ongoing depression, Chicago Edison quadrupled its connected load and watched its sales soar from 2.8 million to 13.7 million kilowatt-hours by 1895, when its operating income exceeded its 1892 capitalization. Yet Insull and his lieutenants still lacked a firm grasp of the economics behind this surge in business, especially the cost of producing light. Two other pivotal experiences added to his education and opened his eyes to fresh possibilities. The Columbian Exposition showed him how far the uses of electricity had come and might go, and how the rotary converter did away with any war between the currents by enabling AC and DC to work together. In August 1897 Insull became one of the first to install a rotary converter. Even more, the fair demonstrated the feasibility of a universal system of distribution.[31]

The second experience came in 1894 when Insull returned to England for a brief vacation in December. To escape London's foggy weather he ventured down to the seaside resort of Brighton, where he was surprised to find nearly every small shop ablaze with electric lights in the evening. His curiosity led him to the local municipal plant and an unassuming young man named Arthur Wright, who attributed the widespread use of electricity in town to a device he had invented called a "demand meter." Unlike other meters, Wright's version recorded not only how much electricity a customer used but also when he used it and the maximum level of his use. As Wright explained how his meter worked, Insull began to see a basis for the policies he had followed on little more than instinct. From this chance encounter flowed one of the major epiphanies in Insull's career. After returning home, he dispatched Ferguson to make a careful study of the meter's use in Brighton and how it might be adapted to Chicago.[32]

The concept was simple yet brilliant. As Wright explained it, producing electricity involved two categories of cost, fixed and operating, both of which varied from one customer to another. A man with a vacation cottage might have only a few lamps that he burned a few weekends each year, yet the fixed costs of providing him power were as high as if he kept his lamps going every day and night. It made no sense to charge him the same price as another customer who had the same number of lamps but used them several hours every night. The station got more income from the latter at a lower price per kilowatt-hour than from the former at four times that price. The trick, therefore, was to measure every customer's usage in two parts: one to see how much equipment he required and the other to see how much he actually used the equipment. The rate charged then consisted of two tiers. The first was a flat charge for use of the equipment and a minimum level of use; the second provided a sliding rate scale for all usage above the minimum. The more he used, the lower it fell.[33]

This revelation did more than excite and energize Insull. It also explained and justified the policies he had already adopted. It confirmed his views that one should enlist as many customers as possible, that isolated plants had no sound economic basis, and that competition drove rates up rather than down as the conventional wisdom asserted. The key element, which had escaped Wright but not Insull, was load factor. All customers used their equipment only a fraction of the time, but not all of them at exactly the same time. Given that fact, a single investment in plant could supply many users, any two of whom could always be supplied more cheaply than one alone. It also followed that two competing stations could not serve these same customers as cheaply as one station because of the higher investment cost. Competition would create higher costs in areas of heavy usage and most likely no service in other areas that lacked enough business to pay. Put another way, monopoly was the surest road to driving down costs and rates.[34]

The beauty of this approach was that it served light and heavy users at once. Residential customers would pay a smaller portion of the company's financial cost because of their lighter peak demand. The company expected to lose money on light users, but that was fine with Insull. "The avowed policy of the company," he said later, "is to do its small business at a loss and make its profits out of its big customers." However, larger users benefited even more because the sliding price scale reduced their unit cost of electricity as they used more of it. Armed with these insights, Insull reorganized his sales organization and set them to getting new customers on the basis of being able to calculate accurately both the cost of electricity and what a given customer would have to pay for it. Ferguson returned from Brighton with a favorable report and was put to work adapting the Wright meter to Chicago. Later Insull, always generous with praise, said, "I do not think it is any exaggeration to say that Mr. Wright first taught us how to sell electricity."[35]

To sell more power, Insull needed to generate more power. The Harrison Street station was a start, but as a DC system its reach was limited. Moreover, the acquisition of so many small companies gave him a glut of small, inefficient power stations that lacked any semblance of coordination. The obvious solution was an AC central station, but that required another large capital investment. Ferguson found a better approach in the form of the rotary converter, which had been displayed at the fair but had not yet entered serious commercial use. Insull acquired two of them and installed one at Harrison Street, where it changed the DC current generated to AC and transmitted it to the Twenty-seventh Street station at high voltage. The second converter, installed at the latter station, stepped down the voltage and reconverted it to DC for distribution. Twenty-seventh Street became in effect a substation with all its power emanating from Harrison Street.[36]

This arrangement went online in August 1897. By then Insull had also pulled off his first major financial coup. Chicago Edison had accumulated a floating debt through its acquisitions and expansion during 1895 and expected more of the same the following year. Insull took the position that the company should use Chicago banks whenever possible for its financial needs. He still bore the emotional scars from his dealings with New York bankers, and he resented the fact that eastern banks insisted on appointing mortgage trustees from New York, Philadelphia, or Boston and paying coupons in those same cities. If Chicago banks could not oblige, Insull looked to London, where he had good contacts. Either option allowed him to bypass New York.[37]

A crisis arose unexpectedly during the summer of 1896 when the nomination of William Jennings Bryan and the Democratic platform endorsement of a bimetallic standard led many banks to stop lending money pending the outcome of the election. This "Bryan panic" slowed business, especially in the Midwest, and caught many central stations and traction companies in a financial squeeze that threw them into bankruptcy. Just before the panic, Insull's board had approved a $6 million bond issue to handle the floating debt and some other obligations such as the acquisition of Chicago Arc Light. Insull protested that the figure was much too small because it left only about $2 million for future expansion, but he could not budge the board. The first batch of $1.2 million in bonds was to be sold in London, but the Bryan panic soured the market there for American securities. Painfully aware that Chicago Edison needed the funds badly, the directors asked Insull to go personally to London.[38]

Insull did so, and within two weeks managed to place the bonds despite the moribund London market. The feat deeply impressed Chicago's financial community. "Such a sale," wrote one editor, "had been thought an impossibility." The transaction, as Insull emphasized, "opened up the London market for me and gave me an outlet for securities for upwards of twenty-five years." It also made Chicago's leading bankers more willing to accommodate Insull's desire to use them for his financial needs. One of them, John J. Mitchell of Illinois Trust and Savings Bank, had joined the Chicago Edison board in February 1896 and soon became a close friend of Insull.[39]

During 1897 and 1898, Insull laid the foundation for what mushroomed into a business empire. A key step involved plunging into the netherworld of Chicago politics. Early in 1897 Charles T. Yerkes, the notorious but capable czar of the city's streetcar companies, moved to get a franchise extension for his properties. Eager to consolidate the city's traction systems, he wanted to issue bonds for that purpose, but several major franchises were due to expire in 1903, and he could not obtain long-term financing without the guarantee of secure franchises. The Chicago City Council was dominated by the voracious "Gray Wolves," who thrived on selling their votes to the highest bidder. Yerkes first tried to skirt the council by supporting a bill in the state legislature that would transfer the right to award franchises from the city to the state. Thwarted in this attempt, he managed to get through the legislature the Allen Act, which gave city councils the power to award fifty-year franchises. Yerkes then tried to get a franchise from the city council but failed, and he ultimately sold out his Chicago interests.[40]

Fortified by the Allen Act, the Gray Wolves moved to play their favorite game of blackmail on Insull. Earlier they had profited handsomely by creating a gas company franchise controlled by themselves and foisting it on the existing gas monopoly, Peoples Gas Light and Coke Company, for an inflated price. When Insull ignored their first overtures, the Wolves created a new paper corporation, Commonwealth Electric, endowed it with a fifty-year franchise, and prepared to turn it into an active competitor to Chicago Edison. Still Insull did not budge, and the reason soon became clear. He held a decisive trump card in the form of his monopoly on electrical equipment. The Wolves could make all the new franchises they wanted, but they could not buy the machinery they needed from anyone except Insull. Four months later they surrendered and sold the franchise to Insull for $50,000. The following year the legislature repealed the Allen Act, leaving Insull with the only fifty-year franchise in the state.[41]

Insull bought Commonwealth in his own name for legal reasons but in practice for the benefit of Chicago Edison. "It was in no sense a personal operation of mine," he insisted, "and I never made or lost a dime on it." He ran the two companies in tandem. Since Chicago Edison utilized DC current, Insull limited its operations to the area within a mile or so of the Loop and turned Commonwealth into an alternating-current company supplying power to the rest of the city. To finance his ambitious expansion plans Insull needed money, and Commonwealth proved an ideal vehicle because its secure franchise made it possible to sell long-term bonds to anxious bankers. Insull had become friends with H. H. Porter, a veteran railroad man, who convinced him that growing businesses with large capital requirements needed large mortgages. The ideal form, he suggested, was one with no set limit. This notion excited Insull, whose vision knew no bounds. He got his trusted legal adviser, William G. Beale, to draw up the necessary papers.[42]

Their work produced a major innovation in corporate financing, the open-end mortgage, which later became standard for many utility companies. Besides having no set limit, the mortgage included a host of new features. The bonds ran for forty-five years instead of the usual twenty and matured all at once rather than serially. Instead of the usual sinking fund to pay them off, they utilized an annual depreciation reserve drawn from earnings along with a provision that they could capitalize no more than 75 percent of existing plant facilities. Somehow Insull managed to win approval of the new mortgage from a board that the previous year had refused to give him more than $6 million in bonds. Well might they have blanched had they been able to peek into the future. Before he was done Insull issued $500 million in bonds under this mortgage.[43]

Having created the mortgage, Insull then had to find ways of selling the bonds. To test the waters, he agreed to acquire several companies on the South

Side that Chicago Edison could not serve. He needed $2 million for this work and, despite the reticence of many Chicago banks, managed to sell enough bonds within two weeks. John Mitchell became Insull's strongest advocate on the board and among Chicago bankers. Together with Beale the two men formulated the strategy that Insull employed in all his financing: They would shun New York banks in favor of Chicago institutions. When Chicago could not serve their needs, they would look to London. This arrangement still left Insull dependent on bankers, but it gave him more flexibility and a more congenial circle of lenders to tap. Later he would come to regret ignoring the New York bankers, who never forgave the slight.[44]

By 1898 Insull was ready to impart his vision to the industry as a whole. Since his arrival in Chicago he had been active in two rival organizations, the National Electric Light Association (NELA) and the Association of Edison Illuminating Companies (AEIC). Gradually he lured old associates, many of whom had worked for him at Edison General Electric, from the AEIC into NELA until he succeeded in replacing competition with cooperation. NELA emerged as the industry's voice in public affairs, AEIC as its forum for private matters and watchdog over the industry's firms. NELA made Insull its president in 1897–98 and provided him a platform to expound his radical views on the industry's future. In his speech at the annual convention on June 7, 1898, Insull outlined the program of action he thought necessary and launched what Harold Platt called "a major crossroads in the energy revolution."[45]

By then Wright and his demand meter were hardly strangers to people in the field. In October 1896 *Electrical World* had run an article describing the system in detail. The following year Insull brought Wright from England to describe his two-tier rate system to the delegates, most of whom regarded it as interesting but purely theoretical. *Electrical World* thought otherwise; in January 1898 it declared flatly that "the Wright demand meter system, or such a modification of it as would be found to best suit Americans conditions, is recommended as the best method for the adjustment of rates and charges." Insull hoped to sway those members who remained dubious about both Wright's system and the need for major changes in the way they did business.[46]

Insull stressed to his audience the importance of selling energy rather than just improving efficiency. The key was ratemaking, which occupied more of his time than any other subject. "The way you sell the current," he insisted, "has more bearing on . . . cost and profit than whether you have the alternating or direct-current system, or a more economical or less economical steam plant." In the discussion that followed, Ferguson reinforced this theme, adding that "there is more money to be made in the intelligent selling of your product than in attempting to introduce further economies in operation." During his speech Insull also stunned the members with his first public statement that the ideal future of the industry was to function as a monopoly under government regulation. "In order to protect the public," he declared, "exclusive franchises should be coupled with the conditions of public control, requiring all charges for services fixed by public bodies to be based on cost plus a reasonable profit."[47]

During the discussion a member asked how small a user he could afford to serve. Insull replied that he would accept a customer who used a single 25-watt bulb if such a bulb existed. "We take any and every customer on the Wright demand meter basis," Ferguson added, but he emphasized that "the intention is to give the low rate to the long-hour." The demand meter could be a powerful incentive for large power users and eventually render the isolated power unit obsolete. However, it would take enormous cuts in the price of electricity for this to happen. If his political and economic approach was used, Insull predicted, the result would be a self-perpetuating cycle of rising consumption and falling rates.[48]

Through these efforts Insull sought nothing less than prodding the industry into an entirely new way of doing business. In May he outlined its contours in a speech at Purdue University, one of many public appearances he made to get across his message. "While the growth of the business has been phenomenal, especially since 1890," he proclaimed, "I think it can be conservatively stated that we have scarcely entered upon the threshold of the development which may be expected in the future." Even more, he proceeded to lead by example. During 1898 Chicago Edison put its version of Wright's rate system into effect, which cut the average residential user's bill by 32 percent. To attract new customers, the company offered to install six lighting outlets free of charge; those who accepted often decided to have their entire home wired as well. Aware that he needed new kinds of data to survey electrical usage, Insull created a statistical department and put E. J. Fowler at its head. The talented Fowler began supplying managers with invaluable reports on a variety of energy-use patterns.[49]

There was a special urgency behind these moves. By 1898 the Welsbach mantle had transformed the gas industry, especially in the areas of heating and cooking, into a powerful competitor for Chicago's energy business. A year earlier the gas interests had managed to complete their consolidation through formation of the Peoples Gas Light and Coke Company. Insull realized that lighting alone could not solve his load dilemma; he needed also to attract large users such as factories, streetcars, and the myriad of users who clung to their isolated plants. As noted earlier, by 1900 electricity supplied less than 5 percent of the power for factories, two thirds of which came from isolated plants.

National streetcar mileage had soared from 8,120 in 1890 to 22,580 in 1902; of the latter figure 21,910 miles used electric power, most of it generated by isolated plants. Chicago alone saw the creation of 614 new isolated plants between 1898 and 1903. Insull wanted all this business; it took him nearly a decade to get it.[50]

He continued to expand his traditional customer base of affluent residents, shopkeepers, and businessmen with a mix of lower rates and special incentives. The two-tier rate system encouraged them to use more electricity, and Insull eased the transition to wired homes by offering to retrofit an entire house at cost. To lure apartment-building owners away from cheaper gas for hall lighting, his salesmen offered the landlord special discounts on hall lights and a 5 percent commission for every tenant he signed up for electric lights. The response to this campaign, observed John Gilchrist, head of the sales department, "was marvelous . . . The attitude of the owners changed from one of cold indifference to one of interested friendliness." Merchants along different streets petitioned Insull for more lighting to compete better with rival commercial areas that already had it. Ferguson obliged with special rates if the merchant groups would guarantee an annual minimum contract of one hundred arc lights. One neighborhood flooded its street with 250 lights, prompting others to ask for a similar number.[51]

In the central business district Insull went after the skyscrapers, hotels, department stores, and other large buildings that used their own plants for light, heating, and power. Their business offered a splendid load for the daylight hours when a central station's generators saw little activity. After studying the problem Insull formed the Illinois Maintenance Company to furnish steam heat through underground pipes to buildings that wished to convert to electricity, and he tossed in some free services along with special rates. His ultimate weapon, however, became the electric elevator, which Harold Platt called "a classic example of a practical solution to a demand awaiting a supply." The elevator required large amounts of energy on an intermittent basis; it had, in short, a terrible load factor and needed special equipment to operate smoothly. Even central station operators tried to avoid elevator contracts, and the problem grew worse as buildings soared ever higher.[52]

Insull and his team thought otherwise. "The electric elevator," declared Ferguson, "is . . . the key to the solution of the much-discussed question, 'central station supply versus the isolated plant.'" The reason lay in the emphasis on spreading the load. According to the gospel of Insull, the liabilities and diseconomies of individual elevators vanished when viewed as a group using large amounts of energy during off-peak daylight hours. Once the electric elevator proved superior to other types, thanks to further development of Tesla's versatile motor, architects began to include in their plans not only elevators but other skyscraper needs that could be handled by electric power, such as water pumps, ventilation fans and blowers, and sewage ejectors. In 1901 Gilchrist asserted that the elevator had done more to promote electrical service in big commercial buildings than any other factor. Why spend huge sums on installing and maintaining a large isolated power plant when everything could be handled by central station service at the special rates offered by Insull?[53]

Factories posed a special problem in part because they had been built around their steam engines, which required such things as reinforced ceilings to bear the weight of heavy shafts, belts, and pulleys. Production machinery could not be organized in the manner best suited to efficient output because the largest machines had to be placed nearest the steam engine. Steam power created a noisy, dirty, dangerous workplace as well as a fire hazard, but conversion to electricity was expensive. Here too the rapid development and obvious versatility of the electric motor sped the process. Insull's salesmen worked zealously to accommodate industries old and new in switching to central station electric power. In 1895 the company provided power to only 3,000 horsepower worth of electric motors; a decade later the figure stood at 69,000 horsepower. Between 1896 and 1902 the use of electricity for power increased about 50 percent faster than the growth of use for lighting.[54]

Insull grasped the value of mass marketing as clearly as he did that of mass production of electricity. In 1901 he formed an advertising department; two years later it began publishing *Electric City*, a free glossy magazine. The department created a model "Electric Cottage" as a mobile showroom that roamed from one neighborhood to another displaying the latest in light fixtures and appliances. Behind it came teams of door-to-door salesmen presenting homeowners with special offers. In one splashy campaign the company gave away ten thousand irons for new customers to try. While the effectiveness of these campaigns cannot be measured, the company scrambled to keep up with the growing demand for its electricity. Between 1898 and 1902 alone consumption increased 152 percent, an average of 38 percent a year. By contrast generating capacity rose only 73.6 percent, an average of 18.4 percent yearly. Having succeeded in stimulating demand with his policies, Insull had to revisit the problem of supply.[55]

By 1902 the Harrison Street station was straining to keep up with demand, and it had become clear that the reciprocating steam engine had reached its capacity. For some years the steam turbine had been under development, especially in Europe. Westinghouse had secured the Parsons patent and installed his first commercial unit in Hartford, Connecticut, in 1898. The previous year a young inventor named Charles G. Curtis had approached General Electric with his own version of a turbine. After two years of dogged experiments, John Kruesi gave the project a thumbs-down. However, the project was handed over to another engineer, William L. Emmet, who painstakingly altered the design to make it commercially viable.[56]

Late in 1901 Charles Coffin tried to interest Insull in the new device shortly before Insull left to spend Christmas with his mother in England. While in Europe Insull visited Frankfurt to inspect the first central station that used steam turbines. Impressed, he asked Ferguson and Sargent to go to Europe and conduct a more thorough study of the turbine. They returned with a favorable report but urged Insull to start slowly with a small unit, perhaps as an annex to Harrison Street. "Their advice was good conservative engineering advice," Insull recalled, "but did not meet the commercial situation." What Insull needed was not small but large power. During 1902 Insull scored a coup by securing a contract to supply electric power to the Lake Street Elevated. He sensed the possibility of scooping the business for the entire elevated system.

"What was needed," Insull concluded, "was production of energy on a very large scale at a very low cost." The high-tension load doubled in 1900 and continued to climb. In April 1901 Sargent advised against expanding Harrison Street even though, as Insull admitted, it would take two years to provide more capacity elsewhere. In December 1901 Insull bought some land on Fisk Street for a new powerhouse for Commonwealth Electric. When Coffin approached him with the idea of trying the experimental 1,000-kilowatt steam turbine in the new plant, Insull shook his head. Already he saw in the turbine the potential to realize his grand vision of a single plant capable of supplying power for the entire city. Forget the 1,000-kilowatt turbine and dynamo, he told Coffin; build instead a unit with a capacity of 5,000 kilowatts. Impossible, said the GE engineers, but Insull dismissed their reaction; they were, he sneered, men who used slide rules to prove anything was impossible. However, his own engineers, including Sargent, voiced similar doubts.[57]

Coffin suggested trying to build a 3,000-kilowatt unit. Not enough, Insull insisted. He needed size, and if GE could not provide it he would try his luck in England with Parsons or someone else. The prospect of a foreign competitor gaining a foothold in the turbine market was enough to spur Coffin. He said that GE would shoulder the risk of building a 5,000-kilowatt unit if Commonwealth Electric agreed to bear the risk of installing it. Insull accepted the offer, and the contracts were signed in April 1902. Although no one realized it, Insull had taken the first step in transforming the delivery of electricity in the United States. Already he had coined a phrase to describe his goal: the "massing of production."[58]

It was a bold venture, and Insull knew it. "I realized as it went along, the risk

I was taking . . . ," he admitted. "I got so nervous that I could not stay in my office. I used to spend most of my time at the Fisk Street Station while it was being built."[59]

The new turbine made its debut in October 1903. Insull was there to watch Sargent steam it up for the first time. His first impression was hardly reassuring. The engines vibrated so wildly at first that the dignitaries were hastily escorted to cover. Sargent shut the unit down, located the problem, fixed it, and prepared to start up again. Noticing that Insull had returned and was watching intently, he suggested that Insull return to his office because he did not know what would happen. "Sam, I think it is all right," he said, "but this thing is shaking like the devil and might blow up. I think I will stay here, but you had better get out."

"There is just as much reason for your leaving here as my leaving here," Insull said.

"No!" replied Sargent. "My being here is in line with my duty."

"Well," said Insull, "I am going to stay. If you are to be blown up, I would prefer to be blown up with you as, if the turbine should fail, I should be blown up anyway."[60]

The turbine ran flawlessly the second time and ushered in a new era. Insull ordered two more units at once. Eighteen months later Insull tore out all three units and replaced them with new ones that had twice the capacity even though they were the same physical size. By 1913 Insull was using generators with 20,000-kilowatt capacity that burned 40 percent less coal than the original 5,000-kilowatt units. The size kept climbing until it reached 75,000 kilowatts in 1924. Later General Electric developed compound machines that combined a high-pressure and a low-pressure turbine with a generator in one unit. In 1928 the company installed a triple cross-compound turbogenerator rated at 208,000 kilowatts. The first turbine units occupied only 10 percent of the space used by the reciprocating steam engines they replaced and weighed only 12.5 percent as much. Their small footprint and superior efficiency sent the reciprocating steam engine into retirement. In 1902 it consumed 6.7 pounds of coal to produce one kilowatt-hour of electricity. By 1928 the new turbogenerators used only 1.76 pounds.[61]

Insull now had the ideal machine for massing the production of electricity. The larger the unit, the more cost-effective it became, and it seemed to have no upper limit as to size. Burning much less coal not only saved money but cut the amount of smoke pollution in the city. When a national coal strike in 1902 threatened his supply of critical fuel, Insull took steps to ensure its availability. Frank S. Peabody of the Peabody Coal Company became his mentor. Chicago Edison joined with Peabody to buy a mine, some coal acreage for

later development, and a small railroad to move the coal. These moves protected his weakest flank. An electric company needed coal every hour of every
day, and these acquisitions provided it.[62]

By 1903 Insull had the tools and the organization needed to master the supply side of electric power. The vision had long been there, and only been refined and elevated by experience. Without missing a beat he set out in earnest
to link his massed production capacity to the realization of his gospel of consumption. He would not only swell the ranks of electricity users to unprecedented size but in the process rationalize the structure of the industry as well.
And he would do it his way.

CHAPTER 20

THE EMPIRE
OF ENERGY

*We owe a great deal to this country. We owe everything to it except
birth and what we all love so much, that liberty which was given
us by the mother country from which we spring. But we owe our
opportunities in life to this country. We do not want to be British-
Americans here; we do not want to be English-Americans or Irish-
Americans or Scotch-Americans. We want to be Americans.*

—SAMUEL INSULL (1916)[1]

*I owe America nothing. She did only one thing for me. She gave me
the opportunity. I did the rest and I repaid America many times
for what she gave me.*

—SAMUEL INSULL (1934)[2]

DURING HIS FIRST TUMULTUOUS decade as head of Chicago Edison, Sam
Insull put his personal life on track as well. In 1896 he paused from his work
long enough to become an American citizen. Three years later, after a lengthy
courtship, he married Margaret Bird, an actress whose stage name was Gladys
Wallis. Small and dainty, with exquisite features, her fragile appearance belied
an inner toughness that enabled her to survive a difficult childhood and prosper in a rough profession that most respectable people looked down on. Unlike the stereotypical actress, Gladys guarded her reputation zealously and
despised both alcohol and sex. Her abstinence from drink pleased Sam Insull
as much as her dislike of sex frustrated him. Both partners had strong wills
and tempers, making for a tempestuous relationship. Yet it endured and produced the great joy of Insull's life, his son, Samuel Insull Jr.[3]

Born in April 1900, Junior remained close to both his parents all his life. As
an only child of devoted parents he basked in constant love and affection

along with strict discipline. On some levels they competed for his attention. Gladys showered him with plays and poetry, hoping to win him to the arts; Sam surrounded him with mechanical toys of every kind, seeing in the boy a miniature version of himself that must be groomed for the family business. It seemed Gladys had the better of it in the boy's early years, but Junior—or Chappie, as Sam called him—decided in his late teens to join his father in business. Although his mother grieved the decision, it proved to be not only wise but crucial to the survival of his parents in later years.

By 1900 Insull had already made his mark on Chicago and the power industry, yet he had barely started the list of his achievements. The creation of his business empire occurred in stages, some of which he initiated and some of which came in response to the moves of others. The potential of the suburbs ' fell into the latter category. A bright, ambitious (some said grasping) lawyer named Charles A. Munroe stumbled into the power business before he understood anything about it by acquiring franchises in suburban villages near a power site he controlled. Then, in 1901, Charles T. Yerkes, the scourge and lightning rod of Chicago business and politics, sold out all his traction holdings and moved to London, where he built a subway. His departure left a vacuum not only in the streetcar but also in the power supply business.[4]

Everyone knew that streetcars and elevated railways gobbled more power than all other users combined. In Chicago, where 1,400 cars served the Loop during rush hour, they consumed three times the energy used by all of Chicago Edison's light and power customers. Munroe figured correctly that Yerkes made money selling power to his own streetcar companies, which meant that their power costs were high. He sought out the president of the Lake Street Elevated and got an informal agreement to sell him all the power Munroe could supply. Munroe lacked the money, power sources, and knowledge of the business to pose a serious threat, but he played a good game of bluff. Insull reacted to the threat of a competitor in typical fashion: by buying him out at once. Munroe soon became another close Insull associate, and Insull secured his entry into the market for delivering power to traction companies.[5]

The streetcar business fit Insull's evolving business model for central stations perfectly. Trolleys swallowed huge amounts of power at the morning and evening rush hours, which fit neatly between the demand for lighting and factory power. Even better, the market had only just began to grow. Of those 1,400 cars serving the Loop in 1902, only 510 used electric power; another 772 relied on cables, and 97 electric cars were attached to cable trains. On the north and west sides of the city horsecars remained a familiar sight. In 1901

Union Traction Company spent as much on horses as on electric power and steam-powered cable cars. All the streetcar lines except the Lake Street Elevated generated their own power.[6]

To wean the traction companies to central station power, Insull resorted to what became known as the Hopkinson system of rates, a version of Wright's two-tier system modified to fit the peculiar usage of streetcars. It created rates so low that the streetcar companies could not afford to ignore the offer. During the next five years Insull landed so many traction company contracts that by 1907 their usage surpassed that of all other Chicago Edison customers combined. The profits from this surge of business enabled Insull to buy the next generation of turbogenerators for Fisk Street and to build a new power plant directly across the Chicago River at Quarry Street with 14,000-kilowatt generators that outstripped Fisk Street in size and efficiency. The traction business with its two rush hours and lighter midday shopper traffic helped the utility company achieve a load factor of 30 to 40 percent, which meant that the equipment remained in use more than a third of the time. Under Insull's formula this reduced costs and allowed still more flexibility in setting rates.[7]

Both Chicago Edison and Commonwealth Electric had grown at a phenomenal pace even before getting the traction business. Their production quadrupled during Insull's first three years at the helm and doubled again during the next three years, jumping from 26 million to nearly 100 million kilowatt-hours. By 1905 annual production had doubled seven times in thirteen years; during the next five years it quintupled, producing more electricity than New York Edison, Brooklyn Edison, and Boston Edison combined. By 1907 the two Chicago companies were sixty times larger than when Insull first arrived and had paid regular 8 percent dividends despite the enormous costs of expansion. Although the two companies had always been run in tandem, Insull decided that they would operate more smoothly and attract funds in the bond market more easily if he could merge them under Commonwealth's fifty-year charter. After some deft political maneuvering he succeeded in uniting the companies as the Commonwealth Edison Company in 1907.[8]

That same year Insull formed an alliance that served him for many years. He was involved in negotiations with three Chicago banks for an issue of $5 million of Chicago Edison debentures, a process that he conceded was "not always of the pleasantest and I clashed more or less with them as to conditions and terms of issues. I used to think that they were rather arbitrary and I presume that they had about the same opinion of me." This time he could not come to terms with the banks and broke off negotiations. When word of the breakdown got out, he received a visit from Harold L. Stuart of N. W. Halsey & Company. A transplanted Rhode Islander, Stuart agreed to take the notes and arranged a syndicate to buy them. For the next decade he continued to handle some of Insull's financing; later he would become the indispensable salesman.[9]

The merger secured Insull's long-term financing. He could now sell bonds in one company with a lengthy charter that reassured investors. Stuart provided another outlet for marketing securities, one that would grow enormously over time. As with the other great pioneers in electricity, Insull's vision and unflagging optimism outran the resources at his disposal. Unlike them, however, he proved adept at solving the financial obstacles that loomed constantly in his path even though he was beginning to work on a scale vastly greater than theirs. His was a different kind of pioneering path, one that wound tortuously through a treacherous landscape of corporate and financial as well as technical intricacies. Well before the merger his horizon had begun to expand beyond the city limits.

A utilities lawyer and friend, Frank J. Baker, called Insull's attention to the suburbs, where a host of companies had sprung up to serve individual, mostly affluent neighborhoods. Baker urged Insull to acquire the plants in Highland Park and Waukegan, Illinois, even though both had unimpressive balance sheets. Insull showed little interest until a conversation with Dr. Emil Rathenau, the head of Germany's electrical giant AEG and Coffin's friend, changed his mind. Rathenau predicted that before long a single central station would serve a radius of fifty miles. Insull expanded this vision to one that targeted a territory of 2,500 square miles. He bought the two small plants, formed a new company called North Shore Electric in 1902, and sold the plants to it for a tidy profit. Then he began buying up other small plants between Evanston and Waukegan and folding them into North Shore, creating an integrated system of generators and substations that provided customers with round-the-clock electric service.[10]

Within a few years the North Shore suburbs had cheap, efficient electrical power. Before Insull got involved, most of the smaller companies provided power part-time at high cost with numerous breakdowns because of undermanned equipment. By 1912 North Shore Electric had expanded its grid to serve 92,500 people in forty-three communities spread across 1,250 square miles. Insull had carried his vision of low rates, load balancing, large central stations powered by huge generators, and transit contracts beyond the dense population of downtown Chicago to the less densely inhabited suburbs. Almost by accident he soon extended it beyond the suburbs to nearby lightly populated rural areas. In 1906 Insull bought a 160-acre farm with a large house near Libertyville, thirty-eight miles northwest of Chicago. Hawthorn Farm became his primary home, supplemented by an apartment in the city; by 1914 he had enlarged it to four thousand acres, erected a mansion, and settled comfortably into the role of country gentleman.[11]

Naturally Insull wanted electric power for his country estate. The obvious solution was to build an isolated plant for the property, but that did not suit Insull; instead he ran a transmission line from a North Shore Electric substation. Then his curiosity took over. If customers in the city could all draw power from a single source, why couldn't transmission lines string together a number of isolated villages with the same economy? Early in 1910 he completed a survey of the region and launched what he called his "Lake County experiment." It embraced twenty-two towns, all with populations under three hundred. Twelve had no electricity, ten had dusk-to-midnight service only, and a few farms drew power from the isolated plant of one well-to-do farmer. Insull could hardly have picked a less promising target locale; one financial writer dismissed it as a "combination of junk piles." Nevertheless, Insull bought the ten small local companies and built a network of transmission lines that furnished power to twenty of the towns and about 125 farms.[12]

The results surprised even Insull. Although the investment per kilowatt-hour more than doubled, the diversity of demand and increased generating efficiency of large units more than compensated for it. Insull soon extended service to the entire area, cut rates, reduced costs, doubled the load factor within two years, and came away with a healthy profit. By 1912 the load factor reached nearly 29 percent for 22,188 customers. In one stroke Insull had demonstrated—with the help of elaborate statistical tables prepared by Fowler—the feasibility of extending electricity to rural areas. "Cheap power," he insisted, "is as essential to the farmer as it is to the manufacturer." The basic formula was simple: Once past the high initial investment in transmission lines, the costs would decline with every new customer and increase in output. The key lay in a grid of powerful central stations feeding a giant network of transmission lines.[13]

While the Lake County experiment proceeded, Insull extended his reach to other regions surrounding Chicago. Southwest of the city Charles Munroe scooped up plants and acquired franchises for fifty-six villages that had no electricity. West of Chicago Insull joined with an old friend, L. E. Myers, who already served Cicero, Berwyn, and Oak Park. To the south he bought out two promoters who made a career of developing properties and then selling them, usually to Insull. Then, in August 1911, Insull swept together all these holdings along with North Shore Electric into a new company, the Public Service Company of Northern Illinois (PSCNI). Its territory covered a population of 302,409, of whom 36,040 used electricity, which invited still more growth of business. Two years later Insull enlarged PSCNI by buying the last major independent energy supplier, Northwestern Gas Light and Coke. This acquisition increased his customer base for gas from sixteen thousand to sixty thousand.[14]

During the 1900s Insull worked steadily at modernizing and upgrading his suburban acquisitions to the level of central station service in Chicago. His new properties served four diverse kinds of customers: residential suburbs, industrial suburbs, small satellite cities, and outlying farms. Once convinced that large power users held the key to obtaining real economies of scale, Insull sought them out relentlessly. Before its merger into PSCNI, North Shore Electric reached a load factor averaging 65 to 70 percent compared with Commonwealth Edison's 40 to 42 percent in the city. As his empire consolidated, the unit cost of producing electricity continued to decline just as he said it would, especially in the suburbs new to his policies.[15]

Through PSCNI Insull forged an impressive regional network that was interconnected with the Commonwealth Edison system. Through it high-tension electricity flowed from Milwaukee on the north to the eastern banks of the Mississippi River on the west, and from Michigan City and South Bend, Indiana, on the east and southeast to the Illinois coal fields south of Chicago. "It was an effort," he said proudly, "by means of operating economies brought about by combination, to bring service to all classes of people residing in communities of vastly different conditions, and to render this service at the lowest possible cost to its users, consistent with an effort to make a proper return to the investors." Even more, he had launched the process by which electricity moved from a luxury enjoyed by the rich to a basic necessity of middle-class life. As rate cuts dropped the price of electricity by 40 percent, the number of households served by Commonwealth Edison soared from fewer than five thousand to nearly eighty thousand, though this was still only 16 percent of the families in Chicago.[16]

During these busy years Insull institutionalized not only his notion of massing production and consumption but also his marketing techniques. As Harold Platt observed, "Insull's spectacular business and political success made his theories seem like immutable laws of economics." In the process he became an institution himself, the guru of electrical utilities. Along with his work in the industry's professional organizations, he gave freely of his time to making speeches that outlined in detail his policies and their results. In Chicago he won the respect and admiration even of the politicians, whom he shrewdly treated as equals, as well as businessmen, bankers, journalists, and his employees. If he had done nothing more in the field, his career would have been acclaimed a success. But Insull was eager to do more, much more.[17]

A personal crisis in 1912 left Insull a changed man and one devoted even more to his work. That January his son came down with scarlet fever and was quarantined at Hawthorn Farm. A distraught Insull accepted the separation; his wife did not. The delicate, pampered, temperamental Gladys entered the house and became Junior's keeper with the help of two devoted nurses and a part-timer. Many people considered scarlet fever a death warrant and would not go near a patient; Gladys and her helpers ignored the risk. For three long months she laboriously scrubbed the house with formaldehyde day after day, antisepsis being the only known safeguard at the time. Early in April Junior's fever spiked, and his heart stopped beating until the head nurse shook and beat him in desperation. Shortly afterward the fever began falling, and the patient had only to get through the last tedious weeks of quarantine.[18]

Through this ordeal Insull could do little but talk to Gladys through the second-story window. He planted trees while Gladys scrubbed and tended to her bleeding hands, rethinking all the while their relationship that had already grown distant because of his devotion to work. When the crisis ended, Gladys had decided to end permanently what she had always found distasteful: the sexual part of their relationship. Already they had not slept together for several months; once the finality of it dawned on Insull, it changed him in unexpected ways. He did not become a philanderer or abusive to the woman who remained his wife in other respects. Instead he found solace in work, elevating his ambitions beyond what he had already created. It was not money he sought but power, the creation of an empire vastly larger than the one he had already fashioned.

When Junior was stricken in January 1912, Insull controlled companies with assets of about $90 million, nearly all of them electric utilities located almost entirely in Chicago and the surrounding region. Five years later he presided over an empire with assets exceeding $400 million, including gas and traction as well as electrical utilities, spread across thirteen states. Each new opportunity came as a separate challenge that he took on because the immediate situation required it and no one else stepped forward to assume the responsibility. Each one also fed his ever hungry ego and his belief that he was capable of meeting any challenge or solving any problem placed before him. In one way or another, the crisis of 1912 became the catalyst for Insull's march toward the erection of a gigantic empire of energy.[19]

AS EARLY AS 1902 Insull had obliged some Chicago bankers by reorganizing an electric light and gas company in New Albany, Indiana. Later he picked up similar holdings in nearby Jeffersonville and extended New Albany's streetcar line to that town. In these cases Insull departed from his usual practice by acquiring control of the companies and giving his younger brother, Martin, the job of managing them. To his surprise Martin proved to be a capable manager with an expansive vision. By the summer of 1911 he presided over a $5 million enterprise of local electric, gas, and streetcar companies. Insull then gave his brother license to replicate the Lake County experiment by extending lines to more surrounding towns. Martin's demands for more capital grew louder just as Insull was putting together PSCNI, which also swallowed money whole.[20]

The logical course was to sell off the Indiana properties at a profit, as Insull had done with peripheral ventures in the past. But Insull decided instead to keep the holdings and expand them still more, which required still more capital. A banker friend and Commonwealth Edison board member, Edward P. Russell, suggested forming a holding company for the Indiana properties. Thus was born in 1912 the Middle West Utilities Company, which raised about $4.5 million through a series of convoluted transactions to repay Insull and finance future expansion. Martin moved to Chicago and became vice president of the new company with Insull as president. Later Insull moved up to chairman as Martin took over the presidency and ran the company.[21]

Middle West became the vehicle through which Insull extended the Lake County model across several midwestern states with the same profitable results. It gobbled up small local plants and turned them into a connected system. Some local plants also provided gas, water, streetcar, even ice service. Insull's company found itself wandering into these businesses as well, often with poor results. Sometimes the purchase of a small utility company also involved a package deal that included other, more distant plants. In this manner Middle West found itself owning two small plants in Missouri, one in New York state, and eleven more in upper New England. The new company grew steadily until by 1914 it served 131,000 electric and 43,000 gas customers in four hundred towns and thirteen states. Most of the customers lived in rural areas and got electricity for the first time at half the price paid by people in other small towns.[22]

During these same years Insull also emerged as the kingpin of Chicago's traction lines. By 1911 a plan had evolved to merge the four elevated and surface companies into one efficient system. The companies needed about $6 million to integrate their lines, but their long history of shady political and financial activities severely damaged their credit. They also needed legislation from the city for the merger, and smart investors doubted it could be obtained. The owners appealed to Insull for help. He agreed to have Commonwealth Edison guarantee the funds in exchange for the stock of a trust created to control the four companies. With this guarantee the companies secured the loans, but the merger never took place. In July 1914 Commonwealth Edison had to assume the loans and put up some working capital as well. In return it held 80 percent of the trust's shares; the electric company owned its largest customers.[23]

Not content to let the traction lines drift, Insull chose the best of the four presidents, Britton I. Budd, to run them. From his efficient management and Insull's encouragement came a startling rise in the quality of service, ranging from cheaper fares to universal transfers to clean stations and polite employees. So well did Budd do his work that writers even from distant Boston hailed him as "Moses for the tractions." The elevateds ceased to be a local joke in Chicago and drew praise from all sides. Later Insull extended his reach to some of the interurban lines such as the Chicago, North Shore & Milwaukee. Budd took charge of it as well and proceeded to modernize the road and its service. Not even Insull could find a long-term solution to the city's transportation problems, but at least he gave the public better service than they had ever seen.[24]

In both cases Insull had taken up the challenge at the request of the owners or bondholders of the properties. Another major responsibility fell his way for the same reason. In 1913 the owners of the beleaguered Peoples Gas Light and Coke Company used a clever ruse to induce Insull to accept the chairmanship of their board. Insull brought with him two of his closest friends, John J. Mitchell and James A. Patten, giving him a majority of the five-member board. He was well aware that Peoples Gas had an unsavory reputation thanks to a notorious past that went back to the Yerkes era and that the pressure was immense for it to reduce rates. Within a short time he managed to ease this pressure and give the company a better public face, but he did not involve himself in the running of the company. Later he would discover that Peoples Gas was in much worse shape than he had suspected.[25]

At the outbreak of World War I, Insull ignored President Woodrow Wilson's dictum to remain neutral and did what he could to aid the British war effort. "At the time it was a penal offense not to be neutral," he admitted, "and it was difficult to carry on some of the activities that I engaged." Once the United States entered the war, the governor of Illinois tapped Insull to head the Illinois State Council of Defense. He promptly galvanized the organization into a forceful advocate of the Allied effort with a brilliant if relentless propaganda campaign, a successful fund-raising effort, and a ruthlessly efficient drive to stamp out war profiteering in the state. Although he won widespread praise for his wartime work, Insull was obliged to neglect his companies during those years. By 1919 he found some of them struggling because of strains imposed by the war.[26]

During the war a political movement gained momentum in Illinois to strip the newly created State Public Utility Commission of its regulatory powers and return them to local authorities in the guise of "home rule." As wartime costs spiraled, utilities found themselves squeezed and pleaded for rate increases that were long delayed or not granted. By January 1919 no fewer than seventy-one transportation companies had gone into receivership, with many others not far behind. The political imbroglio over rates and service affected Insull's companies in different ways. Commonwealth Edison, PSCNI, and Middle West Utilities got through the troubles unscathed, but the elevated railway companies floundered badly, and Peoples Gas stood at the brink of collapse. As Forrest McDonald observed, "Its history could well serve as a manual on how not to run a utility company."[27]

Insull found himself wading into a sea of new problems. Labor posed an ominous difficulty. The war and its aftermath had brought shortages of men, sharp increases in wages, and turbulent relationships. During the war workers demanded better pay and moved to higher-paying jobs elsewhere when they did not receive it. The cost of coal and other commodities soared, as did interest rates. The regulatory battle royal, with its attendant public relations problems, complicated every other issue, as did a huge postwar surge in the demand for power. However, the war had some effects that served Insull well. It produced a coal famine in Chicago and elsewhere, the first energy crisis in American history. Its hardships sounded a death knell for isolated power plants by convincing many businesses and manufacturers to shift to more reliable central station power, especially when the Fuel Administration Agency gave priority to central stations over private users.[28]

The reason was simple. Commonwealth Edison used two and a half pounds of coal to produce one kilowatt-hour of electricity compared to eight or nine pounds for an isolated plant. Manufacturers and other heavy power users began in 1918 to abandon their private plants for central station power. In 1914 large power consumers used only 16.4 percent of Commonwealth Edison's total output, compared to 63.9 percent for traction. Their consumption jumped to 29.4 percent in 1919, 37.4 percent in 1924, and 47.2 percent in 1929, while traction usage fell to 25.2 percent by 1929. Factories accounted for a large part of the soaring demand for power that confronted Insull after the war. To provide it, he had first to solve the problems caused by the war and then find ways to boost capacity. It took him nearly four years to work through these entanglements. In the process he continued to usher the nation into what Harold Platt called the "energy-intensive society," which was the logical outcome of his gospel of consumption.[29]

The war also gave Insull a more paternalistic attitude toward his workers. In struggling through his labor problems he developed a greater sensitivity to their conditions and needs. As he and Gladys drifted apart, he came to regard

Commonwealth Edison's employees as a family. They noticed the change in his attitude; as one remarked, "Now he felt that he belonged to his people." He bestowed on them a wide range of benefits along with an intensive indoctrination program to increase their loyalty to the company and their pride in their work. Insull had learned well from his wartime propaganda activities and put them to good use, exhorting his people to involve themselves in community activities. This demand applied to everyone from himself down to the lowest underling. He made himself more accessible to employees and remained a soft touch for anyone with problems. His personal ledger overflowed with donations to individuals in need. In one of many such cases, he paid a brother and sister leaving his employ fifty dollars a month for the rest of their lives "as a token of my high regard for you and my appreciation of your services to me during many years." He estimated that his private pension list ran to $50,000 a year, most of it to "widows and children of faithful employees."[30]

Insull survived the wartime coal crisis largely through his longstanding arrangement with Frank Peabody, who kept the miners happy by treating them generously while making backroom deals with their leaders. Money was another matter. Interest rates soared to new highs in 1920 just as the demand for power underwent a sudden and huge increase. By 1921 even sound utilities had to pay as much as 10 percent for loans. The pressure was especially great on Middle West Utilities; the combined output of its subsidiary companies skyrocketed from less than 200 million kilowatt-hours in 1914 to nearly a billion in 1923. To meet the crisis Insull borrowed wherever he could, shifted funds from one company to another, used the credit of the strong firms to aid the weaker ones, and relied heavily on his personal credit. Between 1914 and 1921 he borrowed in his own name and loaned to Middle West more than $12 million, $3.5 million of it in 1920 alone. He carried its payroll through the period with his own funds until the money crunch began easing late in 1921.[31]

Once the crisis eased, Insull took steps to prevent it from happening again. Almost by chance he hit on the notion that the customers of his many companies represented an untapped market for securities. He created a securities sales department in every company and later consolidated them into a separate corporation, the Utility Securities Company, owned by its client firms. Everyone from department heads to meter readers pushed stock to friends, neighbors, and customers with astonishing results. Middle West and its subsidiaries had 6,000 stockholders by 1918, 54,000 five years later, and nearly 250,000 in 1928. Other Insull companies did almost as well; by 1930 they boasted together more than a million stockholders.[32]

For bonds and debentures, Insull took a similar tack. The New York investment banks continued to dominate the industrial bond market, but they had been slow to move into utilities and showed little interest in retailing bonds. Insull turned again to Harold Stuart, who had formed his own company in 1911. He responded by developing a system of mass-marketing bonds to a network of smaller outlets. Stuart first showed his mettle in 1922 by selling $27 million in bonds for Insull at slightly over half the interest previously paid. During the next decade Stuart's masterful salesmanship resulted in the financing of almost $1.44 billion in bonds for the Insull empire. Having pioneered in the wholesaling of electricity, Insull and Stuart did the same for the retailing of securities. In the process they loosened New York's stranglehold on the bond market and spread the holders of Insull company debt far and wide. With the stocks and bonds of his companies scattered among so many investors, Insull had little to fear from any threatened takeover.[33]

The gas company posed an even greater challenge. Long the most unpopular utility in the state, it had battled the city of Chicago for years over rates and standards. Instead of compromising with the city, it resorted to the courts with interminable lawsuits. As outrage mounted, the technology of producing gas underwent a profound change. For years manufacturers had relied on "water gas," which required large amounts of oil and coal. It burned more brightly and remained cheap as long as the supply of oil was abundant. An alternative method created "coal gas" from coal, but it produced an illumination of only 22 candlepower, and a city regulation required a minimum of 26 candlepower. The gas mantle eliminated this obstacle by producing incandescent illumination from burning gas. Then the rise of the automobile and wartime demand made oil scarcer and more expensive. Peoples Gas urgently needed to build an efficient new coal gas plant but could not so long as it remained implacable in its fight with the city over a rate dispute that involved as much as $10 million.[34]

During the war Peoples Gas went downhill rapidly. Wartime inflation doubled the cost of building a new plant, and soaring oil prices drove the cost of producing water gas to the point that the company lost money in daily operations. It petitioned frantically for a rate increase but did not receive one until the war was nearly over. During the first eight months of 1918 it lost $1.5 million and saw its credit vanish. That same year the accounting department discovered that many of the company's meter readers had for six months or so been turning in falsified billings without bothering to check the meters. In September 1918 some thirty-eight thousand customers received no bill at all. The beleaguered accounting department responded by walking off the job and dumping their account books into manholes scattered around the city. By April 1919 Peoples Gas had no cash or unpledged assets, lacked credit, and needed $1 million to avoid being sold for unpaid taxes.[35]

Well might Insull have ducked this mess. Neither the company nor its troubles were his, and he had more than enough to do elsewhere. But he could not resist the challenge. As John Mitchell told him, "Sam, do you realize what a black eye it will be for Chicago if we let a utility this size go under?" It was a plea Insull could not ignore. In January 1919 he became president of Peoples Gas and sent four of his top aides to pave the way for his arrival. Charles Munroe and George Mitchell, Insull's private secretary, ruthlessly streamlined the organization, cutting everything except payrolls to boost the morale of a demoralized workforce. Insull and some top executives took hefty pay cuts. Bernard J. Mullaney, Insull's public relations man, somehow managed to convince the employees that all would be well with the company under Insull's direction. Fred Sargent looked into the physical and technical elements for ways to improve them. When Insull formally took command in March, he found a workforce eager to help him turn Peoples Gas around.[36]

Within weeks Insull worked his magic. The pressing, seemingly insoluble problem of taxes mysteriously vanished as the county treasurer lost the company's address. As Charles Munroe later said, "Mr. Insull spoke to some of his political friends." Insull launched a crusade to cut expenses without touching wages. By renegotiating unfavorable fuel and supply contracts he saved $750,000; streamlining procedures knocked off another $1 million, and a rate increase that finally materialized near the war's end added $2 million in income during 1919. For the first time in three years the company managed to break even, but it still hovered on the brink. When a rising tide of inflation threatened to sink the company in 1920, Insull and Munroe came up with a brilliant stroke.[37]

Chicago had many steel and other industrial plants that produced coke oven gas as a by-product and burned most of it off as waste. Insull sent Munroe to see whether there was enough wasted gas to buy, and whether the companies would sell it. Munroe returned with a contract from Wisconsin Steel Company for 8 million to 10 million cubic feet a day, about 12 percent of Peoples Gas's total output, for a price of 38.4 cents per thousand cubic feet, 20 cents below Peoples Gas's manufacturing cost. This contact alone saved Peoples Gas nearly $600,000 a year. Munroe then discovered that most mills closed on weekends but left their ovens running and burned the gas produced as waste. He made the rounds of several companies and landed contracts for this dump gas at an average price of fifteen cents per thousand cubic feet. These contracts gave Peoples enough gas to fill a quarter of its total needs while saving the company $2 million a year.[38]

Through these resourceful moves Insull put Peoples Gas in the black again, but he understood that the best solution was to build a state-of-the-art new gas plant that would yield gas cheaply along with by-products that could be sold. Insull took his plan to H. B. Rust, president of the Koppers Company, the one firm capable of building such a facility. They calculated that the plant would pay for itself; the problem was finding the $20 million to build it. Investors still would not touch any securities with Peoples Gas's name on them. Rust took the problem to his boss, Andrew W. Mellon, one of the richest men in America and a future secretary of the Treasury. Before deciding anything Mellon asked to meet Insull, who obliged by stopping in Pittsburgh on his way to New York. For three hours he and Insull chatted about this topic and that, everything but the subject at hand. The meeting convinced Mellon to back the project personally, and the new plant opened for production in October 1921.[39]

Shortly after the new plant opened, Peoples Gas resumed paying dividends, which reached a steady 8 percent in two years. Insull had turned the company around in spectacular fashion. In November 1921 he took his message to the gas industry. "We have all lost heavily by the war," he said in a speech to the American Gas Association. Inflation had driven costs up faster than rates, and the public clamored for still lower rates. The old days would not return, and in fact the gas industry was to blame for "educating the public on uneconomic lines with relation to rates." Both the gas and streetcar companies erred badly in taking the position of charging the same price per unit of product regardless of the amount of capital needed to supply it or the conditions under which it was used. "Until the public utility men get the idea out of their heads that they can sell service irrespective of the amount of capital required to take care of that service," he scolded, "you are not going to have really fair rates."[40]

Insull then hammered home his familiar themes. Drawing a parallel to the electric utility industry, he noted that the war had taught them the primacy of wholesale to retail business. The former comprised about 75 percent of the business for electric utilities, while the gas companies supplied about 75 percent of their product to retail customers, "the very smallest and least profitable portion of the business." Streetcar companies did even worse; their entire business went to retail customers. In all three industries the biggest item of expense, apart from interest, was labor costs, which hurt the streetcar and gas businesses much more than the electric utilities. "The best thing we can do," Insull insisted, "is to develop our business upon lines that will give us the largest amount of income from the least expenditure of labor."

Nor should the industry look to rate increases for relief. "We are past the day when we can get increased rates," Insull declared. "My belief is that, instead of using every legal means to stop the decrease in rates, our general attitude ought to be sympathetic toward a decrease in rates, provided that we can stand the gaff." The solution was to lower the unit cost of production, which required, among other things, "a very decided improvement in engineering . . . I would suggest that the gas industry needs a first-class scrap pile." It needed to junk not only outmoded equipment but also obsolete ideas. "I think the electric men have beaten you in the race of brains," he said bluntly. Insull offered suggestions on how to improve output and held the transformation of Peoples Gas up as a model for the process. "We have tried to perform every operation mechanically that can be performed mechanically at a saving," he added, "and we have not hesitated at all to take power from our neighbor when we thought it was cheaper to take it from the electric power company than it was to manufacture it ourselves."

Public relations also mattered greatly. "It is more important beyond even the raising of capital for the business," Insull emphasized, "because there is no use in raising the capital . . . unless you establish such connection with the public as to insure the safety of that capital." Every employee from the top executive to the delivery boy should be drilled in the fundamentals of the company and carry to the community its message "that recognition of the duty we owe to the public, which we serve in every household, stands in our minds alongside of our duty to the owners of the properties we manage."

This was Insull at his best, preaching what he practiced to the heathen, trying to drag them into the new era of energy use as outlined by his gospel of consumption and the massing of production. The nation stood on the threshold of a new era of electricity, one that saw it permeate American life to such an extent as to become a prime necessity. The electric utility industry, once the lowly stepchild of the manufacturers, grew at a pace that soon dwarfed them. Having shown the way for this remarkable expansion of energy, Insull proceeded to erect a gigantic empire that would have been inconceivable only a few years earlier.

"THE HOME OF the future," wrote advertising maven Bruce Barton with his usual panache, "will lay all of its tiresome, routine burdens on the shoulders of electrical machines, freeing mothers for their real work, which is motherhood. The mothers of the future will live to a good old age and keep their youth and beauty to the end."[41]

As Insull said in his speech, mechanization held the key to reaching this promised land. Between 1914 and 1929 manufacturers added 350 percent more machine power than manpower. The use of high-speed machines and assembly lines required fewer, more highly skilled workers and resulted in a fourfold increase in labor productivity. Intensive use of heat and refrigeration technology gave Americans a wider, more nutritious variety of foods. Mindful of a

1909 study that showed ice cream makers having one of the highest load factors, Insull pursued the refrigeration market relentlessly. Refrigeration worked
beautifully with lighting; it peaked during the summer as the use of lights
reached seasonal lows. By offering more special rates, Insull began rounding
up contracts from ice-making firms that previously relied on natural cutting
or steam power. By 1917 he had captured 20 percent of the ice-making market;
in five years the figure ballooned to 70 percent. Meat packers and cold-storage
companies also switched from isolated to central stations.[42]

Using electricity to heat and cool food to extreme temperatures proved instrumental in revolutionizing the nation's food supply and eating habits. It
made possible the emergence of the self-serve supermarket, which first appeared in 1916. Refrigerators remained a luxury well into the 1920s; only
twenty thousand existed in the entire country in 1923. An industry study conceded that the appliance cost $465 and used about $50 in electricity a year. Insull went after the appliances market vigorously. His Electric Shops offered a
variety of appliances, including irons ($7.50 to $10), the most popular single
item, vacuum cleaners ($50), coffee percolators ($25), and washing machines
($165 to $190). Consumers could pay with cash or use the newfangled install-
ment plan. By 1926, according to a survey, 85 percent of central station customers had irons, nearly 71 percent had vacuum cleaners, 42 percent had
washing machines, and 31 percent had toasters.[43]

During these same years the automobile, led by Henry Ford's fabulous
Model T, transformed American life as well as the landscape. Insull and his
colleagues played a key role in that process by supplying the electricity that enabled Ford to revolutionize production with his assembly line. But Insull had
more direct ambitions in the field. Between 1917 and 1922 Insull invested heavily in garages to service electric-powered cars and backed the effort with an intensive publicity campaign. He also promoted facilities for an electric truck,
the Walker. Edison liked the idea, especially if Insull could push the costs down
enough to compete with gasoline-powered vehicles. "You have a very good
business proposition," he assured Insull. However, the campaign flopped because Chicagoans, like most Americans, preferred the internal combustion engine to battery power.[44]

Another, more modest electrical appliance became the hit consumer item
of the 1920s. Radios hit the mass market in 1923 and proliferated with the creation of network broadcasting and the production of electric rather than
battery-operated versions. By 1930 some 60 percent of Chicago families owned
a radio; the number reached 94 percent five years later. Other electric appliances also made their debut during the decade, including the toaster, heater,
waffle iron, heating pad, table stove, curling iron, and a host of others. None of

them made a serious impact at first, but their availability fed the desire for a home with electrical power capable of operating the new conveniences. A pattern commenced that came to define the consumer goods economy: Today's luxury items became tomorrow's necessities.[45]

Electric power itself exemplified that pattern. In 1923 Insull opened a model electric house in a suburb next to Oak Park in the territory served by PSCNI. During its sixteen-day life the house welcomed forty-six thousand visitors who came to see not the world of tomorrow but the possibilities of today. Although Insull advertised heavily, he did not need to convince people that electricity made life easier and more convenient. His creation of a six-thousand-square-mile regional power network did much to fuel the flight to new suburbs around the city even before the war. Between 1919 and 1929 the army of suburban families with electricity soared from 97,000 to 275,000. In the city itself, studies showed that improved lighting boosted productivity in offices, as did other labor-saving devices. As office buildings soared higher, they required more electric power along with outside lighting to dazzle the public with their majesty. Few displayed themselves more effectively than the new Wrigley Building in Chicago.[46]

The growing ring of new suburbs around the city not only welcomed electricity but devoured ever more of it. Residents and merchants alike sharply increased their consumption between 1910 and 1930. The flight to the suburbs began in earnest for both groups, posing a threat to downtown businesses, theaters, and other entertainments. By 1926 Marshall Field & Company, the bulwark department store of the Loop, had already established branch stores in three upscale suburbs. Despite Insull's efforts, however, most of rural America remained outside the spreading web of electric power lines. Although Insull's Lake County experiment pointed the way to extending power to small rural villages, few utility companies bothered to risk the investment. As late as 1941 seven out of ten rural inhabitants still lacked electricity. The lack of electric power, and all the amenities it brought, widened the already yawning chasm between urban and rural Americans.[47]

Insull alone straddled that chasm in terms of electric power. In 1923 he extended his reach into northern Indiana, where power development had taken place in a helter-skelter fashion. Both PSCNI and Peoples Gas owned some properties there, as did a company headed by Randall Morgan, a Philadelphia financier. Insull met with Morgan and agreed to form a new holding company to put all their Indiana companies together, fold them into the new Midland Utilities Company, and rationalize their operation. The plan had a special bonus for Insull. He became president but made Junior his assistant, his son's first executive position. Heretofore, Junior had been shuffled through a series of low-level posts as a means of learning the ropes. "It is a great opportunity for you if you are willing to work hard and to take the criticism which alone can mature your judgment," he wrote Junior, who was in Europe at the time. "It is the best opportunity I know of in our business & gives you a better chance than I had when I became Edison's Private Secretary."[48]

The old man wanted nothing more than to have his son serving at his side, yet he agonized over the decision. "Sometimes I wonder if it is the best thing to do," he added. "I am impatient for your judgment to mature and sometimes I wonder whether you can stand the criticism from me which you would expect to take as a matter of course from anyone else." Junior came aboard and did just fine. He had shown his mettle on an earlier trip to Europe with his mother, when he visited the Pirelli Manufacturing Company in Milan and noticed a new type of underground cable that carried an astonishing amount of voltage. Ordinary cable was wrapped in oil-soaked paper and sealed in a lead sheath; it was rated at 33,000 volts but could handle only about 23,000. The Pirelli cable featured a hole in its center through which oil flowed, making it more flexible and impervious to heat and cold. It carried 80,000 volts, and Pirelli claimed it could deliver a version that could handle 138,000 volts.[49]

Junior notified his father of the discovery and sent along a sample. The General Electric engineers assured Insull that it could not possibly work, but Insull persisted with tests that proved the cable's worth. Pirelli licensed GE to manufacture the cable, and within a short time it became the standard in all major American cities. The result was yet another major innovation in electrical supply pioneered by Insull, this one originating with his son. Insull fortified Junior in his new post by surrounding him with able consultants, who dominated the work. "I was nothing but a messenger boy . . . ," Junior recalled. "They were working for me, and they could chase me out of the place." The arrangement lasted about a year until Junior impressed his father by devising a solution to a complicated organizational problem that had stumped Sam and his lawyers. From that moment Insull began giving his son more leeway until he became a full-fledged member of the family executive corps along with brother Martin.[50]

Sam had trained Martin to be his right hand and trusted him completely. Although a bright, talented man, Martin differed from his brother in several key respects. Like their father, he tended to be a man of ideas rather than practical judgment, a theoretician who often missed the obvious nuts and bolts of a problem. As Forrest McDonald put it, "Though he had a Cornell degree in electrical engineering and had brilliant theoretical understanding of the subject, his practical knowledge extended little beyond replacing a burned-out light bulb." Like Junior, he also had the misfortune of dwelling in Sam's enormous shadow. He admired Sam and was deeply devoted to him, but he also wanted to prove his own abilities through independent action. This urge sometimes led him to rash and unwise choices.[51]

Martin had the job of overseeing Middle West Utilities, which grew to enormous proportions during the 1920s. Eventually its lines extended into thirty-two states and serviced five thousand communities. Its expansion mirrored the explosive growth of central stations in general, a phenomenon that transformed utility bonds from a sideshow into a major market that approached $1 billion a year by 1926. Bankers who had previously ignored them gradually discovered that they were one of the most lucrative securities to market. Insull never realized that financing a utility was often more profitable than running it. Harold Stuart grasped this fact early and dominated the market until the light dawned on New York bankers as to what they were missing. Bond houses wedged their way into the field by creating holding companies and investment trusts to control utility firms and capture their financing.[52]

Inevitably this activity caught the attention of the most powerful banking house of all, J. P. Morgan & Company. Brushing aside its longtime aversion to stock operations, Morgan in January 1929 swept together a bundle of major electric utility firms in a new company, United Corporation, which controlled 38 percent of the electric power produced in twelve eastern states and 20 percent of the nation's total supply. "The year 1929," predicted *Time* magazine, "may well be famed as the year in which the House of Morgan became also a Power House." The bankers entered the field as investors, not operators, but their influence continued to spread. No other utility conglomeration came close to rivaling the new Morgan entity except for Insull's empire.[53]

The new financial game involved leveraging and pyramiding properties to obtain maximum control with minimum investment. A new holding company bought utility or other companies, paid for them with bonds, and used the dividends from the power companies to pay their bondholders. Insull had never cared much for the financial game. He was at heart a builder, but he could not ignore what was going on around him. The stock market was advancing almost as fast as the utility business was expanding. During 1928 Martin had led Middle West Utilities into two controversial acquisitions, both of them holding companies of utilities serving fourteen eastern states. The companies were sound but heavily leveraged. In examining the figures Insull said glumly, "It will take us ten years to tear down this pyramid and make this a reasonable investment." But he signed the papers, apparently unconcerned that this new addition to the empire took him for the first time into the heart of the territory the House of Morgan had begun to organize and claimed as its own.[54]

"These New York fellows were jealous of their prerogatives," declared Harold Stuart's brother later, "and if you wanted to get along you had to be deferential of them and keep your opinions to yourself. Mr. Insull wouldn't, and that made bad blood between them. Real bad blood."[55]

During the summer of 1928 Insull became aware that someone was buying large blocks of common stock in all four of his major operating companies. The purchaser turned out to be not Morgan people but Cyrus Eaton, a Cleveland capitalist who had cleverly put together a minor empire of his own in utilities before moving into the financial big leagues. Insull had friendly but not close relations with Eaton. That summer they happened to be on the same liner returning from Europe. Insull already knew of Eaton's purchases and was surprised that "when we met on the steamer he simply passed the time of day and made no reference whatsoever to his accumulation of stock in the operating companies." Although Eaton made no threats or overtures, he had covered his buying cleverly. For the first time Insull had to consider the possibility that someone might be plotting to wrest control of his companies from him.[56]

Times were changing, events were moving fast, and Insull had to move with them. Money was plentiful; for once the bankers were eager to lend him as much as he wanted or more. The growth of his empire had made Insull a business celebrity, one of those giants of whom more was always expected. He had risen to the top, which made him a prime target for raiders. Eaton as yet posed no direct threat, but even the hint of a threat stirred Insull to action. He did not like the new financial game of pyramiding and had long spoken out against it, yet in the face of this perceived threat he created the very kind of structure he disliked, an investment trust. He did it not to take advantage of the feeding frenzy on Wall Street but to defend the empire. In December 1928, after much deliberation, he formed Insull Utility Investments (IUI) and exchanged for its securities all his own and his family's holdings in Commonwealth Edison, Peoples Gas, PSCNI, and Middle West Utilities.[57]

Later he even reluctantly sold it the family firm of Insull, Son & Company. Insull did so only because "I did not want anyone to get the impression that I was trying personally, either for myself or for members of my family, to make money through the use of a separate concern." The new corporation thrust the Insulls deeper into the shark-infested waters of Wall Street on the eve of its most spectacular rise. Insull agonized over what price to set for the new stock and finally settled on $12 a share. On January 17, its first day of trading, IUI common opened at 25, closed at 30, and never looked back. By spring it had surged past 80. During the torrid summer of 1929 it soared to 150 as Commonwealth Edison rose from 202 in January to 450 in August and Middle West from 169 to 529. In the fifty-day stretch ending August 23, Insull securities appreciated at the astounding rate of $7,000 a minute around the clock until their total rise reached $500 million.[58]

This mad summer frenzy of the stock market left Insull dazed. Only a few years earlier he had reckoned his net worth at about $5 million, shockingly low for a man in his position, but he had never chased money hard. That summer his holdings totaled $150 million on paper. "My God," he joked to Stuart, "a hundred and fifty million dollars! Do you know what I'm going to do? I'm going to buy an ocean liner." But the mad rush upward disturbed rather than gratified Insull because he sensed it was a bubble that had to burst. "It is a relief to me that the extreme upward trend should have been broken," he wrote after a sharp market retreat early in August. "I do not like spectacular advances. I would rather see things take a more orderly course." The surge created a thorny problem for him: The inflated prices of Insull company stocks made it difficult for IUI to buy them, which complicated his goal of ensuring control of all the operating companies. Moreover, a technical flaw in the creation of IUI still left the door open for the possible wresting of control from the Insulls.[59]

To remedy these problems Insull and Stuart decided to create yet another giant holding company, the Corporation Securities Company of Chicago (known as Corp), which would be controlled by a voting trust to ensure no outside raider could get hold of it. Its assets consisted of IUI securities given it by the Insulls and by Halsey, Stuart & Company in exchange for Corp securities, which meant in effect that the two securities companies owned a large part of each other. At the same time Insull used the surging bull market to undertake a huge refinancing of Middle West Utilities. Insull wanted the stock held on the widest basis possible, but the enormous surge in price defeated that goal. To reclaim it, Insull split the stock ten for one, retired all outstanding notes and floating debt, called in the outstanding 8 percent preferred stock and issued a lesser amount of 5 percent preferred, and changed the new common stock from a cash to stock dividend basis. When the smoke cleared, Middle West emerged with no debt and hardly any fixed charges.[60]

Corp made its debut on October 5, 1929, eighteen days before the opening of the horror show that became known as the Great Crash. Not even the market's fall shook Insull's strength and standing. Amid the debris of the next few weeks he rescued employees burned by margin calls on their stocks, presided over the opening of Chicago's magnificent new Civic Opera House, and participated in the construction of a new natural gas pipeline from the Texas panhandle to Chicago. As the new year opened, he helped Chicago avoid bankruptcy and bailed out its floundering transportation system. During the dark interlude between the crash and the onset of depression in October 1930, Insull remained a tower of strength on a landscape of crumbling icons. That year, in response to President Herbert Hoover's plea that businesses not cut their spending, Insull's firms poured $197 million into capital projects.[61]

Insull seemed impervious even to the depression, but appearances were deceiving. Important decisions loomed before him, and in nearly every case he made the wrong one. The man whose instincts had guided him so brilliantly in the creation of an empire of energy found himself awash in a sea of troubles. Like many an emperor of old, he discovered that the defense of an empire was far more difficult than the creation of one. In only two years he descended from the pinnacle of the business world to the depths of disgrace that was as complete as it was unwarranted.

APRIL 1934. On the deck of a small steamer called the *Exilona* sat an utterly weary and defeated Sam Insull in the company of Burton Berry, the third secretary of the American Embassy at Istanbul, who was accompanying the old man back to America to stand trial. For two long years Insull had fought extradition only to lose in the end. During the twenty-five-day trip home Insull roused himself to regale the few newsmen and handful of passengers aboard the ship with stories of his career, but much of the time he sat gazing mournfully out to sea. Berry saw him collapse occasionally into tears and heard him talk of suicide. "I have known what it is to have $150,000,000 of property . . . ," Insull said bitterly. "Now I have less than fifty dollars in the world and there is nobody ready to loan me money . . . Throughout my life I have aided thousands of people and contributed to hundreds of funds for the widows of bankrupt statesmen but in my hour of need only one man has come forward with an offer of financial assistance and that man is one who I have not spoken to for fifteen years."[62]

How quickly his life had come apart, and with such brutal finality! Like the interlocking pieces of some perversely complex puzzle, a series of missteps and misfortunes had come together in a pattern of ruin. Foremost among them was his decision to take on a formidable load of debt in the teeth of a depression. It was, he reasoned, the only way to get things done. The market crash had soured the public on stocks and liberated large sums of money from the call market for loans at low rates. The operating companies each increased its funded debt by about 10 percent; IUI issued $60 million in debentures and Corp $30 million in notes to complete their programs of Insull company stock purchases. Worst of all, while Sam was absent on a trip to Europe, Martin decided foolishly to have Middle West issue $50 million in gold notes for activities outside its utilities operation. In one ill-considered stroke he undid the benefits of Sam's refinancing work on Middle West in 1929.[63]

Seasoned financiers scratched their heads at this outpouring of securities while the economy grew more anemic each day. Yet sales went so well that Halsey, Stuart had to ration the offerings among buyers. Early in 1930, however, Cyrus Eaton reentered the picture. Having increased his holdings to eighty-five thousand shares of Commonwealth Edison, sixty thousand of Peoples Gas, and thirteen thousand of PSCNI, he offered to sell the whole lot to Insull for $400 a share or a total of $63.2 million, nearly $12 million above current market prices. If Insull declined, Eaton hinted slyly, the securities would surely find a ready market with New York bankers. Martin liked the idea but worried about the huge cash requirement; Harold Stuart urged Insull to reject the deal. Insull offered a lower price, Eaton refused, and the matter seemed closed.[64]

However, Donald R. McLennan, an insurance man and director of both Commonwealth Edison and Chicago's largest bank, reopened it. Eager to win Eaton's sizable insurance business, he urged Insull four times in the spring of 1930 to buy Eaton's stock, assuring him that his bank stood solidly behind him in the deal. Four times Insull said no; then McLennan came a fifth time, and Insull finally agreed to make the purchase at $350 a share even though he had not consulted Stuart about how to finance the deal. The price tag came to $56 million, $48 million of it in cash and the rest in IUI and Corp stock. All the cash would have to come from bank loans. Once the deal had been made, however, the Chicago bank found it could not handle the entire amount. Insull promptly borrowed $5 million from General Electric but had no time to arrange loans in London. He had no choice but to get money from New York. There followed a chain of complex transactions, at the end of which IUI and Corp owned Eaton's stock and increased their floating debt by $48 million. Of that amount New York banks held $20 million, a seemingly small piece of a $3 billion empire.[65]

However, the stock market continued its downward plunge during late 1930 and 1931. In his biography of Insull, Forrest McDonald makes a persuasive argument that the New York bankers orchestrated an intensive bear campaign to break the Insull stocks. As their market price fell, their value as collateral for the loans shrank, forcing Insull to give the banks ever more stock to cover the difference. The utility companies themselves actually performed well in a dismal economy, but the holding company stocks continued to decline. Still Insull did not fully grasp the forces arrayed against him. Late in August, after going over his account books with Junior, he said, "Well, I never owed so much money in all my life put together; but then, I never had so much collateral." But the available collateral continued to dwindle; by December it had run out. All the securities owned by IUI and Corp lay in the bankers' hands, and loans were coming due. The nightmare had begun.[66]

In mid-December Insull launched a frantic campaign to get the banks to extend their loans or at least accept standstill agreements. He borrowed as much as he could in his own name, shuffled money among the companies, and tried to cut expenses everywhere. To his surprise the bankers proved far less accommodating than they had been in better times. "Bankers," he once snorted, "will lend you umbrellas only when it doesn't look like rain." Now it was pouring, and the bankers wanted their money. The Chicago banks, struggling with their own difficulties, deferred leadership to the New York bankers, who relished the opportunity to gain control of the Insull empire.[67]

At the annual meetings of IUI and Corp on February 16, 1932, Insull defended his actions, admitted his mistakes, fielded questions frankly, and won votes of confidence from the stockholders. However, in March the bankers insisted on giving Arthur Andersen of the accounting firm the power to approve all "extraordinary expenditures." Insull understood that "this practically placed the bankers in control of the three large operating companies, the Middle West Utilities Company, and the two investment companies [IUI and Corp] which heretofore I personally had always occupied." Andersen was then instructed to study the figures of all the Insull companies and report any irregularities. The New York bankers made these moves even though technically they were not creditors but mere holders of the collateral stock.[68]

Using a different accounting system from that used by Insull, Andersen predictably found a series of corporate indiscretions or irregularities, most of them involving loans made between companies during Insull's desperate efforts late in 1931. Rumors leaked out that the Insulls were not only in trouble but had been embezzling through their thick web of interrelated companies. The showdown came during a series of meetings in early April. Insull was trying desperately to raise enough cash to meet a $10 million Middle West note due April 1. At the second of two meetings in the office of Owen D. Young, head of GE, the bankers went into a private session with Young that pointedly excluded the Insulls and Harold Stuart. After about an hour, Young emerged from the room, took time to fill and light his pipe, and then said the decision had been made not to put any more money into Middle West. "Young," asked Insull, "does this mean receivership?" Young replied that it looked that way.[69]

"I think Anderson [sic] has done the three properties great injury," Insull told his son two months later. By then Insull's world had collapsed. The bankers already had IUI, Corp, and Middle West. Early in June the directors of the three operating companies demanded Insull's resignation from all of them and their subsidiaries, more than sixty companies in all. On June 6 Insull issued his resignation and said to the reporters, "Well, gentlemen, here I am, after forty years a man without a job." Eight days later he boarded a train to Montreal, went to Quebec, and then left by ship for Cherbourg, France.[70]

In his absence John Swanson, the state's attorney for Cook County and long an admirer of Insull, nevertheless launched a high-profile investigation into the Insull companies, which made a choice political target during a campaign season. Businessmen had basked in the glory of the robust national economy during the 1920s and cheerfully taken credit for it. The onset of depression with its disheartening chain of bank and business failures soon turned the heroes of good times into pariahs. Few of their number were more conspicuous in the public eye or made a more inviting scapegoat than Insull. In September Franklin D. Roosevelt, stopping in Chicago on his return from a western trip, blasted Insull personally and was soon followed by other Democratic candidates. In October a grand jury indicted Sam and Martin Insull on charges of embezzlement and larceny. Four months later the federal government followed suit on grounds of mail fraud.[71]

Insull and his wife went into exile, planning to stay in Europe until the political storms subsided. Junior remained in Chicago to conduct the fight against efforts to extradite his father. Money soon became a problem. The operating companies had given Insull an annual pension of $50,000 upon his resignation. In October 1932, however, the pension was cut to $21,500, and in November 1933 it was canceled entirely. His own fortune had been swallowed by the collapse, leaving him dependent on his son for support. While Junior scrambled to mend fences in Chicago, Insull fidgeted in Paris, eager for whatever morsels of news he could get. He saw few people he knew and chafed at his imposed life of leisure. "I miss my business and the active life I led," he wrote Junior mournfully, "but above all I miss the association with you." He could not even bring himself to visit a seaside resort. "It would give me too much time to think," he explained, "and thinking gives me the blues as I feel that the finish of my career has brought disaster not only to myself but to so many others . . . I feel it very much indeed."[72]

Above all, he hated the idleness. "What I am suffering from," he confided to his son, "is a lack of something to occupy me and lack of someone to talk to. Sometimes I will go for several days and not speak to a soul except your mother. It is such a strange reversal of my whole life. I do not know whether I will get used to it."[73]

While his father stewed in exile, Junior's life turned into an unrelenting ordeal. He remained assistant to James Simpson, who replaced Insull as chairman of the three operating companies, but his salary got stretched thin supporting his family, his father, and Uncle Martin along with waging the legal battles on behalf of his father. The federal government indicted him twice along with several others, and he had to fend off several civil suits. Worst of all, his wife, Adelaide, died suddenly from an illness, leaving him with a two-year-old son. Through this nightmare Junior showed unflinching courage; he was his father's son in life as well as business. However much he had to endure, Junior also had the task of trying to boost his father's flagging spirits. "I wonder if my day is over," Sam mused in one letter, "and my ideas are permanently in the discard."[74]

In October 1932, as the extradition fight heated up, Insull left Paris and moved to Athens. A month later the U.S. Senate ratified an extradition treaty with Greece, and the fight began in the Greek courts. In May 1933 the federal government added a new charge, violation of the Bankruptcy Act, but still made no headway. Not until the year's end did American pressure force the Greeks to expel Insull. Ordered to leave by March 15, 1934, Insull borrowed enough money to charter a Greek boat and spent two weeks in the Mediterranean pondering his next move. Congress then passed a bill ordering his arrest in any country with a treaty giving the United States extraterritorial rights. Turkey complied by seizing Insull aboard his ship, detaining him for a trial of sorts, and then putting him aboard the *Exilona* for the journey home.[75]

Although the trip home was a welcome respite, it could not slake Insull's growing despair. Junior met him and handed him a hundred dollars to defray his expenses on the crossing; Insull remarked that it was the most money he had seen since leaving Athens. Once back in Chicago, he settled into the Seneca Hotel to await trial. Floyd E. Thompson, the attorney representing both Insull and his son, mounted a brilliant defense. The federal trial opened October 2, 1934, and lasted until November 24. Using Insull's own spellbinding testimony as a cornerstone, the defense utterly dismantled the prosecution's case and won acquittal on all charges. In March 1935 Sam and Martin Insull were acquitted of the state charges after a brief trial; three months later Insull, Junior, and Harold Stuart were tried on the federal Bankruptcy Act charge. After listening to the prosecution, the judge instructed the jury to deliver a verdict of not guilty.[76]

Insull left the last trial a free man, exonerated of all charges, but the barrage of criticism, innuendo, and hostility from politicians, magazine writers, journalists, and others had destroyed his reputation. Those who had defended him showered him with congratulations. "Be assured," wrote the chairman of the Friends of Insull in America, "that no matter what the outcome is, we will still continue to believe in you." Floyd Thompson declared it "a great privilege to represent you." Congratulations also came from a fellow pioneer. "The report that the vicious attacks on you are to be discontinued was most welcome news . . . ," wrote Nikola Tesla. "You passed through a fierce crucible and came out as pure as gold." At the same time he tried to interest Insull in two of his latest inventions, both of them "not only revolutionary but of incalculable importance to the world." Insull accepted the sentiments but declined the invitation to invest.[77]

The operating companies restored Insull's pension, but they could not restore his image. To many people he remained not a pioneer in the field of electric power but an emblem of corporate ruthlessness and dishonesty. Despite his vindication, the ordeal drained most of the life out of Insull. Like Westinghouse and Edison before him, he had been driven from his company by the bankers. Edison he had always adored, yet he also greatly admired Westinghouse, whom he called "one of the great figures in the electrical manufacturing business." Long ago, just after Insull had gone to Schenectady to run the Edison Machine Works, Westinghouse paid him a visit and assured him that, despite their rivalry, he would maintain friendly and cooperative relations with him. It would have pleased Insull immensely to be ranked with these two great men as one of the founding fathers of the industry.[78]

That honor came to him, but not quickly. Since Gladys refused to live in America, Insull returned to Paris. He made occasional trips to Chicago to visit Junior but did not make a public appearance until February 1937, when he served as guest speaker for an Edison dinner at the Sherman Hotel. He found some peace during his last years. "I am really enjoying life over here," he wrote. "It is an idle kind of existence." He relished the fact that "the attitude of people towards me is so different. I am a 'has been' in Chicago but I am 'someone' here . . . How different to the treatment I received from my old institutions— the ones I created in Chicago."[79]

"For his fifty-three years of labor to make electric power universally cheap and abundant," wrote McDonald mordantly, "Insull had his reward from a grateful people: He was allowed to die outside prison." Death claimed him in July 1938 with a heart attack while he stood on the platform of a Paris metro station. His operating companies endured and prospered; none of them ever went into bankruptcy or cost their investors more than a fraction of 1 percent in losses. Even more important, the Insull gospel of cheap, abundant electricity flourished after 1945 as more people received power and ever more uses were found for it.[80]

The age of electricity had arrived in earnest, and no man had done more to make it so than Samuel Insull. What Edison, Tesla, and Westinghouse had done for the technology of electricity, Insull did for its distribution. No man did more to give the American people access to a cheap and abundant supply of electricity—the key power source for every major technology of the twentieth century. The two countries that created the most admired government-owned systems, England and Canada, came to Insull for lessons on how to do it. No amount of tarnish could dim the luster of that achievement.

EPILOGUE

A SHOW OF
POSSIBILITIES:
NEW YORK 1939

Today we are in a position where we must master our inventions . . . Mere mechanical progress is no longer an adequate or practical theme for a world's fair. *Instead we must demonstrate an* American Way of Living. *We must tell the story of the relationships between objects in their everyday use—how they may be used and when purposefully used how they may help us.*

—MICHAEL HARE[1]

ONCE HE FIGURED OUT HOW IT WORKED, the subway took Ned directly to the fairgrounds in Queens. He marveled at the speed with which it hurtled beneath the city, enveloped in a blackness relieved only by occasional tunnel lights and stations. Ned had never been to New York before, had not been any closer to the city than his visit to Philadelphia for the Centennial Exposition when he was a boy of nine. Nor had he been any farther west than Denver, which he chose for one of his rare vacations many years ago. Except for these occasional forays, he had spent his entire seventy-two years in the heartland, a midwesterner to the core. The decision to come to the nation's biggest city was a bold departure from his past, except for its involving a visit to a world's fair. That was a lure he could not resist. The first one in 1876 had opened his eyes to the larger world outside his family's farm; Chicago in 1893 had introduced him to wonders he had not known even existed. What marvelous surprises did New York hold in store for him?

Although still spry for his age, Ned was no longer the wide-eyed boy of Philadelphia or the confident, eager young man of Chicago. Life's sorrows had inevitably begun to crowd in on him. His wife had died four years earlier, and his three children were all grown and married with families of their own. Although they kept in touch and got together on special occasions, not one of them lived within a day's journey of his house. They were all struggling to make a go of their lives, plodding uphill against the weight of the depression that had hung like a millstone over the country for nearly a decade. Ned was fairly comfortable on his modest income, thanks to his even more modest needs. Going to the World's Fair in New York amounted to a huge splurge for him, but he could not resist the temptation. Three world's fairs in one lifetime! That was a wonder in itself.

Now he was alone, and the trip to New York represented a kind of last fling in life. He had never expected to see another world's fair. The world was in such bad shape that he wondered whether it would even be around in a few years. War clouds hung over Europe and Asia alike; blood had already been spilled in Spain and China. At home the New Deal seemed all but dead, a lifeless caricature of its former self, and the president the lamest of lame ducks. He had only another twenty or so months to serve, and nobody knew who would take his place. A vociferous debate had broken out over what to do about the deteriorating world scene. One set of loud voices demanded that America stay out of the world arena and avoid repeating the disastrous course that led the nation into the Great War. Another group insisted just as adamantly that America could not play ostrich in times so perilous and that she must add her considerable weight to the stand against enemies of freedom and democracy.

Ned didn't know what to think about such matters, and he looked forward to the trip to New York as an escape from the pall of gloom that seemed to be gathering about the nation both at home and abroad. He watched eagerly as the subway left the underground and became a surface railway train. Once in Queens, it moved through the Jamaica Yard over the newly built extension to the temporary station at the edge of the fairgrounds. The entire ride from downtown had cost him only a nickel. Entrance to the fair cost him another seventy-five cents, and he had brought extra cash with him, having heard that some of the exhibits and concessions charged admission as well.[2]

Once through the gate, Ned walked past the Star Pylon, which represented the force of electricity, and headed down the Court of Communications to see the most imposing sight of the fair: the 610-foot tall Trylon and the 180-foot wide Perisphere. Sitting atop a thousand piles driven deep into the ground and connected by a graceful bridge, they looked like the discarded toys of some giant. Ned rode partway up the Trylon on what was billed as the world's highest escalator and joined the long line on the bridge waiting to get into the Perisphere. He had read about the show inside, called Democracity, but still did not know what to expect. When his turn finally came, he plunked down his quarter,

stepped inside onto one of the two rings revolving around the exhibit in opposite directions, and gazed down at a sight that amazed him.[3]

Democracity was an enormous diorama depicting eleven thousand square
miles of a future city and suburbs with a population of 1.5 million. Noted news
reporter H. V. Kaltenborn told its story in a six-minute narration recorded to a
musical score by William Grant Still conducted by André Kostelanetz. At its
core lay Centerton, the business district, flanked by Millvilles (industry) and
Pleasantvilles (bedroom communities) interwoven with great stretches of
parks and agricultural land, all tied together with ultramodern expressways
and parkways. In those six enchanted minutes the lighting moved from dawn
to dusk. At the program's end, film projections of happy farmers and workers
escorted the visitors to the Helicline, a 950-foot spiral ramp that took them
back to ground level.[4]

"Building the World of Tomorrow." That was the theme of the fair, and Ned
had to agree. Nothing he saw around him looked much like the world of today.
Behind the Theme Center, as the Trylon/Perisphere area was called, lay the
Court of Power. Ned strolled in that direction, trying to recall the sights that
had most impressed him at the earlier fairs. Scanning the guide he had bought,
he noticed that this fair had no huge buildings like Machinery Hall or Main
Hall. The buildings on the vast grounds lacked the beauty and symmetry of
Chicago in 1893, yet there was an impressiveness to the stark lines of many of
the structures. Nearly all of them hewed to the futuristic theme, and many if
not most of them represented corporations rather than nations or sectors of
industry. Instead of an electricity building housing all the companies, he found
a General Electric Building, a Westinghouse Building, and a Consolidated Edison Building, along with one featuring electrical utilities and another a variety
of electrical products.[5]

Electricity was no longer a novelty, and here it sprang up everywhere in a
myriad of forms. Ned marveled at the variety of futuristic light standards that
utilized clusters of strange, rod-shaped units called fluorescent tubes. A long
row of them stood like sentinels alongside the Court of Power; at night they
glowed a soft but cold white light unlike any Ned had ever seen. The front of the
Consolidated Edison Building glistened with the Water Ballet, a beautiful fountain display designed by Alexander Calder that changed constantly. Ned stepped
inside to see the fabulous City of Light, the world's largest diorama, outdoing
even Democracity. Stretching as long as a city block and standing three stories
high, it was a scale replica of the New York skyline complete with working factories, subways and elevators that moved, and power surging through underground mains. The freestanding Empire State Building stood twenty-two feet
high. Amid brightly colored, shifting lights, the skyline passed through a

twenty-four-hour cycle of time in twelve minutes and culminated in a sky-darkening thunderstorm. "This is the City of Light," intoned an unseen narrator, "where night never comes . . . a world of power at the motion of a hand."[6]

The Electric Utilities Building across the way included one eighty-five-foot wall that was a waterfall and spillway. The modernistic walkway at its edge took him inside to an intriguing exhibit that contrasted a typical street in 1892 with a present-day one to show the advance of electricity. The old-time street featured live performers acting out a day's routine. Ned felt a twinge of nostalgia as he surveyed the gaslights in the shops and homes casting their dim, yellowish pallor on the actors playing policemen, housewives, merchants, and a boy whose kite was tangled in the overhead wires. "Hell," muttered one spectator, "place I come from looks just like this today." He admitted to hailing from the Ozarks.[7]

From the Electric Utilities Building Ned crossed the Plaza of Light to the General Electric Building, where a huge stainless steel sculpture of a lightning bolt greeted him. Inside Steinmetz Hall he jumped when a controller hidden overhead unleashed a 10-million-volt bolt of lightning across a thirty-foot arc. Fascinated, Ned stayed to watch the bolt twice more before moving on to examine an X-ray of a swathed and encased mummy. His gaze drifted upward to a beautiful mural done by Rockwell Kent depicting electricity's role in man's progress from obscurity to enlightenment. While browsing among the GE products on display, Ned discovered a kitchen with talking appliances as well as a House of Magic, which mystified and excited its audience with a bundle of scientific tricks. "Not many in the packed audiences understood the significance of the tricks they saw performed with thyratrons and stroboscopes," observed *Business Week*. "But they came away thrilled, mystified, and soundly sold on the company."[8]

A few steps away stood the Westinghouse Building, its forecourt dominated by the Singing Tower of Light looming up from the center of a fountain. Its show included flashing lights that changed color, climaxing in a burst of fireworks, smoke, and water, all to a musical score. Above the curved entrance to the building large letters proclaimed Westinghouse to be THE NAME THAT MEANS EVERYTHING IN ELECTRICITY. Inside the pavilion Ned found an astonishing array of exhibits. He marveled at how different they were from the GE and Westinghouse wares in Chicago. In 1893 the vast majority of items on display involved electrical equipment and curiosities; here most of the exhibits highlighted products, especially home appliances, that made life easier and more convenient. Before him stood a circular platform with Elektro, a seven-foot robot, and his mechanical dog, Sparko. Ned stayed to watch the demonstration, at which Elektro not only talked and performed tasks but even smoked a cigarette. At a nearby display of appliances he enjoyed what Westinghouse called "a battle of the century" between "Mrs. Drudge," who washed a stack of dishes by hand, and "Mrs. Modern," who did them in an electric dishwasher. The kitchen of the future was a marvel, Ned agreed, but it had come too late for him to enjoy.[9]

The future was everywhere around him, or so the exhibits promised. He wandered across the Bridge of Wings to the Transportation Zone. The General Motors complex of four interconnected buildings used the theme "Highways and Horizons" for the largest individual exhibit at the fair. Ned joined the long line and waited patiently for the Futurama, which had become the biggest hit at the fair. Once through the imposing sloped entrance, he slipped into one of 552 paired seats fitted with individual speakers that carried a soothing narrative. Below him, covering thirty-six thousand square feet, lay a model of a highway system as imagined in a metropolis of 1960, with seven-lane expressways monitored by radio control towers that allowed speeds as high as a hundred miles an hour. As the chairs revolved around the exhibit, Ned gazed down on a city similar to that in the Democracity diorama, one free of pollution, unstained by slums, suffused with parks and civic enclaves, filled with model homes and farms through which traffic flowed easily and unobtrusively.[10]

As the program drew to a close, the scale abruptly enlarged to a projected typical intersection of 1960. Leaving the darkened room, Ned walked out into the sunlight and found himself looking down at a full-sized replica of the same intersection, complete with cars of 1939 but no people. The ride through Futurama, noted one impressed writer, "covers a third of a mile and lasts fifteen minutes. It should be twice as long." At Chicago the automobile had been a mere curiosity, but at this fair it held center stage. Ned moved next door to the elaborate Ford Building, which offered a half-mile spiral Road of Tomorrow that gave visitors, riding in sample cars, shifting views of the fairgrounds. Inside Ned marveled at a huge mural by Henry Billings and, most impressive of all, a hundred-foot revolving turntable that displayed eighty-seven animated groups depicting every step in the building of a car beginning with the extraction of ore from the ground.[11]

After lunch Ned walked over the Bridge of Wheels and headed to the AT&T Building on the Court of Communications. Inside he found an enormous wall map of the United States filled with bulbs that lit the route of a long-distance call when it was placed. Ned entered the lottery for making a free long-distance call anywhere in the country but did not win one. AT&T had another treat in store for its visitors in the form of VODER (Voice Operation Demonstrator), a synthesizer that mimicked human speech. After listening for a while, Ned moved across the street to the RCA Building. The gleaming white structure was shaped like a radio tube. Above it loomed a tall transmission tower.[12]

Here was something entirely new, Ned thought. Radio had no part in the earlier fairs because it had not yet been discovered. How quickly it had become an important part of everyday life!

And there was something still newer. Beyond a wall filled with a giant mural, Ned happened on a newfangled device called a television. He had read about its development but never imagined that he would actually see one. There were versions in the GE and Westinghouse buildings as well, but none so elaborate as the RCA version. Set in a handsome console, its picture was not seen directly but reflected on a mirror in the raised lid. RCA began broadcasting when the fair opened, relaying the signals from an antenna atop the Empire State Building over a fifty-mile range. At six hundred dollars the set was far too expensive for Ned to own one, and what was there to see on it anyway? It only reached the New York area with two hours of regular programming a week along with coverage of occasional special events. Still, it was a sight to behold.[13]

Leaving the building, Ned strolled to the nearby picnic grounds and found a bench to rest his aching legs. As always there was so much to see, but he had the leisure to take it all in. He had given himself three days to wander the grounds and could extend it to four if he felt like it. This fair opened at 9:00 A.M., and the exhibits stayed open until 10:00 P.M. The amusement area, clustered at the far eastern end around Fountain Lake, remained open until 2:00 A.M. Ned had already decided to stay at the fair all day and evening and take his dinner there. He had never seen New York City, but it would be there a long time. The fair was a one-time thing, and he wanted to take as much of it in as possible. There was no hurry to go back home. Nothing waited for him in that small drab midwestern town, certainly none of the enticements offered by the fair. He felt himself in the presence of the future, and it thrilled him. What did it matter that he would most likely not be around to see most of it?[14]

ECONOMIC DEPRESSIONS SEEMED to be the constant companion of world's fairs. The Centennial fair opened midway through what was then the nation's worst depression, and the Great White City saw an even more severe downturn shortly after its gates opened for business. Forty years later Chicago hosted another world's fair in the depths of what became the newest, deepest, and longest depression in American history. That economic trough was entering its ninth year when New York beckoned the world to its fantasy land at Flushing Meadows. In every case the fair offered struggling Americans a beacon of hope, a window of relief, and, most important of all, the promise of a better life.

Development of the fair's concept rested in the hands of an ad hoc committee. It produced a detailed plan for "The Fair of the Future," which, the committee emphasized, should "stress the vastly increased opposite and the developed mechanical means which the twentieth century has brought to the masses for better living . . . Mere mechanical progress is no longer an adequate or practical theme for a World's Fair, we must demonstrate that super-civilization that is based on the swift work of machines, not on the arduous toil of men."[15]

On this basis the Theme Committee discarded the approach of earlier fairs, which organized exhibits around such categories as science, art, manufacturing, and agriculture. Instead it divided the fair into seven major sectors: production and distribution, transportation, communications and business systems, food, medicine and public health, science and education, and community interests. Each one featured a focal exhibit to provide an overview of the sector's theme and tie together the social significance of its exhibits. The Perisphere contained a central theme exhibit designed to summarize the fair's overall message and integrate the seven focal exhibits. From this approach emerged the central message of the fair: The future gleamed with promise, and technology held the key to its realization.[16]

Given the miserable state of affairs at home and abroad, the future certainly held more appeal. The fair set out to display visions of that wonderful future, not only in exhibits but in a myriad of impressive murals, statues, and other forms of artwork created by many of the nation's best artists. Virtually every aspect of the fair exuded the theme that technology made life easier, more convenient, and more enjoyable for people in every walk of life. To that end the exhibits strove mightily to portray life as it might be in the glorious time ahead. The future could be found everywhere at the fair. There was the Town of Tomorrow, the farm of the future, which was called the Electrified Farm because it featured more than a hundred applications of electricity in its exhibit, and even Borden's "Dairy World of Tomorrow."[17]

For all the dazzling displays, lighting itself remained a major attraction as it had been at the Great White City. The *Official Guide Book* sang its praises:

> Visible for miles around, a flood of multi-colored light drenches the sky above the glowing spectacle that is the Fair at night. Light, fire, color, water, and sound have been ingeniously and subtly blended to create a dazzling scene that embraces every band of the spectrum. . . . The city of magic, it might well be called, an enchanting vision hinting at the future in artificial illumination.[18]

Nothing symbolized the fair more strikingly than its signature exhibit, the Trylon and Perisphere. Replicas of it could be found everywhere—on shoes, wallpaper, pianos, cuff links, tiaras, plates, shirts, ties, soap, sunglasses, lamps, and lollipops. One estimate had it appearing on forty thousand pieces of merchandise, with every manufacturer paying a royalty to the fair. For all its emphasis on the future, world harmony, and peace, the fair never lost sight of its need to make money. To that end it hosted a parade of special days for states, counties, organizations, and events.[19]

In opening the fair on April 30, President Franklin Roosevelt had no lever to pull or button to push, no giant engine or banks of machines to start. The power was all in place, its sources neatly hidden and requiring no symbolic start. In showcasing the world of tomorrow, electricity had firmly established its primacy in the world of today. It was no longer a show in itself. Visitors took it for granted even as they admired the spectacles made possible by its presence. Nowhere did the guidebooks or exhibits discuss the steam engines or generators or machinery that produced the power behind virtually every function at the fair. The most basic technology of the century had simply been absorbed into everyday life as people awaited the next marvelous product utilizing its power. Once past the horrors of the next few years, that promise would be realized on a scale that dwarfed even the advances of the past two decades. Electric power had become not only ubiquitous but indispensable to the American way of life.

Scant traces of the pioneers could be found at the fair. Edison's image was illuminated on the Perisphere at his birthday, and Westinghouse's name adorned the building of the company that had long ceased to be his. The name of Samuel Insull, who died the previous year, was nowhere to be found, nor that of Elihu Thomson. Nikola Tesla still lived in New York, but there is no evidence that he ever went to the fair. This absence of recognition, coupled with the invisibility of sources for electric power at the fair, was a vivid demonstration of how far the world had already passed them by.

IT HAD BEEN a long day, one spent largely in the amusement zone sampling its wide variety of attractions. A fair was supposed to be fun, after all, Ned mused, and this fair might have been the most fun of all. Certainly it was by far the most racy he had witnessed. Several exhibits featured scantily clad and sometimes even partially nude women. In one Living Magazine Cover for a publication of the future, a gorgeous blonde posed with nothing on except a modest shield and shoes. Ned's jaw dropped at the sight; he could not imagine any magazine showing something that spicy. Bathing beauties could be found at several attractions, especially Billy Rose's Aquacade, which became the fair's most popular attraction despite its forty-cent admission charge. A bemused

Times reporter depicted the amusement zone exhibits as a war between culture and sex. "Sex, we announce with some regret," he concluded, "seems to have the best of it so far." Mayor La Guardia agreed and promised to crack down on nudity at the fair after the Cuban Village was raided for staging a "Miss Nude of 1939" contest.[20]

The Aquacade had thrilled Ned with its elaborate staging and the performances by Johnny Weissmuller, Eleanor Holm, and Gertrude Ederle. He wandered through Admiral Byrd's Penguin Island, admired the animals in Frank Buck's Jungleland, watched tribesmen wrestle alligators in the Seminole Village, gaped at the bathing beauties in the Arctic Girl's Tomb of Ice, toured the Gay Nineties exhibit with a touch of nostalgia, and looked in at a couple of the freak shows. He sampled some of the rides but decided prudently that he was too old to try the parachute jump. For something different he visited the Merrie England area and was enchanted by the very abridged versions of Shakespeare plays acted out in a reconstructed version of the Globe Theater.[21]

It was all good fun, and he had enjoyed himself immensely even though he had spent more money than he had planned. He didn't mind the expense; it was a one-time splurge. Soon it would be time to go home, back to his familiar world where every day seemed the same as the last one. Yet Ned knew this was not so, or certainly would not be for much longer. The world was changing, moving ever faster. Things looked so bad in Europe that it seemed war would surely break out before the year's end; some said it would happen before summer's end. Given the hard times and the uncertainty of things overseas, what better place to spend a few days than in the world of tomorrow?

The title was a good one, Ned decided. The fair had opened his eyes to all sorts of new things. He didn't understand all or even most of them, but he found their possibilities to be exciting and hoped he might live long enough to see some of them realized. The fair in its brazen way promised him what American life had always promised—a better tomorrow. He understood that, like it or not, the world he had grown up in was gone forever. Even more, he sensed that a scary new pattern of life was emerging, one in which change itself was becoming a constant more than ever before. Electricity and the machines it spawned had created a way of life in which change came ever faster, and obliterated the past more ruthlessly than ever before. Sitting on a bench at the close of a busy day, the Trylon and Perisphere looming before him in the distance, he wondered how anyone could ever come to feel comfortable or secure in such a fast-moving world.

ELECTRICAL CIRCUITS

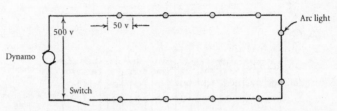

The above diagram illustrates a circuit wired in series. It is connected to a high-voltage, constant-current dynamo. The same current passes through all the lights, and the voltage depends on the number of arc lights in the circuit. Closing one switch or the failure of one light turns off all the lights.

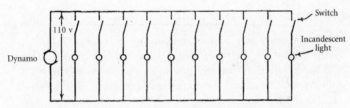

In this diagram incandescent lights are connected via a parallel circuit. The dynamo has low and constant voltage. The amount of current depends on the number of lights in the circuit. Most important, the closing of any one switch turns off only one light. This type of connection did much to refute the longstanding belief that light could not be subdivided.

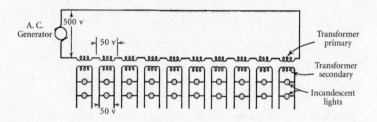

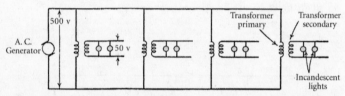

These drawings illustrate the difference between the Gaulard-Gibbs system and the one devised by Westinghouse and his engineers. In the first, the transformer primaries are connected in series and receive 50 volts across each one with no voltage transformation. The secondaries also have 50 volts across each one.

The Westinghouse system connects the primaries in parallel, which puts 500 volts across each one. The transformer has a voltage transformation of 10 to 1, and the secondaries still have 50 volts across each one.

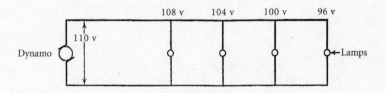

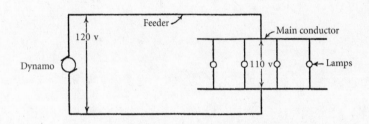

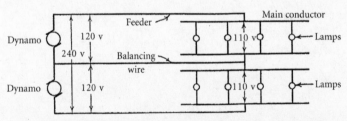

These drawings illustrate three versions of a dynamo-lamp network. In the first, an ordinary parallel network, the lamps have uneven illumination and the voltage drops on the lamps farther from the dynamo.

In Edison's feeder-and-main network most of the voltage drop occurs in the feeder conductor and the main conductors operate at a nearly constant voltage. Edison's three-wire feeder-and-main version improved on the earlier version by doubling the voltage without increasing it at the lamps.

NOTES

ABBREVIATIONS

CFC   *Commercial and Financial Chronicle*
DAB   *Dictionary of American Biography*
EE   *Electrical Engineer*
EW   *Electrical World*
EW&E   *Electrical World and Engineer*
FM   Forrest McDonald papers
NCAB   *National Cyclopedia of American Biography*
SI   Samuel Insull papers
TAE   Thomas A. Edison papers
WSJ   *Wall Street Journal*

INTRODUCTION

1. David M. Potter, *People of Plenty: Economic Abundance and the American Character* (Chicago, 1954).

2. Herbert Croly, *The Promise of American Life* (New York, 1964 [1909]), 10. For a more detailed account of the relationship between materialism and industrialization see Maury Klein, *The Genesis of Industrial America, 1870–1920* (New York, 2007).

PROLOGUE: A SHOW OF POWER: PHILADELPHIA 1876

1. Although the character of Ned is fictional, it is based on the narrative of an actual nine-year-old boy who went to the Centennial Exhibition. See *Prairie Farmer*, December 16, 1876.

2. *New York Times*, August 18, 1876.

3. Details on the Corliss engine are taken from *New York Herald*, May 22, 1876, *Philadelphia Inquirer*, May 10, 1876, J. S. Ingram, *The Centennial Exposition* (New York, 1976 [1876]), 697, 699, and Robert C. Post (ed.), *1876: A Centennial Exhibition* (Washington, 1976), 15, 31. See also *New York Herald*, May 15, 1876.

4. *New York Herald*, May 22, 1876.

5. Ibid., May 15 and May 22, 1876.

6. William Dean Howells, "A Sennight of the Centennial," *Atlantic Monthly* 38 (July 1876): 96.

7. Post, *1876*, 31; Dee Brown, *The Year of the Century: 1876* (New York, 1966), 130; *New York Times*, May 14, 1876.

8. *New York Times*, May 15 and August 18, 1876; *New York Herald*, May 22 and August 14, 1876; Ingram, *Centennial Exposition*, 699–700; Howells, "Sennight," 96.

9. *New York Times*, May 14 and May 24, 1876.

10. Ibid., May 29, 1876.

11. *New York Tribune*, May 10, 1876.

12. Ibid. This discussion is drawn from Ingram, *Centennial Exposition*, 21–40, and from an insert entitled "The Growth of Expositions" that came with my copy of Post, *1876*. I have not been able to find the original source.

13. Ingram, *Centennial Exposition*, 47–51; Post, *1876*, 69.

14. Howells, "Sennight," 104.

15. Post, *1876*, 15, 17, 67.

16. Details on opening day are taken from *New York Times*, May 11, 1876; *New York Herald*, May 11, 1876; *New York Tribune*, May 11, 1876; *Philadelphia Inquirer*, May 10, 1876; *Chicago Tribune*, May 11, 1876; and Ingram, *Centennial Exposition*, 73–98, 753–60.

17. For details on the Corliss valve, see Richard L. Hills, *Power from Steam: A History of the Stationary Steam Engine* (Cambridge, Eng., 1989), 178–81.

18. *Philadelphia Evening Bulletin*, August 8, 1876; Post, *1876*, 29–31. For a jaded view of the Corliss offer, see *Philadelphia Inquirer*, August 3, 1876.

19. *Philadelphia Inquirer*, May 11, 1876; *New York Herald*, May 11, 1876.

20. Ingram, *Centennial Exposition*, 755. On the Sunday opening controversy, see, for example, *New York Herald*, May 7, 1876, and Post, *1876*, 21. Some Web sites claim that the Corliss engine fell silent on Sundays because Corliss himself did not wish it to operate on the Sabbath. In fact, it was the Sunday closing policy that prevailed.

21. Sarah Orne Jewett, *The Country of the Pointed Firs and Other Stories* (Garden City, N.Y., 1956). This is a paperback reprint of the 1925 collection.

22. *New York Herald*, May 12 and August 14, 1876; *New York Times*, May 14, 1876; *New York Sun*, May 16, 1876; *Philadelphia Evening Bulletin*, October 24, 1876.

23. *New York Times*, June 12, June 17, and October 5, 1876; Matthew Josephson, *Edison: A Biography* (New York, 1959), 104; Ingram, *Centennial Exposition*, 297–98, 303; Post, *1876*, 41–43, 80, 99, 145.

24. *New York Times*, August 18 and August 21, 1876; Post, *1876*, 39.

25. Ingram, *Centennial Exposition*, 686–87, 706–8.

26. Ibid., 701–2; William Peirce Randel, *Centennial: American Life in 1876* (Philadelphia, 1969), 289; *New York Times*, May 15, 1876.

27. Howells, "Sennight," 96, 107.

CHAPTER 1: THE MACHINE THAT CHANGED THE WORLD

1. Quoted in Hills, *Power from Steam*, 1.

2. H. W. Dickinson, *James Watt: Craftsman and Engineer* (Cambridge, Eng., 1936), 1–28.

3. Ibid., 29–35. Unless otherwise indicated, the descriptions of Watt's experiments are drawn from this source.

4. Isaac Asimov, *Biographical Encyclopedia of Science and Technology* (Garden City, N.Y., 1982), 194–96. This is the second revised edition. See also Dickinson, *James Watt*, 35–36, and Eric Robinson and A. E. Musson (eds.), *James Watt and the Steam Revolution* (New York, 1969), 39–40.

5. Dickinson, *James Watt*, 35–36; H. W. Dickinson, *A Short History of the Steam Engine* (New York, 1939), 66–69; Hills, *Power from Steam*, 51–54; Robert H. Thurston, *A History of the Growth of the Steam Engine* (Ithaca, 1939 [1878]), 80–88.

6. This description is drawn from Ruth Schwartz Cowan, *A Social History of American Technology* (New York, 1997), 29–39.

7. John Lord, *Capital and Steam-Power, 1750–1800* (New York, 1965), 21–24. This is the second edition of a work originally published in 1923.

8. Ibid., 25–34.

9. Thurston, *History*, 4–9; Hills, *Power from Steam*, 13.

10. Thurston, *History*, 18–24; Dickinson, *Short History*, 13–16.

11. Hills, *Power from Steam*, 14; Thurston, *History*, 29–30.

12. Hills, *Power from Steam*, 14–16. For an explanation of Boyle's law, see Asimov, *Biographical Encyclopedia*, 135–36.

13. Ibid., 16–20; Thurston, *History*, 30–44; Dickinson, *Short History*, 18–28. I have modernized the spelling in the quotations.

14. Hills, *Power from Steam*, 21–22; Dickinson, *Short History*, 31–32; Thurston, *History*, 56–58.

15. Dickinson, *Short History*, 29–43, Thurston, *History*, 57–62, and Hills, *Power from Steam*, 20–30, discuss the technical details of Newcomen's engine and include diagrams of it.

16. Hills, *Power from Steam*, 20–30; Dickinson, *Short History*, 29–42.

17. Hills, *Power from Steam*, 31–40.

18. Ibid., 41–50.

19. Ibid., 2.

20. Ibid., 49–53; Dickinson, *Short History*, 54–65. Both works give more detailed explanations of how the atmospheric steam engine works. See also the brief but clear explanation at www.egr.msu.edu/~lira/supp/steam/.

21. H. W. Dickinson, *Matthew Boulton* (Cambridge, Eng., 1937), 38, 79–80.

22. Ibid., 80–82. Boulton's letter can also be found in Robinson and Musson, *Watt and the Steam Revolution*, 62–63.

23. Dickinson, *James Watt*, 38–43, 67–77. See also Watt's patent application of 1769 in Robinson and Musson, *Watt and the Steam Revolution*, 60–61.

24. Quoted in Dickinson, *James Watt*, 57.

25. Dickinson, *Matthew Boulton*, 76, 83; Dickinson, *James Watt*, 79.

26. Dickinson, *Matthew Boulton*, 83–84; Dickinson, *James Watt*, 81–85.

27. Dickinson, *Matthew Boulton*, 36, 75–76, 84–88; Thurston, *History*, 103.

28. Dickinson, *James Watt*, 85–94, 114–15.

29. Sanford P. Bordeau, *Volts to Hertz: The Rise of Electricity* (Minneapolis, 1982), 32; Dickinson, *Short History*, 71–74. For more detail on how Boulton & Watt structured their payments, see Hills, *Power from Steam*, 75, 87–88.

30. Dickinson, *Short History*, 80–82; Hills, *Power from Steam*, 32, 59–60; Thurston, *History*, 105.

31. Hills, *Power from Steam*, 60–62; Dickinson, *Short History*, 79–80; Robinson and Musson, *Watt and the Steam Revolution*, 88. By "fire-engine" Boulton refers to engines heated by fire, not machines to put out fires.

32. Hills, *Power from Steam*, 63; Dickinson, *Short History*, 82.

33. Hills, *Power from Steam*, 63–66; Dickinson, *Short History*, 83–85.

34. Hills, *Power from Steam*, 66–69; Dickinson, *Short History*, 82.

35. Hills, *Power from Steam*, 69.

36. Ibid., 85–86; Dickinson, *Short History*, 83; Asimov, *Biographical Encyclopedia*, 208–9.

37. Hills, *Power from Steam*, 85–86; Thurston, *History*, 123–24, 417; Dickinson, *Short History*, 83–85.

38. Hills, *Power from Steam*, 89; Asimov, *Biographical Encyclopedia*, 208–9. Desaguliers adopted the figure of 27,500 and Smeaton 22,916.

39. Hills, *Power from Steam*, 70.

CHAPTER 2: CONQUERING THE WATERS

1. Greville Bathe and Dorothy Bathe, *Oliver Evans: A Chronicle of Early American Engineering* (Philadelphia, 1935), 200.

2. Dickinson, *Short History*, 92–93.

3. Ibid., 91; Hills, *Power from Steam*, 97.

4. Dickinson, *James Watt*, 139–40; Hills, *Power from Steam*, 101–3; Thurston, *History*, 138, 159, 174–79. For Murdock's steam carriage models, see www.cottontimes.co.uk/murdocho .html. For Trevithick, see also www.cottontimes.co.uk/trevithicko.html.

5. Dickinson, *Short History*, 174–75.

6. Hills, *Power from Steam*, 120.

7. Ibid., 120–29; Dickinson, *Short History*, 93–97; Thurston, *History*, 170. Dickinson includes a photograph of Trevithick's engine and boiler.

8. Bathe and Bathe, *Oliver Evans*, 1–4. This is the fullest and almost the only significant source on Evans's life.

9. Ibid., 4. For more on the pattern of creative minds, see Maury Klein, *The Change Makers* (New York, 2003), chapter 1.

10. Ibid., 5–8.

11. Ibid., 9–10.

12. Ibid., 11–27. Evans's own description of the old manual process of producing flour is given ibid., 12.

13. Cowan, *Social History*, 70–72. A drawing of the Evans mill, taken from his own book, is in Bathe and Bathe, *Oliver Evans*, 32; www.greenbankmill.org/oliverevans/html.

14. Cowan, *Social History*, 72; Bathe and Bathe, *Oliver Evans*, 28, 32–33, 48. Evans wrote his own comparison between his engine and that of Watt. See Bathe and Bathe, *Oliver Evans*, 300–302.

15. Dickinson, *Short History*, 94; Bathe and Bathe, *Oliver Evans*, 15, 34–40, 50–63.

16. Dickinson, *Short History*, 94–95, 108, 119–20; Bathe and Bathe, *Oliver Evans*, 66; Hills, *Power from Steam*, 97.

17. Bathe and Bathe, *Oliver Evans*, 101–2, 108–9.

18. Ibid., 97–112, 156, 165, 303–9. Evans himself wrote that the Oruktor circled Center Square. The Bathes took him at his word but noted that not a single Philadelphia newspaper reported the incident. It is hard to believe that so spectacular an event could have passed unnoticed by the press.

19. Ibid., 120–23, 141–42, 159–64, 166–68, 180, 188; Cowan, *Social History*, 73–74. Evans and his sons took over the flour mill after another partner's death in 1812.

20. Bathe and Bathe, *Oliver Evans*, 97 and passim; www.greenbankmill.org/oliverevans/html.

21. Hills, *Power from Steam*, 98–99; Dickinson, *Short History*, 119. Drawings that show the evolution of Evans's boiler designs are in Bathe and Bathe, *Oliver Evans*, opposite 140.

22. Quoted in Bathe and Bathe, *Oliver Evans*, 140.

23. Ibid., 157–58.

24. Ibid., 172, 267–71. Evans's last original patent was filed in 1811.

25. Louis C. Hunter, *Steamboats on the Western Rivers: An Economic and Technological History* (New York, 1993 [1949]), 22. This is a corrected reprint of the original edition of this classic work.

26. Bathe and Bathe, *Oliver Evans*, 57–58; Hunter, *Steamboats*, 5.

27. Bathe and Bathe, *Oliver Evans*, 76–78.

28. This summary of Fitch's early life is drawn from Frank D. Prager (ed.), *The Autobiography of John Fitch* (Philadelphia, 1976), 19–61, and Andrea Sutcliffe, *Steam: The Untold Story of America's First Great Invention* (New York, 2004), 19–26. See also Thomas Boyd, *Poor John Fitch: Inventor of the Steamboat* (New York, 1935), and Thompson Westcott, *Life of John Fitch, the Inventor of the Steam-Boat* (Philadelphia, 1857).

29. Fitch, *Autobiography*, 113–14.

30. Ibid., 156–57.

31. Sutcliffe, *Steam*, 1–9. Five other states refused Rumsey's application. Bath is now Berkeley Springs.

32. Ibid., 13–16, 35–37; Jack L. Shagena, *Who Really Invented the Steamboat?* (Amherst, N.Y., 2004), 130–44.

33. Fitch, *Autobiography*, 158–74; Sutcliffe, *Steam*, 38–47; Shagena, *Who Really*, 184–94. The quotation about Voigt is on 168.

34. Fitch, *Autobiography*, 172–79; Sutcliffe, *Steam*, 48–49; Shagena, *Who Really*, 189–95.

35. Sutcliffe, *Steam*, 14–15, 35–38; 53–54; Shagena, *Who Really*, 145–47.

36. Sutcliffe, *Steam*, 54–56; Shagena, *Who Really*, 147–49. Sutcliffe says this first cruise lasted "almost two hours." Shagena says it lasted three hours.

37. Sutcliffe, *Steam*, 56–60; Shagena, *Who Really*, 148–50.

38. Fitch, *Autobiography*, 181; Sutcliffe, *Steam*, 61–67; Shagena, *Who Really*, 151–53.

39. Sutcliffe, *Steam*, 78–87, 133–43; Shagena, *Who Really*, 154–59.

40. Fitch, *Autobiography*, 182–84; Sutcliffe, *Steam*, 67–69; Shagena, *Who Really*, 192–95.

41. Fitch, *Autobiography*, 184.

42. Ibid., 187–91; Sutcliffe, *Steam*, 76–78, 89–91; Shagena, *Who Really*, 198–99.

43. Fitch, *Autobiography*, 190–91; Sutcliffe, *Steam*, 91; Shagena, *Who Really*, 199–201.

44. Fitch, *Autobiography*, 192–200; Sutcliffe, *Steam*, 93–99, 147–59; Shagena, *Who Really*, 200–203.

45. Fitch, *Autobiography*, 193.

46. Most of the details on Fulton's life are taken from Cynthia Owen Philip, *Robert Fulton* (New York, 1985). For other versions of Fulton's life see H. W. Dickinson, *Robert Fulton, Engineer and Artist: His Life and Works* (London, 1913), Alice C. Sutcliffe, *Robert Fulton* (New York, 1915), and Kirkpatrick Sale, *The Fire of His Genius: Robert Fulton and the American Dream* (New York, 2001). See also the relevant passages in Hunter, Shagena, and Andrea Sutcliffe.

47. Philip, *Fulton*, 142. Contrary to legend, the boat was not named the *Clermont* but had no name. Later it was named the *North River of Clermont*, which got shortened to the latter name.

48. The figures are taken from Sutcliffe, *Steam*, 184–85, 191, 197.

49. Details on the trip down the river are taken from ibid., 201–3.

50. Hunter, *Steamboats*, 15–19, 128. A Fulton boat with a low-pressure engine, the *Aetna*, also made four trips from New Orleans to Louisville but did not reach Pittsburgh.

51. Ibid., 62–63, 123–41. Hunter provides ample detail on the development of the high-pressure engine as well as other aspects of the western steamboat.

52. Ibid., 17, 33–34, 101–5.

53. All quotations are taken from ibid., 27–28.

CHAPTER 3: THE GREATEST ENGINE OF ALL

1. Quoted in Archibald Douglas Turnbull, *John Stevens: An American Record* (New York, 1928), 358–59.

2. Albro Martin, *Railroads Triumphant* (New York, 1992), 33–34.

3. Klein, *Genesis of Industrial America*, 57.

4. Marc McCutcheon, *The Writer's Guide to Everyday Life in the 1800s* (Cincinnati, 1993), 68–69.

5. United States Bureau of the Census, *Historical Statistics of the United States from Colonial Times to 1970* (Washington, 1975), 2:728, 731; Albert A. Hopkins and A. Russell Bond, *Scientific American Reference Book* (New York, 1905), 117.

6. Thurston, *History*, 172–73.

7. Ibid., 145–46; Hills, *Power from Steam*, 143, 163.

8. Thurston, *History*, 157–60, 174–76; Dickinson, *Short History*, 95–96; Alfred W. Bruce, *The Steam Locomotive in America* (New York, 1952), 19.

9. Hills, *Power from Steam*, 140–42. Hills estimated that at start-up a textile mill engine needed only about 15 percent of its overall power.

10. Thurston, *History*, 183–90.

11. Ibid., 191–99; Bruce, *Steam Locomotive*, 20–21.

12. The struggles between Stevens and Fulton and Livingston are detailed in Philip, *Robert Fulton*, and less cogently in Turnbull, *John Stevens*.

13. Thurston, *History*, 178–80; Bruce, *Steam Locomotive*, 21; Turnbull, *John Stevens*, 359–85, 473–83.

14. For the Erie Canal's story, see Peter L. Bernstein, *Wedding of the Waters: The Erie Canal and the Making of a Great Nation* (New York, 2005). The broader issue of competing forms of internal improvements can be followed in John Lauritz Larson, *Internal Improvement: National Public Works and the Promise of Popular Government in the Early United States* (Chapel Hill, 2001).

15. James D. Dilts, *The Great Road: The Building of the Baltimore & Ohio, the Nation's First Railroad, 1828–1853* (Stanford, Calif., 1993), 1–97. Dilts recounts the building of the B&O in loving detail and is by far the best study of its construction.

16. Ibid., 97; John F. Stover, *The Life and Decline of the American Railroad* (New York, 1970), 11–12; John F. Stover, *American Railroads* (Chicago, 1961), 14–15.

17. Stover, *American Railroads*, 15–19; Stover, *Life and Decline*, 14–17. The quotation is from the latter source. The four states lacking track were Arkansas, Missouri, Tennessee, and Vermont.

18. John F. Stover, *Life and Decline*, 27; Maury Klein, *Unfinished Business: The Railroad in American Life* (Hanover, N.H., 1994), 70. The trip by rail from coast to coast required several changes of trains as the traveler moved from one rail company to another. No one American railroad has ever traversed the entire continent.

19. Stover, *Life and Decline*, 16–17; Stover, *American Railroads*, 17–18.

20. Stover, *American Railroads*, 32–33; Stover, *Life and Decline*, 21.

21. Martin, *Railroads Triumphant*, 18; Stover, *Life and Decline*, 34.

22. Larson, *Internal Improvements*, 231.

23. John K. Brown, *The Baldwin Locomotive Works, 1831–1915* (Baltimore, 1995), 3–4.

24. Bruce, *Steam Locomotive*, 23–24; Stover, *Life and Decline*, 19; Brown, *Baldwin Locomotive Works*, 5–6. A drawing of *Old Ironsides* can be found in Brown, 6. The company published a notice saying that "the locomotive-engine, built by M. W. Baldwin of this City, will depart daily, when the weather is fair, with a train of passenger-cars. On rainy days horses will be attached"; www.famousamericans.net/matthiaswilliambaldwin.

25. Bruce, *Steam Locomotive*, 24–25; Stover, *Life and Decline*, 20; John H. White Jr., *A History of the American Locomotive: Its Development, 1830–1880* (New York, 1968), 33–57. The 4-4-0 designation refers to the wheel alignment beneath the locomotive. Locomotives are classified by their wheel arrangements. The first number represents the leading wheels, the second the driving wheels, and the final number the trailing wheels. Thus the 4-4-0 has two sets of leading wheels, two sets of driving wheels, and no trailing wheels.

26. Bruce, *Steam Locomotive*, 34, 36; Brown, *Baldwin Locomotive Works*, 4. For details on the air brake, see Henry G. Prout, *A Life of George Westinghouse* (New York, 1922), 21–76. Details on the various components of the locomotive can be found in White, *American Locomotive*, 93–235.

27. Bruce, *Steam Locomotive*, 29, 31, 35, 36, 76, 92; White, *American Locomotive*, 239–442. A huge Union Pacific "Big Boy" can be seen on display at Kenefick Park in Omaha, Nebraska.

28. White, *American Locomotive*, 4–6, 443–44.

29. Bruce, *Steam Locomotive*, 81–82, 95, 98, 152–54, 159.

30. For figures on locomotive production, see ibid., 67, 423. For an introduction to the replacement of steam locomotives by diesel engines, see Klein, *Unfinished Business*, 143–54. On the broader impact of the railroad's role in American life, see Maury Klein, *The Flowering of the Third America* (Chicago, 1993).

31. Thurston, *History*, 264–86; Dickinson, *Short History*, 127–28. The *Savannah* needed sails because it burned wood rather than coal for fuel.

32. Thurston, *History*, 284–88.

33. Ibid., 288–89, 414–16. The *Great Eastern* was never a commercial success. After sitting idle for several years, it was broken up for scrap in 1889–90.

34. Unless otherwise indicated, the following discussion is drawn from Chandler's seminal article, "Anthracite Coal and the Beginnings of the Industrial Revolution in the United States," *Business History Review* 46 (Summer 1972): 141–81.

35. These figures are taken from tables in ibid., 155, 157.

36. Quoted in Reynold M. Wik, *Steam Power on the American Farm* (Philadelphia, 1953), 3. The following discussion is drawn largely from this excellent source.

37. Ibid., 4–12.

38. Ibid., 17–21.

39. Ibid., 21–28.

40. Ibid., 36–41.

41. These figures are calculated from those given in ibid., 99–101.

42. Post, *1876*, 39; Hills, *Power from Steam*, 147–61, 173.

43. Hills, *Power from Steam*, 173–82, explains the workings of the Corliss improvements in detail. The purpose of the valve gear is to regulate the movement of valves that control the emission of steam to and from the cylinders.

44. Ibid., 183–88; Post, *1876*, 29, 31; Thurston, *History*, 505. This last portion of Thurston's book, entitled "Supplementary History to 1939," was written not by Thurston but by W. N. Barnard. It embraces pages 500–45.

45. Thurston, *History*, 506–12; Hills, *Power from Steam*, 193–203; Dickinson, *Short History*, 141–42. Hills provides the most technically detailed account.

46. Thurston, *History*, 506–12; Hills, *Power from Steam*, 193–203.

47. Hills, *Power from Steam*, 208–12.

48. Dickinson, *Short History*, 145–47.

49. Hills, *Power from Steam*, 231–32.

50. Thurston, *History*, 526–27.

CHAPTER 4: IN SEARCH OF THE MYSTERIOUS ETHER

1. Marc J. Seifer, *Wizard: The Life and Times of Nikola Tesla* (New York, 1998), 98.

2. Edmund Whittaker, *A History of the Theories of Aether and Electricity* (New York, 1989), 37–39. This is a reprint of a two-volume work originally published in 1951 and 1953. Whittaker's work is the fullest and most technical discussion of the ether and its history.

3. Asimov, *Biographical Encyclopedia*, 161–62, 222–26.

4. Hills, *Power from Steam*, 164–65.

5. Dickinson, *Short History*, 176–77; Asimov, *Biographical Encyclopedia*, 242–44; Thurston, *History*, 434–36; Hills, *Power from Steam*, 167. Asimov explains Thompson's experiment in detail.

6. Hills, *Power from Steam*, 167–68.

7. Ibid., 165–66; Dickinson, *Short History*, 177. Carnot's book was entitled *Reflections on the Motive Power of Fire*.

8. Hills, *Power from Steam*, 167; Dickinson, *Short History*, 179. Italics are in the original. Carnot did not formulate the second law of thermodynamics. It was deduced from his work and articulated by Rudolf J. E. Clausius in 1850.

9. Asimov, *Biographical Encyclopedia*, 398–400; Dickinson, *Short History*, 178.

10. Hills, *Power from Steam*, 168–69; Dickinson, *Short History*, 178–79; *EE*, November 1889, 500–501. Joule's measure for the BTU was later revised to 778 pounds.

11. Hills, *Power from Steam*, 6–7.

12. Ibid., 6–7, 165–72.

13. Ibid., 170–72; Asimov, *Biographical Encyclopedia*, 414–15.

14. Herbert W. Meyer, *A History of Electricity and Magnetism* (Cambridge, Mass., 1971), 1–4.

15. Ibid., 8–10; Asimov, *Biographical Encyclopedia*, 61–62, 91; Bordeau, *Volts to Hertz*, 2–11; Jill Jonnes, *Empires of Light* (New York, 2003), 18–20. The full title of Gilbert's work, translated into English, was *On the Magnet, Magnetick Bodies Also, and on the Great Magnet the Earth; A New Philosophy Demonstrated by Many Arguments and Experiments*.

16. Jonnes, *Empires of Light*, 96; Philip Dray, *Stealing God's Thunder* (New York, 2005), 46.

17. Whittaker, *Theories of Aether and Electricity*, 1–8.

18. Meyer, *History*, 11–18; Bordeau, *Volts to Hertz*, 16–18; Asimov, *Biographical Encyclopedia*, 173–75; Jonnes, *Empires of Light*, 20–24. The two scientists were E. G. von Kleist of Pomerania, Germany, and Pieter van Musschenbroek of Leyden. Credit for being the first to discover the Leyden jar has gone back and forth between them. Basically it is a glass jar coated with metal foil on both sides. Later it was found that storage of electricity required only two conductive coatings separated by an insulator such as glass. One end of a brass wire dangled into the water-filled jar while the other end, with a small brass ball atop it, stood above a tight nonconducting cork lid.

19. I. Bernard Cohen, *Benjamin Franklin's Science* (Cambridge, Mass., 1990), 14–19, 61–65; J. L. Heilbron, *Electricity in the 17th and 18th Centuries: A Study of Early Modern Physics* (Berke-

ley, Calif., 1979), 324–39. Cohen's book offers masterful and lucid summaries of the major issues surrounding Franklin's scientific career. Meyer and others attribute Franklin's interest in electricity to having read one of Watson's books, but Heilbron's more detailed study shows that Franklin's interest was aroused by a report on some German experiments along with the glass tube that accompanied the journal in which the report appeared.

20. Benjamin Franklin, *Autobiography* (New Haven, 1964), 196, 240–41; Cohen, *Benjamin Franklin's Science*, 24–27; Dray, *God's Thunder*, 37–47; Heilbron, *Electricity*, 324–39; Carl van Doren, *Benjamin Franklin* (New York, 1938), 156–73; Bordeau, *Volts to Hertz*, 19–20.

21. Van Doren, *Benjamin Franklin*, 158–59; Michael Brian Schiffer, *Draw the Lightning Down: Benjamin Franklin and Electrical Technology in the Age of Enlightenment* (Berkeley, Calif., 2003), 67–68; Cohen, *Benjamin Franklin's Science*, 7.

22. Cohen, *Benjamin Franklin's Science*, 66–109; Heilbron, *Electricity*, 339–72; Meyer, *History*, 22–27; van Doren, *Benjamin Franklin*, 164–68. The famous kite episode remains steeped in controversy. Franklin left no firsthand account, and some scholars doubt that it ever occurred. Van Doren, Franklin's most thorough biographer, says that "the chances that he did not do it at all are nearly as good as that he did it when proof was no longer called for" (168). However, Cohen offers a persuasive argument that Franklin did indeed perform the experiment, most likely in June 1752 (66–81, 93–100).

23. Cohen, *Benjamin Franklin's Science*, 6–9.

24. Bordeau, *Volts to Hertz*, 20–21; Heilbron, *Electricity*, 468–70, 490–91.

25. Heilbron, *Electricity*, 468–77; Meyer, *History*, 30–31; Bordeau, *Volts to Hertz*, 22–26.

26. The following account of Galvani and Volta is drawn from Heilbron, *Electricity*, 491–94; Meyer, *History*, 28–30, 34–44; Bordeau, *Volts to Hertz*, 44–56. For more detail on the electroscope, see Bordeau, 50–51.

27. The electrophorus "worked on the principle that a charge may be repeatedly induced on a conductor and transferred to a storage device such as a Leyden Jar, without depleting the original inducing charge." Bordeau, *Volts to Hertz*, 48.

28. The pile consisted of a zinc disk, then a silver one, and then the cardboard soaked in liquid. This pattern was repeated for the entire pile, which could be built in many sizes.

29. Bordeau, *Volts to Hertz*, 54; Meyer, *History*, 152; Asimov, *Biographical Encyclopedia*, 284–85; Jonnes, *Empires of Light*, 35. Asimov says that Davy's giant battery contained 250 metal plates; Bordeau gives the figure of "500 voltaic cells," and Jonnes "two thousand pairs of plates." The Asimov and Bordeau numbers can be reconciled with one another, but not with that of Jonnes. None of the three gives any source for the figure.

30. Meyer, *History*, 46–48; Bordeau, *From Volts to Hertz*, 60–67; Harold I. Sharlin, *The Making of the Electrical Age* (New York, 1963), 9.

31. Bordeau, *Volts to Hertz*, 67; L. Pearce Williams, *Michael Faraday: A Biography* (New York, 1965), 142–43.

32. Bordeau, *Volts to Hertz*, 70–76; Williams, *Michael Faraday*, 143–45.

33. Williams, *Michael Faraday*, 145–49; Asimov, *Biographical Encyclopedia*, 306, 310, 403.

34. Williams, *Michael Faraday*, 149–51.

35. Bordeau, *Volts to Hertz*, 77–85, 117. Bordeau gives the mathematical equations for Ampère's several laws. Ampère also established the concept of the complete electric circuit. See Percy Dunsheath, *A History of Electrical Power Engineering* (Cambridge, Mass., 1969), 58–59.

36. Bordeau, *Volts to Hertz*, 86–100. Bordeau gives a detailed account of Ohm's career and work.

37. Ibid., 26–27, 39–41, 56–57, 69, 84, 106.

38. Ibid., 108–16; Meyer, *History*, 52–55; John Meurig Thomas, *Michael Faraday and the Royal Institution* (Bristol, Eng., 1991), 4–5.

39. Thomas, *Michael Faraday*, 5–8.

40. Williams, *Michael Faraday*, 1–106; Frank A. J. L. James, *The Correspondence of Michael Faraday*, vol. 1 (London, 1991), 591. The quotation about Faraday's soul is in Williams, 6. For more detail on Faraday and his religion, see Geoffrey Cantor, *Michael Faraday: Sandemanian and Scientist* (New York, 1991).

41. Bordeau, *Volts to Hertz*, 116–17. Controversy continues over whether Faraday was present during the experiment. See for example the differing views in David Gooding and Frank A. J. L. James (eds.), *Faraday Rediscovered: Essays on the Life and Work of Michael Faraday, 1791–1867* (New York, 1985), 44, 85, 110–20, and Williams, *Michael Faraday*, 153.

42. Gooding and James, *Faraday Rediscovered*, 110–20; Williams, *Michael Faraday*, 151–60.

43. Bordeau, *Volts to Hertz*, 118–19; Thomas, *Michael Faraday*, 29–30; Dunsheath, *History of Electrical Power Engineering*, 60. For more detail on the relationship between Faraday and Davy, see the essay by David M. Knight in Gooding and James, *Faraday Rediscovered*, 33–49.

44. Bordeau, *Volts to Hertz*, 118–19; Williams, *Michael Faraday*, 170–73. Faraday's contributions to the science of chemistry were considerable in themselves. See Thomas, *Michael Faraday*, 23.

45. Williams, *Michael Faraday*, 151.

46. Bordeau, *Volts to Hertz*, 119–22; Sharlin, *Electrical Age*, 136–39; Asimov, *Biographical Encyclopedia*, 212, 286, 295.

47. This discussion of Faraday's great discovery is drawn from Bordeau, *Volts to Hertz*, 119–22, Sharlin, *Electrical Age*, 136–39, Williams, *Michael Faraday*, 179–202, and Thomas, *Michael Faraday*, 40–45.

48. The quotations are from Sharlin, *Electrical Age*, 137, 139.

49. Ibid., 134, 139–41; Bordeau, *Volts to Hertz*, 122–23. Sharlin adds that Faraday "had discovered the principle which became known as Faraday's Law[:] . . . The size of the voltage induced in a conductor is a measure of the rate with which a conductor cuts lines of flux."

50. Sharlin, *Electrical Age*, 133, 139, 192; Bordeau, *Volts to Hertz*, 121.

51. A useful summary of Faraday's accomplishments can be found in Thomas, *Michael Faraday*, 45–94. For Faraday's work in field theory, see Williams, *Michael Faraday*, 408–64, and the essay by Nancy J. Nersessian in Gooding and James, *Faraday Rediscovered*, 175–87. For field theory, see R. A. R. Tricker, *The Contributions of Faraday and Maxwell to Electrical Science* (New York, 1966), 75–91.

52. Jonnes, *Empires of Light*, 45; Williams, *Michael Faraday*, 465–505.

53. Thomas, *Michael Faraday*, 116–33; Williams, *Michael Faraday*, 465–505.

54. Asimov, *Biographical Encyclopedia*, 454–55; Tricker, *Faraday and Maxwell*, 92–96.

55. Sharlin, *Electrical Age*, 85; Tricker, *Faraday and Maxwell*, 96–97.

56. Williams, *Michael Faraday*, 506–9; Sharlin, *Electrical Age*, 83; Tricker, *Faraday and Maxwell*, 107–9.

57. Williams, *Michael Faraday*, 510–11; Sharlin, *Electrical Age*, 77.

58. This discussion of Maxwell's work is drawn from Williams, *Michael Faraday*, 509–11, Sharlin, *Electrical Age*, 76–86, Tricker, *Faraday and Maxwell*, 101–31, and Bordeau, *Volts to Hertz*, 190–202.

59. Bordeau, *Volts to Hertz*, 192; Sharlin, *Electrical Age*, 80.

60. Of the sources listed in note 48, Sharlin's is the clearest and most succinct. Bordeau and

Tricker provide more detailed descriptions of the experiments, including illustrations, and Tricker supplies many of the mathematical formulas.

61. The clearest explanation of this part of Maxwell's paper is in Bordeau, *Volts to Hertz*, 194–95.

62. Ibid., 195–96. Emphasis is in the original.

63. Ibid., 197–202. Later the equations were refined into four equations that became known simply as "Maxwell's equations." Bordeau discusses each one, as does Tricker, *Faraday and Maxwell*, 125–29.

64. Ibid., 85–86; Tricker, *Faraday and Maxwell*, 126. Tricker, 226–86, reprints a substantial portion of Maxwell's third paper.

65. Sharlin, *Electrical Age*, 87.

66. Ibid., 73, 86–87; Tricker, *Faraday and Maxwell*, 97–99. For details on the prize see Bordeau, *Volts to Hertz*, 223–24, 230. Hertz died in 1894 at the age of 36. See *EW*, January 27, 1894, 197–98.

67. Bordeau, *Volts to Hertz*, 206; *EW*, February 10, 1894, 174. See also *EW*, February 17, 1894, 204–5, and March 17, 1894, 355–57.

68. *EE*, May 18, 1889, 226–28; *EW*, August 26, 1893, 143.

69. *EW*, December 2, 1893, 423.

70. Seifer, *Wizard*, 102–10; Asimov, *Biographical Encyclopedia*, 540–41.

71. Asimov, *Biographical Encyclopedia*, 154–55, 540–42.

72. Ibid. Several Web sites have useful explanations and diagrams of the Michelson-Morley experiments. See, for example, http://scienceworld.wolfram.com/physics/Michelson-MorleyExperiment.html; Michael Fowler, "The Michelson-Morley Experiment," http://galileoandeinstein.physics.virginia.edu/lectures/michelson.html; http://planetquest.jpl.nasa.gov/technology/michelson.cfm.

73. Asimov, *Biographical Encyclopedia*, 674–75.

CHAPTER 5: LET THERE BE LIGHT

1. Quoted in David E. Nye, *Electrifying America: Social Meanings of a New Technology, 1880–1940* (Cambridge, Mass., 1992), 3.

2. Sharlin, *Electrical Age*, 134–35.

3. Maury Klein, "What Hath God Wrought?" *Invention & Technology*, Spring 1993, 37; Bordeau, *Volts to Hertz*, 83–84; Dunsheath, *History of Electrical Power Engineering*, 67–68.

4. Sharlin, *Electrical Age*, 9–10, 133.

5. Taliaferro P. Shaffner, "Early Electric Telegraphs" and "Steinheil's Electric Telegraph," in George Shiers (ed.), *The Electric Telegraph: An Historical Anthology* (New York, 1977), 135–37, 157–78; Geoffrey Hubbard, *Cooke and Wheatstone and the Invention of the Electric Telegraph* (New York, 1968 [1965]), 6–14. The rather bizarre Shiers anthology reprints several useful older sources but gives the whole volume no consecutive pagination. As a result, the page numbers cited here refer only to the specific articles.

6. Dunsheath, *History of Electrical Power Engineering*, 69–74; Hubbard, *Cooke and Wheatstone*, 27–33.

7. Taliaferro P. Shaffner, "History of the English Electric Telegraph," in Shiers, *Electric Telegraph*, 179–215; Hubbard, *Cooke and Wheatstone*, 38–86; Sharlin, *Electrical Age*, 10, 12.

8. Albert E. Moyer, *Joseph Henry: The Rise of an American Scientist* (Washington, 1997), 9–22; Thomas Coulson, *Joseph Henry: His Life and Work* (Princeton, 1950), 3–22.

9. Coulson, *Joseph Henry*, 25–39; Moyer, *Joseph Henry*, 23–60, 126–33. Coulson and Moyer differ on when and how Henry first learned of Sturgeon's magnet. Coulson says it occurred on an 1826 trip to New York City; Moyer says it happened in January 1827.

10. Coulson, *Joseph Henry*, 39–68; Moyer, *Joseph Henry*, 61–70; Asimov, *Biographical Encyclopedia*, 335–36; Bordeau, *Volts to Hertz*, 154–55.

11. Bordeau, *Volts to Hertz*, 154–55.

12. Differing versions of this episode can be found in Coulson, *Joseph Henry*, 65–95, and Moyer, *Joseph Henry*, 78–111.

13. Coulson, *Joseph Henry*, 65–95; Moyer, *Joseph Henry*, 86–96, 162–64. For an explanation of self-induction, see Bordeau, *Volts to Hertz*, 159, Coulson, 89–91, and Moyer, 99–101. Scientists later referred to the original discovery of Faraday and Henry as "mutual induction" because it involved two circuits as opposed to the single circuit involved in self-induction.

14. Coulson, *Joseph Henry*, 69–71; Moyer, *Joseph Henry*, 72–74; Bordeau, *Volts to Hertz*, 156–57.

15. Coulson, *Joseph Henry*, 63–64; Moyer, *Joseph Henry*, 69–70; William B. Taylor, "An Historical Sketch of Henry's Contribution to the Electro-Magnetic Telegraph," in Shiers, *Electric Telegraph*, 32–34.

16. Coulson, *Joseph Henry*, 64, 107–8; Moyer, *Joseph Henry*, 143–44; Taylor, "Historical Sketch of Henry's Contribution," 33–34, 63–66, 77–79, 99–100.

17. Moyer, *Joseph Henry*, 178.

18. Ibid., 236. For Henry's trip to Europe and his Smithsonian career, see ibid., 205–68, and Coulson, *Joseph Henry*, 114–207.

19. This summary of Morse's life is drawn from Carleton Mabee, *The American Leonardo: A Life of Samuel F. B. Morse* (New York, 1943), 3–147, and Kenneth Silverman, *Lightning Man: The Accursed Life of Samuel F. B. Morse* (New York, 2003), 3–123. Mabee's work has more detail. Silverman's includes some new material, but his source notes could hardly be arranged more inconveniently for the reader.

20. Robert Luther Thompson, *Wiring a Continent: The History of the Telegraph Industry in the United States, 1832–1866* (Princeton, 1947), 8; Mabee, *American Leonardo*, 148–54, 196–98; Silverman, *Lightning Man*, 152–59. Mabee and Silverman recount the later dispute that arose over whether Morse or a fellow passenger, Charles T. Jackson, came up with the idea and the design.

21. Mabee, *American Leonardo*, 155–56, 181–82; Silverman, *Lightning Man*, 124–32.

22. Mabee, *American Leonardo*, 183–93; Silverman, *Lightning Man*, 148–50; Sharlin, *Electrical Age*, 12–13. All three sources have pictures of Morse's original apparatus.

23. Sharlin, *Electrical Age*, 12; Klein, "What Hath God Wrought?" 39; Taylor, "Historical Sketch of Henry's Contribution," 43–44. For more detail on the code, see Mabee, *American Leonardo*, 201–5. Morse himself never approved of operators interpreting the messages by ear.

24. Taylor, "Historical Sketch of Henry's Contribution," 41–43; Mabee, *American Leonardo*, 189–91.

25. Taylor, "Historical Sketch of Henry's Contribution," 45–47; Silverman, *Lightning Man*, 160–63. Vail had attended seminary before going to the university. For background on the optical telegraph see Daniel R. Headrick, *When Information Came of Age: Technologies of Knowledge in the Age of Reason and Revolution, 1700–1850* (New York, 2000), 193–203.

26. Taylor, "Historical Sketch of Henry's Contribution," 63–65; Mabee, *American Leonardo*, 201–25, 249–50; Silverman, *Lightning Man*, 164–95.

27. Mabee, *American Leonardo*, 251–72; Silverman, *Lightning Man*, 211–32; Klein, "What Hath God Wrought?" 41–42.

28. Mabee, *American Leonardo*, 272–74. Mabee attributes the idea for stringing the wire on poles to Vail. Robert Charles Post, *Physics, Patents, and Politics: A Biography of Charles Grafton Page* (New York, 1976), 66, credits the idea to Page. Silverman, *Lightning Man*, 232, says Morse had described the method six years earlier when first applying for the appropriation.

29. Mabee, *American Leonardo*, 274–75; Silverman, *Lightning Man*, 234–36.

30. Mabee, *American Leonardo*, 276–78; *New York Herald*, June 4, 1844; Silverman, *Lightning Man*, 243.

31. Mabee, *American Leonardo*, 281–377; Silverman, *Lightning Man*, 274–440. For the dispute with Henry see Coulson, *Joseph Henry*, 214–34, and Moyer, *Joseph Henry*, 243–61.

32. Frederic Hudson, *Journalism in the United States, from 1690 to 1872* (New York, 1872), 603.

33. Arthur A. Bright Jr., *The Electric-Lamp Industry: Technological Change and Economic Development from 1800 to 1947* (New York, 1949), 19–21.

34. Sharlin, *Electrical Age*, 133–41.

35. Ibid., 141–43; Harold C. Passer, *The Electrical Manufacturers, 1875–1900* (Cambridge, Mass., 1953), 12. A commutator is, in Sharlin's words, "a rotating switch that reverses the connection between the armature winding and the outside circuit each time the current changes direction in the winding. In this arrangement the external current always flows in the same direction."

36. Sharlin, *Electrical Age*, 141–45; Dunsheath, *History of Electrical Power Engineering*, 99–104.

37. Dunsheath, *History of Electrical Power Engineering*, 104–6; Sharlin, *Electrical Age*, 145.

38. Dunsheath, *History of Electrical Power Engineering*, 107–13; Sharlin, *Electrical Age*, 145–47; Bordeau, *Volts to Hertz*, 138–41. The principle of self-excitation was developed by three different inventors at about the same time: Charles Wheatstone and C. F. Varley of England and William Siemens of Germany. Henry Wilde of England was one of the first to use electromagnets for field excitation in a machine he patented in 1864. In this machine part of the current generated in the coils of the rotating armature was fed back into the coils of the electromagnet, increasing the magnetic strength.

39. Elihu Thomson, "Electricity in 1876 and in 1893," *Engineering* 6:444–45; Sharlin, *Electrical Age*, 146.

40. Dunsheath, *History of Electrical Power Engineering*, 38, 123–25.

41. Ibid., 126; Sharlin, *Electrical Age*, 149–50; Bordeau, *Volts to Hertz*, 147–48; Bright, *Electric-Lamp Industry*, 21–29. Bright lists the early inventors who contributed to the arc light.

42. Jonnes, *Empires of Light*, 47; Josephson, *Edison*, 177.

43. Charles F. Brush, "Some Reminiscences of Early Electric Lighting," *Journal of the Franklin Institute* 206 (July 1928): 3–4, and *NCAB* 21:1.

44. Brush, "Reminiscences," 4–5; Passer, *Electrical Manufacturers*, 14–15. Brush's original dynamo was later exhibited at the Paris Exposition of 1900.

45. Brush, "Reminiscences," 6; Passer, *Electrical Manufacturers*, 15.

46. Brush, "Reminiscences," 6–7; Passer, *Electrical Manufacturers*, 16–17.

47. Bright, *Electric-Lamp Industry*, 31; Brush, "Reminiscences," 9–11; Passer, *Electrical Manufacturers*, 17–18. Petroleum coke was also called "still coke" because it was the residue at the bottom of the paraffin stills at the Standard Oil refinery where Brush got his supply. Brush noted that the volume of carbon sales eventually reached 200 million.

48. Brush, "Reminiscences," 7–9, 14; Passer, *Electrical Manufacturers*, 16–18.

49. Brush, "Reminiscences," 11–13; Passer, *Electrical Manufacturers*, 18–20; *Scientific American*, December 25, 1880, 400.

50. *Scientific American*, April 10, 1880, 232.

51. Brush, "Reminiscences," 12–14; Passer, *Electrical Manufacturers*, 20–21.

52. Brush, "Reminiscences," 8–9.

53. Ibid., 14.

54. David O. Woodbury, *Beloved Scientist: Elihu Thomson, a Guiding Spirit of the Electrical Age* (New York, 1944), 82–83.

CHAPTER 6: A COVEY OF COMPETITORS

1. Franklin L. Pope, *Evolution of the Electric Incandescent Lamp* (Elizabeth, N.J., 1889), iii.

2. David O. Woodbury, *A Measure for Greatness: A Short Biography of Edward Weston* (New York, 1949), 1–24.

3. Ibid., 25–45.

4. http://electrochem.cwru.edu/ed/encycl; Sharlin, *Electrical Age*, 145.

5. Woodbury, *Measure for Greatness*, 46–65.

6. Ibid., 66–84; Passer, *Electrical Manufacturers*, 31–33.

7. Woodbury, *Measure for Greatness*, 85–92; Passer, *Electrical Manufacturers*, 32–33.

8. Woodbury, *Measure for Greatness*, 93–100; *Scientific American*, June 26, 1880, 399.

9. This account of Thomson's early years is drawn primarily from W. Bernard Carlson, *Innovation as a Social Process* (New York, 1991), 16–65, and Woodbury, *Beloved Scientist*, 3–61. Carlson's study, which does not purport to be a full biography, is the more authoritative and is fully documented. Woodbury's biography was commissioned by the Thomson family, has no documentation, and must be used with caution, but it contains many useful details along with some serious errors.

10. For background on Central High School, see Franklin Spencer Edmonds, *History of the Central High School of Philadelphia* (Philadelphia, 1902).

11. Woodbury, *Beloved Scientist*, 62, and Carlson, *Innovation as a Social Process*, 57–63, give radically different interpretations of this episode. Woodbury tells the story in a way that leads him to conclude that "the discovery was made almost wholly by Elihu Thomson, with Houston only looking on, and later making the report." Carlson assigns Houston a much larger role in the whole process. For a version favoring Edison, see Francis Jehl, *Menlo Park Reminiscences*, vol. 1 (Dearborn, Mich., 1936), 80–93.

12. *New York Herald*, December 2, 1875. For more detail, see Paul Israel, *Edison: A Life of Invention* (New York, 1998), 110–14, and Josephson, *Edison*, 127–29. In his experiment Edison used an electromagnet with a steel armature placed horizontally across its core. Whenever he stopped the flow of current to the electromagnet, a bright spark jumped between the core and armature.

13. For the controversy, see *Scientific American*, December 25, 1875, 400, January 1, 1876, 2, January 8, 1876, 17, January 29, 1876, 69, February 5, 1876, 89, and February 12, 1876, 101.

14. The best description of Thomson's second experiment is in Carlson, *Innovation as a Social Process*, 61–62.

15. For the situation at Central High School, see ibid., 141–42.

16. Ibid., 66–79; Elihu Thomson, "Personal Recollections of the Development of the Electrical Industry," *Engineering Magazine*, July 1905, 565.

17. Carlson, *Innovation as a Social Process*, 80–87; Woodbury, *Beloved Scientist*, 88–90.

18. This discussion is drawn from Carlson, *Innovation as a Social Process*, 86–95. This splendid work also includes details on the laws proposed by Thomson and Houston in their 1879 paper.

19. Ibid., 96–108.

20. Thomson, "Personal Recollections," 566. For a romanticized version of these events, see Woodbury, *Beloved Scientist*, 98–104.

21. Thomson, "Personal Recollections," 566; Carlson, *Innovation as a Social Process*, 116–27. Carlson includes more technical detail on the open-coil design, which utilized three coils.

22. Carlson, *Innovation as a Social Process*, 124–32.

23. For more technical detail on the regulator, see ibid., 135–36.

24. Ibid., 137–46.

25. Ibid. Garrett and McCollin then organized the Thomson-Houston Light Company of Philadelphia and sold arc-lighting systems to individual companies.

26. This account of Farmer is taken from Bright, *Electric-Lamp Industry*, 44–46; *DAB* 6:279–80; *NCAB* 7:361–62.

27. Post, *1876*, 64–65; Bright, *Electric-Lamp Industry*, 46. Post includes a picture of the Wallace-Farmer dynamos on display.

28. *New York Sun*, September 16 and October 20, 1878.

29. Ibid., September 10, 1878; Jehl, *Reminiscences* 1:210–11; Josephson, *Edison*, 178; Frank Lewis Dyer and Thomas Commerford Martin, *Edison: His Life and Inventions*, vol. 1 (New York, 1910), 246–48. The Dyer and Martin biography was authorized by Edison and must be handled with care. The *Sun* article and many other important documents are reproduced in Theresa M. Collins and Lisa Gitelman, *Thomas Edison and Modern America* (Boston, 2002).

30. Josephson, *Edison*, 1–33; Israel, *Edison*, 1–18; Dyer and Martin, *Edison* 1:9–113. Italics are in the original. For Edison's own version of his deafness, see Dagobert D. Runes (ed.), *The Diary and Sundry Observations of Thomas Alva Edison* (New York, 1968 [1948]), 44–55.

31. Josephson, *Edison*, 33–64; Israel, *Edison*, 19–47; Runes, *Edison Diary*, 47–48.

32. Josephson, *Edison*, 65–155; Israel, *Edison*, 48–118.

33. Jehl, *Reminiscences* 1:220, 223–24; Dyer and Martin, *Edison* 1:269.

34. Dyer and Martin, *Edison* 1:279–82; Jehl, *Reminiscences* 1:284–85, 2:515–19.

35. Josephson, *Edison*, 139–41; Israel, *Edison*, 131–40; Dyer and Martin, *Edison* 1:114–202; Sharlin, *Electrical Age*, 36–51.

36. Josephson, *Edison*, 159–73; Israel, *Edison*, 142–56; Dyer and Martin, *Edison* 1:203–33.

CHAPTER 7: THE LIGHT DAWNS

1. Josephson, *Edison*, 158.

2. Bright, *Electric-Lamp Industry*, 36–42.

3. Ibid., 38–41. Bright lists twenty inventors working on incandescent lights before 1879. For more detail on Swan, see Dunsheath, *History of Electrical Power Engineering*, 129–34.

4. Christopher J. Castaneda, *Invisible Fuel: Manufactured and Natural Gas in America, 1800–2000* (New York, 1999), 3–9. Castaneda provides an excellent and concise history of the industry.

5. Ibid., 13–44; George B. Cortelyou, *The Gas Industry* (New York, 1933), 7–8.

6. Bright, *Electric-Lamp Industry*, 42–56; Sharlin, *Electrical Age*, 151–53.

7. *New York Sun*, September 16, 1878; Francis R. Upton, "Edison's Electric Light," *Harper's Monthly*, February 1880, 534; Robert D. Friedel and Paul Israel, *Edison's Electric Light: Biography of an Invention* (New Brunswick, N. J., 1986), 8–9.

8. Josephson, *Edison*, 184–89; Israel, *Edison*, 173; Passer, *Electrical Manufacturers*, 84–85; Jean Strouse, *Morgan: American Financier* (New York, 1999), 182–83.

9. *New York Sun*, October 20, 1878; Samuel Insull, *The Memoirs of Samuel Insull*, ed. Larry Plachno (Polo, Ill., 1992), 33; Passer, *Electrical Manufacturers*, 86. Details on the organization of the company can also be found in Payson Jones, *A Power History of the Consolidated Edison System, 1878–1900* (New York, 1940), 25–31. See also the correspondence reproduced in Collins and Gitelman, *Edison and Modern America*, 85–87. According to Jones, $290,000 of the capitalization was paid in by January 16, 1879, the rest by November 11.

10. Jehl, *Reminiscences* 2:691–92, 699.

11. Ibid., 862.

12. Morgan to Walter Burns, October 30, 1878, in Collins and Gitelman, *Edison and Modern America*, 86–87. For the fall in gas stocks, see also *New York Herald*, October 12, October 14, and October 17, 1878.

13. Ibid., 82; Josephson, *Edison*, 186; Jehl, *Reminiscences* 1:217–18, 231–32; Dyer and Martin, *Edison* 1:242. The author of the 1879 book was Paget Higgs.

14. *New York Herald*, September 24 and October 12, 1878.

15. Ibid., October 14, October 17, October 18, and November 29, 1878; *New York Tribune*, December 26, 1878.

16. *New York Herald*, October 14, October 17, October 18, and November 29, 1878; *New York Tribune*, October 18, 1878.

17. *New York Herald*, October 17, 1878.

18. Passer, *Electrical Manufacturers*, 86–88; Jehl, *Reminiscences* 1:214–15. Passer includes more financial details. Before the new additions, Edison had only three buildings at Menlo Park: the laboratory, a carpenter shop, and the carbon shed. See Jehl, *Reminiscences* 1:214.

19. Jehl, *Reminiscences* 1:215; *New York Herald*, December 11, 1878; Josephson, *Edison*, 181, 185; *New York Tribune*, September 28, 1878; Israel, *Edison*, 169; Friedel and Israel, *Edison's Electric Light*, 29, 66–68; Dyer and Martin, *Edison* 1:263–66. "His notebooks relating specifically to 'electricity vs. gas as general illuminant' covered an astounding range of inquiry and comment," recalled Francis Jehl. Friedel and Israel include details on Edison's calculations of comparative costs between gas and electricity.

20. Friedel and Israel, *Edison's Electric Light*, 33–37. Descriptions and pictures of all these men can be found scattered through all three of Jehl's volumes.

21. For details on Edison's expenditures, see Jones, *Power History*, 32–34.

22. Passer, *Electrical Manufacturers*, 83; Josephson, *Edison*, 195–96; Jehl, *Reminiscences* 1:266–68. For Ohm's law and Joule's law, see chapter 4. "The ohm," wrote Jehl, "was the only unit of electricity we at Menlo Park could utilize with facility or to the highest degree of accuracy." For the problem of subdivision of light as then understood, see *New York Tribune*, December 26, 1878.

23. Dyer and Martin, *Edison* 1:250–52; Friedel and Israel, *Edison's Electric Light*, 116–17.

24. Jehl, *Reminiscences* 1:241–44, 269–70, 2:819–20; Josephson, *Edison*, 193–96; Friedel and Israel, *Edison's Electric Light*, 55–56. Friedel and Israel call Edison's grasp of the high-resistance principle a "major breakthrough" that "began to distinguish the work at Menlo Park from what had gone before."

25. Israel, *Edison*, 168–70; Friedel and Israel, *Edison's Electric Light*, 76.

26. Josephson, *Edison*, 183–84, 198–99; Israel, *Edison*, 171–72; Bright, *Electric-Lamp Industry*, 61–63; Friedel and Israel, *Edison's Electric Light*, 44–49. For Edison's coining of the term "filament," see Jehl, *Reminiscences* 1:346–47.

27. Josephson, *Edison*, 199; Israel, *Edison*, 172; Dyer and Martin, *Edison* 1:255–56.

28. Josephson, *Edison*, 192; Israel, *Edison*, 177–78.

29. *New York Sun*, November 28, 1879; Israel, *Edison*, 174–83; Friedel and Israel, *Edison's Electric Light*, 42–43; Josephson, *Edison*, 207–10.

30. Jehl, *Reminiscences* 1:292–95. Jehl includes a photograph of a replica of the wooden drum.

31. Israel, *Edison*, 182–83; Friedel and Israel, *Edison's Electric Light*, 67–75; Jehl, *Reminiscences* 1:298–312; Bright, *Electric-Lamp Industry*, 68; Jones, *Power History*, 65; *New York Herald*, December 10, 1878.

32. *New York Sun*, January 3, 1880; Israel, *Edison*, 180–83; Josephson, *Edison*, 199–200; Friedel and Israel, *Edison's Electric Light*, 51–54, 61–62, 81–82, 88, 116; Jehl, *Reminiscences* 1:250–54, 324–27; Dyer and Martin, *Edison* 1:256–57.

33. *New York Herald*, December 11, 1878, and January 17, 1879; *New York Tribune*, November 16, 1878.

34. *New York Herald*, March 27, 1879; Josephson, *Edison*, 201–3.

35. Dyer and Martin, *Edison* 1:289.

36. Ibid., 292–97; *New York Herald*, April 25–27, 1879.

37. Friedel and Israel, *Edison's Electric Light*, 76–81.

38. Israel, *Edison*, 184–86; Josephson, *Edison*, 214–16; Friedel and Israel, *Edison's Electric Light*, 94–98; Jehl, *Reminiscences* 1:330–43. For a sample of the work involved, see the documents in Collins and Gitelman, *Edison and Modern America*, 91–100. For the early telephone experiments see Jehl, *Reminiscences* 1:112–34. Friedel and Israel, 89–91, contains an excellent discussion of Edison's relationship with carbon.

39. Jehl, *Reminiscences* 1:344–46. "This outline," wrote Jehl, "does not convey a fair idea of the tedious, back-breaking and heart-breaking delays experienced as we went through the various processes."

40. Israel, *Edison*, 186–87; Josephson, *Edison*, 218–20; Friedel and Israel, *Edison's Electric Light*, 100–106; Jehl, *Reminiscences* 1:351–57. Jehl perpetuates the forty-hour myth, from which October 21 came to be celebrated as "Edison Lamp Day." Friedel and Israel show that the breakthrough came more gradually.

41. *New York Sun*, November 11 and December 22, 1879; *New York Herald*, December 21, 1879; Josephson, *Edison*, 221–23; Israel, *Edison*, 187; Jehl, *Reminiscences* 1:380–94. Edison had allowed the *Herald* reporter to take full notes for his story. See Francis Upton to his father, December 21, 1879, in Collins and Gitelman, *Edison and Modern America*, 102–3.

42. Friedel and Israel, *Edison's Electric Light*, 108–11; Jehl, *Reminiscences* 2:740.

43. *New York Sun*, December 23–24, 1879.

44. *New York Herald*, December 28, 1879.

45. Ibid., December 29–31, 1879.

46. Jones, *Power History*, 87; *New York Times*, December 28, 1879.

47. *New York Times*, December 29, 1879, and January 4, 1880.

48. *New York Herald*, January 1, 1880; Jehl, *Reminiscences* 2:487, 713. Jehl estimated that about a hundred lamps were lit during the demonstration.

49. *New York Herald*, January 1, 1880.

50. *New York Herald*, January 1 and January 4, 1880; *New York Sun*, January 3, 1880; Israel,

Edison, 187–88; Josephson, *Edison*, 223–27; Thomas Commerford Martin, *Forty Years of Edison Service* (New York, 1922), 21–22.

51. *Operator and Electrical World*, May 19, 1883, 309. Unless otherwise indicated, this account of Sawyer's life is drawn from Charles D. Wege and Ronald G. Greenwood, "William E. Sawyer and the Rise and Fall of America's First Incandescent Electric Light Company, 1878–1881," *Business and Economic History* 13 (1984): 31–48. For the Sawyer-Man lamp see Bright, *Electric-Lamp Industry*, 50–53.

52. Pope, *Evolution*, 6.

53. Hiram S. Maxim, *My Life* (New York, 1915), 120–22. In Maxim's book Sawyer is referred to simply as "Mr. D."

54. For Sawyer's early work, see Pope, *Evolution*, 7–9. Sawyer claimed that he was given the option of accepting Maxim as a joint inventor or leaving the firm and that he chose to leave. See Jehl, *Reminiscences* 2:708.

55. Pope, *Evolution*, 8–12, 30–37; Jehl, *Reminiscences* 1:255; Bright, *Electric-Lamp Industry*, 51–52.

56. S. L. Griffin to Lowrey, November 1, 1878, and Lowrey to Edison, December 10, 1878, in Collins and Gitelman, *Edison and Modern America*, 87–90; *New York Tribune*, November 16, 1878; *Scientific American*, December 7, 1878, 351.

57. *Scientific American*, January 8, 1876, 20, and February 5, 1876, 89.

58. *New York Sun*, December 22, 1879.

59. Ibid., December 23, 1879; *New York Herald*, December 24, 1879; Jehl, *Reminiscences* 1:395–97. For the Menlo Park episode, see Josephson, *Edison*, 225–26.

60. *New York Sun*, December 27, 1879; *Scientific American*, April 10, 1880, 230. Josephson, *Edison*, 225, identifies Sawyer as the drunk shouting curses at Edison. The *New York Sun*, January 1, 1880, alludes to an "inebriated pseudo scientist." For the exhibit of the Sawyer lamp, see *New York Times*, March 17, 1880.

61. For a less than reliable version of Sawyer's demise, see Maxim, *My Life*, 127–28.

CHAPTER 8: THE PEARL STREET SYSTEM

1. Josephson, *Edison*, 263–64. The quotation comes from Edison's testimony in a trial.

2. For details on the cost analysis, see Friedel and Israel, *Edison's Electric Light*, 119–25.

3. *New York Herald*, January 1, January 4, January 9, January 10, and January 23, 1880; Jehl, *Reminiscences* 2:537–40.

4. *New York Herald*, January 9, January 10, and January 23, 1880; Jehl, *Reminiscences* 2:492–94.

5. *New York Sun*, January 16 and February 27, 1880.

6. Israel, *Edison*, 191–96; Friedel and Israel, *Edison's Electric Light*, 129–30; *New York Herald*, December 30, 1879. For profiles of some of the new men, see Jehl, *Reminiscences* 1:401–8.

7. Jehl, *Reminiscences* 2:460–66, 500–507.

8. Josephson, *Edison*, 97–101, 104, 134–35; Israel, *Edison*, 73–75, 122–24; Robert Conot, *A Streak of Luck: The Life and Legend of Thomas Alva Edison* (New York, 1979), 46–48, 53, 55, 63, 66, 77–78, 87, 112, 121–22, 127, 136, 139, 162.

9. Josephson, *Edison*, 99–101; Conot, *Streak of Luck*, 87, 122.

10. Friedel and Israel, *Edison's Electric Light*, 146–47.

11. Israel, *Edison*, 197, 214, 314; Collins and Gitelman, *Edison and Modern America*, 122–23; Friedel and Israel, *Edison's Electric Light*, 140–47; Jones, *Power History*, 89–92; Jehl, *Reminiscences* 2:557–64, 747–49.

12. Israel, *Edison*, 196–97, 202–3; Friedel and Israel, *Edison's Electric Light*, 132–34, 156–57; Jones, *Power History*, 80; Dyer and Martin, *Edison* 1:262–63, 299–317; Josephson, *Edison*, 233–36; Jehl, *Reminiscences* 2:614–33. Bamboo was used in Edison lamps from July 1880 to 1894.

13. Friedel and Israel, *Edison's Electric Light*, 158–71.

14. Israel, *Edison*, 197; Friedel and Israel, *Edison's Electric Light*, 171–74; Jehl, *Reminiscences* 2:714–15.

15. Bright, *Electric-Lamp Industry*, 68; Josephson, *Edison*, 230–31; Dyer and Martin, *Edison* 1:341, 386; Jehl, *Reminiscences* 2:736–37, 821–24.

16. *New York Sun*, November 28, 1879; Bright, *Electric-Lamp Industry*, 69; Friedel and Israel, *Edison's Electric Light*, 64–65; Dyer and Martin, *Edison* 1:250; Jehl, *Reminiscences* 1:287–89, 2:637–45, 653–69, 819. Edison claimed to have perfected a meter as early as December 1878. See *New York Herald*, December 3, 1878.

17. Jones, *Power History*, 35; Passer, *Electrical Manufacturers*, 98; Josephson, *Edison*, 248.

18. Passer, *Electrical Manufacturers*, 92–93; Josephson, *Edison*, 248; Israel, *Edison*, 199; Jones, *Power History*, 60; Jehl, *Reminiscences* 2: 673, 786–813. Jehl provides useful detail on several aspects of the manufacturing process. Batchelor originally oversaw the factory, but health issues forced him to relinquish the post to Upton in January 1881. A memoir by Clarke can be found in Jehl, *Reminiscences* 2:855–63.

19. Jones, *Power History*, 100–103; Jehl, *Reminiscences* 2:713–26, 846–48; Friedel and Israel, *Edison's Electric Light*, 178–80.

20. Jehl, *Reminiscences* 2:727–28.

21. Passer, *Electrical Manufacturers*, 89–90, 94–95; Jones, *Power History*, 28. Passer says that six men formed Edison Illuminating, five of them from Edison Electric. Passer gives no source for his figure, and Jones names the individuals involved. A map of the First District can be found as front matter in the Jones book and in Friedel and Israel, *Edison's Electric Light*, 206.

22. Jones, *Power History*, 111–17; *New York Evening Post*, December 17, 20–21, 1880, January 11, 1881.

23. Josephson, *Edison*, 250–51; Israel, *Edison*, 205; Jehl, *Reminiscences* 1:402–5; Dyer and Martin, *Edison* 1:359–60.

24. Passer, *Electrical Manufacturers*, 94; Josephson, *Edison*, 249; Jehl, *Reminiscences* 2:743–45, 753–69.

25. Jehl, *Reminiscences* 2:505–6, 531–45, 738–39; Jones, *Power History*, 93–95; Friedel and Israel, *Edison's Electric Light*, 205–7.

26. Friedel and Israel, *Edison's Electric Light*, 181–82, 207; Jehl, *Reminiscences* 2:778–85.

27. Dyer and Martin, *Edison* 1:393; Friedel and Israel, *Edison's Electric Light*, 207–10.

28. Dyer and Martin, *Edison* 1:395–96; Friedel and Israel, *Edison's Electric Light* 210–12. Friedel and Israel note that in fact Edison, having defined the First District, had few options on locating his power station since he had to be within half a mile of any building he planned to light.

29. Jehl, *Reminiscences* 2:676–77, 866–94; Friedel and Israel, *Edison's Electric Light*, 198–99, 212–14; Jones, *Power History*, 139–46. For more detail on the Jumbo generators, see Martin, *Forty Years of Edison Service*, 46–50.

30. Friedel and Israel, *Edison's Electric Light*, 215–22; Dyer and Martin, *Edison*, 400–406.

31. *New York Herald*, September 5, 1882.

32. Josephson, *Edison*, 256.

33. Ibid., 264.

34. Jones, *Power History*, 177, 292. Batchelor, Jehl, and some other assistants were in Europe at the time.

35. Herbert L. Satterlee, *J. Pierpont Morgan: An Intimate Portrait, 1837–1913* (New York, 1939), 190, 207; Strouse, *Morgan*, 182–83.

36. Jones, *Power History*, 177–78, 295; Friedel and Israel, *Edison's Electric Light*, 222.

37. Jones, *Power History*, 178–81; *New York Times*, September 5, 1882.

38. *New York Herald, New York Sun, New York Tribune*, and *New York World*, September 5, 1882; *New York Times*, September 10, 1882; Jones, *Power History*, 182.

39. Jones, *Power History*, 154–55, 182–90. Jones provides both data on customers and lamps used and a list of first-year customers.

40. Ibid., 35–36; Passer, *Electrical Manufacturers*, 97–99.

41. Satterlee, *Morgan*, 208, 212–14.

42. Ibid., 216.

43. Dyer and Martin, *Edison* 1:373–74. The Vanderbilts later electrified their house through a central station.

44. Passer, *Electrical Manufacturers*, 112–14.

45. Jehl, *Reminiscences* 2:787, 790–815; Passer, *Electrical Manufacturers*, 93–94.

46. Jehl, *Reminiscences* 2:816; Passer, *Electrical Manufacturers*, 95.

47. Jehl, *Reminiscences* 2:676–78; Passer, *Electrical Manufacturers*, 94–95; Jones, *Power History*, 139–41; Israel, *Edison*, 214.

48. Israel, *Edison*, 204–5.

49. Ibid., 214–17; Jehl, *Reminiscences* 2: 897.

50. Jehl, *Reminiscences* 2:897–98.

51. Israel, *Edison*, 208–9.

CHAPTER 9: THE COWBIRD, THE PLUGGER, AND THE DREAMER

1. Carlson, *Innovation as a Social Process*, 1.

2. *Scientific American*, June 5, 1880, 355, and June 12, 1880, 368; *New York Herald*, May 15 and July 23, 1880. For Dr. Scott's magical brush, see the ads in *Scientific American* during late 1879 and 1880 and the *New York Evening Post*, November 29, 1880.

3. Jehl, *Reminiscences* 2:491, 700–705. For the rival versions, see *Scientific American*, April 10, 1880, 230; April 17, 1880, 245; August 21, 1880, 116; and October 23, 1880, 256.

4. Maxim, *My Life*, 1–141; *DAB* 12:436–37.

5. Maxim, *My Life*, 120; Jehl, *Reminiscences* 2:611–13, 705–8.

6. Jehl, *Reminiscences* 2:611–13, 705–8; Maxim, *My Life*, 140–41; *Scientific American*, October 23, 1880, 256; *New York Express*, November 26, 1880; Passer, *Electrical Manufacturers*, 147–48; Bright, *Electric-Lamp Industry*, 47–49. In his memoir Maxim is extremely vague about his newly invented lamp and neglects to mention his visit to Edison's laboratory or his luring Boehm away.

7. *New York Evening Express*, November 26, 1880; *New York Evening Post*, November 22, 1880.

8. Jehl, *Reminiscences* 2:705–8; Passer, *Electrical Manufacturers*, 147–48; *New York Evening Post*, November 12, 1880; Pope, *Evolution*, 76–80; Conot, *Streak of Luck*, 199.

9. *Scientific American*, November 27, 1880, 336; Jehl, *Reminiscences* 2:467–76; Bright, *Electric-Lamp Industry*, 53–56.

10. *Scientific American*, November 27, 1880, 336; Jehl, *Reminiscences* 2: 474; Bright, *Electric-Lamp Industry*, 55, 72–73, 105.

11. Woodbury, *Measure for Greatness*, 114–23. See also the Web site of the Weston Electrical Instrument Corporation at http://weston.ftldesign.com.

12. Woodbury, *Measure for Greatness*, 109–10, 124–47; Bright, *Electric-Lamp Industry*, 72–73; Passer, *Electrical Manufacturers*, 32.

13. Woodbury, *Measure for Greatness*, 148–55; Israel, *Edison*, 260–61.

14. Bright, *Electric-Lamp Industry*, 71–73; Passer, *Electrical Manufacturers*, 144–45.

15. Woodbury, *Beloved Scientist*, 108–9.

16. Carlson, *Innovation as a Social Process*, 147–54. Houston published numerous articles in *Electrical World* and elsewhere. He remained active in professional organizations. In 1894 he left his teaching post at Central High School and joined Arthur Kennelly in a consulting firm. See *EW*, February 24, 1894, 262.

17. Carlson, *Innovation as a Social Process*, 172–73, 207.

18. Ibid., 157–70. Carlson provides some technical detail on the improvements and accessories.

19. Ibid., 171–75; Thomson, "Personal Recollections," 567. Carlson includes a list of the customers who bought systems from American Electric.

20. Carlson, *Innovation as a Social Process*, 176–81.

21. Ibid., 181–82. Woodbury, *Beloved Scientist*, 131–41, offers a more romanticized version of these events but with no documentation.

22. Ibid., 185; Passer, *Electrical Manufacturers*, 51–52.

23. Carlson, *Innovation as a Social Process*, 186–87; Passer, *Electrical Manufacturers*, 51–52.

24. Carlson, *Innovation as a Social Process*, 187–92.

25. Ibid., 192–93; Thomson, "Personal Recollections," 567–68.

26. Carlson, *Innovation as a Social Process*, 193–96. Although the charter increased the capital stock to $1 million, the officers issued only $250,000 worth at first.

27. For a sketch of Coffin, see ibid., 206–11, and Charles E. Wilson, *Charles A. Coffin (1844–1926): Pioneer Genius of General Electric Company* (New York, 1946), 11–14.

28. Carlson, *Innovation as a Social Process*, 197–99.

29. Ibid.; Passer, *Electrical Manufacturers*, 144–46.

30. Carlson, *Innovation as a Social Process*, 203, 222.

31. Ibid., 199–200; Woodbury, *Beloved Scientist*, 144–47, 158–59.

32. Israel, *Edison*, 230–33; Conot, *Streak of Luck*, 218–19.

33. Israel, *Edison*, 230–33.

34. Jehl, *Reminiscences* 3:1121–37.

35. Conot, *Streak of Luck*, 226–27.

36. Ibid., 215; Jehl, *Reminiscences* 2:825–27, 3:1092–94; Passer, *Electrical Manufacturers*, 99, 178–79; Bright, *Electric-Lamp Industry*, 68. Bright describes the three-wire system this way: "With the two larger wires acting as the conducting mains and the smaller third wire serving as a neutral wire, lamps were connected in parallel between either of the outer wires and the third wire." A diagram of the the three-wire system is in Passer, 179.

37. Israel, *Edison*, 222–23; Jehl, *Reminiscences* 3:930.

38. Insull, *Memoirs*, 28–31; Forrest McDonald, *Insull* (Chicago, 1962), 3; Dyer and Martin, *Edison* 1:368–70; Jehl, *Reminiscences* 3:987; Israel, *Edison*, 220–21; Alfred O. Tate, *Edison's Open Door* (New York, 1938), 264.

39. Jehl, *Reminiscences* 3:1096–113; Dyer and Martin, *Edison* 1:425–28. Jehl includes a reproduction of Edison's twelve pages of handwritten notes.

40. Jehl, *Reminiscences* 3:1116–20; Dyer and Martin, *Edison* 1:438–42; Israel, *Edison*, 223, 497.

41. Conot, *Streak of Luck*, 215–17.

42. Ibid., 217–19; Josephson, *Edison*, 295; Runes, *Edison Diary*, 12.

43. Josephson, *Edison*, 294; Conot, *Streak of Luck*, 219–20; Israel, *Edison*, 225–26.

44. Josephson, *Edison*, 295–98; Israel, *Edison*, 225–27; Conot, *Streak of Luck*, 220–21; *New York Tribune*, October 26, 1884.

45. McDonald, *Insull*, 31; Conot, *Streak of Luck*, 221; Strouse, *Morgan*, 231.

46. Israel, *Edison*, 228.

47. Ibid.; Conot, *Streak of Luck*, 221–22; Josephson, *Edison*, 298–99; McDonald, *Insull*, 31–32; Vincent P. Carosso, *The Morgans: Private International Bankers, 1854–1913* (Cambridge, Mass., 1987), 271.

48. Israel, *Edison*, 228–29; Conot, *Streak of Luck*, 222; Passer, *Electrical Manufacturers*, 100–101.

49. Israel, *Edison*, 228.

50. Maury Klein, *The Change Makers* (New York, 2003), 88.

51. Stover, *American Railroads*, 152–53; Prout, *Westinghouse*, 21–86; Francis E. Leupp, *George Westinghouse: His Life and Achievements* (Boston, 1919), 59–60.

52. Prout, *Westinghouse*, 1–8; Leupp, *Westinghouse*, 10–17. There is no recent biography of Westinghouse.

53. Prout, *Westinghouse*, 8–9, 301; Leupp, *Westinghouse*, 10–11, 24, 45–46.

54. Prout, *Westinghouse*, 90, 248; Leupp, *Westinghouse*, 131; Passer, *Electrical Manufacturers*, 129; Charles A. Terry, *The Early History of the Westinghouse Electrical and Manufacturing Company* (East Pittsburgh, 1925), 9, 100. The Terry volume, a series of booklets, was done for the company's educational department and used in an extension course for employees.

55. Prout, *Westinghouse*, 212–23; Passer, *Electrical Manufacturers*, 130–32; Terry, *Early History*, 10–11. There is no biography of Stanley. For basic information on him, see Laurence A. Hawkins, *William Stanley (1858–1916): His Life and Work* (n.p., 1951), and http://web.mit.edu/Invent/iow/stanley.html.

56. Passer, *Electrical Manufacturers*, 130–31; Prout, *Westinghouse*, 224–26; Leupp, *Westinghouse*, 119–30.

57. Passer, *Electrical Manufacturers*, 131.

58. Ben Johnston (ed.), *My Inventions: The Autobiography of Nikola Tesla* (Williston, Vt., 1982), 18, 28. For differing versions of Tesla's childhood, see John J. O'Neill, *Prodigal Genius: The Life of Nikola Tesla* (Hollywood, Calif., 1978), 9–38, Margaret Cheney, *Tesla: Man out of Time* (Englewood Cliffs, N.J., 1981), 6–18, and Seifer, *Wizard*, 1–14. The literature on Tesla is extensive but must be approached with caution. The earliest biography by O'Neill, originally published in 1944, is hagiographic and often unreliable; see the comments by Ben Johnston in Tesla, *My Inventions*, 21–23. Cheney is better but perpetuates some of the myths spawned by O'Neill. Seifer provides the fullest, most authoritative, and best documented study, but the book has some curious quirks as well as numerous errors of fact. Taken as a whole, the bibliography on Tesla is almost as strange as its subject.

59. Tesla, *My Inventions*, 35–36.

60. Ibid., 33, 48–54; Seifer, *Wizard*, 11–14.

61. Tesla, *My Inventions*, 56; Seifer, *Wizard*, 15–16.

62. Tesla, *My Inventions*, 57; Seifer, *Wizard*, 16.

63. Seifer, *Wizard*, 17–20.

64. Ibid., 20–21; Tesla, *My Inventions*, 65.

65. Tesla, *My Inventions*, 59–60.

66. Ibid., 60–61, 65.

67. Ibid., 62–63; Seifer, *Wizard*, 22–24.

68. Tesla, *My Inventions*, 66–67; Seifer, *Wizard*, 24, 27–29.

69. Tesla, *My Inventions*, 67–68; Seifer, *Wizard*, 29–30.

70. Tesla, *My Inventions*, 70–71.

71. Ibid.

CHAPTER 10: THE ALTERNATIVE SYSTEM

1. Thurston, *History*, 3.

2. Passer, *Electrical Manufacturers*, 120–21; Carlson, *Innovation as a Social Process*, 217–18; Andre Millard, *Edison and the Business of Innovation* (Baltimore, 1990), 88; Jones, *Power History*, 209–12. Millard says that Pearl Street lost money for five years, but Jones and Passer show that it became profitable in 1884.

3. Passer, *Electrical Manufacturers*, 117–21.

4. Ibid., 118–23.

5. Ibid., 151–54.

6. Jones, *Power History*, 141; Jehl, *Reminiscences* 3:1003–4; Josephson, *Edison*, 299–300; Conot, *Streak of Luck*, 228.

7. Josephson, *Edison*, 301–2; Israel, *Edison*, 234–44; Conot, *Streak of Luck*, 228–30.

8. Josephson, *Edison*, 303–5; Israel, *Edison*, 244–45; Conot, *Streak of Luck*, 231–32.

9. Runes, *Edison Diary*, 14–15; Israel, *Edison*, 245–46.

10. Runes, *Edison Diary*, 17, 54–55; Israel, *Edison*, 246–47.

11. Runes, *Edison Diary*, 37; Israel, *Edison*, 247–53; Conot, *Streak of Luck*, 236–41.

12. Israel, *Edison*, 253–55.

13. Ibid., 256–57; Conot, *Streak of Luck*, 247; Josephson, *Edison*, 311.

14. Israel, *Edison*, 253–54; Conot, *Streak of Luck*, 243–44; Insull, *Memoirs*, 46.

15. Israel, *Edison*, 254; Conot, *Streak of Luck*, 244; McDonald, *Insull*, 38–39; Insull, *Memoirs*, 48. Alfred O. Tate became Edison's private secretary after the move, but Insull continued to handle Edison's financial affairs.

16. Jones, *Power History*, 36, 141; Jehl, *Reminiscences* 2:675, 836; Israel, *Edison*, 257.

17. Woodbury, *Measure for Greatness*, 151–53; Millard, *Edison and the Business of Innovation*, 7; Israel, *Edison*, 260–61.

18. Carlson, *Innovation as a Social Process*, 204, 218; Passer, *Electrical Manufacturers*, 27.

19. Carlson, *Innovation as a Social Process*, 218–23, 240.

20. Woodbury, *Beloved Scientist*, 159.

21. Carlson, *Innovation as a Social Process*, 209–10, 234; Wilson, *Coffin*, 14–16.

22. Carlson, *Innovation as a Social Process*, 211–12; Passer, *Electrical Manufacturers*, 27–28.

23. Carlson, *Innovation as a Social Process*, 212–14, 223; Passer, *Electrical Manufacturers*, 28–29.

24. Carlson, *Innovation as a Social Process*, 216–17; Passer, *Electrical Manufacturers*, 28–29.

25. Carlson, *Innovation as a Social Process*, 217–18.

26. William Stanley, "Alternating-Current Development in America," *Journal of the Franklin Institute*, June 1912, 561–62; Prout, *Westinghouse*, 89–90. The water main analogy was used in Harold L. Platt, *The Electric City: Energy and the Growth of the Chicago Area, 1880–1930* (Chicago, 1991), 71.

27. Carlson, *Innovation as a Social Process*, 249.

28. Sharlin, *Electrical Age*, 192.

29. Carlson, *Innovation as a Social Process*, 249–50. Carlson includes a reproduction of Thomson's sketch as well as the text.

30. Ibid., 251–53.

31. Hawkins, *Stanley*, 9–12; *DAB* 17:514–15.

32. Hawkins, *Stanley*, 12–14.

33. Stanley, "Alternating-Current Development," 565–67.

34. Ibid., 567.

35. Ibid., 563–64; Bright, *Electric-Lamp Industry*, 98–99; Sharlin, *Electrical Age*, 192–93; Passer, *Electrical Manufacturers*, 133–35; Prout, *Westinghouse*, 100–101; Thomas P. Hughes, *Networks of Power: Electrification in Western Society, 1880–1930* (Baltimore, 1983), 86–97. Passer includes a drawing of the difference between the series and parallel connections. He notes that Gaulard and Gibbs "never viewed the transformer as a voltage-changing device." Hughes provides the most detailed analysis of the Gaulard-Gibbs transformer. The three Hungarian pioneers were Károly (Charles) Zipernowsky, Miksa (Max) Déri, and Ottó T. Bláthy. For brief profiles of them see *EW*, April 16, 1892, 258.

36. *EW*, October 24, 1885, 167; Passer, *Electrical Manufacturers*, 131–32; Terry, *Early History*, 10.

37. Passer, *Electrical Manufacturers*, 132; Prout, *Westinghouse*, 101–3.

38. Seifer, *Wizard*, 46; Stanley, "Alternating-Current Development," 569–70; Hughes, *Networks of Power*, 101–3.

39. Stanley, "Alternating-Current Development," 567–68.

40. Ibid., 568; Woodbury, *Measure for Greatness*, 141; *DAB* 15:75–76; *EW*, October 19, 1895, 424.

41. Stanley, "Alternating-Current Development," 570–72; Jonnes, *Empires of Light*, 134.

42. Stanley, "Alternating-Current Development," 570–72; Jonnes, *Empires of Light*, 133–35. In his article Stanley erroneously reported the date of his first turning on the system as March 6.

43. Jonnes, *Empires of Light*, 135–36.

44. Stanley, "Alternating-Current Development," 573–75; Passer, *Electrical Manufacturers*, 137–38; Hughes, *Networks of Power*, 104–5.

45. Prout, *Westinghouse*, 111–13; Leupp, *Westinghouse*, 136–39; Terry, *Early History*, 10–11.

46. Leupp, *Westinghouse*, 139–40; Prout, *Westinghouse*, 114; Stanley, "Alternating-Current Development," 575.

47. Stanley, "Alternating-Current Development," 573; Prout, *Westinghouse*, 107.

48. Tesla, *My Inventions*, 71–72.

49. Ibid., 72; Seifer, *Wizard*, 37–38. For a different version of Edison's remark, see Hughes, *Networks of Power*, 112.

50. Tesla, *My Inventions*, 72; Seifer, *Wizard*, 38–39; Tate, *Edison's Open Door*, 149; Conot, *Streak of Luck*, 213.

51. Josephson, *Edison*, 233; Tesla, *My Inventions*, 72; Seifer, *Wizard*, 40–41.

52. Seifer, *Wizard*, 41–43.

CHAPTER 11: EVENTFUL CURRENTS

1. Quoted in Carlson, *Innovation as a Social Process*, 206.

2. This discussion of the gas industry is drawn from Passer, *Electrical Manufacturers*, 195–203, and Castaneda, *Invisible Fuel*, 57–65.

3. The data can be found in Passer, *Electrical Manufacturers*, 197.

4. Martin, *Forty Years*, 11.

5. Prout, *Westinghouse*, 200, 301–2; *EW*, March 21, 1914, 637.

6. Prout, *Westinghouse*, 5–6. The quotation is attributed to an "eminent engineer" who first came to work for Westinghouse in 1888. Charles F. Scott fits both of those criteria and is probably the source.

7. Ibid., 290, 293.

8. Terry, *Early History*, 19.

9. Leupp, *Westinghouse*, 287–89.

10. Jonnes, *Empires of Light*, 122.

11. Leupp, *Westinghouse*, 246–58; Prout, *Westinghouse*, 294–300.

12. Passer, *Electrical Manufacturers*, 138, 141.

13. Ibid., 138–39, 141–42; Terry, *Early History*, 24–25. For a brief early history of meters, including that of Shallenberger, see *EW*, May 30, 1891, 404–5. For a brief profile of Shallenberger, see *EW*, June 23, 1894, 827.

14. Passer, *Electrical Manufacturers*, 138–40; Terry, *Early History*, 17, 35.

15. Passer, *Electrical Manufacturers*, 141; Terry, *Early History*, 30–34; Prout, *Westinghouse*, 114. Terry gives no further information about the kodak. It is not clear whether it preceded or followed George Eastman's famous 1888 camera or whether the name was borrowed from Eastman.

16. Passer, *Electrical Manufacturers*, 147; Prout, *Westinghouse*, 96–97; *EW*, February 2, 1889, 53. For a description and pictures of the early Westinghouse AC system, see *EW*, September 3, 1887, 125–27.

17. Jonnes, *Empires of Light*, 118; Carlson, *Innovation as a Social Process*, 255.

18. Carlson, *Innovation as a Social Process*, 255–56.

19. Ibid., 234–47.

20. Harold C. Passer, "Development of Large-Scale Organization," *Journal of Economic History* 12, no. 4 (Fall 1952): 379.

21. Carlson, *Innovation as a Social Process*, 225–35, 246.

22. Ibid., 257; Passer, *Electrical Manufacturers*, 145–46. Passer has additional details on the agreement and a supplementary one signed in December 1887.

23. Passer, *Electrical Manufacturers*, 146.

24. Ibid., 146; *EE*, November 1889, 462; Carlson, *Innovation as a Social Process*, 212, 298. Brush had also just developed its own AC system. See *EE*, August 1889, 338–41.

25. Carlson, *Innovation as a Social Process*, 257; Passer, *Electrical Manufacturers*, 146–50.

26. Carlson, *Innovation as a Social Process*, 259–61.

27. This sketch of the new facility is drawn from Israel, *Edison*, 260–76, and Millard, *Edison and the Business of Innovation*, 10–21.

28. Israel, *Edison*, 264, 266; Conot, *Streak of Luck*, 273–75.

29. Israel, *Edison*, 266–70; Millard, *Edison and the Business of Innovation*, 45, 55.

30. Millard, *Edison and the Business of Innovation*, 55; Tate, *Edison's Open Door*, 140.

31. Millard, *Edison and the Business of Innovation*, 60–87; Israel, *Edison*, 272, 277–302. For a profile of Kennelly, see *EW*, March 24, 1892, 391.

32. Israel, *Edison*, 305.

33. Ibid., 310–17; Millard, *Edison and the Business of Innovation*, 89–90.

34. Millard, *Edison and the Business of Innovation*, 91–92.

35. Sharlin, *Electrical Age*, 160–61.

36. Ibid., 161–62; *EW*, April 11, 1891, 270. Sharlin provides more detail on all these developments.

37. Millard, *Edison and the Business of Innovation*, 90; Israel, *Edison*, 219–22.

38. Millard, *Edison and the Business of Innovation*, 96; Passer, *Electrical Manufacturers*, 172; Israel, *Edison*, 325–26.

39. Josephson, *Edison*, 346.

40. Millard, *Edison and the Business of Innovation*, 96; Israel, *Edison*, 324; Josephson, *Edison*, 345; Conot, *Streak of Luck*, 252.

41. Millard, *Edison and the Business of Innovation*, 97; Israel, *Edison*, 303–4.

42. Millard, *Edison and the Business of Innovation*, 99–100; Israel, *Edison*, 303–5.

43. Millard, *Edison and the Business of Innovation*, 99.

44. Israel, *Edison*, 321–22.

45. Tesla, *My Inventions*, 72; Seifer, *Wizard*, 42–43; O'Neill, *Prodigal Genius*, 66–67.

46. Seifer, *Wizard*, 43–44. See also the obituary in *EW*, May 24, 1924, 1100.

47. Seifer, *Wizard*, 44–45; Thomas Commerford Martin, *The Inventions, Researches, and Writings of Nikola Tesla* (New York, 1995 [1894]), 5, 9.

48. Seifer, *Wizard*, 47–48; Martin, *Inventions, Researches, and Writings*, 9–25. For the paper see *Electrician*, June 15, 1888, 173–77.

49. Seifer, *Wizard*, 48.

50. Carlson, *Innovation as a Social Process*, 244; Passer, *Electrical Manufacturers*, 277. Woodbury, *Beloved Scientist*, 180, turns this story on its head: "Tesla was present in New York at Thomson's lecture and was so inspired by him that he abandoned Edison's direct-current work and began experimenting in the new field." He also declares that "Tesla respected and admired Professor Thomson and considered he had given him his start. But this did not prevent him from accepting credit for later inventions which were clearly not his." These statements do not bear close scrutiny. Tesla and Thomson crossed swords on several other issues later in their careers.

51. Passer, *Electrical Manufacturers*, 277–78.

52. Ibid.; Seifer, *Wizard*, 49.

53. Seifer, *Wizard*, 50.

54. Ibid., 51–52; Nikola Tesla, "Tribute to George Westinghouse," *EW&E*, March 21, 1914, 637.

55. Seifer, *Wizard*, 52; Passer, *Electrical Manufacturers*, 278. The terms given are taken from Passer, who cites a July 7, 1888, memorandum. Seifer's version, drawn from a July 11, 1888, memorandum, says that Tesla received $60,000 instead of $50,000 and two hundred shares of stock in Westinghouse Electric as well. Seifer also describes the royalty as being "$2.50 per watt," which is doubtless a slip of the pen.

CHAPTER 12: GAINING TRACTION

1. Frank J. Sprague, "The Electric Railway," *Century* 70 (July and August 1905): 435. This is a two-part article spread across two issues.

2. Sharlin, *Electrical Age*, 170–73. See, for example, the editorial in *EW*, December 12, 1896, 713.

3. Passer, *Electrical Manufacturers*, 213.

4. Ibid., 214; Sharlin, *Electrical Age*, 182; Nye, *Electrifying America*, 90; David M. Young, *Chicago Transit: An Illustrated History* (DeKalb, Ill., 1998), 14–16; Frank Rowsome Jr., *Trolley Car Treasury* (New York, 1956), 17–34; John Anderson Miller, *Fares, Please! A Popular History of Trolleys, Horsecars, Streetcars, Buses, Elevateds, and Subways* (New York, 1960 [1941]), 1–34;

Charles W. Cheape, *Moving the Masses: Urban Public Transit in New York, Boston, and Philadelphia, 1880–1912* (Cambridge, Mass., 1980), 1–5.

5. Passer, *Electrical Manufacturers*, 214; Rowsome, *Trolley Car Treasury*, 49–64; Miller, *Fares*, 35–53; Cheape, *Moving the Masses*, 5–6.

6. Passer, *Electrical Manufacturers*, 211–12; Jehl, *Reminiscences* 2:565–74; Sprague, "Electric Railway," 435–36; *Scientific American*, June 12, 1880, 368.

7. Passer, *Electrical Manufacturers*, 216–17. As Passer explains, "The ratio between the motor speed and the vehicle speed had to be chosen, and the position and manner of mounting the motor had to be determined."

8. Jehl, *Reminiscences* 2:576–79; Sprague, "Electric Railway," 437.

9. Jehl, *Reminiscences* 2:581–85; *EW&E*, June 10, 1899, 797–801; Dyer and Martin, *Edison* 1:454–56.

10. Jehl, *Reminiscences* 2:586; *Scientific American*, June 5 and June 12, 1880; Dyer and Martin, *Edison* 1:450, 458–59. For publicity about the railroad, see *New York Herald*, May 15, July 23, and July 24, 1880; *New York Graphic*, July 27, 1880.

11. Dyer and Martin, *Edison* 1:460–61; Passer, *Electrical Manufacturers*, 219–20; Josephson, *Edison*, 242.

12. Passer, *Electrical Manufacturers*, 220; Dyer and Martin, *Edison* 1:465–67; Sprague, "Electric Railway," 436–37, 440.

13. Passer, *Electrical Manufacturers*, 223–24. Frank Sprague thought that the "fundamental error" of both Edison demonstrations "was the use of the running rails only as the working conductors." He also thought that "Mr. Edison, engrossed in his electric light activities, simply missed his opportunity." See "Mr. Sprague Discusses Early Steps in the Development of Electric Traction," *American Electrical Railway Association*, February 1932, 849–50.

14. Passer, *Electrical Manufacturers*, 225–26; Rowsome, *Trolley Car Treasury*, 72–73.

15. Passer, *Electrical Manufacturers*, 227–29.

16. Ibid., 230–31; Rowsome, *Trolley Car Treasury*, 77–78; Sprague, "Electric Railway," 440. For a brief profile of Van Depoele, see *EW*, March 26, 1892, 210.

17. Passer, *Electrical Manufacturers*, 231–33; Sprague, "Electric Trolley," 444.

18. Passer, *Electrical Manufacturers*, 233.

19. Ibid., 233–36; Rowsome, *Trolley Car Treasury*, 78–80.

20. Harold C. Passer, "Frank Julian Sprague," in William Miller (ed.), *Men in Business: Essays in the History of Entrepreneurship* (Cambridge, Mass., 1952), 216–17.

21. Ibid., 217; Harriet Sprague, *Frank J. Sprague and the Edison Myth* (New York, 1947), 7–8.

22. Passer, "Frank Julian Sprague," 217–18.

23. Ibid., 219–20; Passer, *Electrical Manufacturers*, 237–38; Dyer and Martin, *Edison* 1:427. Sprague already knew the Hopkinson three-wire system well from his time in London.

24. Passer, "Frank Julian Sprague," 220; Dyer and Martin, *Edison* 1:427–29, 440; Passer, *Electrical Manufacturers*, 238; Israel, *Edison*, 222–23; Sprague, "Electric Railway," 445–46; "Mr. Sprague Discusses Early Steps," 851.

25. Passer, *Electrical Manufacturers*, 238–40.

26. Sprague, "Electric Railway," 446–47.

27. Ibid., 447–49; Passer, *Electrical Manufacturers*, 240–42.

28. Passer, "Frank Julian Sprague," 223; Passer, *Electrical Manufacturers*, 242–43; Sprague, "Electric Railway," 512; Sharlin, *Electrical Age*, 179. The contract stipulated that Sprague was to

complete the job within ninety days after track construction had gone far enough for electrical installation work to begin. Most of the material in Passer's earlier article on Sprague is repeated, often verbatim, in his later book.

29. Sprague, "Electric Railway," 512–13; Passer, *Electrical Manufacturers*, 243; *EW*, February 12, 1898, 221. The long Sprague quotation is taken from the latter source; a similar but modified version of it appears in the 1905 article cited.

30. Sprague, "Electric Railway," 513–14; Passer, *Electrical Manufacturers*, 243–44; Sharlin, *Electrical Age*, 182.

31. Sprague, "Electric Railway," 514.

32. Ibid.; Passer, "Frank Julian Sprague," 224–25.

33. Sprague, "Electric Railway," 515–16; Passer, "Frank Julian Sprague," 225.

34. Sprague, "Electric Railway," 516–17; Passer, "Frank Julian Sprague," 225–26.

35. Sprague, "Electric Railway," 516–19; Passer, "Frank Julian Sprague," 226. Passer, *Electrical Manufacturers*, 245, observes that "the real significance of the Richmond installation was that it contained the technical features which permitted successful operation on a large scale and set the pattern for electric street railway development." The *New York Times*, September 29, 1889, reported that "for a year or more the [Richmond] Union Company did an immense traffic, and no public complaint was made of the inefficiency of the system."

36. Cheape, *Moving the Masses*, 6; *New York Herald*, April 21, 1888; *EW*, October 18, 1890, 274; *New York Times*, March 31, 1890.

37. Passer, *Electrical Manufacturers*, 228, 249–50.

38. Ibid., 252; Sprague, "Electric Railway," 519; Passer, "Frank Julian Sprague," 228. For a comparison of the Sprague and Thomson-Houston systems, see *EW*, November 1, 1890, 312.

39. Passer, *Electrical Manufacturers*, 256–57; *WSJ*, October 12, 1889.

40. Passer, *Electrical Manufacturers*, 257–59; Terry, *Early History*, 57; *EW*, September 20, 1890, 197. Westinghouse also joined with George Pullman to explore the creation of an electric Pullman Palace Car. See *EW*, October 25, 1890, 294.

41. Passer, *Electrical Manufacturers*, 258–59.

42. *WSJ*, July 19 and September 20, 1890.

43. Israel, *Edison*, 322; Josephson, *Edison*, 352; Insull, *Memoirs*, 44.

44. Israel, *Edison*, 322.

45. Henry Villard, *Memoirs of Henry Villard*, vol. 2 (Boston, 1904), 325–26; Carosso, *Morgans*, 272–73; Conot, *Streak of Luck*, 276; Strouse, *Morgan*, 312. Drexel, Morgan took only a $600,000 share in the offering.

46. Israel, *Edison*, 323; Carosso, *Morgans*, 272–73; Passer, *Electrical Manufacturers*, 103; *EW*, January 12, 1889, 15, May 18, 1889, 286–87, May 18, 1889, 287, and December 7, 1889, 369; *EE*, February 1889, 43; *New York Times*, January 12, 1890. Passer says the new company incorporated in January; Carosso gives the date as May. Israel provides the exact date. The incorporation papers were filed January 4.

47. Passer, *Electrical Manufacturers*, 221.

48. Passer, "Frank Julian Sprague," 229–31; *New York Times*, January 12, 1890; *EW*, February 1, 1896, 128–31. For other work by Sprague, see *EW*, February 21, 1891, 143–46, and June 18, 1892, 415–17.

49. Passer, *Electrical Manufacturers*, 231–37; Sprague, "Electric Railway," 521–25; *EW*, May 1, 1897, 551, and August 7, 1897, 155–58.

CHAPTER 13: COMPETITION AND ELECTROCUTION

1. George Westinghouse Jr., "A Reply to Mr. Edison," *North American Review* 149 (December 1889): 654.

2. Alfred D. Chandler Jr., *The Visible Hand: The Managerial Revolution in American Business* (Cambridge, Mass., 1977).

3. Carlson, *Innovation as a Social Process*, 260.

4. Israel, *Edison*, 326; Mark Essig, *Edison & the Electric Chair* (New York, 2003), 114. Essig's fine book is succinct yet thorough in its research.

5. Passer, *Electrical Manufacturers*, 121, 149; Carlson, *Innovation as a Social Process*, 218; Conot, *Streak of Luck*, 254.

6. Conot, *Streak of Luck*, 254–55; Essig, *Edison & the Electric Chair*, 115.

7. Jonnes, *Empires of Light*, 147–48; *New York Times*, July 10 and August 12, 1888, and June 23, 1889; *EE*, April 1889, 155–56, and July 1889, 295–96. The copper corner can be followed in the *Times* and other papers during 1888 and 1889. For a summary, see *New York Times*, December 14, 1889, and *EE*, December 1889, 507.

8. Essig, *Edison & the Electric Chair*, 85–87; *Scientific American*, January 8, 1876, 16. In 1879 the *New York Herald* researched "the best substitute for the process of hanging criminals. See *New York Herald*, November 22, 1879.

9. Essig, *Edison & the Electric Chair*, 87–89.

10. Ibid., 89–94.

11. Ibid., 94–99; Terry S. Reynolds and Theodore Bernstein, "Edison and 'The Chair,'" *IEEE Technology and Society Magazine*, March 1989, 20.

12. Essig, *Edison & the Electric Chair*, 116; Reynolds and Bernstein, "Edison and 'The Chair,'" 20.

13. Essig, *Edison & the Electric Chair*, 116–18; Reynolds and Bernstein, "Edison and 'The Chair,'" 21; Israel, *Edison*, 327–28. Reynolds and Bernstein erroneously give the date of Edison's letter as December 9. The original can be viewed online at TAE along with Southwick's letters.

14. A copy of this pamphlet can be read online at TAE. Jonnes, *Empires of Light*, 150, makes far more out of this publication than the document warrants. She attributes it to Edison although the company issued it and Edward Johnson signed it. I know of no evidence that shows Edison to be the author, and Jonnes provides none. Nor is it likely that Johnson was merely fronting for Edison. At the time their relationship was rather strained. Israel, *Edison*, 330, attributes it to the company, and Josephson, *Edison*, 347, to Johnson. Jonnes writes that Edison "lashed out publicly, issuing what surely stands as America's longest and most splenetic howl of corporate outrage . . . the official public salvo in one of the most unusual and caustic battles in American corporate history." This description exaggerates and to some extent distorts the pamphlet's content, which is as much a defense of the Edison system as it is an attack on its rivals.

15. "A Warning from the Edison Electric Light Company," TAE, 3–25, 28–31, 66–69. The quotation is on p. 17.

16. Ibid., 26–27, 45–55, 58–60. The quotations are on p. 26. Emphasis in the original.

17. Passer, *Electrical Manufacturers*, 166.

18. *New York Times*, March 13 and June 6, 1888; *New York Tribune*, April 17, 1888; *New York Sun*, April 28 and June 6, 1888; *New York Herald*, April 30 and June 6, 1888; Jonnes, *Empires of Light*, 141–43; Leupp. *Westinghouse*, 143. For the storm, see *New York Evening Post*, March 12–16, 1888.

19. Essig, *Edison & the Electric Chair*, 136–39, has the most succinct account of the campaign to get the wires buried. See also *New York Times*, May 12, June 1, June 6, June 9, and June 10, 1888; *New York Sun*, June 1, 1888; *New York World*, June 16–18 and June 20, 1888; *New York Herald*, June 19, 1888. For the United States Illuminating Company suit, see *New York Times*, November 28, 1888, January 3, January 4, and January 8, 1889; *New York World*, July 4, July 17, and July 20, 1888; *New York Herald*, July 6 and September 25, 1888; *New York Sun*, July 20, 1888; *EE*, February 1889, 66.

20. *EW*, July 30, 1887, 56, August 6, 1887, 70, October 1, 1887, 186, November 19, 1887, 265, December 3, 1887, 289, April 27, 1889, 241, May 4, 1889, 255, 263, June 8, 1889, 333–34; *New York World*, June 28 and July 11, 1888; *New York Herald*, July 31 and August 31, 1888; *New York Sun*, September 1–2, 1888; *New York Times*, April 15–18, April 21, and May 8, 1889; *EE*, March 1889, 82, 140–46, April 1889, 179, May 1889, 230, 243–44, June 1889, 270, July 1889, 321. The history of this fascinating fight remains to be written. Its travails are best followed in the New York newspapers 1887–90. For background to the board, see the exposé in *New York Herald*, September 23, 1888.

21. Essig, *Edison & the Electric Chair*, 122–23; Israel, *Edison*, 328; *New York Times*, June 5, 1888; *New York Evening Post*, June 5, 1888; *New York Sun*, June 5–6, 1888. Emphasis is in original. Brown's letter was dated May 24 but was not published until June 5.

22. Israel, *Edison*, 328–29; Thomas P. Hughes, "Harold P. Brown and the Executioner's Current: An Incident in the AC-DC Controversy," *Business History Review* 32 (Summer 1958): 146–47. Brown's letter and the responses to it are reprinted in *EW*, July 28, 1888, 40–46.

23. Westinghouse to Edison, June 7, 1888, and Edison to Westinghouse, June 12, 1888, both in TAE.

24. *New York World*, June 24, 1888; *New York Herald*, June 24, 1888; *EE*, February 1889, 74–75.

25. Hughes, "Executioner's Current," 148; *New York World*, July 31, 1888.

26. *New York World*, July 31, 1888; *New York Times*, July 31, 1888; Essig, *Edison & the Electric Chair*, 146–47.

27. *New York World*, July 31, 1888; *New York Times*, July 31, 1888; *New York Sun*, August 4, 1888; Brown to Kennelly, August 4, 1888, TAE.

28. *New York Times*, December 6, 1888; *New York Sun*, November 15, 1888; Reynolds and Bernstein, "Edison and 'The Chair,'" 22; Hughes, "Executioner's Current," 151.

29. *New York Times*, December 6 and December 13, 1888; *New York Sun*, December 12–13, 1888; Essig, *Edison & the Electric Chair*, 152–53.

30. Millard, *Edison and the Business of Innovation*, 104–6; Israel, *Edison*, 330; Essig, *Edison & the Electric Chair*, 148–49.

31. *New York Evening Post*, December 12, 1888; *New York Times*, December 13 and December 18, 1888; *New York Sun*, December 13 and December 18, 1888; *EW*, January 5, 1889, 13.

32. Essig, *Edison & the Electric Chair*, 147, 160–61; Conot, *Streak of Luck*, 256.

33. Josephson, *Edison*, 349.

34. Essig, *Edison & the Electric Chair*, 192–93.

35. Ibid., 163–73; *New York Times*, October 10, 1889; Thomson to Coffin, May 16, 1889, in Harold J. Abrahams and Marion B. Savin (eds.), *Selections from the Scientific Correspondence of Elihu Thomson* (Cambridge, Mass., 1971), 69–70.

36. *EE*, April 1889, 155–56, and December 1889, 507; *New York Times*, May 8, May 18, June 26, and December 14, 1889; Hughes, "Executioner's Current," 156–57. Kemmler and his companion had run off together from unhappy marriages and settled in New York under assumed names.

37. Josephson, *Edison*, 333–37; Hughes, "Executioner's Current," 156–57; Passer, *Electrical Manufacturers*, 169; Jonnes, *Empires of Light*, 178–79; Conot, *Streak of Luck*, 282–86; *New York Tribune*, August 4, 1889; *EE*, July 1889, 295–96. The testimony and discussion of it can be followed in *New York Times*, July 10, July 12, July 16–17, July 19–20, July 24–26, and August 1–2, 1889; *New York Tribune*, July 16–17, July 20, July 23–24, and July 26, 1889; *EW*, July 13, 1889, 25–26, July 20, 1889, 43–44, July 27, 1889, 57, and August 3, 1889, 63, 76; *EE*, August 1889, 335, 366. For an adulatory article on Edison, entitled "Wizard Edison at Home," see *New York World*, November 17, 1889.

38. *New York Sun*, August 25, 1889; *EW*, August 31, 1889, 153–54; *EE*, September 1889, 377–78; Essig, *Edison & the Electric Chair*, 190–99. Essig provides considerable detail on this episode.

39. *New York Sun*, August 23, 1889.

40. *New York Times*, October 7, 1889; *New York Herald*, October 7, 1889.

41. Josephson, *Edison*, 354–56; Conot, *Streak of Luck*, 280–82.

42. Essig, *Edison & the Electric Chair*, 206.

43. *New York Times*, September 18 and October 10, 1889; *New York Tribune*, October 10, 1889; *New York Herald*, October 10, 1889; *New York World*, October 10 and October 12, 1889; *EW*, October 19, 1889, 268. For the Feeks episode, see *New York Times*, *New York Sun*, *New York Tribune*, and *New York World*, October 12, 1889. Essig, *Edison & the Electric Chair*, 212–16, draws his narrative largely from these sources.

44. *New York Times*, June 28, September 3, September 14, and October 9, 1889; *New York Tribune*, September 3, September 6–7, September 14, September 16–17, and October 9, 1889; *New York World*, September 3, September 6, September 14, and September 17, 1889. There were also some nonfatal electric accidents. See *New York Times*, September 20, 1889; *New York Tribune*, September 20 and September 22, 1889.

45. *New York Tribune*, October 10–11, 1889; *New York Times*, October 10, 1889; *New York World*, October 10–11, 1889; *New York Herald*, October 10, 1889. The *World*, October 12, 1889, listed those people killed or injured by electricity since January 1887.

46. *New York Times*, October 12, 1889; *New York World*, October 12, 1889, *New York Tribune*, October 12, 1889, *New York Herald*, October 12, 1889, *New York Sun*, October 12, 1889; Essig, *Edison & the Electric Chair*, 215.

47. *New York World*, October 12–13, 1889; *New York Times*, October 12, 1889.

48. *New York Tribune*, October 12, 1889.

49. Essig, *Edison & the Electric Chair*, 216–19; *New York World*, October 13–15, 1889; *New York Times*, October 13–15, 1889; *New York Tribune*, October 13–15, 1889; *New York Herald*, October 13–15, 1889.

50. *New York Tribune*, October 16–22, October 25, and October 27, 1889; *New York Herald*, October 16–22, October 26, and October 30, 1889; *New York Times*, October 16–22, October 25, and October 29–30, 1889; *New York World*, October 16–22, October 26, and October 29–30, 1889; *EW*, October 19, 1889, 268, October 26, 1889, 284, November 2, 1889, 289; *EE*, November 1889, 462.

51. Passer, *Electrical Manufacturers*, 169–71; *New York Times*, October 24, 1889; *EE*, November 1889, 498.

52. Harold P. Brown, "The New Instrument of Execution," *North American Review* 149 (November 1889): 586, 589, 593; Thomas A. Edison, "The Dangers of Electric Lighting," *North American Review* 149 (November 1889): 625.

53. Edison, "Dangers of Electric Lighting," 625–29.

54. Ibid., 630–34.

55. Westinghouse, "Reply to Mr. Edison," 653–55; *New York World*, December 1, 1889.

56. Westinghouse, "Reply to Mr. Edison," 655–58.

57. Ibid., 658–60.

58. Ibid., 661–64.

59. Essig, *Edison & the Electric Chair*, 220–23; *New York Times*, November 30–December 4 and December 12–16, 1889; *New York World*, November 30, December 2, December 10–11 and December 14–15, 1889; *New York Herald*, November 30, December 10–11, and December 14–15, 1889; *New York Tribune*, December 3 and December 10–15, 1889; *EE*, December 1889, 505–6, 520–21, 528.

60. *New York Tribune*, December 16–18, December 28, and December 31, 1889; *New York World*, December 16–18, December 20–22, December 26, and December 30, 1889; *New York Herald*, December 16, December 18, December 20–22, and December 28, 1889; *New York Times*, December 17, December 20–21, December 25, December 29, and December 31, 1889.

61. Essig, *Edison & the Electric Chair*, 224–33; *New York World*, December 27 and December 31, 1889; *New York Tribune*, December 31, 1889; *New York Times*, January 1 and February 12, 1890. A drawing of Fell's chair is in Essig, 239.

62. Essig, *Edison & the Electric Chair*, 234–42; *New York World*, April 30 and May 2, 1890; *New York Times*, March 22, April 29, April 30–May 2, May 6, and May 24, 1890; *New York Herald*, December 31, 1889, April 30 and May 4, 1890.

63. *New York Times*, June 25 and August 4–6, 1890. The death scene is well told in Essig, *Edison & the Electric Chair*, 243–54. See also all the major New York newspapers for August 7, 1890.

64. Essig, *Edison & the Electric Chair*, 254–56; *New York Herald*, August 7–8, 1890.

65. Essig, *Edison & the Electric Chair*, 256–57; *New York Times*, August 7, 1890.

66. Essig, *Edison & the Electric Chair*, 258–64, 330; *EW*, August 16, 1890, 97, 99–100; *New York Times*, August 17–18, October 9, and November 11, 1890.

67. For the Baring failure and its effects, see Carosso, *Morgans*, 298–302.

CHAPTER 14: MONEY, MERGERS, AND MOTORS

1. Edward W. Byrn, *The Progress of Invention in the Nineteenth Century* (New York, 1900), 38.

2. For a small example, see the price reduction advertisement by Edison in *New York Times*, September 2, 1890.

3. For background on the industrial securities market, see Thomas R. Navin and Marian V. Sears, "The Rise of a Market for Industrial Securities, 1887–1902," *Business History Review* 29 (June 1955): 105–38. See also Strouse, *Morgan*, 310, and *EW&E*, April 29, 1899, 534.

4. Navin and Sears, "Rise of a Market for Industrial Securities," 116–19, 125. Preferred stocks were considered a more conservative investment because they paid dividends ahead of common stock. Thomson-Houston obtained a charter amendment that authorized it to issue $5 million in preferred stock and to increase its common stock from $1 million to $10 million. See *EE*, April 1889, 190.

5. Carlson, *Innovation as a Social Process*, 259–61.

6. Ibid., 216, 264.

7. Ibid., 264–66.

8. Ibid., 262–63.

9. Ibid., 266–68; Woodbury, *Beloved Scientist*, 195–99, 202; Malcolm MacLaren, *The Rise of the Electrical Industry During the Nineteenth Century* (Princeton, 1943), 158–59.

10. Passer, *Electrical Manufacturers*, 175, 206.

11. Insull to Edison, July 16, 1890, TAE; Israel, *Edison*, 332–33.

12. Israel, *Edison*, 333–34; Edison to Henry Villard, February 8, 1890, TAE.

13. Passer, *Electrical Manufacturers*, 321–22; Carlson, *Innovation as a Social Process*, 291–92; Josephson, *Edison*, 359–60.

14. Edison to Villard, February 8, 1890, TAE.

15. Ibid.; Carlson, *Innovation as a Social Process*, 293.

16. Passer, *Electrical Manufacturers*, 149; Leupp, *Westinghouse*, 156–57; Passer, "Development," 389–90. The Pittsburgh factories made alternating-current equipment, arc-lighting equipment, and electric railway apparatus. New York manufactured direct-current equipment for arc, incandescent, and power use, while both it and Newark produced incandescent lamps. Newark also made some DC generators for electric railway use even though the motors were made in Pittsburgh.

17. *New York Times*, May 8, 1893.

18. Leupp, *Westinghouse*, 242.

19. Ibid., 239–40.

20. Ibid., 156–58; Arthur S. Dewing, *Corporate Promotions and Reorganizations* (Cambridge, Mass., 1930), 167–68. Westinghouse had earlier acquired the charter of a defunct company for the purpose of consolidating his several firms. See *New York Tribune*, July 12, 1889.

21. Leupp, *Westinghouse*, 156–58. Leupp errs in placing the new stock issue and corporate name change after the Baring collapse rather than before it.

22. Ibid., 157–58; Dewing, *Corporate Promotions and Reorganizations*, 168–69.

23. Leupp, *Westinghouse*, 159–60.

24. Clarence W. Barron, *More They Told Barron* (New York, 1973 [1931]), 38–39.

25. Ibid., 39–40.

26. Leupp, *Westinghouse*, 160–61; Dewing, *Corporate Promotions and Reorganizations*, 169.

27. Dewing, *Corporate Promotions and Reorganizations*, 169.

28. Ibid., 169–73; *CFC*, February 21, 1891, 322, May 16, 1891, 762, July 18, 1891, 98, and December 5, 1891, 846. The preferred stock was to receive priority in payment of dividends, with the "assenting" stock next in line. Some creditors balked at taking preferred stock for their claims. Westinghouse paid them off with his own personal note for $1 million secured by the preferred stock. For their pains the bankers received $125,000 in cash and a like amount of preferred stock.

29. Dewing, *Corporate Promotions and Reorganizations*, 172–74.

30. Navin and Sears, "Rise of a Market for Industrial Securities," 124–25; Leupp, *Westinghouse*, 160–61; Carlson, *Innovation as a Social Process*, 290–91, 293; Edward C. Kirkland, *Charles Francis Adams, Jr., 1835–1915: The Patrician at Bay* (Cambridge, Mass., 1965), 175–76.

31. Prout, *Westinghouse*, 275.

32. Seifer, *Wizard*, 58–59; Benjamin Garver Lamme, *An Autobiography* (New York, 1926), 60. Seifer offers the best vanilla version of this episode. Jonnes, *Empires of Light*, 227–29, recounts in toto an elaborate scene depicted in O'Neill, *Prodigal Genius*, 79–82, which is moving but apocryphal and lacking any source reference.

33. Terry, *Early History*, 57–58; Prout, *Westinghouse*, 161; Lamme, *Autobiography*, 37–38, 43–47; *WSJ*, February 28, 1891.

34. Lamme, *Autobiography*, 48, 209; Charles F. Scott, "Early Days in the Westinghouse Shops," *EW*, December 20, 1924, 587.

35. Prout, *Westinghouse*, 161–62; Seifer, *Wizard*, 59. Lamme, *Autobiography*, 47–48, said that Westinghouse had to scrap between two hundred and three hundred of the old motors. This comparison between Edison and Westinghouse is drawn in part from Steven W. Usselman, "From Novelty to Utility: George Westinghouse and the Business of Invention During the Age of Edison," *Business History Review* 66 (Summer 1992): 251–304. This excellent article is one of the few modern studies of Westinghouse.

36. Prout, *Westinghouse*, 6, 162, 255; Usselman, "From Novelty to Utility," 273–74.

37. Carlson, *Innovation as a Social Process*, 298.

38. Josephson, *Edison*, 360; *CFC*, February 21, 1891, 322. Emphasis is in the original.

39. Dietrich G. Buss, *Henry Villard: A Study of Transatlantic Investments and Interests, 1870–1895* (New York, 1978), 215–17; Villard, *Memoirs* 2:342–43, 357–58; *EW*, December 26, 1891, 462.

40. Josephson, *Edison*, 360; Carlson, *Innovation as a Social Process*, 293–94; Strouse, *Morgan*, 313.

41. Conot, *Streak of Luck*, 295; Villard, *Memoirs* 2:326; Edison to Villard, February 24, 1891, TAE.

42. This version of the merger is drawn from existing accounts of the negotiations, which are frustratingly sketchy and often contradictory. See Buss, *Henry Villard*, 218–19; Josephson, *Edison*, 362–63; Carlson, *Innovation as a Social Process*, 294–95; Conot, *Streak of Luck*, 301–2; Strouse, *Morgan*, 313; Woodbury, *Beloved Scientist*, 203–4; and Israel, *Edison*, 336.

43. The quotation is from Strouse, *Morgan*, 313. Conot, *Streak of Luck*, 301, put the situation this way: "Villard was anxious to sell out, Twombly to buy, Coffin to merge, and J. P. Morgan to end 'ruinous competition.' "

44. Passer, *Electrical Manufacturers*, 322; *Cleveland Plain Dealer*, February 10, 1892; *New York Journal*, February 13, 1892; *EW*, February 13, 1892, 100, and March 5, 1892, 148; *New York Times*, February 13 and February 18, 1892; Villard, *Memoirs* 2:326; Insull, *Memoirs*, 55. Villard's memoirs are conspicuously silent on the events leading up to the merger. Several historians, using a dubious source by Hugh R. Fraser, have portrayed the event as one in which Villard did much of the negotiating only to be shoved aside by Morgan at the end. Jonnes, *Empires of Light*, 240–43, sidesteps the issue and relies on the version in Josephson, *Edison*, 362–63. However, Josephson relies on the same flawed Fraser source, the weaknesses of which are outlined in Carlson, *Innovation as a Social Process*, 296–97. As Carlson indicates, there is little evidence that Villard played any significant role in the negotiations. Insull recalled that "Villard was not really anxious to see the consolidation go through."

45. Insull, *Memoirs*, 55–56.

46. *New York Times*, February 21, 1892.

47. Passer, *Electrical Manufacturers*, 322–24, lists the officers of the new company and their previous affiliation. See also the profile in *EW*, May 14, 1892, 331. Of the other Edison men, John Kruesi remained in charge of the Schenectady Works and Francis Upton of the East Newark lamp factory. J. P. Ord, Edison's comptroller, became second vice president of the new firm.

48. Carlson, *Innovation as a Social Process*, 311; Tate, *Edison's Open Door*, 260–62; Conot, *Streak of Luck*, 302; Insull, *Memoirs*, 56; Israel, *Edison*, 336–37.

49. *New York Times*, February 21, 1892; Israel, *Edison*, 336–37; Tate, *Edison's Open Door*, 278.

50. Tate, *Edison's Open Door*, 278–79.

51. Prout, *Westinghouse*, 122; Terry, *Early History*, 46; Sharlin, *Electrical Age*, 185.

52. Prout, *Westinghouse*, 122; Passer, *Electrical Manufacturers*, 281.

53. Prout, *Westinghouse*, 124; Lamme, *Autobiography*, 60.

54. Hughes, *Networks of Power*, 127–28. Hughes lists 25, 30, 40, 50, 60, 66⅔, 83⅓, 125, and 133⅓ as the cycles in use.

55. Terry, *Early History*, 48–49; Passer, *Electrical Manufacturers*, 280–81. Passer provides a more technical explanation of the phase issue.

56. Lamme, *Autobiography*, 59–65; Hughes, *Networks of Power*, 121; Prout, *Westinghouse*, 131–32; Passer, *Electrical Manufacturers*, 300–301.

57. Prout, *Westinghouse*, 131; *EW*, November 27, 1897, 630, February 11, 1899, 165–66.

58. Hughes, *Networks of Power*, 143, 161; Lamme, *Autobiography*, 26. Scott became chief electrician at Westinghouse in 1896 and later professor of electrical engineering at Yale.

59. Scott, "Early Days in the Westinghouse Shops," 537.

CHAPTER 15: A SHOW OF LIGHTS: CHICAGO 1893

1. Trumbull White and William Igleheart, *The World's Columbian Exposition, Chicago, 1893* (Philadelphia, 1893), 301. Barrett, the head of Chicago's electrical department, wrote the chapter on the electrical exhibits for this book.

2. Donald L. Miller, *City of the Century: The Epic of Chicago and the Making of America* (New York, 1996), 181; Robert Muccigrosso, *Celebrating the New World: Chicago's Columbian Exposition of 1893* (Chicago, 1993), 83.

3. David F. Burg, *Chicago's White City of 1893* (Lexington, Ky., 1976), 132; John E. Findling, *Chicago's Great World's Fairs* (New York, 1994), 18; Harold M. Mayer and Richard C. Wade, *Chicago: Growth of a Metropolis* (Chicago, 1969), 194. Mayer and Wade have a clear map of the exposition grounds. The statue was done by Daniel Chester French and the fountain by Frederick William MacMonnies. For more detail on the fountains, see *EW*, December 17, 1892, 388, and January 21, 1893, 42–43.

4. Burg, *Chicago's White City*, 139, 214.

5. Ibid., 216–23; Findling, *Chicago's Great World's Fairs*, 27. Most historians of the fair mention the performance of "Little Egypt, the Darling of the Nile," but, as Miller notes, Little Egypt made her debut not at the fair but afterward at Coney Island.

6. Burg, *Chicago's White City*, 224; Miller, *City of the Century*, 496–97; Muccigrosso, *Celebrating the New World*, 176–77. The wheel itself was 250 feet high and was set on a 15-foot platform.

7. White and Igleheart, *World's Columbian Exposition*, 575–77; Miller, *City of the Century*, 496–97; Muccigrosso, *Celebrating the New World*, 176–77. For a detailed description of the wheel's construction, see *Chicago Tribune*, June 18, 1893.

8. Burg, *Chicago's White City*, 204–5.

9. *EE*, August 16, 1893, 153; White and Igleheart, *World's Columbian Exposition*, 139–49; J. R. Cravath, "Electricity at the World's Fair," *Review of Reviews* 8 (July 1893): 36. Most sources refer to the giant steam engine as having 2,000 horsepower. However, both White and *EE* give the figure as 3,000 horsepower.

10. Platt, *Electric City*, 62–63; Burg, *Chicago's White City*, 204; F. Herbert Stead, "An Englishman's Impressions at the Fair," *Review of Reviews* 8 (July 1893): 34. Between September 1892 and June 1893, *Electrical World* ran articles on electricity at the fair in almost every issue.

11. *EW*, May 6, 1893, 334–38; Nye, *Electrifying America*, 37–39. For this temporary system the third rail was laid to the side of the other two rails rather than between them.

12. Murat Halstead, "Electricity at the Fair," *Cosmopolitan* 15 (September 1893): 578.

13. White and Igleheart, *World's Columbian Exposition*, 310–14; Cravath, "Electricity at the Fair," 36. For detailed descriptions of the underground facilities, see *EW*, July 16, 1892, 37, and November 26, 1892, 337. The main entrance to the subway was in Machinery Hall just north of the Westinghouse dynamos. Floor plans of the Electricity and other buildings can be found in *Chicago Tribune*, August 27, 1893.

14. *EW*, May 6, 1893, 333, and May 20, 1893, 380; White and Igleheart, *World's Columbian Exposition*, 319–23; McDonald, *Insull*, 61. Different sources give the number of lamps on the Tower of Light as being five thousand to eighteen thousand. White and Igleheart, 323, give the lowest figure. So too do figures for the cut glass vary from thirty thousand to forty thousand. Moses Farmer came to Chicago to help prepare the exhibit but caught pneumonia and died. See *EW*, June 3, 1893, 402, July 31, 1897, 128–29. For a photo of the Tower of Light illuminated, see *EW*, April 8, 1893, 263.

15. *EE*, August 16, 1893, 153–55; *EW*, July 22, 1893, 60; White and Igleheart, *World's Columbian Exposition*, 316–18; Seifer, *Wizard*, 119–20; Martin, *Inventions, Researches, and Writings of Nikola Tesla*, 477–85.

16. *New York Times*, June 10, 1893.

17. Burg, *Chicago's White City*, 198–201.

18. Ibid., 170–72, 184.

19. Ibid., 113.

20. Ibid., 75–76; Elizabeth Stevenson, *Park Maker: A Life of Frederick Law Olmsted* (New York, 1977), 395.

21. Burg, *Chicago's White City*, 75–88; Stevenson, *Park Maker*, 396–98; Miller, *City of the Century*, 380–85. For details on the electrical work, see *EW*, July 30, 1892, 69.

22. Burg, *Chicago's White City*, 83–85; Muccigrosso, *Celebrating the New World*, 72–74; Findling, *Chicago's Great World's Fairs*, 17.

23. Estimates on the construction costs vary from source to source. The figure given is taken from the official report of the auditor as reprinted in *CFC*, December 2, 1893, 939.

24. Jonnes, *Empires of Light*, 248–49; Charles H. Baker, *Life and Character of William Taylor Baker* (New York, 1908), 158.

25. Leupp, *Westinghouse*, 163; Terry, *Early History*, 82; *Chicago Tribune*, April 29 and May 17, 1892; *Chicago Times*, April 23, 1892; *Chicago Inter Ocean*, April 27, 1892. The *Chicago Tribune*, May 24, 1892, gives the total original bid as $1,713,567.

26. *Chicago Inter Ocean*, April 13, 1892; *Chicago Times*, April 23, 1892.

27. *Chicago Times*, April 23, 1892; Baker, *Life and Character of William Taylor Baker*, 159.

28. *Chicago Inter Ocean*, April 24 and April 26, 1892; Leupp, *Westinghouse*, 163.

29. *Chicago Inter Ocean*, April 27, 1892; *Chicago Tribune*, April 29, 1892.

30. Chicago *Tribune*, May 3 and May 7, 1892; *Chicago Inter Ocean*, May 3–4, 1892; *New York Times*, May 7, 1892.

31. *Chicago Times*, May 17, 1892.

32. Terry, *Early History*, 82–83; *Chicago Times*, May 17, 1892; *Chicago Tribune*, May 17, 1892.

33. Westinghouse's letter is reprinted in *Chicago Inter Ocean*, May 24, 1892.

34. Ibid., May 17, 1892.

35. Ibid., May 18 and May 20–21, 1892; *Chicago Times*, May 18, 1892.

36. Chicago *Inter Ocean*, May 21 and May 24, 1892; *Chicago Tribune*, May 24, 1892; *Chicago Times*, May 24, 1892.

37. *Chicago Inter Ocean*, May 21 and May 24, 1892; *Chicago Tribune*, May 24, 1892; *Chicago Times*, May 24, 1892.

38. Jonnes, *Empires of Light*, 256–57.

39. Ibid.

40. Leupp, *Edison*, 164–65; Passer, *Electrical Manufacturers*, 142–43, 154–55; Prout, *Westinghouse*, 135–37; *EW*, July 18, 1891, 39, July 25, 1891, 68, September 12, 1891, 171, May 14, 1892, 339, October 22, 1892, 267, January 7, 1893, 2, and January 28, 1893, 71, 77.

41. Passer, *Electrical Manufacturers*, 142–43; Leupp, *Westinghouse*, 165–69; Prout, *Westinghouse*, 136–37.

42. *Chicago Inter Ocean*, April 23 and May 3, 1892; *Chicago Tribune*, May 10, 1892; Cravath, "Electricity at the World's Fair," 37–38; *New York Times*, June 4, 1892; John Winthrop Hammond, *Men and Volts: The Story of General Electric* (Philadelphia, 1941), 213–15. This company history, though informative, must be used with care. In discussing the fair, for example, Hammond never mentions that Westinghouse received the lighting contract. In listing General Electric's chief competitors of that time, he neglects even to mention Westinghouse.

43. Passer, *Electrical Manufacturers*, 281; Terry, *Early History*, 84; Lamme, *Autobiography*, 61; Prout, *Westinghouse*, 138–39; Leupp, *Westinghouse*, 169.

44. *EE*, August 16, 1893, 153; *EW*, May 20, 1893, 374; Leupp, *Westinghouse*, 169; Terry, *Early History*, 84.

45. Terry, *Early History*, 85; Prout, *Westinghouse*, 139; *EE*, August 16, 1893, 153.

46. Martin, *Inventions, Researches, and Writings of Nikola Tesla*, 477–85; Seifer, *Wizard*, 121.

47. Seifer, *Wizard*, 119; White and Igleheart, *World's Columbian Exposition*, 319–29. For details on the operation of Gray's teleautograph machine, see *Chicago Tribune*, August 6, 1893.

48. Burg, *Chicago's White City*, 109–11; *Chicago Tribune*, May 2, 1893; *Chicago Times*, May 2, 1893; White and Igleheart, *World's Columbian Exposition*, 301.

49. *Pittsburgh Commercial Gazette*, May 2, 1893.

50. *Chicago Tribune*, May 1, May 2, May 5, May 8, May 11, May 13, and May 16, 1893.

51. Ibid., May 14 and May 25, 1893.

52. Burg, *Chicago's White City*, 112, 180; Muccigrosso, *Celebrating the New World*, 80, 87. According to Muccigrosso, the investors netted $1.4 million profit after expenses of $67 million.

53. Leupp, *Westinghouse*, 170.

54. Burg, *Chicago's White City*, 286–88; Miller, *City of the Century*, 531–32.

55. William E. Cameron, *The World's Fair: Being a Pictorial History of the Columbian Exposition* (New Haven, 1894), 318; Lamme, *Autobiography*, 66; Terry, *Early History*, 49. For more detail on Westinghouse's electric railway exhibits, see *EE*, August 16, 1893, 154.

56. Prout, *Westinghouse*, 123; Sharlin, *Electrical Age*, 185–87; Hughes, *Networks of Power*, 110–19. For an explanation of an induction motor, see Hughes, 111. For a profile of Ferraris, see *EW*, January 2, 1892, 3.

57. Sharlin, *Electrical Age*, 188; Hughes, *Networks of Power*, 125.

58. For more detail on this concept, see Maury Klein, "Replacement Technology: The Diesel as a Case Study," *Railroad History*, Spring 1990, 109–20.

59. Prout, *Westinghouse*, 140.

60. This section is drawn entirely from Thomson, "Electricity in 1876 and in 1893," 442–55. For a comparison of the Chicago fair with other world's fairs, see *EW*, November 11, 1893, 373.

61. For a profile of Helmholtz, see *EW*, October 6, 1894, 329–30.

62. This account of Tesla's appearance is drawn from *Chicago Tribune*, August 22, 1893, and Seifer, *Wizard*, 120–21.

63. White and Igleheart, *World's Columbian Exposition*, 95–97; Burg, *Chicago's White City*, 97.

64. White and Igleheart, *World's Columbian Exposition*, 98–138; Thomas J. Schlereth, *Victorian America: Transformations in Everyday Life, 1876–1915* (New York, 1991), 171.

65. Muccigrosso, *Celebrating the New World*, 93; Burg, *Chicago's White City*, 184, 199, 212; Findling, *Chicago's Great World's Fairs*, 29.

CHAPTER 16: THE NIAGARA FALLOUT

1. Essig, *Edison & the Electric Chair*, 254–64, 286–94; Richard Moran, *Executioner's Current: Thomas Edison, George Westinghouse, and the Invention of the Electric Chair* (New York, 2002), 213–31. As late as 1895 charges lingered that the victims died not from electrocution but from the "surgeon's knife on the post-mortem table." *EW*, June 8, 1895, 657.

2. *New York Times*, January 7, January 17, February 16, February 18, February 20, March 2, March 6–7, March 15, March 23, April 1, April 20, May 19, June 13, July 9, September 20, and October 18, 1890, January 21, 1892; *EW*, July 5, 1890, 1, March 14, 1891, 217, March 21, 1891, 23, April 11, 1891, 269, April 23, 1892, 275–76, December 1, 1894, 563, December 29, 1894, 659–60, May 18, 1895, 577–78, January 30, 1897, 159, 163–65.

3. Thomson-Houston also had the Brush AC system after buying that company in November 1889. See *EE*, August 1889, 338–41, and November 1889, 462–63.

4. Passer, *Electrical Manufacturers*, 150; Strouse, *Morgan*, 313; *EW*, February 13, 1892, 100.

5. Carlson, *Innovation as a Social Process*, 301–4.

6. Passer, *Electrical Manufacturers*, 151–54. For the McKeesport case, see *EW*, February 23, 1889, 96, March 9, 1889, 149, and March 23, 1889, 177. For the Canadian decision see *EW*, December 7, 1889, 368, and *EE*, February 1889, 41, and April 1889, 199–202.

7. Passer, *Electrical Manufacturers*, 154–61; *EW*, October 12, 1889, 243, supp. 1–2, October 19, 1889, 259, and October 26, 287; *EE*, June 1889, 286–93, and November 1889, 461, 487.

8. Passer, *Electrical Manufacturers*, 161–64; *EE*, June 1889, 293; *WSJ*, March 29, April 22–23, and May 16, 1893; *CFC*, April 1, 1893, 538, and April 23, 1893, 668; *New York Times*, May 8, 1893. The American patent law contained a clause to the effect that if an invention had received a foreign patent earlier than its American counterpart, the American patent would expire on the same date as the foreign one. Edison's Canadian patent expired on November 14, 1894, and with it the American version as well. See Passer, 153–54, 160.

9. Passer, *Electrical Manufacturers*, 256–62; Lamme, *Autobiography*, 43–48; Terry, *Early History*, 50.

10. Hughes, *Networks of Power*, 123–25.

11. Passer, *Electrical Manufacturers*, 164–65; *EW*, December 4, 1897, 657.

12. *WSJ*, May 4, 1893; *EW*, December 21, 1895, 662. Van Depoele died in March 1892, but he had assigned his patents to Thomson-Houston, and therefore to General Electric.

13. Usselman, "From Novelty to Utility," 269–71; Prout, *Westinghouse*, 307–8.

14. Usselman, From Novelty to Utility," 271–73.

15. Ibid., 273.

16. Passer, *Electrical Manufacturers*, 164–66.

17. Ibid., 282–83. For the early history of Niagara Falls and attempts to harness its power, see Edward Dean Adams, *Niagara Power: History of the Niagara Falls Power Company, 1886–1918*, vol. 1 (Niagara Falls, N.Y., 1927), 5–114. This two-volume set is a treasure trove of information on the Niagara project.

18. Adams, *Niagara Power*, 1:18–19, 32; Passer, *Electrical Manufacturers*, 282. Adams includes figures from several estimates of the falls' power made between 1841 and 1911.

19. Adams, *Niagara Power* 1:101–7; Passer, *Electrical Manufacturers*, 283; Steven Lubar, "Transmitting the Power of Niagara: Scientific, Technological, and Cultural Contexts of an Engineering Decision," *IEEE Technology and Society Magazine*, March 1989, 12.

20. Adams, *Niagara Power* 1:115–16; Passer, *Electrical Manufacturers*, 283.

21. Adams, *Niagara Power* 1:115–29; Passer, *Electrical Manufacturers*, 283–84; Satterlee, *Morgan*, 325.

22. Adams, *Niagara Power* 1:130–35.

23. Ibid., 141–42; Charles F. Scott, "Personality of the Pioneers of Niagara Power," 1. I am indebted to Robert Dischner of National Grid for furnishing me a copy of this article.

24. Adams, *Niagara Power* 1:142–44; Jonnes, *Empires of Light*, 281.

25. Adams, *Niagara Power* 1:144–52.

26. Coleman Sellers to Edward Dean Adams, October 5, 1893, TAE. See also the more detailed report by Sellers in Sellers to Adams, December 17, 1893, TAE.

27. Adams, *Niagara Power* 1:153–74. In addition to Lord Kelvin the commission consisted of Coleman Sellers, Professor Eleuthère Mascart, Lieutenant Colonel Theodore Turrettini, and Professor William Cawthorne Unwin.

28. Ibid., 162.

29. Ibid., 163–92, 363. All fourteen entries got a prize of some kind. See ibid., 444.

30. Ibid., 444–45; Passer, *Electrical Manufacturers*, 286–87. For the water turbine issue, see Adams, *Niagara Power* 2:85–115.

31. Adams, *Niagara Power* 2:178–80; Hughes, *Networks of Power*, 130–35; *EW*, January 9, 1892, 25–26. For a copy of Brown's paper, see Adams, *Niagara Power* 2:426–30. Soon afterward Brown left Oerlikon and became an independent consultant.

32. Quoted in Charles F. Britton, "An Early Electric Power Facility in Colorado," *Colorado Magazine* 49 (1972): 194.

33. Charles F. Scott, "Long Distance Transmission for Lighting and Power," *EW*, June 18, 1892, 419. Scott includes more technical detail and declares that "the simplicity of the system, and the perfection of its operation, both from a theoretical and practical standpoint, are noteworthy."

34. This account of the Telluride installation is drawn from Britton, "Early Electric Power Facility in Colorado," 186–95, and *EW*, March 21, 1891, 223, January 23, 1892, 32–33, and June 18, 1892, 420–21.

35. Terry, *Early History*, 47; *EW*, January 23, 1892, 151. The quotation is from *EW*, March 21, 1891, 223. A synchronous motor is basically an AC generator run in reverse. Since a synchronous motor lacks starting torque, it requires a starting motor to bring it up to speed.

36. Scott, "Long Distance Transmission for Lighting and Power," 421.

37. *EW*, January 23, 1892, 32–33. For a less optimistic view about alternating current, see *EW*, March 5, 1892, 151.

38. Adams, *Niagara Power* 2:190–91; *EW*, November 7, 1891, 346, and October 1, 1892, 205.

39. Lamme, *Autobiography*, 55–67. The kodak was a direct-current generator that was directly coupled to high-speed engines.

40. Ibid., 66; Adams, *Niagara Power* 2:193–94.

41. Adams, *Niagara Power* 2:195; Prout, *Westinghouse*, 253; Usselman, "From Novelty to Utility," 278.

42. Adams, *Niagara Power* 2:225; Lamme, *Autobiography*, 64; Jonnes, *Empires of Light*, 287–88.

43. Adams, *Niagara Power* 2:222–28. The letter of invitation with its stipulations is reprinted here.

44. Ibid., 173; William Stanley, "Notes on the Distribution of Power by AC," *EW*, July 6, 1889, 9, and February 6, 1892, 88; Lamme, *Autobiography*, 80.

45. Seifer, *Wizard*, 100–101; Jonnes, *Empires of Light*, 291.

46. Adams, *Niagara Power* 2:237; Terry, *Early History*, 49–50; Lamme, *Autobiography*, 80; Passer, *Electrical Manufacturers*, 313; Jonnes, *Empires of Light*, 291.

47. Adams, *Niagara Power* 2:194–95; Jonnes, *Empires of Light*, 293.

48. Jonnes, *Empires of Light*, 293–94; Tesla to Edward Dean Adams, January 9, February 2, February 6, March 12, and March 22, 1893. These letters may be viewed on the Web at www .electrotherapymuseum.com/TeslaArchive/Adams. A more complete collection of fourteen letters, including these five, is housed in the Niagara Mohawk Power archives now belonging to National Grid. I am grateful to Robert Dischner for furnishing me copies of these letters.

49. Tesla to Adams, March 12 and March 22, 1893, National Grid Archives.

50. Ibid., March 21 and March 22, 1893, National Grid Archives.

51. Adams, *Niagara Power* 2:228–29, 235. For Forbes's objections to the generator design, see George Forbes, "Harnessing Niagara," *Blackwood's*, September 1895, 439.

52. Adams, *Niagara Power* 2: 233–34.

53. Adams to General Electric Co., May 11, 1893, TAE.

54. Jonnes, *Empires of Light*, 297–98; *Saturday Review*, August 3, 1895, 135. For all the detail Adams lavished on other aspects of the Niagara project, he was conspicuously reticent on this episode. He said only that "after a careful analysis of the designs submitted and with the approval of its scientific staff the Cataract Company declined all proposals and proceeded under the guidance of Professor Forbes in the preparation of the plans and designs for a new type of generator." Adams, *Niagara Power* 2:229.

55. *Chicago Tribune*, May 6–7, 1893; *EW*, May 13, 1893, 352.

56. *EW*, September 16, 1893, 208.

57. *Chicago Tribune*, May 7, 1893.

58. Ibid.; *EW*, May 13, 1893, 352, and May 20, 1893, 371.

59. *EW*, May 20, 1893, 371; *EE*, June 14, 1893, 587.

60. Tesla to Adams, May 11, 1893, National Grid Archives.

CHAPTER 17: HARD TIMES

1. Adams, *Niagara Power* 2:256.

2. Ibid., 229–33; Forbes, "Harnessing Niagara," 431–32, 439–40; Jonnes, *Empires of Light*, 301–3. Jonnes mistakenly attributes the August 10 letter to Coleman Sellers. Not surprisingly, Forbes portrays himself as a giant among midgets in the development of machinery for the Niagara project.

3. Adams, *Niagara Power* 2:233; Passer, *Electrical Manufacturers*, 288–89; Jonnes, *Empires of Light*, 303. Italics are in the original. For a brief profile of Sellers see *EW*, February 17, 1894, 227.

4. Jonnes, *Empires of Light*, 303–4; Passer, *Electrical Manufacturers*, 289; Adams, *Niagara Power* 2:409–12.

5. Forbes, "Harnessing Niagara," 440.

6. *EW*, October 5, 1895, 361, 378.

7. Jonnes, *Empires of Light*, 304; Adams, *Niagara Power* 2:237.

8. Passer, *Electrical Manufacturers*, 290–91; Adams, *Niagara Power* 1:164.

9. *EW*, September 16, 1893, 208.

10. Adams, *Niagara Power* 2:237–38; Jonnes, *Empires of Light*, 304–5.

11. Passer, *Electrical Manufacturers*, 292; Adams, *Niagara Power* 2:238–39; *EW*, November 4, 1893, 356; *WSJ*, November 11, 1893.

12. Adams, *Niagara Power* 2:241–43; Jonnes, *Empires of Light*, 305. Technical details on the construction and breakthroughs are given by Lamme in Adams, *Niagara Power* 2:411–18. Forbes also delivered a paper on the Niagara installation that inspired ridicule. See *EW*, December 9, 1893, 439, January 13, 1894, 42, January 20, 1894, 92.

13. Jonnes, *Empires of Light*, 305–6; Adams, *Niagara Power* 2:245; Prout, *Westinghouse*, 155.

14. Adams, *Niagara Power* 2:245–46; Lamme, *Autobiography*, 69–70; Terry, *Early History*, 87–90; *EW*, February 3, 1894, 141–43, October 5, 1895, 361, 378.

15. Adams, *Niagara Power* 2:65.

16. Ibid., 116–17, 248–49, 442. General Electric also supplied the first five alternators for the Canadian powerhouse.

17. Ibid., 308–9; Passer, *Electrical Manufacturers*, 293; Charles C. Carr, *Alcoa: An American Enterprise* (New York, 1952), 12–90. Adams offers two different dates for the first power usage by Pittsburgh Reduction: August 5 and August 26, 1895.

18. Adams, *Niagara Power* 2:311–13; Pierre Berton, *Niagara: A History of the Falls* (New York, 1997), 168–69.

19. Adams, *Niagara Power* 2:307; *EW*, March 6, 1897, 341.

20. Adams, *Niagara Power* 2:307–23.

21. Ibid., 37, 170; *EW*, February 13, 1892, 101–3, November 21, 1896, 618, 621–22; Sharlin, *Electrical Age*, 195–96. For the Buffalo project, see also *EW*, November 28, 1896, 653–55, December 5, 1896, 685–86, December 12, 1896, 717–20, December 19, 1896, 749–52.

22. Passer, *Electrical Manufacturers*, 293; Adams, *Niagara Power*, 2:257–59; *EW*, March 9, 1895, 298, 301–2.

23. Adams, *Niagara Power* 2:367–68.

24. *WSJ*, March 3, May 23, July 26, August 14, October 17, and December 19, 1894, April 4, April 18, and May 9, 1896; *EW*, August 11, 1894.

25. *WSJ*, June 9 and June 14, 1893, April 10 and December 6, 1894, January 25, 1895; *EW*, February 17, 1894, 225, October 20, 1894, 414, June 22, 1895, 712; Cheape, *Moving the Masses*, 115–16.

26. *EW*, August 19, 1893, 140, December 30, 1893, 503.

27. Carlson, *Innovation as a Social Process*, 304; Passer, *Electrical Manufacturers*, 328; *WSJ*, February 28, 1893.

28. *WSJ*, March 29, April 13, and April 20, 1893; *CFC*, April 15, 1893, 625–27. The committee consisted of Henry L. Higginson and T. Jefferson Coolidge representing the Boston interests and Hamilton Twombly and Charles H. Coster the New York interests.

29. *WSJ*, April 20–21, May 5, May 11, May 20, May 24–26, May 29, and May 31, 1893; *CFC*, May 13, 1893, 792, May 27, 1893, 886.

30. Hammond, *Men and Volts*, 221–23; *WSJ*, July 9, July 21–22, and August 21, 1893; *CFC*, April 7, 1894, 600. In order to buy equipment from GE, the local electric companies obtained bank loans with GE endorsement and used them as partial payment. The endorsement made GE liable for the loans if the local company defaulted on payment.

31. *EW*, August 5, 1893, 107, August 19, 1893, 140.

32. Ibid., August 5, 1893, 107; Hammond, *Men and Volts*, 223–24; *WSJ*, March 11, July 29,

August 1, August 10, August 14, and August 16, 1893; *CFC*, August 5, 1893, 217, August 19, 1893, 298, April 7, 1894, 600. The last source has details on the plan.

33. Carlson, *Innovation as a Social Process*, 304; *Chicago Tribune*, August 13–14, 1893; *WSJ*, August 14, August 18, August 31, and September 1, 1893; *EW*, September 9, 1893, 204, September 16, 1893, 221. According to Carlson, the syndicate members interested in GE included the following: Morgan took $945,000, Twombly $780,000, Higginson $600,000, T. Jefferson Coolidge $556,000, Frederick L. Ames $556,000, and Coffin himself $272,000, a total of $3.709 million.

34. *WSJ*, September 5, September 7, September 13, September 28, October 27, and November 13, 1893, January 18, 1894; *CFC*, September 9, 1893, 422; *EW*, September 30, 1893, 260. Details on the plan are in *CFC*, October 28, 1893, 722, and May 19, 1894, 866. Once the market for the deposited securities improved, the trustees gradually sold them and used the proceeds to buy back the stock in the new company from those who took it.

35. *WSJ*, November 16–17 and November 21–25, 1893; *CFC*, November 25, 1893, 893; *EW*, October 7, 1893, 279, December 9, 1893, 451.

36. *WSJ*, December 1, December 4–6, December 12, December 14–16, and December 21, 1893; *CFC*, December 9, 1893, 979, December 16, 1893, 1039, December 30, 1893, 1122; *EW*, October 14, 1893, 301, November 18, 1893, 402, November 25, 1893, 419, December 2, 1893, 435, December 23, 1893, 485.

37. *WSJ*, December 6, December 22, and December 28, 1893, January 6, January 11, January 27, January 29, and May 28, 1894; *CFC*, December 30, 1893, 1122, January 6, 1894, 43, April 7, 1894, 600; *EW*, December 16, 1893, 469, January 6, 1894, 4, January 13, 1894, 65, February 10, 1894, 194, July 27, 1895, 103.

38. *EW*, February 24, 1894, 257, March 10, 1894, 346, March 17, 1894, 380, March 31, 1894, 446, April 7, 1894, 482, April 21, 1894, 554, April 28, 1894, 586.

39. For more detail, see the annual report in *EW*, April 14, 1894, 512–15, and *CFC*, April 7, 1894, 600–602.

40. *WSJ*, February 12, February 27, March 12, March 16, March 20, April 9, May 22–23, and September 4, 1894.

41. Carlson, *Innovation as a Social Process*, 304–6; Passer, *Electrical Manufacturers*, 328–29; *WSJ*, June 6–9, August 31, and October 13, 1894, January 12 and January 18, 1895; *CFC*, April 7, 1894, 600, May 4, 1895, 796; *EW*, August 27, 1898, 201–2.

42. *WSJ*, April 12, June 26, August 25, August 27, October 19, October 23, October 26, November 5, and December 26, 1894, February 2, February 8, February 13, February 27, March 9, April 9, June 14, July 1, July 15, August 28, September 7, November 9, and December 18, 1895; *CFC*, November 10, 1894, 835; *EW*, May 26, 1894, 724, June 2, 1894, 755, June 9, 1894, 787, June 16, 1894, 819, June 30, 1894, 882, August 14, 1894, 114, October 6, 1894, 349, November 3, 1894, 482, November 10, 1894, 506, November 17, 1894, 533, November 24, 1894, 558.

43. *WSJ*, April 3, April 10, and May 29, 1894, January 7–8, January 23, and May 1, 1895; *CFC*, April 14, 1894, 636, May 26, 1894, 902.

44. *WSJ*, April 8, May 16, 1893 ; *CFC*, May 27, 1893, 874, June 3, 1893, 932, May 19, 1894, 867; *EW*, May 20, 1893, 381; *Pittsburgh Commercial Gazette*, September 12, 1893.

45. *WSJ*, May 18, 1893, January 15, January 26, April 16, and May 10, 1894; *CFC*, May 12, 1894, 814–15; *EW*, June 24, 1893, viii; *Pittsburgh Commercial Gazette*, April 4, 1893.

46. Prout, *Westinghouse*, 308; *WSJ*, November 11 and December 27, 1893, April 18, 1894.

47. *WSJ*, April 26, May 9, May 12, June 26, July 24, July 28, August 14, and October 19, 1894. For progress on the new building, see *EW*, February 3, 1894, 162, February 17, 1894, 235, July 14,

1894, 41, August 18, 1894, 161, October 20, 1894, 414, December 15, 1894, 630, January 12, 1895, 64, February 23, 1895, 252.

48. *WSJ*, November 22, 1894; *EW*, February 23, 1895, 252, September 29, 1894, 321. For a detailed description of the new plant, complete with excellent photographs, see *EW*, June 22, 1895, 713–20, June 29, 1895, 745–52.

49. *WSJ*, April 30, May 1, June 5, and July 5–6, 1895; *CFC*, July 6, 1895, 25–26; *EW*, July 13, 1895, 60, July 27, 1895, 119.

50. *EW*, October 20, 1894, 414; *WSJ*, October 31 and November 1, 1895. For an advertisement boasting the Tesla polyphase system, see *EW*, June 29, 1895, xii.

51. Wilson, *Coffin*, 11–20.

52. Ibid., 21.

53. Carlson, *Innovation as a Social Process*, 306–8; *EW*, August 21, 1897, 209–10.

54. Carlson, *Innovation as a Social Process*, 308–11; Passer, *Electrical Manufacturers*, 296–305.

55. Hammond, *Men and Volts*, 63–64, 210.

56. Ibid., 210–12.

57. Passer, *Electrical Manufacturers*, 302–5.

58. Ibid., 306–7.

59. Ibid., 307–9; Hawkins, *William Stanley*, 24–25. Passer includes more technical details on the differences between the synchronous and inductor generators. See also *EW&E*, August 5, 1899, 186.

60. Passer, *Electrical Manufacturers*, 308–9; Hawkins, *William Stanley*, 26–27.

61. Passer, *Electrical Manufacturers*, 308–9.

62. Ibid., 333–34; *WSJ*, August 17 and October 3, 1894; *EW*, August 18, 1894, 161, September 8, 1894, 245, December 4, 1897, 661. The other investors in Walker Company were Roswell P. Flower, J. W. Hinkley, and Perry Belmont.

CHAPTER 18: THE FUTURE ARRIVES

1. Passer, *Electrical Manufacturers*, 297, 313–16. Unless otherwise indicated, the discussion of phase and frequency is drawn from Passer's excellent summary.

2. *EW*, February 9, 1895, 162, May 25, 1895, 611.

3. Ibid., May 25, 1895, 611.

4. Passer, *Electrical Manufacturers*, 314–15.

5. Ibid., 329–31; *EW*, November 23, 1895, 563, December 21, 1895, 662.

6. *WSJ*, July 17 and July 20, 1893, January 8, January 9, January 24, May 10, September 14, September 19, and November 14, 1894; *Chicago Tribune*, July 21, 1893; *CFC*, July 22, 1893, 144, September 15, 1894, 472.

7. *WSJ*, March 18 and March 21–22, 1895; *EW*, March 23, 1895, 381, March 30, 1895, 409.

8. *WSJ*, March 23, March 25–28, April 4–6, April 8–11, April 17–18, 1895; *CFC*, March 30, 1895, 561; *EW*, March 30, 1895, 385, April 6, 1895, 414, April 13, 1895, 443.

9. *WSJ*, April 18, April 23, May 8, and May 11, 1895; *EW*, April 20, 1895, 473, April 27, 1895, 497, May 4, 1895, 545.

10. *WSJ*, May 31, June 3–5, June 8, June 17, August 9, and August 13–14, 1895; *EW*, May 25, 1895, 620, August 24, 1895, 224.

11. *WSJ*, August 14–15, 1895; *EW*, June 8, 1895, 649, June 15, 1895, 700.

12. *WSJ*, April 12, May 10, May 25, June 11, July 5, and November 9, 1894, April 25, May 18 and July 26, 1895; *EW*, July 27, 1895, 103.

13. *EW*, July 27, 1895, 103, August 24, 1895, 207; *WSJ*, July 27 and August 21, 1895, February 4, April 22, and April 28, 1896.

14. *WSJ*, November 6, November 8, November 11, November 13, and November 20, 1895, January 21–22, January 30, February 15, February 29, March 3–4, March 9, and March 11, 1896; *CFC*, November 16, 1895, 871; *EW*, November 23, 1895, 563, February 29, 1896, 219.

15. *WSJ*, March 12–14 and April 22, 1896; *CFC*, March 14, 1896, 502, April 4, 1896, 635; *EW*, March 12, 1896, 306, March 21, 1896, 306–7, 323, April 25, 1896, 451. The patents excluded from the agreement pertained to cables and underground trolley material. The patent committee consisted of Coffin and his chief counsel, Frederick Fish, Westinghouse and his chief counsel, Paul D. Cravath, and E. B. Thomas, president of the Erie Railroad. See *WSJ*, April 22, 1896, and *CFC*, June 6, 1896, 1040.

16. *WSJ*, April 6–8, April 12, June 23, July 3, July 14, July 18, July 27, October 30, November 10–13, and November 17, 1896, January 7–8, February 13, and November 22, 1897; *EW*, December 18, 1897, 794, September 24, 1898, 309–10, April 1, 1899, 397; Passer, *Electrical Manufacturers*, 330–34.

17. Passer, *Electrical Manufacturers*, 315–16.

18. Ibid., 310.

19. Ibid., 389; Dickinson, *Short History*, 186–95, Hills, *Power from Steam*, 283; Abbot Payson Usher, *A History of Mechanical Inventions* (Cambridge, Mass., 1954), 382–96. Usher has the most detailed description of how turbines work.

20. Dickinson, *Short History*, 194–95; Passer, *Electrical Manufacturers*, 310; Rollo Appleyard, *Charles Parsons: His Life and Work* (London, 1933), 1–46. The quotation is taken from Dickinson.

21. Dickinson, *Short History*, 195–212; Passer, *Electrical Manufacturers*, 310; Hills, *Power from Steam*, 285–87. For a later version of the Laval turbine, see *EW*, February 10, 1894, 189.

22. Passer, *Electrical Manufacturers*, 310–11; *EW*, October 12, 1895, 408, June 25, 1898, 770; Prout, *Westinghouse*, 185.

23. Passer, *Electrical Manufacturers*, 311; Prout, *Westinghouse*, 185–86, 202–4; Hammond, *Men and Volts*, 275–82.

24. Sharlin, *Electrical Age*, 214–15. For the significance of the Hopkinsons' paper, see ibid., 161–63.

25. Passer, *Electrical Manufacturers*, 311–13; Hills, *Power from Steam*, 230–31; Lamme, *Autobiography*, 114; Prout, *Westinghouse*, 206.

26. Passer, *Electrical Manufacturers*, 311–13; Prout, *Westinghouse*, 201; *WSJ*, July 24, 1909.

27. Passer, *Electrical Manufacturers*, 341; Nye, *Electrifying America*, 186. The Sears material is drawn from the index in a facsimile 1902 Sears catalog in my possession. The Heidelberg belt is listed on 475–76.

28. Hawkins, *William Stanley*, 28–34; Hammond, *Men and Volts*, 348–49, 382, 384.

29. Carlson, *Innovation as a Social Process*, 311–43.

30. Ibid., 343; Woodbury, *Beloved Scientist*, 343.

31. Josephson, *Edison*, 474–85. For Edison's later years, see Josephson, *Edison*, 367–485, and Conot, *Streak of Luck*, 308–472.

32. Seifer, *Wizard*, 148–49; *EW*, March 23, 1895, 357; *New York Sun*, March 14, 1895. The *Sun* called the fire "more than a calamity. It is a misfortune to the whole world."

33. *EW*, April 14, 1894, 489–90. See also the analysis of Tesla's work in the same issue, 496–99.

34. O'Neill, *Prodigal Genius*, 81; Jonnes, *Empires of Light*, 354.

35. Seifer, *Wizard*, 167–70, 182–83.

36. Ibid., 192–203. For Tesla's almost mystical explanation of what he called "teleautomatics," see *My Inventions*, 93–110.

37. Ibid., 204–35. As early as 1896 Tesla talked of communicating with Mars as well as around the globe. He claimed to have "had the scheme under consideration for five or six years." See *EW*, April 4, 1896, 369. For other projects see *EW*, November 19, 1898, 527, and November 26, 1898, 552.

38. Ibid., 236–38.

39. Ibid., 245–73; Strouse, *Morgan*, 395; Tesla, *My Inventions*, 79–91.Tesla's account includes a photograph of the tower.

40. Seifer, *Wizard*, 274–79.

41. *EW*, January 25, 1896, 95.

42. Seifer, *Wizard*, 283–335; Jonnes, *Empires of Light*, 262–64. O'Neill, *Prodigal Son*, 214, asserts that Morgan continued to advance funds to Tesla "almost up to the time of his own death" in 1913, but the evidence contradicts him. Seifer suggests that Morgan and others were part of a vast conspiracy to block Tesla's invention because it threatened their interests in GE as well as copper holdings and other investments. Neither his argument nor his evidence is convincing. Seifer's handling of business history throughout his book is shaky; the careful reader will find numerous errors of fact.

43. Seifer, *Wizard*, 336–42.

44. O'Neill, *Prodigal Genius*, 316–17.

45. Ibid., 281–317; Seifer, *Wizard*, 378–445.

46. *EW*, September 23, 1893, 225.

47. For the copper issue and inventory examples, see *WSJ*, March 19, March 27–28, and May 7, 1908.

48. *WSJ*, January 14, February 18, March 14, and October 1, 1907; *EW*, July 24, 1897, 105, August 13, 1898, 168, September 10, 1898, 258, December 3, 1898, 605–7; *EW&E*, August 5, 1899, 214, October 7, 1899, 543–47. A list of the seventeen companies controlled by Westinghouse is in the January 14 issue of *WSJ*.

49. *WSJ*, March 14, 1907; Dewing, *Corporate Promotions and Reorganizations*, 178. Dewing has the most detailed and cogent account of the Westinghouse crisis.

50. *WSJ*, March 14, April 26–27, and May 16, 1907; Dewing, *Corporate Promotions and Reorganizations*, 178. The remaining $200,000 was lost through depreciation in the value of other securities held in the company's treasury.

51. Dewing, *Corporate Promotions and Reorganizations*, 179, 182.

52. *WSJ*, April 2 and April 26–27, 1907.

53. Ibid., April 26–27, May 16, and July 8, 1907; Leupp, *Westinghouse*, 206–7. Earlier, in 1897, Westinghouse had increased the capital stock from $10 million to $15 million and taken $3 million of the new shares for his own Security Investment Company. See *EW*, July 3, 1897, 21.

54. *WSJ*, July 8, 1907; Dewing, *Corporate Promotions and Reorganizations*, 182. The stock offering raised only about $2.5 million.

55. *WSJ*, June 10, June 17, July 1, and July 8, 1907.

56. Ibid., July 11, July 19, and July 22, 1907; *EW*, May 19, 1894, 665, 689–90; Leupp, *Westinghouse*, 208–9.

57. *WSJ*, July 19, July 22, and July 30, 1907; Leupp, *Westinghouse*, 208–9.

58. *WSJ*, July 30 and October 24, 1907, April 30 and July 25, 1908. Westinghouse owned all $6 million of the Security Investment Company stock and included many of his own holdings in other companies as part of its assets.

59. Ibid., July 30 and October 15, 1907; Strouse, *Morgan*, 573.

60. *WSJ*, August 7, August 10, August 12, August 23, August 28, and September 28, 1907, January 1, 1908.

61. Ibid., October 1, 1907; *Historical Statistics* 2:1023; Strouse, *Morgan*, 574.

62. For details on the crisis and Morgan's role, see Strouse, *Morgan*, 574–96, and Carosso, *Morgans*, 535–49.

63. Dewing, *Corporate Promotions and Reorganizations*, 182–83; Carosso, *Morgans*, 536–37; Leupp, *Westinghouse*, 207–8.

64. Leupp, *Westinghouse*, 209.

65. Ibid., 209–10; *WSJ*, October 24, 1907.

66. Dewing, *Corporate Promotions and Reorganizations*, 178–83; Leupp, *Westinghouse*, 211–12; *WSJ*, October 16, October 24, October 26, and November 14, 1907. For the company's balance sheet as of October 23, 1907, see *WSJ*, December 2, 1907.

67. Dewing, *Corporate Promotions and Reorganizations*, 183–85; *WSJ*, December 3 and December 23, 1907, January 21, 1908; *CFC*, January 25, 1908, xv, 233, February 29, 1908, 550.

68. Dewing, *Corporate Promotions and Reorganizations*, 185–86. The bankers did agree to extend Westinghouse's personal liabilities, which amounted to $8 million, by requiring him to put up additional collateral of $10 million. They stipulated that the liabilities could be paid off in equal installments over three years. See *CFC*, March 14, 1908, 672, March 21, 1908, 726.

69. Dewing, *Corporate Promotions and Reorganizations*, 186–87; Leupp, *Westinghouse*, 212.

70. Dewing, *Corporate Promotions and Reorganizations*, 187–89; Leupp, *Westinghouse*, 212–13; *WSJ*, April 10, 1908. Both Dewing and Leupp give more details on the plan, especially the bankers' provisions. See also *CFC*, April 11, 1908, 922.

71. Dewing, *Corporate Promotions and Reorganizations*, 190–91; Leupp, *Westinghouse*, 213; *WSJ*, April 30 and May 22, 1908. The stockholders were asked to buy one additional share for every four shares they owned. See *WSJ*, May 18 and October 2, 1908.

72. *WSJ*, April 11, April 30, May 11, May 13, May 18, May 22, May 27, May 30, June 12, June 15, and June 17, 1908; *CFC*, May 2, 1908, 1105, May 16, 1908, 1228, June 6, 1908, 1413, and June 13, 1908, 1474; Leupp, *Westinghouse*, 213–14.

73. *WSJ*, June 18, June 23, June 25, July 3–4, July 8, July 11, July 30, September 16–17, September 28, September 30, October 1–3, October 14, October 20, October 31, November 7, and November 13, 1908; *CFC*, July 4, 1908, 43, July 11, 1908, 101, September 5, 1908, 617, September 26, 1908, 815, October 3, 1908, 875, October 17, 1908, 1014; Dewing, *Corporate Promotions and Reorganizations*, 193–94; Leupp, *Westinghouse*, 214–15.

74. *WSJ*, November 23 and November 25, 1908; *CFC*, October 31, 1908, 1163, November 21, 1908, 1361, November 28, 1908, 1425, December 5, 1908, 1483, December 12, 1908, 1531, 1538; Dewing, *Corporate Promotions and Reorganizations*, 194–97; Leupp, *Westinghouse*, 215. Dewing described it as a "truly remarkable reorganization."

75. *WSJ*, June 25 and December 8, 1908; Dewing, *Corporate Promotions and Reorganizations*, 198–99. The bankers on the committee of six were Albert H. Wiggin of Chase National Bank, Thomas W. Lamont of J. P. Morgan & Company, and James S. Kuhn of Pittsburgh First National Bank. The merchandise representatives were Joseph W. Marsh of Standard Underground Cable and George M. Verrity of American Rolling Mill.

76. *CFC*, January 9, 1909, 236; Jonnes, *Empires of Light*, 344. For a brief sketch of Mather, see *NCAB* 17:351.

77. Leupp, *Westinghouse*, 249–52.

78. Dewing, *Corporate Promotions and Reorganizations*, 199–200; *WSJ*, July 24, 1909. Dewing credited Mather with showing "a far sighted wisdom in the direction of the Company's affairs," and regarded his death in 1911 as "a real calamity to the Company." For the March 1909 annual report, see *CFC*, July 24, 1909, 220. Mather did not make the first dividend payment on the preferred stock until October 1909. See *CFC*, October 2, 1909, 850.

79. Leupp, *Westinghouse*, 274–75.

80. Ibid., 216–17; Jonnes, *Empires of Light*, 344–46.

81. *WSJ*, November 13, 1909.

CHAPTER 19: MASTERING THE MYSTERIES OF DISTRIBUTION

1. William Eugene Keily (ed.), *Central-Station Electric Service: Its Commercial Development and Economic Significance as Set Forth in the Public Addresses (1897–1914) of Samuel Insull* (Chicago, 1915), 39–40.

2. All figures are taken from data in *Historical Statistics* 2:818, 821, 827, and Passer, *Electrical Manufacturers*, 343.

3. Insull, *Memoirs*, 3–18; McDonald, *Insull*, 1–19.

4. Insull, *Memoirs*, 7–8; McDonald, *Insull*, 4–8.

5. Insull, *Memoirs*, 27–30; McDonald, *Insull*, 19–20.

6. Insull, *Memoirs*, 30–31; McDonald, *Insull*, 21–22; Burton Y. Berry, "Mr. Samuel Insull," SI, box 17, folder 5, 7, 33–34. This typescript is the memoir of a man who, as third secretary of the American Embassy in Istanbul, served as guard and companion for Insull on his forced return to America to face trial in April and May 1934. "Mr. Insull's conversations with me during these twenty-four days," he wrote, "I recorded each night before I retired. What he said appears in the following pages, unaltered in order or textual content." Berry, "Insull," 1.

7. Insull, *Memoirs*, 31, 37; McDonald, *Insull*, 23.

8. McDonald, *Insull*, 35–36; Tate, *Edison's Open Door*, 53.

9. Insull, *Memoirs*, 46–48; McDonald, *Insull*, 37–38.

10. Insull, *Memoirs*, 49–50; McDonald, *Insull*, 38–39.

11. Insull to Alfred O. Tate, June 4, September 1–2, and October 6, 1887, TAE; Insull to Edison, October 6, 1887, TAE. The Edison papers have many more examples of Insull's money chase.

12. Insull to Edison, August 3, 1887, TAE; McDonald, *Insull*, 38–39.

13. McDonald, *Insull*, 41–42.

14. Ibid., 42–52; Insull, *Memoirs*, 54–59.

15. Insull, *Memoirs*, 58.

16. Ibid., 58–62; McDonald, *Insull*, 52–53; Edison to Insull, May 2, 1898, TAE.

17. Insull, *Memoirs*, 62–63, 66; McDonald, *Insull*, 56; Fish to Insull, June 25, 1892, SI; Coffin to Insull, June 25, 1892, SI.

18. Insull, *Memoirs*, 62–65; McDonald, *Insull*, 53–54.

19. Platt, *Electric City*, 40–58. Platt's excellent study is the fullest and most lucid account of the electrical industry in Chicago during the Insull era.

20. Ibid., 68–71; Keily, *Central-Station Electric Service*, 112, 320.

21. Platt, *Electric City*, 75–78; McDonald, *Insull*, 59; Chicago Edison executive committee

minutes, July 14, 1891, FM. Platt lists the companies acquired by Insull and their purchase price, which totaled only $511,000 for the fourteen smaller stations. All executive committee minutes references are taken from the notes made by Forrest McDonald and can be found in FM, box 2, folder 2.

22. Insull, *Memoirs*, 76; McDonald, *Insull*, 58; Platt, *Electric City*, 76–77.

23. Chicago Edison executive committee minutes, November 8, 1893, FM; Insull, *Memoirs*, 76; McDonald, *Insull*, 58–60; Keily, *Central-Station Electric Service*, 112. The two generators from the fair were still operating in 1909. The engines installed at Harrison Street station were marine-type triple-compound engines that required large amounts of water for condensation.

24. McDonald, *Insull*, 60–62.

25. Ibid., 63; Keily, *Central-Station Electric Service*, 58.

26. This discussion is drawn from McDonald, *Insull*, 64, and several of the essays in Keily, *Central-Station Electric Service*.

27. McDonald, *Insull*, 63–65.

28. Keily, *Central-Station Electric Service*, 24–29. In this talk Insull provides more explanation on these points. Load factor can be defined as the ratio of average power demand over a given period of time to the peak or maximum demand for that same period of time.

29. Ibid., 27–30; McDonald, *Insull*, 65.

30. Insull, *Memoirs*, 73–76; McDonald, *Insull*, 66; Platt, *Electric City*, 80, 312. Platt provides more detail on the terms of the contract.

31. McDonald, *Insull*, 66; Platt, *Electric City*, 80–81; Insull, *Memoirs*, 70–71.

32. McDonald, *Insull*, 67–68; Insull, *Memoirs*, 89; Platt, *Electric City*, 84; William Eugene Keily (ed.), *Public Utilities in Modern Life: Selected Speeches (1914–1923) by Samuel Insull* (Chicago, 1924), 347. Insull gives the date of his trip as 1896 in his memoirs and 1897 in his speech, but McDonald presents a convincing argument that it occurred in 1894.

33. McDonald, *Insull*, 67–68; Platt, *Electric City*, 84–85.

34. McDonald, *Insull*, 68.

35. Ibid., 68–69; Keily, *Central-Station Electric Service*, 41, 59–60, 217; Platt, *Electric City*, 84–85.

36. McDonald, *Insull*, 69–70; Platt, *Electric City*, 81.

37. Insull, *Memoirs*, 83–84; McDonald, *Insull*, 71.

38. Insull, *Memoirs*, 84–85; McDonald, *Insull*, 71–72; Chicago Edison board of directors minutes, June 5 and August 18, 1896, FM; Chicago Edison executive committee minutes, August 31, 1896, FM. McDonald writes that Insull "let his directors stew for two months. Then, late in August, he summarily announced to the executive committee that he was going to London to sell the bonds himself." Neither the minutes nor Insull's memoirs confirm this interpretation. Insull wrote that a large stockholder, Byron L. Smith, urged him to go to London.

39. Insull, *Memoirs*, 86–87; McDonald, *Insull*, 72–73; Chicago Edison board of directors minutes, February 18, 1896, FM.

40. The fullest account of these episodes is in John Franch, *Robber Baron: The Life of Charles Tyson Yerkes* (Urbana, Ill., 2006), 235–64. See also the briefer but more incisive account in McDonald, *Insull*, 82–90, which portrays Yerkes more as victim than villain in the fight. Prior to the Allen Act, the council was empowered to award only twenty-year franchises.

41. McDonald, *Insull*, 88–89; Platt, *Electric City*, 81–82, 313. McDonald gives the sale price at $50,000; Platt notes that a city report in 1907 put the figure at $170,000. The franchise was a bargain at either price. Insull also had to fend off an effort by newly created Chicago Union Traction Company to buy control of Chicago Edison. See *EW*, August 26, 1899, 328, December 2, 1899, 876. For background on this company, see Franch, *Yerkes*, 265–70.

42. Insull, *Memoirs*, 81–83; McDonald, *Insull*, 91–92. Insull put the stock in his name because the law did not permit one corporation to own stock in another. He did this with other purchases as well. See, for example, Chicago Edison board of directors minutes, April 19, 1898, FM. For the division of territory between the two companies, see the board of directors minutes, July 25, 1898, FM.

43. McDonald, *Insull*, 91–92.

44. Ibid., 93–95; Insull, *Memoirs*, 82–85.

45. McDonald, *Insull*, 102–3; Platt, *Electric City*, 85–86; *EW* June 12, 1897, 770–74.

46. *EW*, October 24, 1896, 505–6, January 20, 1898, 141–42.

47. Platt, *Electric City*, 86–87; Keily, *Central-Station Electric Service*, 45. Keily, 34–47, reprints Insull's entire speech but not the discussion afterward, which is summarized in Platt.

48. Platt, *Electric City*, 87, 315.

49. Ibid., 87, 89, 95; Keily, *Central-Station Electric Service*, 32.

50. Platt, *Electric City*, 89–91, 95–96; Passer, *Electrical Manufacturers*, 342–43.

51. Platt, *Electric City*, 98–101. Platt's account of Insull's sales campaign provides more detail.

52. Ibid., 102–4.

53. Ibid., 104–5. See the map showing downtown elevator installations in ibid., 106.

54. Ibid., 105, 107. For example, in 1907 United States Steel, the nation's largest company, placed an order worth $5.5 million with Allis-Chalmers. It was the largest order for AC generators yet made in the United States. The *Wall Street Journal* reported that "the use of electrical power in modern steel mills has now been extended to cover practically all the machinery of the plant; the various systems of teleferage, conveyers, and appliances for handling hot metal being now most frequently driven by individual motors." *WSJ*, May 28, 1907.

55. Platt, *Electric City*, 108–11. The figures are taken from the table on 110.

56. Hammond, *Men and Volts*, 275–79; "Story of the Curtis Turbine," 1, SI, box 19, folder 12. See also chapter 18.

57. Insull, *Memoirs*, 78–79; Commonwealth Edison board of directors minutes, February 19 and April 16, 1901, FM; Commonwealth Edison executive committee minutes, January 7, 1902, FM; McDonald, *Insull*, 98–99; Platt, *Electric City*, 114; Keily, *Central-Station Electric Service*, 137–38. McDonald says that Sargent told Insull so large a unit "could not be done." Platt declares that Ferguson and Sargent urged Insull "to insist on something larger still."

58. Commonwealth Edison board of directors minutes, January 7, 1902; Insull, *Memoirs*, 79; McDonald, *Insull*, 99. Insull used the phrase "massing of production" as early as the turn of the century; see McDonald, 98.

59. Insull, *Memoirs*, 79.

60. Ibid., 79–80; "Commonwealth Edison Company," SI, box 19, folder 6.

61. Insull, *Memoirs*, 80; Hammond, *Men and Volts*, 282; John W. Hammond, "Steam Turbine Development in the United States," 3–4, SI, box 19, folder 12.

62. Insull, *Memoirs*, 67–69; Platt, *Electric City*, 119, 322.

CHAPTER 20: THE EMPIRE OF ENERGY

1. Keily, *Public Utilities in Modern Life*, viii.

2. Berry, "Insull," 14.

3. These paragraphs on Insull's personal life are drawn from McDonald, *Insull*, 74–82, and my own research in SI. See also Insull, *Memoirs*, 66, 100–103.

4. McDonald, *Insull*, 96–97; Franch, *Yerkes*, 265–310.

5. McDonald, *Insull*, 97; Platt, *Electric City*, 119–20, 186–87. These two sources differ on several points and even spell Munroe's name differently. I have used McDonald's spelling; Platt uses Monroe. Platt provides considerably more detail.

6. Platt, *Electric City*, 119–20. In 1901 Union Traction spent $322,000 for electric power along with $240,000 for its cable system and $562,000 on horses.

7. Ibid., 120–22.

8. Ibid., 132–36; McDonald, *Insull*, 103; *WSJ*, June 18, 1907. At the time of the merger Chicago Edison had assets totaling $25.3 million compared to nearly $21 million for Commonwealth Electric. Chicago Edison earned 9.12 percent on its $13.6 million in stock compared to 3.94 percent by Commonwealth Electric on its $10 million in stock. See *WSJ*, May 28, 1907.

9. Insull, *Memoirs*, 160–61; McDonald, *Insull*, 205. For details on Stuart and the notes, see the handwritten memorandum in SI, box 19, folder 10.

10. Insull, *Memoirs*, 95, 98; Keily, *Central-Station Electric Service*, 136; Platt, *Electric City*, 172–73.

11. Platt, *Electric City*, 164; McDonald, *Insull*, 135–36.

12. Insull, *Memoirs*, 95–96; McDonald, *Insull*, 137–39; Keily, *Central-Station Electric Service*, 358–62.

13. Insull, *Memoirs*, 96–97; Keily, *Central-Station Electric Service*, 362–66; McDonald, *Insull*, 137–42. McDonald has a good discussion of the factors involved in the Lake County experiment.

14. Platt, *Electric City*, 181–90, 342–43; McDonald, *Insull*, 144; "Public Service Co. of Northern Illinois," 1–3, SI, box 19, folder 2.

15. Platt, *Electric City*, 164–78.

16. Ibid., 147; Insull, *Memoirs*, 98–99. Platt estimated that about half of Chicago's middle-class families had installed electricity by 1912.

17. Platt, *Electric City*, 97. For a sample of his speeches during these years, see the collection in Keily, *Central-Station Electric Service*. More can be found in SI, boxes 20–22.

18. This story is drawn from McDonald, *Insull*, 146–49. Although McDonald inferred some of his observations, he offers a persuasive argument for them.

19. One of Insull's other investments was in the Illinois Electric Vehicle Company, of which he served as president. See *EW*, July 1, 1899, 31.

20. Insull, *Memoirs*, 103; McDonald, *Insull*, 149. For Martin's background see *EW*, August 13, 1898, 171, April 22, 1899, 530.

21. Insull, *Memoirs*, 104–5; McDonald, *Insull*, 150–51. McDonald offers an explanation of the convoluted transactions and the reasons for them.

22. McDonald, *Insull*, 152–56.

23. Ibid., 156–57.

24. Ibid., 158; Insull, *Memoirs*, 180–81; "Chicago Rapid Transit Company," SI, box 19, folder 4; "Chicago, North Shore & Milwaukee Railroad Company," SI, box 19, folder 4.

25. Insull, *Memoirs*, 172–73; McDonald, *Insull*, 158–61.

26. Insull, *Memoirs*, 110–26; McDonald, *Insull*, 162–77; Keily, *Public Utilities in Modern Life*, 166–77.

27. McDonald, *Insull*, 177–88.

28. Ibid., 188–89; Platt, *Electric City*, 201–12.

29. Platt, *Electric City*, 201–17. Platt has an especially good discussion of the war's impact on energy usage.

30. McDonald, *Insull*, 192–97; Berry, "Insull," 28; Insull to Frank and Annie Rose, December 20, 1919, SI, box 2, folder 6. The money was paid to them jointly. For a survey of Insull's charitable activities, see McDonald, *Insull*, 241–45.

31. McDonald, *Insull*, 197–203.

32. Ibid., 203–4.

33. Ibid., 204–5; Insull, *Memoirs*, 161–62; "Halsey, Stuart & Co. Issues Statement on Bill, Four Hundred Million of Insull Securities," undated, SI, box 19, folder 13.

34. Insull, *Memoirs*, 172–73; McDonald, *Insull*, 206–7; Platt, *Electric City*, 226–29.

35. McDonald, *Insull*, 207–8; Platt, *Electric City*, 229–30.

36. McDonald, *Insull*, 208–9.

37. Ibid., 209–10.

38. Ibid., 210–11; *Chicago Tribune*, February 11, 1921. Peoples owned several large gas holders, which it filled with the dump gas on weekends and drained throughout the week.

39. Insull, *Memoirs*, 173–75; McDonald, *Insull*, 211–12. Mellon and his brother put up securities owned by them to back the project. Mellon's own holdings included the Aluminum Corporation of America, Gulf Oil, Koppers, the family banks, and numerous other companies. For a brief portrait of Mellon, see Maury Klein, *Rainbow's End: The Crash of 1929* (New York, 2001), 68–71.

40. The speech is reprinted in Keily, *Public Utilities in Modern Life*, 252–62. I have also used a typescript copy of the speech in SI, box 20, folder 8.

41. Quoted in Platt, *Electric City*, 237.

42. Ibid., 220–24.

43. Ibid., 224–26, 237–41; Nye, *Electrifying America*, 275–76.

44. Platt, *Electric City*, 243.

45. Ibid., 243–44, 251; Insull to Edison, September 27, 1921, SI, box 3, folder 3. Edison returned the letter to Insull with his reply scribbled on it in pencil.

46. Platt, *Electric City*, 245, 247, 253. Platt notes that "exterior lighting added a new dimension to the decorative design of buildings and became almost as important as ensuring adequate interior illumination" (247).

47. Ibid., 249–67. As Platt indicates, Insull opened a demonstration farm in Lake County in 1928 that drew more than fifty thousand visitors in eighteen months. He also launched a five-year plan to finance rural electrification that doubled the size of his rural distribution network. See ibid., 269–70.

48. McDonald, *Insull*, 230–31; Insull to "Chappie," June 3 and June 24, 1923, SI, box 1, folder 9. Midland Utilities was originally called Public Service Investment Company and is called that in the June 24 letter. It was owned jointly by Morgan's United Gas Improvement Company of Philadelphia, Commonwealth Edison, Peoples Gas, PSCNI, Middle West Utilities, and Insull himself.

49. McDonald, *Insull*, 229–30.

50. Ibid., 230–32.

51. Ibid., 223–25.

52. Ibid., 228, 248–49. The investment trust was an early version of the mutual fund. For more detail on its origins, see Klein, *Rainbow's End*, 129.

53. Ibid., 249–51; *Time*, September 23, 1929; Klein, *Rainbow's End*, 168; Ron Chernow, *The House of Morgan: An American Banking Dynasty and the Rise of Modern Finance* (New York, 1990), 308–9.

54. McDonald, *Insull*, 251–52.

55. Ibid., 247.

56. Ibid., 278–79; Insull, *Memoirs*, 188, 192.

57. McDonald, *Insull*, 280–81; Insull, *Memoirs*, 189. For some details on the creation of IUI, see Samuel Insull to Martin J. Insull, August 27, 1928, and Samuel Insull to Samuel Insull Jr., August 27, 1928, both in SI, box 1, folder 2, and "Insull Utility Investments," January 15, 1929, SI, box 8, folder 1.

58. McDonald, *Insull*, 281–82; Insull, *Memoirs*, 190; Klein, *Rainbow's End*, 155, 165–89.

59. McDonald, *Insull*, 282; Insull, *Memoirs*, 191; Insull to Insull, Son & Co., Ltd., August 5 (two letters), August 6, and August 7, 1929, SI, box 1, folder 2; Insull to Samuel Insull Jr. and F. J. McEnroe, August 20, 1929, SI, box 1, folder 9. The technical flaw involved preemptive rights that went with the IUI stock.

60. McDonald, *Insull*, 282–83; Insull, *Memoirs*, 191; Insull to Samuel Insull Jr., August 20, 1929, SI, box 1, folder 9; H. L. Stuart to Insull, August 29, 1929, SI, box 8, folder 1; Samuel Insull Jr. to Insull, September 11, 1929, SI, box 7, folder 3. Insull wrote that "we wanted Insull Utility Investment Inc. to own as much Corporation Securities Company of Chicago stock as possible and wanted Corporation Securities Company of Chicago to own as much Insull Utility Investments Inc. stock as we could possibly carry." Untitled handwritten notes, SI, box 19, folder 10.

61. McDonald, *Insull*, 283–85.

62. Ibid., 318; Insull to Samuel Insull Jr., April 16, 1934, SI, box 2, folder 1; Berry, "Insull," 1–3, 5–6, 8–9, 27.

63. McDonald, *Insull*, 285–86.

64. The Eaton episode is detailed in ibid., 287–89, and Insull, *Memoirs*, 192–97.

65. Insull, *Memoirs*, 196–98; McDonald, *Insull*, 288–89. The Chicago bank was the new Continental Illinois National Bank and Trust Company, the product of a recent merger between Continental National Bank and John J. Mitchell's Illinois Merchants Trust Company. Mitchell had died in an automobile accident in 1927.

66. Insull, *Memoirs*, 198–203; McDonald, *Insull*, 290–94; Insull to E. Ogden Ketting, January 26, February 14, June 20, July 25, and November 27, 1931, SI, box 2, folder 4; Samuel Insull Jr. to Insull, April 23 and April 28, 1931, SI, box 7, folder 3.

67. Insull, *Memoirs*, 203–8; McDonald, *Insull*, 210, 294–96. The complicated dealings with the banks can be tracked in the Insull papers as well. See for example, "Statement by Mr. Samuel Insull, Jr.," undated box 7, folder 4, and untitled memorandum, December 14, 1931, box 19, folder 1. For Insull's version of a controversial $5 million loan, see Insull to Samuel Insull Jr., September 4, 1933, SI, box 1, folder 2.

68. Insull, *Memoirs*, 208–11; McDonald, *Insull*, 296–98.

69. Insull, *Memoirs*, 211–15; McDonald, *Insull*, 298–301; memorandum by Harold L. Stuart, August 22, 1934, SI, box 19, folder 3. Insull's version of the April 8 meeting in his memoirs is taken almost word for word from Stuart's memorandum.

70. Insull, *Memoirs*, 215–32; McDonald, *Insull*, 301–7; Chicago *Tribune*, June 7, 1932; Insull to Samuel Insull Jr., July 10 and August 30, 1932, SI, box 1, folder 10; "Statement by Mr. Samuel Insull for publication in morning newspapers of June 7, 1932," SI, box 8, folder 5. The latter statement includes a list of the companies from which Insull resigned. In the August 30 letter Insull says frankly that he was " 'fired' from the three companies."

71. McDonald, *Insull*, 307–15.

72. Insull, *Memoirs*, 232–33; Insull to Samuel Insull Jr., June 30, July 15, July 18, and August 4, 1932, SI, box 1, folder 10, February 3, 1933, SI, box 2, folder 1, and November 23, 1933, SI, box 8, folder 1. The quotations are from the July 15 letter.

73. Insull to Samuel Insull Jr., August 4, 1932, SI, box 1, folder 10.

74. Ibid., August 2, 1932, SI, box 1, folder 10; McDonald, *Insull*, 314–15.

75. McDonald, *Insull*, 315–17; Insull, *Memoirs*, 233–65; Insull to Samuel Insull Jr., June 5 and September 4, 1933, April 12 and April 16, 1934, SI, box 2, folder 1.

76. McDonald, *Insull*, 318–33. The trials can be followed in detail in the papers in SI.

77. Ibid., 334; A. Howard Walker to Insull, November 19, 1934, SI, box 8, folder 3; Floyd E. Thompson to Insull, June 21, 1935, SI, box 8, folder 2; Tesla to Insull, March 18, March 25, March 31, and June 17, 1935, SI, box 8, folder 2.

78. Insull, *Memoirs*, 140–41; *Chicago Tribune*, July 31, 1935; *Chicago Daily News*, July 30, 1935.

79. Insull to Howard, August 23, 1937, SI, box 2, folder 4.

80. McDonald, *Insull*, 333, 337–39; N. M. Thomson Jr. to Insull, February 12, 1937, SI, box 3, folder 4.

EPILOGUE: A SHOW OF POSSIBILITIES: NEW YORK 1939

1. Quoted in Queens Museum, *Dawn of a New Day: The New York World's Fair, 1939/40* (New York, 1980), 6. Emphasis is in the original. Hare was secretary of the Fair of the Future Committee and had served on its Board of Design.

2. Stanley Appelbaum, *The New York World's Fair, 1939/1940, in 155 Photographs* (New York, 1977), xiv.

3. Ibid., 2–4; Queens Museum, *Dawn of a New Day*, 24, 50. The Trylon was so named because it was a triangular pylon.

4. Appelbaum, *New York World's Fair*, 3; Queens Museum, *Dawn of a New Day*, 14, 29, 33, 62.

5. For the layout of the fair, see the useful map in Appelbaum, *New York World's Fair*, just before the introduction.

6. Ibid., 12, 14, 48–49; Queens Museum, *Dawn of a New Day*, 22, 47, 52–53, 97; David Hillel Gelernter, *1939: The Lost World of the Fair* (New York, 1995), 265–66.

7. Appelbaum, *New York World's Fair*, 57; *New York Times*, May 7, 1939.

8. Appelbaum, *New York World's Fair*, xvii, 58–59; Queens Museum, *Dawn of a New Day*, 25, 96; *New York Times*, May 3, May 7, and May 10, 1939; *Business Week*, November 4, 1939, 23.

9. Appelbaum, *New York World's Fair*, 60–61; Queens Museum, *Dawn of a New Day*, 13, 47, 74, 98; *New York Times*, May 5, May 7–9, May 17, and June 15, 1939; Joseph J. Corn and Brian Horrigan, *Yesterday's Tomorrows: Past Visions of the American Future* (Baltimore, 1984), 46–47.

10. Appelbaum, *New York World's Fair*, 18–20; Queens Museum, *Dawn of a New Day*, 14, 25, 51, 100–101; *New York Times*, May 17, June 4, and June 13, 1939.

11. Corn and Horrigan, *Yesterday's Tomorrows*, 46–49; *Business Week*, November 4, 1939, 22; *New Yorker*, May 13, 1939, 86; Appelbaum, *New York World's Fair*, 22–25; Queens Museum, *Dawn of a New Day*, 104, 108–9. Corn and Horrigan have a photograph of the full-sized intersection replica.

12. Appelbaum, *New York World's Fair*, 30–31, 42; *New York Times*, May 8, May 10, and May 16, 1939; Queens Museum, *Dawn of a New Day*, 25, 83–85.

13. Queens Museum, *Dawn of a New Day*, 54, 82–83; Appelbaum, *New York World's Fair*, 43–44; *New Yorker*, June 3, 1939, 66–68; Alva Johnston, "Television's Here," *Saturday Evening Post*, May 6, 1939, 8–9, 42, 45–46, 48, and "Now What Can We Do with Television?" May 20, 20–21, 103–7. The last two sources have the fullest account of available televisions, their prices

($199.50 to $600) and features, and the development of the medium. The largest screen, for example, was eight by ten inches. The Crosley Appliance exhibit also featured a version of television.

14. Appelbaum, *New York World's Fair*, xiv.

15. Queens Museum, *Dawn of a New Day*, 3–5; Gelernter, *1939*, 342–43; *New Yorker*, April 23, 1939, 16.

16. Queens Museum, *Dawn of a New Day*, 5–10; Gelernter, *1939*, 25–26; Appelbaum, *New York World's Fair*, xiv.

17. *New York Times*, March 5, 1939; Appelbaum, *New York World's Fair*, 9; Queens Museum, *Dawn of a New Day*, 21.

18. Quoted in Queens Museum, *Dawn of a New Day*, 46.

19. *New Yorker*, April 29, 1939, 16. For examples of special days see *New York Times*, May 2–4, May 9–10, May 17–19, June 3–7, June 10, June 13–15, June 17–20, June 24, June 25, and June 29, 1939.

20. Queens Museum, *Dawn of a New Day*, 108, 113; *New York Times*, May 14, June 3, and June 13, 1939. Some Aquacade seats went as high as ninety-nine cents. With its huge cast of five hundred, and elaborate sets, the show grossed more than $4.3 million during its two-year run at the fair.

21. Appelbaum, *New York World's Fair*, 65, 128–29, 131, 133–44; Queens Museum, *Dawn of a New Day*, 25, 108, 110, 113; *New York Times*, May 6, May 9–10, and May 16–19, 1939.

BIBLIOGRAPHY

GENERAL

Anderson, Oscar E. *Refrigeration in America: A History of a New Technology and Its Impact* (Princeton, 1953).

Barron, Clarence W. *More They Told Barron* (New York, 1973 [1931]).

Berton, Pierre. *Niagara: A History of the Falls* (New York, 1997).

Bliss, Charles A. *The Structure of Manufacturing Production* (New York, 1939).

Bluhm, Andreas, and Louise Lippincott. *Light! The Industrial Age, 1750–1800* (New York, 2001).

Byrn, Edward W. *The Progress of Invention in the Nineteenth Century* (New York, 1900).

Carosso, Vincent P. *The Morgans: Private International Bankers, 1854–1913* (Cambridge, Mass., 1987).

Castaneda, Christopher J. *Invisible Fuel: Manufactured and Natural Gas in America, 1800–2000* (New York, 1999).

Chandler, Alfred D., Jr. *The Visible Hand: The Managerial Revolution in American Business* (Cambridge, Mass., 1977).

Clark, Victor S. *History of Manufactures in the United States: Volume 3: 1893–1928* (New York, 1929).

Cortelyou, George B. *The Gas Industry* (New York, 1933).

Cowan, Ruth Schwartz. *A Social History of American Technology* (New York, 1997).

Debeir, Jean-Claude. *In the Servitude of Power: Energy and Civilisation Through the Ages* (London, 1991).

Dewing, Arthur S. *Corporate Promotions and Reorganizations* (Cambridge, Mass., 1930).

Giedion, Siegfried. *Mechanization Takes Command* (New York, 1948).

Headrick, Daniel R. *When Information Came of Age: Technologies of Knowledge in the Age of Reason and Revolution, 1700–1850* (New York, 2000).

Hills, Richard L. *Power from Wind: A History of Windmill Technology* (Cambridge, Eng., 1994).

———. *Power in the Industrial Revolution* (New York, 1970).

Hunter, Louis C. *Waterpower*, vol. 1 of *A History of Industrial Power in the United States, 1780–1830* (Charlottesville, Va., 1979).

Kendrick, John W. *Productivity Trends in the United States* (Princeton, 1961).

Klein, Maury. *The Change Makers: From Carnegie to Gates, How the Great Entrepreneurs Transformed Ideas into Industries* (New York, 2003).

Lawson, Thomas W. *Frenzied Finance* (New York, 1968 [1905]).

Leach, William R. *Land of Desire: Merchants, Power, and the Rise of a New American Culture* (New York, 1993).

Lorant, Stefan. *Pittsburgh: The Story of an American City* (New York, 1964).

Luckiesh, Matthew. *Artificial Light: Its Influence upon Civilization* (New York, 1920).

———. *Light and Work* (New York, 1924).

Lynd, Robert S., and Helen M. Lynd. *Middletown: A Study in American Culture* (New York, 1929).

———. *Middletown in Transition: A Study in Cultural Conflicts* (New York, 1937).

Maril, Nadja. *American Lighting, 1840–1940* (West Chester, Pa., 1995).

Marvin, Carolyn. *When Old Technologies Were New* (New York, 1988).

McCabe, James D., Jr. *New York by Gaslight* (New York, 1984 [1882]).

Mumford, Lewis. *Technics and Civilization* (New York, 1934).

O'Dea, William Thomas. *The Social History of Lighting* (London, 1958).

Pierce, Bessie Louise. *A History of Chicago* (New York, 1937–57), 3 vols.

Rice, W. G. C. *Seventy-five Years of Gas Service in Chicago* (Chicago, 1925).

Schlereth, Thomas J. *Victorian America: Transformations in Everyday Life, 1876–1915* (New York, 1991).

Schurr, Sam H., and Bruce C. Netschert. *Energy in the American Economy, 1850–1975* (Baltimore, 1977 [1960]).

Strasser, Susan. *Never Done: A History of American Housework* (New York, 1982).

Tarr, Joel, and Gabriel Dupuy, eds. *Technology and the Rise of the Networked City in Europe and America* (Philadelphia, 1988).

Thompson, Robert Luther. *Wiring a Continent: The History of the Telegraph Industry in the United States, 1832–1866* (Princeton, 1997).

Trescott, Martha Moore. *The Rise of the American Electrochemicals Industry, 1880–1910* (Westport, Conn., 1981).

Usher, Abbott Payson. *A History of Mechanical Inventions* (Cambridge, Mass., 1954).

STEAM POWER

Briggs, Asa. *The Power of Steam: An Illustrated History of the World's Steam Age* (Chicago, 1982).

Brown, John K. *The Baldwin Locomotive Works, 1831–1915* (Baltimore, 1995).

Bruce, Alfred W. *The Steam Locomotive in America: Its Development in the Twentieth Century* (New York, 1952).

Dickinson, H. W. *A Short History of the Steam Engine* (New York, 1939).

Evans, Oliver. *The Abortion of the Young Steam Engineer's Guide* (Philadelphia, 1805).

Flexner, James T. *Steamboats Come True: American Inventors in Action* (New York, 1944).

Hills, Richard L. *Power from Steam: A History of the Stationary Steam Engine* (Cambridge, Eng., 1989).

Hunter, Louis C. *Steamboats on the Western Rivers: An Economic and Technological History* (New York, 1993 [1949]).

Lord, John. *Capital and Steam-Power, 1750–1800* (New York, 1965 [1923]).

Martin, Albro. *Railroads Triumphant* (New York, 1992).

Parsons, Charles A. *The Steam Turbine* (Cambridge, Eng., 1911).

Savery, Thomas. *The Miner's Friend, or An Engine to Raise Water by Fire* (London, 1702).

Shagena, Jack L. *Who Really Invented the Steamboat?* (Amherst, N.Y., 2004).

Sutcliffe, Andrea. *Steam: The Untold Story of America's First Great Invention* (New York, 2004).

Thurston, Robert H. *A History of the Growth of the Steam Engine* (Ithaca, 1939 [1878]).

White, John H., Jr. *A History of the American Locomotive: Its Development, 1830–1880* (New York, 1968).

Wik, Reynold M. *Steam Power on the American Farm* (Philadelphia, 1953).

ELECTRICITY

Adams, Edward Dean. *Niagara Power: History of the Niagara Falls Power Company, 1886–1918* (Niagara Falls, N.Y., 1927), 2 vols.

Behrend, B. A. *The Induction Motor* (New York, 1921).

Bordeau, Sanford P. *Volts to Hertz: The Rise of Electricity* (Minneapolis, 1982).

Brandon, Craig. *The Electric Chair: An Unnatural American History* (Jefferson, N.C., 1999).

Bright, Arthur A., Jr. *The Electric-Lamp Industry: Technological Change and Economic Development from 1800 to 1947* (New York, 1949).

Brittain, James E., ed. *Turning Points in American Electrical History* (New York, 1976).

Brown, D. Clayton. *Electricity for Rural America: The Fight for the REA* (Westport, Conn., 1980).

Brown, Harold P. "The New Instrument of Execution." *North American Review* (November 1889): 586–94.

Childs, Marquis. *The Farmer Takes a Hand: The Electric Power Revolution in Rural America* (New York, 1952).

Clement, William E. *Selling Electric Service, 1879–1954* (New Orleans, 1954).

Cohen, I. Bernard. *Benjamin Franklin's Science* (Cambridge, Mass., 1990).

Coyle, David C., ed. *Electric Power on the Farm: The Story of Electricity, Its Usefulness on Farms, and the Movement to Electrify Rural America* (Washington, 1936).

DuBoff, Richard B. *Electric Power in American Manufacturing, 1889–1958* (New York, 1979 [1964]).

———. "The Introduction of Electric Power in American Manufacturing." *Economic History Review* 20 (December 1967): 509–18.

Dunsheath, Percy. *A History of Electrical Power Engineering* (Cambridge, Mass., 1969).

Edison, Thomas A. "The Dangers of Electric Lighting." *North American Review* 149 (November 1889): 625–35.

———. "The Success of the Electric Light." *North American Review* 131 (October 1880): 295–301.

Edison Electrical Institute. *Networks of Electric Power* (New York, 1962).

Essig, Mark. *Edison & The Electric Chair: A Story of Light and Death* (New York, 2003).

Fleming, John Ambrose. *The Alternate Current Transformer* (New York, 1899).

Foster, Abram John. *The Coming of the Electrical Age to the United States* (New York, 1979).

Friedel, Robert D., and Paul Israel. *Edison's Electric Light: Biography of an Invention* (New Brunswick, N.J., 1986).

Gerry, Elbridge T. "Capital Punishment by Electricity." *North American Review* 149 (September 1889): 321–26.

Gilbert, Chester G., and Joseph E. Pogue. *America's Power Resources* (New York, 1921).

Gould, Jacob M. *Output and Productivity in the Electric and Gas Utilities, 1899–1942* (New York, 1946).

Hammond, John Winthrop. *Men and Volts: The Story of General Electric* (Philadelphia, 1941).

Hammond, Robert. *The Electric Light in Our Homes* (New York, 1884).

Heilbron, J. L. *Electricity in the 17th and 18th Centuries: A Study of Early Modern Physics* (Berkeley, Calif. 1979).

Hirsh, Richard F. *Technology and Transformation in the American Electric Utility Industry* (New York, 1989).

Hogan, John. *A Spirit Capable: The Story of Commonwealth Edison* (Chicago, 1986).

Houston, Edwin J. *Electricity in Every-day Life* (New York, 1905), 3 vols.

Hubbard, Geoffrey. *Cooke and Wheatstone and the Invention of the Electric Telegraph* (New York, 1968 [1965]).

Hughes, Thomas P. "Electrification of America: The System Builders." *Technology and Culture* 20, no. 1 (1979): 124–61.

———. "Harold P. Brown and the Executioner's Current: An Incident in the AC-DC Controversy." *Business History Review* 32 (Summer 1958): 143–65.

———. *Networks of Power: Electrification in Western Society, 1880–1930* (Baltimore, 1983).

Hyman, Leonard S. *America's Electric Utilities: Past, Present, and Future* (Arlington, Va., 1983).

Jones, Payson. *A Power History of the Consolidated Edison System, 1878–1900* (New York, 1940).

Jonnes, Jill. *Empires of Light: Edison, Tesla, Westinghouse, and the Race to Electrify the World* (New York, 2003).

Klein, Maury. "What Hath God Wrought?" *Invention & Technology*, Spring 1993, 34–42.

Lodge, Oliver J. *Modern Views of Electricity* (New York, 1889).

Lubar, Steven. "Transmitting the Power of Niagara: Scientific, Technological, and Cultural Contexts of an Engineering Decision." *IEEE Technology and Society Magazine*, March 1989, 11–18.

MacLaren, Malcolm. *The Rise of the Electrical Industry During the Nineteenth Century* (Princeton, 1943).

Marsh, Charles. *Trade Unionism in the Electric Light and Power Industry* (Urbana, Ill., 1928).

Martin, Thomas Commerford. *Forty Years of Edison Service* (New York, 1922).

McDonald, Forrest. *Let There Be Light: The Electric Utility Industry in Wisconsin, 1881–1955* (Madison, 1957).

McMahon, A. Michal. *The Making of a Profession: A Century of Electrical Engineering in America* (New York, 1984).

Metzger, Th. *Blood and Volts: Edison, Tesla, and the Electric Chair* (Brooklyn, 1996).

Meyer, Herbert W. *A History of Electricity and Magnetism* (Cambridge, Mass., 1971).

Miller, Raymond C. *Kilowatts at Work: A History of the Detroit Edison Company* (Detroit, 1957).

Moran, Richard. *Executioner's Current: Thomas Edison, George Westinghouse, and the Invention of the Electric Chair* (New York, 2002).

Nye, David E. *Electrifying America: Social Meanings of a New Technology* (Cambridge, Mass., 1992).

Parsons, R. H. *The Early Days of the Power Station Industry* (Cambridge, Eng., 1939).

Passer, Harold C. "Development of Large-Scale Organization: Electrical Manufacturing Around 1900." *Journal of Economic History* 12, no. 4 (Fall 1952): 378–93.

———. "The Electric Science and the Early Development of the Electric Manufacturing Industry," *Annals of Science* 7 (December 1951): 382–92.

———. *The Electrical Manufacturers, 1875–1900* (Cambridge, Mass., 1953).

Platt, Harold L. *The Electric City: Energy and the Growth of the Chicago Area, 1880–1930* (Chicago, 1991).

Pope, Franklin Leonard. "The Electric Motor and Its Applications." *Scribner's Magazine* 3 (March 1888): 306–21.

———. *Evolution of the Electric Incandescent Lamp* (Elizabeth, N.J., 1889).

Preece, W. H. "Electric Lighting in America." *Electrical World*, December 27, 1884, 265.

Ramsay, M. L. *Pyramids of Power: The Story of Roosevelt, Insull, and the Utility Wars* (Indianapolis, 1937).

Reader, W. J. *A History of the Institution of Electrical Engineers, 1871–1971* (London, 1987).

Reich, Leonard S. "Lighting the Path to Profit: GE's Control of the Electric Lamp Industry, 1892–1941." *Business History Review* 66 (Summer 1992): 305–34.

Reynolds, Terry S., and Theodore Bernstein. "Edison and 'The Chair.'" *IEEE Technology and Society Magazine*, March 1989, 19–28.

Rowbottom, Margaret, and Charles Susskind. *Electricity and Medicine: The History of Their Interaction* (London, 1984).

Rudolph, Richard, and Scott Ridley. *Power Struggle: The Hundred-Year War Over Electricity* (New York, 1986).

Schiffer, Michael Brian. *Draw the Lightning Down: Benjamin Franklin and Electrical Technology in the Age of Enlightenment* (Berkeley, Calif., 2003).

Schivelbusch, Wolfgang. *Disenchanted Night: The Industrialization of Light in the Nineteenth Century* (Berkeley, Calif., 1989).

Sharlin, Harold I. *The Making of the Electrical Age* (New York, 1963).

Shiers, George, ed. *The Electric Telegraph: An Historical Anthology* (New York, 1977).

Stanley, William. "Alternating-Current Development in America." *Journal of the Franklin Institute*, June 1912, 561–80.

Steinmetz, Charles. *AC Phenomena* (New York, 1900).

Terry, Charles A. *The Early History of the Westinghouse Electric and Manufacturing Company* (East Pittsburgh, 1925).

Thompson, Carl D. *Confessions of the Power Trust* (New York, 1932).

Thomson, Sir William. "Electric Lighting and Public Safety." *North American Review* 150 (February 1890): 189–97.

Tricker, R. A. R. *The Contributions of Faraday and Maxwell to Electrical Science* (New York, 1966).

Usselman, Steven W. "From Novelty to Utility: George Westinghouse and the Business of Innovation During the Age of Edison." *Business History Review* 66 (Summer 1992): 251–304.

Wainwright, Nicholas B. *History of the Philadelphia Electric Company, 1881–1961* (Philadelphia, 1961).

Westinghouse, George, Jr. "A Reply to Mr. Edison." *North American Review* 149 (December 1889): 653–65.

———. "Sir William Thomson and Electric Lighting." *North American Review* 150 (March 1890): 321–30.

Whittaker, Edmund. *A History of the Theories of Aether and Electricity* (New York, 1989 [1951, 1953]).

TRACTION INDUSTRY

Cheape, Charles W. *Moving the Masses: Urban Public Transit in New York, Boston, and Philadelphia, 1880–1912* (Cambridge, Mass., 1980).

Hilton, George W. *The Cable Car in America* (Pasadena, Calif., 1970).

Hilton, George W., and John F. Due. *The Electric Interurban Railways in America* (Stanford, Calif., 1960).

Miller, John Anderson. *Fares, Please! A Popular History of Trolleys, Horsecars, Streetcars, Buses, Elevateds, and Subways* (New York, 1960 [1941]).

Rowsome, Frank, Jr. *Trolley Car Treasury: A Century of American Streetcars—Horsecars, Cable Cars, Interurbans, and Trolleys* (New York, 1956).

Sprague, Frank J. "The Electric Railway." *Century Magazine* 70 (July and August 1905): 434–50, 512–27.

Young, David M. *Chicago Transit: An Illustrated History* (DeKalb, Ill., 1998).

BIOGRAPHY, AUTOBIOGRAPHY

Abrahams, Harold J., and Marion B. Savin, eds. *Selections from the Scientific Correspondence of Elihu Thomson* (Cambridge, Mass., 1971).

Anderson, L. *Nikola Tesla: On His Work with Alternating Currents and Their Application to Wireless Telegraphy, Telephony, and Transmission of Power* (Denver, 1992).

Appleyard, Rollo. *Charles Parsons: His Life and Work* (London, 1933).

Asimov, Isaac. *Biographical Encyclopedia of Science and Technology* (Garden City, N.Y., 1976).

Baker, Charles H. *Life and Character of William Taylor Baker* (New York, 1908).

Baldwin, Neil. *Edison: Inventing the Century* (New York, 1995).

Bathe, Greville, and Dorothy Bathe. *Oliver Evans: A Chronicle of Early American Engineering* (Philadelphia, 1935).

Beckhard, A. *Electrical Genius Nikola Tesla* (New York, 1959).

Boyd, Thomas. *Poor John Fitch: Inventor of the Steamboat* (New York, 1935).

Brush, Charles F. "Some Reminiscences of Early Electric Lighting." *Journal of the Franklin Institute* 206 (July 1928): 3–15.

Buss, Dietrich G. *Henry Villard: A Study of Transatlantic Investments and Interests, 1870–1895* (New York, 1978).

Campbell, Lewis, and William Garnett. *The Life of James Clerk Maxwell* (London, 1882).

Cantor, Geoffrey. *Michael Faraday: Sandemanian and Scientist: A Study of Science and Religion in the Nineteenth Century* (New York, 1991).

Carlson, W. Bernard. *Innovation as a Social Process: Elihu Thomson and the Rise of General Electric, 1870–1900* (New York, 1991).

Caro, Robert A. *The Power Broker: Robert Moses and the Fall of New York* (New York, 1975).

Cheney, Margaret. *Tesla: Man out of Time* (Englewood Cliffs, N.J., 1981).

Clark, Ronald. *Edison: The Man Who Made the Future* (New York, 1977).

Collins, Theresa M., and Lisa Gitelman. *Thomas Edison and Modern America* (New York, 2002).

Conot, Robert. *A Streak of Luck: The Life and Legend of Thomas Alva Edison* (New York, 1979).

Coulson, Thomas. *Joseph Henry: His Life and Work* (Princeton, 1950).

Dickinson, H. W. *James Watt: Craftsman and Engineer* (Cambridge, Eng., 1936).

———. *Matthew Boulton* (Cambridge, Eng., 1937).

———. *Robert Fulton, Engineer and Artist: His Life and Works* (London, 1913).

Dickinson, H. W., and Rhys Jenkins. *James Watt and the Steam Engine* (Oxford, 1927).

Dray, Philip. *Stealing God's Thunder: Benjamin Franklin's Lightning Rod and the Invention of America* (New York, 2005).

Dyer, Frank Lewis, and Thomas Commerford Martin. *Edison: His Life and Inventions*, 2 vols. (New York, 1910).

Franch, John. *Robber Baron: The Life of Charles Tyson Yerkes* (Urbana, Ill., 2006).

Franklin, Benjamin. *Autobiography* (New Haven, 1964).

Gooding, David, and Frank A. J. L. James (eds.). *Faraday Rediscovered: Essays on the Life and Work of Michael Faraday, 1791–1867* (New York, 1985).

Hawkins, Laurence A. *William Stanley (1858–1916): His Life and Work* (n.p., 1951).

Insull, Samuel. *The Memoirs of Samuel Insull*. Ed. Larry Plachno (Polo, Ill., 1992).

Israel, Paul. *Edison: A Life of Invention* (New York, 1988).

James, Frank A. J. L. *The Correspondence of Michael Faraday*, vol. 1 *1811–1831* (London, 1991).

Jehl, Francis. *Menlo Park Reminiscences* (Dearborn, Mich., 1936–41), 3 vols.

Johnson, Robert Underwood. *Remembered Yesterdays* (Boston, 1923).

Josephson, Matthew. *Edison: A Biography* (New York, 1959).

Keily, William Eugene (ed.). *Central-Station Electric Service: Its Commercial Development and Economic Significance as Set Forth in the Public Addresses (1897–1914) of Samuel Insull* (Chicago, 1915).

————. *Public Utilities in Modern Life: Selected Speeches (1914–1923) by Samuel Insull* (Chicago, 1924).

Kirkland, Edward C. *Charles Francis Adams, Jr., 1835–1915: The Patrician at Bay* (Cambridge, Mass., 1965).

Lamme, Benjamin Garver. *An Autobiography* (New York, 1926).

Lanier, Charles D. "Thomas A. Edison, Greatest of Inventors." *Review of Reviews* 8 (July 1893): 41–53.

Leonard, Jonathan N. *Loki: The Life of Charles Proteus Steinmetz* (Garden City, N.Y., 1928).

Leupp, Francis E. *George Westinghouse: His Life and Achievements* (Boston, 1919).

Loth, David. *Swope of G.E.* (New York, 1958).

Mabee, Carleton. *The American Leonardo: A Life of Samuel F. B. Morse* (New York, 1943).

Marsden, Ben. *Watt's Perfect Engine: Steam and the Age of Inventions* (New York, 2002).

Martin, Thomas Commerford. *The Inventions, Researches, and Writings of Nikola Tesla* (New York, 1995 [1894]).

————. "Nikola Tesla." *Century* 47 (February 1894): 582–86.

Maxim, Hiram S. *My Life* (New York, 1915).

McDonald, Forrest. *Insull* (Chicago, 1962).

Millard, Andre. *Edison and the Business of Innovation* (Baltimore, 1990).

Morgan, John S. *Robert Fulton* (New York, 1977).

Moyer, Albert E. *Joseph Henry: The Rise of an American Scientist* (Washington, 1997).

Nye, David E. *The Invented Self: An Anti-Biography, from Documents of Thomas A. Edison* (Odense, Denmark, 1983).

O'Neill, John J. *Prodigal Genius: The Life of Nikola Tesla: Inventor Extraordinary* (Hollywood, 1978).

Passer, Harold C. "Frank Julian Sprague." In William Miller (ed.), *Men in Business: Essays in the History of Entrepreneurship* (Cambridge, Mass., 1952), 213–37.

Philip, Cynthia Owen. *Robert Fulton* (New York, 1985).

Prager, Frank D., ed. *The Autobiography of John Fitch* (Philadelphia, 1976).

Prout, Henry G. *A Life of George Westinghouse.* (New York, 1921).

Pupin, Michael. *From Immigrant to Inventor* (New York, 1923).

Ratzlaff, John T., ed. *Tesla Said* (Millbrae, Calif., 1984).

Ratzlaff, John T., and Leland I Anderson. *Dr. Nikola Tesla Bibliography, 1884–1978* (Palo Alto, Calif., 1979).

Raucher, Elizabeth, and Toby Grotz, eds. *Tesla 1984: Proceedings of the Tesla Centennial Symposium* (Colorado Springs, 1984).

Robinson, Eric, and A. E. Musson, eds. *James Watt and the Steam Revolution* (New York, 1969).

Runes, Dagobert D., ed. *The Diary and Sundry Observations of Thomas Alva Edison* (New York, 1968 [1948]).

Sale, Kirkpatrick. *The Fire of His Genius: Robert Fulton and the American Dream* (New York, 2001).

Satterlee, Herbert L. *J. Pierpont Morgan: An Intimate Portrait, 1837–1913* (New York, 1939).

Scott, Charles F. "Early Days in the Westinghouse Shops." *Electrical World*, September 20, 1924, 585–87.

Seifer, Marc J. *Wizard: The Life and Times of Nikola Tesla* (New York, 1998).

Silverberg, Robert. *Light for the World: Edison and the Power Industry* (Princeton, 1967).

Silverman, Kenneth. *Lightning Man: The Accursed Life of Samuel F. B. Morse* (New York, 2003).

Smiles, Samuel. *Lives of the Engineers: Boulton and Watt* (London, 1878).

Sprague, Harriet. *Frank J. Sprague and the Edison Myth* (New York, 1947).

Stevenson, Elizabeth. *Park Maker: A Life of Frederick Law Olmsted* (New York, 1977).

Strouse, Jean. *Morgan: American Financier* (New York, 1999).

Sutcliffe, Alice C. *Robert Fulton* (New York, 1915).

Tate, Alfred O. *Edison's Open Door* (New York, 1938).

Tesla, Nikola. *My Inventions: The Autobiography of Nikola Tesla*. Ed. Ben Johnston (Williston, Vt., 1982).

Thomas, John M. *Michael Faraday and the Royal Institution* (New York, 1991).

Thompson, Silvanus P. *Michael Faraday: His Life and Work* (New York, 1898).

Thomson, Elihu. "Personal Recollections of the Development of the Electrical Industry." *Engineering Magazine*, July 1905, 563–72.

Turnbull, Archibald Douglas. *John Stevens: An American Record* (New York, 1928).

Upton, Francis R. "Edison's Electric Light." *Scribner's Monthly*, February 1880, 531–44.

Van Doren, Carl. *Benjamin Franklin* (New York, 1938).

Villard, Henry. *Memoirs of Henry Villard* (Boston, 1904), 2 vols.

Wachhorst, Wyn. *Thomas Alva Edison: An American Myth* (Cambridge, Mass., 1981).

Westcott, Thompson. *Life of John Fitch, the Inventor of the Steam-Boat* (Philadelphia, 1857).

Williams, L. Pearce. *Michael Faraday: A Biography* (New York, 1965).

Wilson, Charles E. *Charles A. Coffin (1844–1926): Pioneer Genius of General Electric Company* (New York, 1946).

Woodbury, David O. *Beloved Scientist: Elihu Thomson, a Guiding Spirit of the Electrical Age* (New York, 1944).

———. *A Measure for Greatness: A Short Biography of Edward Weston* (New York, 1949).

Wrege, Charles D., and Ronald G. Greenwood. "William E. Sawyer and the Rise and Fall of America's First Incandescent Electric Light Company, 1878–1881." *Business and Economic History* 13 (1984): 31–48.

MISCELLANEOUS

Baum, L. Frank. *The Master Key: An Electrical Fairy Tale, Founded upon the Mysteries of Electricity and the Optimism of Its Devotees* (Indianapolis, 1901).

Carlson, W. Bernard and A. J. Millard. "Defining Risk Within a Business Context: Thomas A. Edison, Elihu Thomson, and the A.C.-D.C Controversy, 1885–1900." In Branden B. Johnson and Vincent T. Covello, eds., *The Social and Cultural Construction of Risk* (Norwell, Mass., 1987), 275–93.

Chandler, Alfred D., Jr. "Anthracite Coal and the Beginnings of the Industrial Revolution in the United States." *Business History Review* 46 (Summer 1972): 141–81.

WORLD'S FAIRS

Appelbaum, Stanley. *The Chicago World's Fair of 1893: A Photographic Record . . .* (New York, 1980).

———. *The New York World's Fair 1939/1940 in 155 Photographs* (New York, 1977).

Badger, Reid. *The Great American Fair: The World's Columbian Exposition* (Chicago, 1979).

Bancroft, Hubert Howe. *The Book of the Fair* (Chicago, 1893).

Barrett, J. P. *Electricity at the Columbian Exposition* (Chicago, 1894).

Bergamini, John D. *The Hundredth Year: The United States in 1876* (New York, 1976).

Bolotin, Norm. *The World's Columbian Exposition* (Washington, 1992).

Brown, Dee. *The Year of the Century: 1876* (New York, 1966).

Brown, Julie K. *Contesting Images: Photography and the World's Columbian Exposition* (Tucson, 1994).

Bruce, Edward C. *The Century: Its Fruits and Its Festival* (Philadelphia, 1877).

Burg, David F. *Chicago's White City of 1893* (Lexington, Ky., 1976).

Burnham, Daniel H. *The Final Official Report of the Director of Works of the World's Columbian Exposition* (New York, 1989).

Cameron, William E. *The World's Fair: Being a Pictorial History of the Columbian Exposition* (New Haven, 1894).

Campbell, James W. *America in Her Centennial Year, 1876* (Washington, 1980).

Corn, Joseph J., and Brian Horrigan. *Yesterday's Tomorrows: Past Visions of the American Future* (Baltimore, 1984).

Cravath, J. R. "Electricity at the World's Fair." *Review of Reviews* 8 (July 1893): 35–39.

Findling, John E. *Chicago's Great World's Fairs* (New York, 1994).

Gelernter, David Hillel. *1939: The Lost World of the Fair* (New York, 1995).

Giberti, Bruno. *Designing the Centennial: A History of the 1876 International Exhibition in Philadelphia* (Lexington, Ky., 2002).

Halstead, Murat. "Electricity at the Fair." *Cosmopolitan* 15 (September 1893): 577–83.

Hobbs, Susan. *1876: American Art of the Centennial* (Washington, 1876).

Howells, William Dean. "A Sennight of the Centennial." *Atlantic Monthly* 38 (July 1876): 92–107.

Hunter, Thomas. *Memorial of the International Exhibition at Philadelphia* (Philadelphia, 1876).

Ingram, J. S. *The Centennial Exposition* (New York, 1976 [1876]).

Johnson, Rossiter, ed. *History of the World's Columbian Exposition* (New York, 1897), 4 vols.

Mattie, Erik. *World's Fairs* (New York, 1998).

Mayer, Harold M., and Richard C. Wade. *Chicago: Growth of a Metropolis* (Chicago, 1969).

Miller, Donald L. *City of the Century: The Epic of Chicago and the Making of America* (New York, 1996).

Muccigrosso, Robert. *Celebrating the New World: Chicago's Columbian Exposition of 1893* (Chicago, 1993).

Post, Robert C. (ed.). *1876: A Centennial Exhibition* (Washington, 1976).

Queens Museum, *Dawn of a New Day: The New York World's Fair, 1939/40* (New York, 1980).

Randel, William Peirce. *Centennial: American Life in 1876* (Philadelphia, 1969).

Rydell, Robert W. *All the World's a Fair: Visions of Empire at American International Exposi-
    tions, 1876–1916* (Chicago, 1984).
———. *World of Fairs: The Century-of-Progress Expositions* (Chicago, 1993).
Stead, F. Herbert. "An Englishman's Impressions at the Fair." *Review of Reviews* 8 (July 1893):
    30–34.
Thomson, Elihu. "Electricity in 1876 and 1893." *Engineering* 6:442–55.
Truman, Benjamin C. *History of the World's Fair: Being a Complete and Authentic Description
    of the Columbian Exposition from Its Inception* (New York, 1976).
White, Trumbull, and William Igleheart. *The World's Columbian Exposition, Chicago 1893*
    (Philadelphia, 1893).
World's Columbian Exposition. *Official Views of the World's Columbian Exposition* (Chicago,
    1893).

DOCUMENTS

Thomas A. Edison papers, Rutgers University. I used the online edition, http:// edison. rutgers
    .edu.
Samuel Insull papers, Loyola of Chicago University Archives.
Forrest McDonald papers, Loyola of Chicago University Archives.
Nikola Tesla letters, www.electrotherapymuseum.com.

A

Abbott, Gordon, 359

Abortion of the Young Steam Engineer's Guide, The (Evans), 36

AC (alternating current)
  Columbian Exposition as demonstration of, 319
  conversion to/from DC, 210, 241–42, 298–99
  deaths from, 270–71, 275
  discovery of, 111
  Edison's attacks on, 233, 257, 261–62, 266–67, 271–75
  for electrocution, 260, 261–62, 264–66
  for electromotive force, 212, 237
  for Jablochkoff candles, 113, 126–27
  overview, 209
  and polyphase motors, 296, 297
  Stanley's fascination with, 212
  Tesla's rotating magnetic field, 218
  See also frequencies; phase

AC generators. *See* alternators

AC motors
  for Columbian Exposition, 312–13
  General Electric's, 363–64
  Stanley's model, 364
  Tesla's, 197–200, 235–37, 252, 290, 296–98, 336, 339–40
  Westinghouse's need for, 217, 297

AC systems
  advantages overcoming disadvantages, 329–30
  for arc lighting, 224, 263
  Brush's, 494n3
  central power station in Manhattan, 361
  for Columbia, SC, textile mill, 363
  danger of, 260, 262–63, 271–75
  Edison's consideration of, 232–33
  in Germany, 335
  for lighting, in Portland, OR, 335–36
  meter for, 223
  potential of, 337
  powering lights and motors, 336
  Stanley's, 237
  for suburban areas, 257, 420–22, 433

Tesla's, 235–37

Thomson's, 209–11, 226, 237

three-phase, 340, 341, 363, 366–67, 367

transmission to consumption voltage ratio, 223–24

Westinghouse's, 216–17, 261–62, 310–11

AC traction systems, 252.

Accum, Friedrich Christian, 137

Acheson, Edward Goodrich, 350, 351

Adams, Charles Francis, Jr., 289, 290, 359

Adams, Edward Dean
  biographical information on, 332–33
  chicanery over generator design, 342, 496n54
  on direct current in cities, 345
  and hydraulic power, 333–34
  as mediator for Edison and Westinghouse, 267
  and Niagara project plant design, 349
  on Niagara project success, 352, 497n17
  and Niagara project technology, 338–42, 345–47, 348

Adams Street, Chicago, IL, power station, 401, 402

aether of Aristotle, 76

Aetna (steamboat), 461n50

agricultural industry, 17–18, 65–67

air brake for railroad cars, 4, 60, 193, 285, 329–30

Allegheny, Pennsylvania, 244, 313, 337–38

Allen, John F., 68, 69

Allen Act, 408–9, 504n40

Allis-Chalmers, 68, 70, 303, 316, 505n54

Allis steam engine, 303, 491n9

alternating current. *See entries starting with* "AC"

alternators (AC generators)
  for Columbian Exposition, 314
  hand-driven, 110–11
  inductor-type, 127, 364
  limitations of, 371–72
  for Niagara project, 342, 345–47, 348–49
  nickel steel for, 349
  with parallel transformers, 213
  Siemens alternators, 213, 214
  slotted-type armatures for, 337

alternators (*continued*)
  Stanley's model, 215, 364
  synchronous, 364
  as synchronous motor, 495*n*35
  Thomson and Houston working on, 127–28
  Thomson's, for incandescent lamps, 211
  with transformers in series, 212–13
  turboalternators, 374–75
  US Steel order for, 505*n*54
Aluminum Company of America (Niagara, NY), 350
aluminum production, 350
amber (*elektron*), 76
American 4-4-0 locomotive, 60, 61, 463*n*25
American Electric Company, 129–30, 182–84, 185
American Gas Association, 430
American Gaslight Association, 141
American Gaslight Journal, 141
American Institute of Electrical Engineers (AIEE), 235
American Journal of Science, 102
American Neurological Association, 276
American Notes (Dickens), 58
American Philosophical Society (APS), 44, 122
American Society of Mechanical Engineers, 393
American Typographic machine, 11
American Watch Company, 11
Ames, Frederick L., 356
Ampère, André-Marie, 82–84, 87, 99, 465*n*35
Ampère's law, 83, 84
amplifier to boost transmission signals, 198
Andersen, Arthur, 440
Anderson, John, 15
Andrews, William, 189
anesthesia, 259
Annapolis Junction, Maryland, 107
Anthony, William, 235
anthracite coal, 37, 64–65
Appleton Mill (Lowell, MA), 69
appliances, 432–33
Aquacade, NY World's Fair, 451–52, 510*n*20
Arago, Dominique, 87
Arago's disk, 87, 89
Archambault, A. L., 66
arc lighting
  AC system for, 224, 263, 363
  arc of light demonstration, 81, 112
  Brush's development of, 113–17
  Chicago Arc Light and Power, 401
  competition in market, 352
  death of dry-good clerk from, 275
  differential control magnet, 117
  double-carbon, 115, 186
  generators for, 112
  Jablochkoff's candles, 113, 126–27
  for lighthouses, 111–12
  limitations of, 135
  problems with, 136
  Tesla's system, 218

Thomson-Houston system, 127–28, 182–83
  two-tire rating system and, 412
  Van Depoele's system, 244
  Weston's business in, 120
arc of light demonstration, 81
Argo, Dominique-François-Jean, 82
Arkwright, Richard, 22
Arlington & Sims steam engine, 176
armatures
  design improvements, 112
  drum-style, 146–47
  for giant "C" model dynamo, 168–69
  for hand-driven AC generator, 110–11
  laminated, 120
  low-resistance, 242
  in rotating magnetic field AC motor, 199
  slotted-type, for alternators, 337
  spherical, 129
  for street railway motors, 249, 250
  windings staggered 90 degrees, 314
Armington & Sims dynamo, 247
Ashton, H. F., 343, 347–48
Asimov, Isaac, 465*n*29
asphalt filaments, 230–31
Associated Press, 108
Association of Edison Illuminating Companies (AEIC), 410
Astor, John Jacob, IV, 377, 378, 379
Atkins, Edward F., 392
atmospheric electricity, 79
atmospheric steam engine, 21–23, 26, 31, 33
AT&T Building, NY World's Fair, 448–49
August Belmont & Company, 288–89
automatic cutoff valve, 8–9, 68
automatic regulator, 128–29
automatic shunt, 115
automobiles, 432, 448
autotransformer, 216
Avery, William, 70

B
Babcock, George, 67
Bacon, Francis, 78–79
Baetz, William, 147
Baker, Frank J., 420
balanced load concept, 231
Baldwin, Matthias, 59–60, 463*n*24
Baldwin Locomotive Works, 59–60
Baltimore, Maryland, 56, 107–8, 138
Baltimore Brush Electric Company, 361
Baltimore & Ohio (B&O) railroad, 56, 59, 462*n*15
bamboo filaments, 163, 165, 475*n*12
banks. *See* financing
Baring Brothers failure, 278, 286, 293
Barker, George, 131
Barrett, J. P., 300, 491*m*1
Barron, Clarence, 287–88
Bartholdi, Frédéric-Auguste, 3
Barton, Bruce, 431

Barton, Silas A., 184, 185
Batchelor, Charles
   driving first electric train in America, 242
   and Edison, 143, 146–47, 192–93
   and Edison Machine Works, 165, 202–3, 205,
      475n18
   and Tesla, 199–200, 217–18
batteries
   arc of light from, 81
   Brush's, 212
   Davy's, 81, 465n29
   electric, 78, 81, 304
   in electric cars, 432
   and electromagnetic induction, 86, 87–89
   Farmer's use of, 130, 131
   intensity, 106
   Julien battery for electric streetcars, 243
   magnetic effect from, 82–84
   for storage, 212
   for telegraphs, 101, 106
   voltaic, 81, 82, 88, 101, 111, 465n29
   Volta's, 80–81, 99
Beale, William G., 409, 410
beam engines, 2–3, 21, 23–24, 31, 68
Belfield, Reginald, 214, 216, 222
Belgium, 12
Bell, Alexander Graham, 11, 96, 134, 316. See also
   telephone
Bell Telephone Company, 134, 203, 229, 304
Belmont, August, 359
Belmont & Company, 288–89
Bennett, James Gordon, 172
Bentley, Edward M., 244
Bentley-Knight Electric Railway Company, 244, 251
Bergmann, Sigmund, 167, 179, 203, 254
Bergmann & Company, 206, 229, 233–34, 253–55
Berlin Industrial Exhibition, 241
Berry, Burton Y., 438, 503n6
Best Friend of Charleston (locomotive), 55, 56
Billy Rose's Aquacade, NY World's Fair, 451–52,
   510n20
bipolar magnets, 147
bituminous coal, 18, 32
Black, Joseph, 15–16
Blake, William, 30
Bláthy, Ottó T., 233, 480n35
Board of Electrical Control for New York, 263,
   264, 271, 272, 275
Board of Fire Underwriters, 360
Boehm, Ludwig, 147, 178–79
boilers
   cast-iron, 32
   at Centennial Exhibition, 3
   high pressure and, 21, 31–32, 37
   horizontal fire-tube, 60
   horizontal tubular, 55
   for steamboats, 43, 44, 45, 46, 60, 67
   tubular, 63
   water-tube, 67

bonds
   Coffin's handling of, 208
   electric companies trading service for, 356,
      385–86
   Insull's, 408, 409–10, 419–20, 428, 504n38
   and NY investment banks, 427–28
   and overextended money supply, 386–87
   value of utility bonds, 435
   Westinghouse's, 389–90
Bosanquet, R. H. M., 231
Boston Post Office Building, 174
Boulton, Matthew, 23, 24–29
Boulton & Watt, 25, 28–29, 45, 48
Boyle, Robert, 20
Bracken, William, 243
Bradley, Charles S., 298
Brady, Anthony N., 365
British thermal unit (BTU), 74, 464n10
Brockton, MA, power station, 190
Brown, Alfred S., 218, 234–35
Brown, Charles E. L., 335
Brown, Harold P., 263–64, 265–66, 267, 268, 269,
   272, 273
Brown & Sharpe machine tool company, 249
Bruce, Alfred, 60
Brugnatelli, Luigi, 119
Brush, Charles F., 113–17, 212, 469n47
Brush Electric Company, 114–16, 181, 196, 227,
   244, 361, 494n3
Brush Electric Light Company, 116, 166
Bryan panic, 408, 504n38
Budd, Britton I., 425
Buffalo, New York, 216, 351–52
Buffalo General Electric Company, 328
"Building the World of Tomorrow," 446. See also
   New York World's Fair
burner. See filaments
Burnham, Daniel, 311
Burnham & Root, architects, 306–7
Byllesby, Henry M., 236, 237, 264
Byrn, Edward W., 279

C
cable cars, 240–41, 353, 418
calcium carbide, 351
caloric theory of heat, 72–73, 74
Camden & Amboy Railroad, 57
Campbell, Henry R., 60
Campbell Printing Press and Manufacturing
   Company, 5
canals for transportation, 56, 57, 59
canals for water energy, 330–32, 363
capital punishment debate, 259. See also
   electrocution of criminals
carbon, 64, 111–13, 115, 137, 145, 149–50
carbon filament lamps, 150, 230, 296–97
carbon lamp, 137, 146, 148, 155–56, 474n54
carbon monoxide for capital punishment, 259
carbon transmitter, 134, 203

Carborundum Company (Niagara, NY), 350–51
Carey, George R., 177
Carlson, W. Bernard, 122, 129, 470n9, 470n11
Carnegie, Andrew, 306, 360
Carnot, Nicolas L. S., 73–74, 75, 464n8
Cataract Construction
    AC selected as current for, 341–42
    establishment of, 332
    and Forbes, 338–39, 341–42, 345–49, 496n2,
        497n12
    Forbes' generator design, 341–42, 345–47
    international commission on developing,
        334–35
    inviting bids on building the generator,
        345–47, 348
    request for proposals, 339
    Westinghouse generator-building process,
        348–59
    See also Adams, Edward Dean
Cauchy, Augustin-Louis, 83
Centennial Exhibition, Philadelphia (1876), 1–13
central power stations
    AC station in NYC, 361
    alternating current vs. direct current, 275
    building in small towns to promote business,
        189–90
    Coffin's belief in, 208
    DC-based, 209, 258, 401, 407
    and demand for electricity, 328–29, 374–75
    Edison and, 141, 162, 188, 257–58
    Edison's underground wiring for, 139, 141, 160,
        163, 165, 167–68
    efficiency of, 284–85
    and electric railways, 248–51, 328, 419
    in Europe, 200
    factories converting to, 361, 412–13, 426, 432
    fixed vs. operating costs, 403–4
    funding for, 403–4
    Insull's goal to monopolize and lower prices,
        403, 406–7
    marketing strategies, 202, 406–7, 410–13
    multiphase capability, 362
    overview, 201–2
    and postwar coal shortage, 426, 427
    rotary converters allow substations, 407
    stand-alone, 361, 363–64, 412–13, 426, 432
    for suburban and rural areas, 421–22
    and Westinghouse, 224, 361, 364
    See also feeder-and-main systems
centrifugal governor, 27–28, 68–69
Chandler, Alfred D., Jr., 64, 256
Charleston & Hamburg Railroad, 56
Cheever, Charles, 157
chemistry, 72–73, 81, 85, 350–51
Chesapeake & Ohio Railroad, 61
Chesney, Cummings C., 365
Chicago, Illinois, 400, 418. See also Columbian
    Exposition
Chicago Arc Light and Power Company, 401, 402

Chicago Edison Company, 400–406, 408, 409,
    419, 504n38, 504n41, 506n8
Chicago Railway Exposition (1883), 243, 244
Chicago River, 402
Chicago Times, 324
Chicago Tribune, 300, 356, 366
Chicago Union Traction Company, 419, 504n41,
    506n6
Chicago World's Fair of 1893. See Columbian
    Exposition
chronometric telegraph, 100
Churchill, Frederick H., 129, 182
circuits
    Ampère and, 83, 465n35
    for customer types, 328
    and electromagnetic force, 92, 101–2
    high-frequency, 235
    and internal vs. external resistance, 147
    low-voltage, 313
    magnetic, 212–13, 216, 231–32
    and Morse code, 105
    mutual induction vs. self-induction, 468n13
    and Ohm, 84
    and potential, 210, 249
    for rotating magnetic field AC motor, 199
    short circuits, 117, 171, 211
    single-phase vs. multiphase, 297, 314, 339–40,
        340, 341
    single series vs. parallel, 128, 144, 148–49
    and subdivision of light, 126–28
    Tesla's multiple-circuit motor, 199
Clarke, Charles L., 165, 166–67
Clarke, William, 343–44
Claudius, Hermann, 167, 189, 246
Clausius, Rudolf J. E., 75, 464n8
Clermont estate, 47
Clermont (steamboat), 48, 461n47
Cleveland Telegraph Supply Company, 114, 115
coal, 18, 32, 37, 64–65, 414–15, 426, 427
coal gas, 110, 137, 138, 428
coal mining, 18–22, 25, 54
Cockran, W. Bourke, 268
Coffin, Charles A.
    biographical information on, 362
    and Brown conspiracy, 268, 269
    Columbian Exposition bids, 308
    and Edison General Electric, 284, 292–95
    on electric railways, 251
    funding sources, 326
    and GE's financial problems, 354–56, 357–58
    and Insull, 414
    retirement from GE, 375–76
    selling product trumps safety concerns, 281–82
    and Thomson, 185–86, 281–83
    and Thomson-Houston, 207–9, 284, 292–95
    and Westinghouse, 227, 287–88, 292, 368–71
    Westinghouse designs theft, 343–44
    See also General Electric Company;
        Thomson-Houston Electric Company

Cohen, I. Bernard, 78, 465n22
coke oven gas, 429
Coleridge, Samuel Taylor, 85
Colorado Springs, Colorado, 378–79
Columbia, SC, textile mill AC power system, 363
Columbian Exposition, Chicago (1893), 300–323
  bids for power station and lighting, 308–12
  closing day, 317–18
  designing and building, 306–7
  electricity and lighting for, 312–16
  exhibits, 301–6, 307, 318, 322–23, 492n14
  Insull's purchase of generators from, 402, 504n23
  investors' profit, 493n52
  opening day and grand illumination, 316–17
  overview, 300–302, 323
  universal system of distribution for, 405
Columbia (steamship), 162, 163
common stocks, 360, 436, 488n4
Commonwealth Edison Company, 419–22, 424–27, 433, 436–41, 506n8
Commonwealth Electric Company, 408–10, 414–15, 419, 504n41, 505n42, 506n8
commutators
  carbon vs. metallic brushes for, 250
  description of, 469n35
  disadvantages of, 236, 237, 250, 252, 297
  Forbes' plans for, 346
  induction coil with, 212
  purpose of, 111, 112, 197–98, 241
  Tesla's plan to do away with, 198–200
Company for Isolated Lighting, 172
compass needle and magnetism, 76, 82
compensated alternating motor, 364
competition
  as barrier to progress, 366–67
  companies left standing, 279, 293, 325–26, 360
  between Edison and Westinghouse, 261, 264–67, 352–54
  between electricity and gas, 138, 141–42, 219–21
  for electric lamp design and manufacturing, 176, 179–81, 196, 224–25
  for electrocution market, 259
  Franklin Institute's, for dynamos, 114–15, 126
  between General Electric and Westinghouse, 325–26
  as incentive for inventions, 233, 256–58, 284–85
  as incentive for product improvement, 230
  between steamboats and stagecoach, 45–46
  between Thomson-Houston and Sprague, 251–52, 254–55
  Thomson-Houston strategy for overcoming, 281–83
  See also patent disputes
compound engines, 67
Compton, Karl T., 376
condenser, 16, 25
conductors
  comparison to magnets, 82

copper, 99, 144, 163–64, 232
  electrostatic charges on surface of, 91
  and generator's power output, 374
  for incandescent light, 144
  for long-distance signaling, 99
  and polarization, 93
  rate of decomposition, 89
  relative motion and electric current, 88–89
  size required for DC systems, 209–10
conservation of energy principle, 74–75
"Conservation of Force, The" (Brush), 114
Consolidated Edison Building, NY World's Fair, 446–47
Consolidated Electric Light Company, 181, 186, 209, 226–27, 269–70, 289
constant-voltage generator, 144
consumer goods economy, 432–33
Continental Illinois National Bank and Trust Company, 439, 508n65
convergence principle, 195–96
Cooke, William, 99–100
Cooke-Wheatstone telegraph, 100
cooking with gas, 220
Coolidge, T. Jefferson, 294, 368, 369
Cooper, Peter, 56
copper
  coating carbon electrodes with, 115
  cost considerations, 144, 163–64, 210, 263–64, 382
  monopolization of, 258, 268, 387
  polarity relationship with other metals, 80–81
  price increases, 258, 485n7
  reducing need for, 188, 201, 210–11
  stabilization of prices, 268
copper conductors, 99, 144, 163–64, 232
copper wires, 101, 144, 263, 336, 341
Corliss, George H., 8–9
Corliss steam engine, 2–4, 9
Cornell, Ezra, 107
Corporation Securities Company of Chicago (Corp), 437–40, 508n60
cosmos, Descartes on, 77
Coster, Charles H., 193, 254, 357
cotton gins, 65–66
cotton thread filaments, 150
Coulomb, Charles-Augustin de, 79, 82, 84
Craig, John, 14–15
Croly, Herbert, xii
Crookes, Sir William, 137, 147
Cugnot, Nicolas-Joseph, 54
Cunard Line, 63
currents
  displacement current concept, 93
  Edison's description of, 273
  for rotating magnetic field AC motor, 199
  See also entries beginning with "AC" *and* "DC"
Curtis, Charles G., 413–14
Custer, George, 10

customer service, importance of, 431
Cutting, Robert L., 148
cylinders
  boring with inadequate tools, 21–22, 59
  compound, 61
  in compound engines, 67
  controlling amount of steam in, 8–9, 27,
    463n43
  double-acting, 43
  heating and cooling in, 15–16, 21, 22–23
  indicator for measuring effectiveness of, 28
  on locomotives, 55
  maintaining heat in, 23–24, 27
  pistons in, 20, 23–24, 27
  and rollers, in Walter press, 5

D
"Dangers of Electric Lighting,
  The" (Edison), 273
Davy, Sir Humphrey, 81, 85–87, 136, 465n29
DC (direct current)
  AC changed to before entering homes, 274
  Brown shocking a dog with, 265
  conversion to/from AC, 210, 241–42, 298–99
  generators, 102, 111
  overview, 209
  roasting meat with, 274
  and rotary converters, 405
  Westinghouse's interest in, 194
  See also generators
DC generators. See generators
DC motors, 236, 241, 252, 297, 335, 401
DC systems
  for arc lighting, 224
  in Chicago pre-Insull, 401, 409
  Edison's vision for, 209, 261
  efficiency of, 330
  for electric railways, 252, 315, 489n16
  limitations of, 209–10, 215, 224, 252, 328
  power stations, 209, 258, 401, 407
  Tesla's denunciation of, 341
  See also feeder-and-main systems
De Fonvielle, Wilfred, 160
Dean, Charles, 175
Delaware River, 45–46, 51
demand meters, 406
Democracity exhibit, 445–446
demonstration farm in Lake County, 420–22,
  433, 507n47
depressions. See economic depressions
Déri, Miksa "Max," 480n35
Descartes, René, 77
Deutsche Bank, 254
Dewing, Arthur S., 503n78
Dickens, Charles, 58
Dickinson, H. W., 31, 35–36
Dickson, W. L., 217
differential control magnet, 117
differential regulator, 127

direct current. See entries starting with "DC"
displacement current concept, 93
distribution
  AC system as answer to copper costs, 232
  with alternating current, 212, 224
  to axles of railroad cars, 241
  with belting and pulleys, 69
  with gears and line shafts, 69
  Insull's contribution to, 443
  natural gas system, 195–96
  overhead vs. underground, 163–64, 262–64
  See also central power stations; Insull, Samuel,
    in Chicago; transmission
Documents tending to prove the superior
  advantages of Railways . . . (Stevens), 55
Dolivo-Dobrowolsky, Michael, 318
double-acting rotative engines, 27
double-carbon arc lamps, 115, 186
double-reduction-gear type motor with carbon
  brushes, 252–53, 290, 291
dredging scow, steam-powered, 36–37
Drexel, Anthony, 139
Drexel, Morgan firm, 139, 170–71, 192, 234, 254,
  355, 484n45. See also Morgan, J. P.
drum armature, 146–47
Du Fay, Charles, 77
Dumoncel, Count, 160
dynamical theory of heat, 74
"Dynamical Theory of the Electromagnetic
  Field, A" (Maxwell), 93–94
dynamo. See generators
dynamoelectric machine, self-excited, 112

E
earthquake of 1811, Mississippi Valley, 49
Eastman, George, 481n15
East Newark lamp factory, 175, 203, 204–5,
  489n16, 490n47
East Pittsburgh, PA, Westinghouse factory,
  359–61, 383, 392
Eaton, Cyrus, 436, 438, 439
Eaton, Sherbourne B., 190, 191–93
economic depressions
  and Centennial Exhibition, 6
  and Columbian Exposition, 308, 317–18
  due to credit shortage, 386–87
  GE's financial woes, 354–59
  Great Depression, 438
  Weston's success during, 119
  world's fairs and, 449
economies of scale, 404, 412–16, 422, 430–31, 443,
  505n58
Edison, Marion "Dot" (daughter), 161–62,
  187–88, 204, 205
Edison, Mary (wife), 161–62, 187–88
Edison, Mina Miller (wife), 204–5
Edison, Thomas A.
  and Batchelor, 143, 146–47, 192–93
  battle against alternating current, 273

biographical information on, 132–33, 161–62, 172, 187–88, 203–4, 296, 398–99
and Brown conspiracy, 269
on capital punishment, 260–61
at Centennial Exhibition, 10–11
companies as public corporations, 233–34
consolidating his companies, 206, 234, 253–55, 268, 284
display on, at Columbian Exposition, 315
and Edison General Electric/Thomson-Houston merger, 284–85, 295–96
and electric wire panic, 271–75
on electrocution, 260–61, 271–75, 277, 485n13
and Faraday, 147
and Insull, 226, 229, 397–98, 399–400, 479n15
inventions of, 376
Jehl on, 140, 161, 165, 472n19
and Johnson, 143, 167, 173–74, 253–54
North American Review article, 273
as public icon, 268–69, 272, 377
and Sprague, 242, 246–47, 483n13
and Standardizing Bureau, 230
and Thomson, 124–25, 182, 226
and Upton, 133, 138
and Villard, 229
and Westinghouse, 232, 264–68, 270, 352–54
Westinghouse compared to, 285–86, 291–92, 490n35
West Orange, NJ, home, 204–5
Edison, Thomas A., as inventor
of AC system, 284
and central power stations, 141, 162, 188
and counter electric motive force, 242
on discovery vs. invention, 136
of dynamos, 146–47, 152
of electrical devices, 133, 134, 145
on electricity vs. gas as general illuminant, 472n19
and electric railways, 242–43, 254–55, 483n13
and electromagnets, 470n12
"etheric" force theory, 123–25, 156
Farmer's influence on, 131–32
and filament material, 145–46, 147, 149–50, 162–63, 230–31
of generators, 146–47, 152, 168–69, 175, 217, 230–31
and high-resistance principle, 472n23
improvements to telegraph, 134
incandescent lighting, 138–40, 143–50
insulation for house wiring, 230
patents held by, 189
on Pearl Street station, 159
perfection of meter, 164, 475n16
Sawyer's criticism of, 156–57
and subdivision of light, 139, 144
and Tesla, 198, 200, 217–18
Westinghouse compared to, 291, 490n35
and West Orange laboratory, 206, 228–30, 233–34, 281

See also feeder-and-main systems; Menlo Park, New Jersey; Pearl Street station; *entries beginning with* "DC"
Edison effect, 376
Edison Electric Illuminating Company of Boston, 354
Edison Electric Illuminating Company of New York, 165–68, 170–74, 354, 475n21
Edison Electric Light Company
board members' lack of confidence and/or interest, 143, 146, 148, 188, 191–92
central power stations installed by, 257–58
generators from Edison shop, 175
and Hungarian ZBD AC system, 233
incorporation and financing, 139–40, 165, 472n9
Isolated Lighting subsidiary, 173–74, 206
Johnson as manager, 202
as "leadened collar" on Edison, 234
patent problems, 190–91
payments to support Edison's laboratory, 229, 233–34
policy against manufacturing, 164
salesman for, 363
Edison Electric Tube Company, 167, 176, 398
Edison General Electric Company, 254, 268, 283–85, 398–99, 502n75
Edison Lamp Company
East Newark factory, 175, 203, 204–5, 489n16, 490n47
Edison resolves problems at, 204
merge into Edison General Electric, 234, 253–55
patent disputes, 259, 260, 269–70, 283, 326
payments to support Edison's laboratory, 229, 233–34
and United Edison Manufacturing, 206
Upton as manager of, 165, 175, 203, 204–5
Edison Lamp Day, 473n40
Edison Machine Works
Batchelor as manager, 202–3
establishment of, 167
Goerck Street facility, 168–69, 175–76, 217, 489n16
merge into Edison General Electric, 234, 253–55
move to Schenectady, 205–6
payments to support Edison's laboratory, 229, 233–34
and United Edison Manufacturing, 206
workers request union representation, 205
Edison Medal, 381, 393
Edison Ore Milling Company, 229
Edison Phonograph Company, 229
Edison Pioneers, 376
efficiency
of 1913 generators, 415
and cost of fuel, 64
of DC systems, 330

efficiency (*continued*)
    of Edison's vs. Westinghouse's power stations,
        284–85
    of generators in early 1900s, 415
    heating and cooling steam cylinder vs., 22–23
    and temperature of the engine, 75
    Watt's improvements to, 15–16, 23–29
Egg of Columbus display, 315
Einstein, Albert, 96–97
electric batteries, 78, 81, 304
Electric Bond and Share Company, 386
electric cars, 432
electric circuit and magnetic field, 83
Electric City (Chicago Edison free magazine), 413
electrical charge, 77–78
Electrical Construction Company, 308–9
Electrical Engineer, 287
electrical industry
    capital requirements, 382–85
    espionage in, 342–44, 347–48
    growth and volatility of, 319–21, 375–76
    Hopkinson system of rates, 419
    imprecision of, 231
    marketing aspect of, 202, 395, 406–7, 410–13
    Niagara Falls Power as breakthrough, 350–51, 352
    Niagara project as breakthrough, 350–51, 352
    NY World's Fair exhibits, 446
    overview, 395–96
    placement of top firms, 279, 293, 325–26, 360
    problems related to demand for electricity,
        328–29
    standardization of frequency, 366–67, 371
    Wright's meter and two-tier rate system, 410
    See also competition
Electrical Properties Company, 386
Electrical Securities Corporation, 386
Electrical Trust. *See* General Electric Company
Electrical World
    on alternating current, 382
    on ether concept, 95
    on Forbes, 346–47
    on GE and Westinghouse pooling patents, 368
    on Hopkinsons' paper, 232
    Martin as editor, 235
    on Niagara project, 324, 332
    in Sprague's elevated railway system, 255
    on steam turbines, 373
    on Tesla, 336, 377
    on transformers, 213
    on Wright's meter and two-tier rate system,
        410
Electrician, 159
electricity
    Ampère's law, 83, 84
    as electrochemical interaction, 81
    extending to rural areas, 421–22
    gas vs., as general illuminant, 472n9
    gospel of consumption concept, 404, 412–13,
        416, 430–31, 443

    for heating and cooling food, 432
    and lightning, 78–79, 465n22
    and magnetism, 75–76
    massing production concept, 414–15, 422,
        430–31, 505n58
    measuring, 164
    Ohm's law, 84, 144
    overview, 69–70, 77–81
    redefining abundance with, xiii
    static electricity, 79–80
    steam engine as basis for, xi
    term coined by Gilbert, 76
    versatility of, 240
electricity conversion
    from AC to DC or DC to AC, 210, 241–42,
        298–99
    current produces magnetic field and force, 82,
        89
    Joule on, 74–75, 464n10
    magnetism to electricity, 88–89, 99, 110
    telegraph as, 89
electric light law, Edison's, 145
electric motors, 153, 163, 239–41, 247, 251–53,
    290–91
electric polarities, 77
Electric Railway Company of the United States,
    243
electric railways
    Bentley-Knight contracts for, 244
    in Chicago, 408, 418–19
    compensated alternating motor for, 364
    competition, 327
    early years, 241–43
    Edison's, 242–43, 254–55, 483n13
    expansion requires converting AC to DC,
        298–99
    generators for, 337
    Manhattan Elevated railway, 374
    merger of four companies in Chicago, 424–25
    multiple-unit control scheme, 255
    overhead-trolley system, 245, 249–50, 328, 370
    problems related to demand for electricity,
        328–29
    Richmond installation, 248–51, 483n28, 484n35
    Siemen's commercial line, 243
    third rail for power, 304, 314, 491n11
    Van Depoele's success with, 244–45
    See also elevated electric railways; urban
        transportation
Electric Utilities Building, NY World's Fair, 447
electric wire panic, 271–75
electrochemical industry, 81, 350–51
electrocution, accidental, 270–71, 275
electrocution of animals, 264–66, 275
electrocution of criminals
    adoption by states, 325
    Auburn prison tests, 275
    bill passes in New York, 263
    debates and debacles, 324–25, 494n1

demonstrations with animals, 264–66, 275
horror stories from, 324–25
issues, 259
Kemmler as first, 276–77
New York commission inquiry on, 258–59, 260
electrodynamic charges, 91
Electro-Dynamic Light Company, 155, 156
electrodynamics, Ampère's theory of, 83, 84, 87
electrolysis, 89
electrolytic meter, 164, 475n16
electrolytic processes, 351
electromagnetic induction, 86, 87–89
electromagnetic motor, 86–87, 102, 111–12
electromagnetic relay, 103
electromagnetic telegraph, 99, 102
electromagnetic theory of light, 93, 94
electromagnetism
    intensity magnets, 101, 106
    Maxwell's theory on electromagnetic waves, 231
    Ohm's study of, 84
    quantity magnets, 101
    Sturgeon's magnet, 100–101, 468n9
    theories of, 94–95
    and Wilde's field excitation machine, 469n38
electromagnets, 231
electromotive force, 80, 84, 92, 212, 214, 237
electrophorus, 80–81, 465nn27–29
electroplating, 111, 119–20
electroscope, 80
electrostatic charges, 91, 94
electrostatic law of inverse squares, 79
electrostatic machinery, 77, 122
electrothermic processes, 351
electrotonic state, 92
electrotyping, 111
elevated electric railways
    in Chicago, 255, 418–19, 424–25
    financial problems, 425–26
    intramural railway at Columbian Exposition, 300, 304, 314
    Lake Street Elevated, 414
    Manhattan Elevated railway, 374
    motor for, 247–48, 255
    prediction of, 242–43
    providing power for, 328
    Van Depoele's overhead trolley patent, 245, 328, 367, 370
    See also electric railways; urban transportation
elevated steam railways, 243, 247–48
elevators, 255, 412–13
Ellsworth, Annie, 107
Emmet, William L., 414
energy, 73, 74–75
Enterprise (steamboat), 49
epicyclic gear, 26
Erie Canal, 56, 59, 360, 370
"Ether, Electricity, and Ponderable Matter" (Thomson), 95

ether as heavens (aether) to Greeks, 76
ether as scientific concept, 77, 83, 91, 95–97, 123–25, 198. See also electricity
Evans, George, 49
Evans, Oliver (1755–1819)
    biographical information on, 32–34, 39
    dredging scow, 36–37
    flour-milling technology, 34–35
    Mars Works iron foundry, 37
    Oruktor of, 37, 40, 54, 460n18
    patent problems, 35, 38–39
    on steam-driven carriages, 30
    steamship on the Mississippi River, 40
Evershed, Thomas, 331
execution, forms of, 259, 260. See also electrocution of criminals
Experimental Researches in Electricity (Faraday), 92
expositions
    Centennial Exhibition, Philadelphia (1876), 1–13
    initiation of international shows, 6–7
    International Electrical Congress, 321–22
    International Electrical Exhibition in Philadelphia, 207
    International Electrical Exposition in Frankfurt, 335
    in London, 6–7, 68, 207, 246
    in Paris, 126, 169, 300, 302, 305, 306, 469n44
    world's fairs in United States, 444–45
    See also Columbian Exposition; New York World's Fair

F
Fabbri, Egisto, 139, 150, 151, 165, 193
Fahrenheit, Gabriel, 15
Fairchild, Charles, 289, 293
"Fair of the Future," 450–51, 452. See also New York World's Fair
Fall River, MA, power station, 190
Faraday, Michael
    biographical information on, 84–86, 89–90
    converting magnetism to electricity, 88–89, 110
    on Cooke's electric telegraph, 99–100
    as discoverer of induction, 101–2
    electromagnetic motor discovery, 86–87
    as Holmes' advisor for lighthouse trials, 111
    influence on Edison, 147
    mathematics deficiency, 86, 90–91
    Maxwell's mathematical proof, 92
    polarization defined by, 93
Faraday's laws of electrolysis, 89, 466n49
Farmer, Moses G., 130–31, 137, 138, 246
farming. See agricultural industry
feeder-and-main systems
    description of, 477n36
    Edison's invention of, 188–90
    faltering sales as AC system dominates, 258
    feeder wires, 164

feeder-and-main systems (*continued*)
 Hopkinson's, 231–32, 246, 483n23
 limitations of, 224
 patent disputes over, 327, 368
 refining, 230–32
 underground wiring for, 247
Feeks, John, 270, 271–72
Fell, George, 275
Ferguson, Louis A., 402, 405, 407, 410–11, 412, 414, 505n57
Ferraris, Galileo, 236–37, 318
Ferris wheel, 302, 491n6
Field, Marshall, 401
Field, Stephen D., 243
field coils, 249–50
field magnets, 147, 231, 232
filaments
 asphalt, 230–31
 bamboo, 163
 carbon, 150, 230, 296–97
 Edison's search for, 145–46, 147, 149–50, 162–63, 230–31
 hydrocarbon vapor treatment, 179
 overview, 137
 Tamidine, 180–81
financial markets
 common stocks, 360, 436, 488n4
 preferred stocks, 281, 289, 437, 488n4, 489n28, 503n78
 See also bonds; market fluctuations
financing
 for building a gas plant in Chicago, 430
 for central power stations, 403–4
 dynamoelectric machine, 112
 incandescent light, 138, 139–40
 industrial sector of market, 279, 280–81
 Insull borrows money in NY during depression, 439–41
 Insull's contacts in Chicago, 419
 and Insull's destruction, 438–41
 inventors' need for, xii, 33–34
 London market as source of funds, 407–8
 open-coil generator and ring-clutch lamp, 114
 with open-end mortgages, 409–10
 organizing factories as, 65
 profitability of utility financing, 435
 railroad's need for, 62
 Sawyer's carbon lamp, 155
 shareholders giving up value for future guarantees, 289
 transportation sector of market, 280
 Westinghouse's opinion of bankers, 288
financing utilities, lucrative nature of, 435
fire-alarm system, electric, 130
fire-based pumping engines, 15–16, 23–29, 36
first law of thermodynamics, 74–75
First Street, Chicago, IL, power station, 414–15
Fish, Frederick P., 219, 294, 313

Fisk Street, Chicago, IL, power station, 414–15, 419
Fitch, John, 40–47
Flint, Charles R., 179, 184
flour-milling industry, 34–35
fluorescent lighting, 376, 378, 379–80, 446
Fontaine, Hippolyte, 141
Forbes, George, 338–39, 341–42, 345–49, 496n2, 497n12
Forbes generator, 346–47
Ford, Henry, 188, 376, 432
Fort Wayne Electric Company, 308, 355, 358–59, 401–2
Fothergill, John, 24
Fowler, E. J., 411
frame for locomotives, 61, 68
Frankfurt, Germany, 335, 414
Franklin, Benjamin, 44, 51–52, 77–79, 464n19, 465n22
Franklin Institute (Philadelphia, PA), 114, 122, 126
frequencies
 designers and engineers choices of, 297–98
 effect of competition, 366
 for Forbes generator, 346, 348
 for light circuits, 339
 and phase, 297–98
 for single-phase AC system, 296–97
 single-phase vs. multiphase motors, 340
 standardization of, 298, 366–67, 371
Fresnel, Augustin-Jean, 83
friction draft gear for railroad cars, 193
friction pulley, 242
Fulton, Robert, 40, 47–49, 55, 63
funding. *See* financing
Futurama, NY World's Fair, 448

G
Gale, Leonard D., 106
Galileo Galilei, 20
Galvani, Luigi, 80
galvanometer telegraph, 99
galvonometer, 88, 99
Ganz & Company (Budapest, Hungary), 213
Garrett, George S., 127–28, 129
gas
 coal gas, 110, 137, 138, 428
 coke oven gas, 429
 natural gas, 138, 195–96, 285, 330
 water gas, 219, 428
gas industry
 in Chicago, 401
 Chicago companies' fights with the City, 428
 competing with electric lighting, 116, 137, 138, 166, 219–21, 472n19
 first commercial company, 138
 Insull's solutions for, 430–31
 and oil prices, 428
 Peoples Gas Light and Coke, 409, 411, 425–26, 428–31, 439, 507n38
 as poor model for electrical industry, 403

resistance to incandescent light, 141–42
and Welsbach mantle, 220, 404, 411
See also Peoples Gas Light and Coke Company
gas lamps, 110, 116, 128, 138
Gates, John W., 386
gauge of "horsepower," 28
Gaulard, Lucien, 212–13
Gaulard-Gibbs AC system, 212–13, 214, 225, 232
Gaulard-Gibbs transformers, 212–13, 216, 227
Geissler pump, 147
General Electric Building, NY World's Fair, 447
General Electric Company
    AC power system for Columbia, SC, textile
        mill, 363
    buying smaller companies, 371
    Columbian Exposition bids, 309–12
    committee on accounts, 354, 497n28
    consolidating operations in Schenectady, NY,
        356, 357, 359, 369–70
    constant-current transformer, 362–73
    financial challenges, 353–59, 497n28, 497n30,
        498nn33–34
    incorporation of, 295–96
    Insull's request for 5,000-kilowatt units, 414–15
    and Niagara project, 340, 341, 348
    and Pirelli cable, 434
    polyphase AC systems for industrial plants,
        363–64
    pooling patents with Westinghouse, 368–71,
        500n15
    searchlights for Columbian Exposition, 314
    and Stanley, 364–65
    and steam turbines, 373
    theft of proposals and designs from
        Westinghouse, 342–44, 347–48
    Tower of Light, 305, 492n14
General Electric Company of Berlin, 254
General Motors complex, NY World's Fair, 448
generators (dynamos)
    Armington & Sims, 247
    for Columbian Exposition, 314
    constant-voltage, 144
    development of, 111–17
    direct-current, 197
    Edison's, 146–47, 152, 168–69, 175, 217, 230–31
    for electroplating, 119–20, 146
    factors in power output, 374
    Gramme model, 126, 146, 197
    hand-driven, 110–11
    for incandescent light, 146–47, 152, 163
    for intramural railway at Columbian
        Exposition, 320
    invention of, 110
    Jumbo generators, 168, 169, 175, 231
    kilowatt capacity increases, 415
    kodak, 495n39
    low-internal-resistance with high-resistance
        field, 242
    motors vs., 239–40

open-coil, 114
for Pearl Street station, 168–69
polyphase, 340
for railway service, 337
self-excited dynamoelectric machine, 112
self-exciting, 131
and Tesla motor, split-phase version, 336
Thomson-Houston collaboration, 126, 127–28
three-phase, 335
turbogenerators, 372–75, 415, 419
See also alternators; frequencies; phase; steam
    turbines
Gerry, Elbridge T., 259–60
Gibbs, John D., 212–13
Gilbert, William, 76
Gilchrist, John, 412
Gilliland, Ezra, 203–4
glassblowing, 147
glass globes for arc lights, 363
Glenmont laboratory, 206, 228–30, 281
Glenmont (West Orange, NJ), 204–5
Goerck Street manufacturing facility, 168–69,
    175, 217, 489n16
Goff, Edward H., 208
Gold King Milling and Mining Company
    (Telluride, CO), 336, 339
gospel of consumption concept, 404, 412–13, 416,
    430–31, 443
Gould, Jay, 247–48
government
    corruption in, 6
    funding for telegraph, 106–7, 108
    indictments against Insulls, 441–43
    as public utility regulator, 425–26
governors (engine part), 27–28, 68–69, 176
Gramme, Zénobe-Théophile, 112, 119, 241
Gramme dynamo, 126, 146, 197
Grant, Hugh L., 263, 271
Grant, Ulysses S., 7, 8, 9
"grasshopper" engine, 36–37
Gray, Elisha, 316, 321
Gray, Stephen, 77
Great Barrington, Massachusetts, 214–15
Great Blizzard of 1888, New York, 262–63
Great Crash, 437–38
Great Depression, 438
Great Eastern (steamship), 63–64
Great Western (steamship), 63
Great White Way, 166
Green, Norvin, 139–40
Greensburg, Pennsylvania, 216
Griffin, Eugene, 310, 311–12
gristmills, 17, 34–35, 65–66
Guericke, Otto von, 77
guide pulleys, 69

H
Hale, Matthew, 260
Hall, Charles M., 350, 351

Hammond, John Winthrop, 493n42
Harrison, Carter, 317–18
Harrison Street, Chicago, IL, power station, 402, 407, 413–14, 504n23
Hastings, Frank, 266
Hawley, Joseph R., 7
Hawthorne Farm (Libertyville, IL), 420–21
Hayes, Rutherford B., 10
heat, 16, 32, 71–75, 144
heating with gas, 220
Hebard, George W., 181
Hefner-Alteneck, Friedrich von, 112
Heinrichs, Ernest, 393
Helicline ramp, 446
Helmholtz, Hermann von, 96, 321
Helmont, Jan Baptista van, 137
Henry, Joseph, 87, 101–3, 210, 468n9
Herr, Herbert T., 291–92, 388
Hertz, Heinrich, 94, 95, 231
Herz, Cornelius, 152
Hewitt, Abram S., 263
Higginson, Henry, 285, 293
high-frequency circuits, 235
high-frequency coil, 316
high-frequency magnetic waves, 125
Highland Park, Illinois, 420
high-pressure engines
   boilers and, 21, 31–32, 37
   and caloric theory of heat, 73
   condenser, in power station, 402
   direct-acting, 49, 63
   Evans' use of, 35–38
   portability of, 21
   for railroad locomotives, 54, 56
   Savery's, 21
   for water travel, 40, 49–50
high-pressure turbines, 415
high-resistance lamps, 144, 145–46, 149
high-resistance principle, 472n23
high-voltage systems. See AC systems
Hill, David B., 259
Hills, Richard L., 27
holding companies, 386
Holmes, Frederick Hale, 111
Holtz electrostatic machine, 122
Hooke, Robert, 71
Hoover, Herbert, 438
Hopkinson, Edward, 232
Hopkinson, John, 231–32, 483n23
Hopkinson system of rates, 419
horizontal fire-tube boiler, 60
horizontal reciprocating engines, 36
horizontal tubular boiler, 55
horsecar trolleys, 241, 418
horsepower measures, 28, 84
horses and mules, 240, 241, 418
Houston, Edwin J., 120, 122–23, 124–29, 182, 470n11, 477n16. See also Thomson-Houston
Howells, William Dean, 3, 7, 12

Hudson River, 47–48
Hughes, Thomas, 319
Hunter, Louis C., 40
Hussey, Obed, 66
Hyde, Henry B., 359
hydraulic power, 73, 334–35. See also Niagara Falls Power Company
hydrocarbon vapor filaments, 179

I
Illinois Maintenance Company, 412
illumination (lighting), 109–10. See also arc lights; incandescent light
impulse turbines, 372
incandescent lamps
   development of, 138–40, 143–50
   Farmer's version of, 131
   high- vs. low-resistance, 144–45
   Sawyer-Man incandescent lamps, 226–27
   stopper lamp, 313, 326–27, 360
   See also Edison Lamp Company; filaments
incandescent lighting
   competition in market, 196, 352
   gas companies' resistance to, 137, 138
   generator for, 146–47
   high-frequency requirements of, 341
   NY World's Fair description of, 450
   overview, 136
   sales figures, 283
   stand-alone systems, 162, 163, 169, 170–74, 176, 217
   subdivision of light, 127, 137, 139, 141–42, 144
   system components, 143–44
   Thomson-Houston's system, 209
   Thomson's work on, 186
   vacuum pump for, 147
   and Westinghouse, 227, 313
Indiana electric service, 423–24, 433
indicator for steam engines, 28
induction, 80, 89, 124–25, 199, 468n13
induction coils, 127, 210–11, 212. See also transformers
induction motors, 296–98, 315, 318–19, 336, 364
inductor alternator, 364
industrialization, xi, 16–17, 65
innovate or perish maxim, 329
insulation, 107, 164, 230, 337
insulators (dielectrics), 89, 93, 94, 101, 165, 464n18
Insull, Martin (brother), 423–24, 434–, 438
Insull, Samuel
   on AC as viable venture, 283–84
   on American patriotism, 417
   biographical information on, 396–97, 417–18, 422–23
   and Coffin, 414
   and Eaton, 191–93
   and Edison, 226, 229, 397–98, 399–400, 479n15
   on experiments in individual companies, 233
   on incorporation of Edison Electric Light, 139

as Machine Works manager, 206, 398–99
and merger of Edison General Electric with
　　Thomson-Houston, 294–95, 296
retirement in France, 443
and Sprague, 253
and Tesla, 442–43
and Westinghouse, 443
Insull, Samuel, in Chicago
and Chicago Edison, 400–406, 408, 409, 419,
　　504n38, 504n41, 506n8
and Chicago electric railways, 424–25, 426
and Commonwealth Edison, 419–22, 424–27,
　　433, 436–41, 506n8
and Commonwealth Electric, 408–10, 414–15,
　　419, 504n41, 505n42, 506n8
and cooperation between competing
　　associations, 410
and Corporation Securities Company of
　　Chicago, 437–40, 508n60
destruction of, 438–41, 503n6, 508n70
electricity for railways in Chicago, 418–19
federal trial of, 442
funding sources, 407–8, 419–20, 428
Hawthorne Farm as home, 420–21
indictments by federal government, 441–43
and Insull Utility Investments, 436–37, 438–40,
　　508nn59–60
Lake County experiment, 420–22, 433, 507n47
on marketing electricity, 395, 404, 406–7,
　　410–13, 416, 443
massing production concept, 414–15, 422,
　　430–31, 505n58
and Middle West Utilities, 424, 427, 435, 437,
　　438
and Midland Utilities, 433–34, 507n48
as NELA president, 410
Peoples Gas rescued by, 428–31
response to takeover attempts, 435–38
selling securities to customers, 427
World War I support efforts, 425
Insull, Samuel, Jr., 417–18, 422–23, 433–34, 441–43
Insull Utility Investments (IUI), 436–37, 438–40,
　　508nn59–60
intensity batteries, 106
intensity magnets, 101, 106
interference fringes, search for, 96
interferometer, 96
interlocking signal system, 329–30
internal combustion engine, 69–70
internal resistance of generators, 147
International Electrical Congress (1893), 321–22
International Electrical Exhibition in
　　Philadelphia (1884), 207
International Electrical Exposition in Frankfurt
　　(1891), 335
International Exhibition in London (1862), 68
International Niagara Commission, 334, 495n27
investment banking, 280
iron foundries, 37, 64–65

iron ore, extracting, 230
Isolated Lighting Company, 173–74, 206
Israel, Paul, 134, 160–61
Ives, Brayton, 359

J
J. I. Case, 67
J. P. Morgan & Company, 139–40, 192, 435. See
　　also Morgan, J. P.
Jablochkoff, Paul, 113
Jablochkoff candles, 113, 126–27
Jarvie, James N., 389
Jehl, Francis
　　on Edison, 140, 161, 165, 472n19
　　on idea stealers, 160
　　on Maxim's copy of Edison's lamp, 179
　　on Menlo Park, 176, 472n22, 473n39
Jenks, W. J., 230
Jewett, Sarah Orne, 10
Johnson, Edward H.
　　and Edison, 143, 167, 173–74, 253–54
　　as Edison Electric manager, 192–93, 202, 253
　　on electrocution for capital punishment,
　　　261–63, 485n14
　　and Insull, 397
　　and Sprague, 247, 249, 253
Johnson bar, 61
Jonnes, Jill, 485n14
Josephson, Matthew, 132
Joule, James Prescott, 74–75, 84, 464n10
Julien, E., 243
Julien Electric Company, 243
Jumbo generators, 168, 169, 175, 231

K
Kaltenborn, H. V., 446
Kelly, John F., 365
Kemble, Fanny, 58
Kemmler, William, 268, 270, 276–77, 283
Kennelly, Arthur, 230, 265, 284, 381
kerosene, 110
Kleist, E. G. von, 464n18
Knickerbocker Trust Company, 387, 388
Knight, Walter H., 244, 343
kodak generator, 481n15, 495n39
Kruesi, John, 143, 146–47, 242, 398, 490n47

L
labor
　　Edison's workers request union
　　　representation, 205
　　increasing productivity of, 430, 431
　　and Insull post-World War I, 426–27
　　Westinghouse's, 222–23, 392
Lackawanna & Wyoming Valley Rapid Transit
　　Company, 385–86
Lake County demonstration farm, 420–22, 433,
　　507n47
Lake Street Elevated (Chicago, IL), 414, 418–19

Lamme, Benjamin Carver, 290–91, 297–98, 299, 314, 337, 338, 349
Lancaster Turnpike Company, 37
Land Ordinance (1785), 41
Lane-Fox, St. George, 138, 181
Lansing, Michigan, electric railway system, 253
Lauffen experiment, 335
Laval, Gustav de, 373
Lavoisier, Antoine, 72
Lawrence, MA, power station, 190
Lawrenceville, Pennsylvania, 216
laws of inverse squares, 79
Lee, Higginson & Company, 281
lethal injections, 260
Leyden jar, 77, 78, 464*n*18, 465*n*27
light, xi–xii, 81, 83, 91, 93–96
light circuit phase and frequency, 339
lighthouse lights, 111–12
lighting (illumination), 109–10. *See also* arc lights; incandescent light
lightning, Franklin and, 78–79, 465*n*22
lightning rod, 78–79
lightning simulations, 315
Livingston, Robert, 47–48, 55
load factor, 231, 403–5, 419, 422, 431–32, 504*n*28
Locksteadt, Charles F., 308, 309
locomotives. *See* steam locomotives
London Crystal Palace Electrical Exhibition (1882), 246
London *Engineer*, 347
London Exposition (1851), 6–7
London Inventions Exhibition (1885), 207
London market as source of funds, 407–8
Lontin, Dieudonné-François, 126, 127
Lorain Steel Company of Cleveland, 365, 371
low-internal-resistance dynamo with high-resistance field, 242
low-pressure engines
    atmospheric, 22–23
    beam engines, 2–3, 21, 23–24, 31, 68
    and caloric theory of heat, 73
    double-acting rotative, 27
    for farm work, 65–66
    "grasshopper," 36–37
    and railroad, 54
    reciprocating steam engines, 19, 20–25, 36, 70, 372, 374–75, 415
    rotative, 25–29
    in steamboats for eastern rivers, 48, 49, 50
    See also high-pressure engines; steam engines
low-pressure turbines, 415
low-resistance armature generator, 242
low-resistance lamps, 144, 145–46, 155, 164, 313
Lowrey, Grosvenor P., 139–40, 142–43, 155, 192, 313
low-voltage systems. *See* DC systems
luminiferous ether, 83
Lynn, Massachusetts, 184–87, 211, 292, 343, 356–57

M
Mach, Ernst, 198
Machinery Building, 302
MacMonnies, Frederick William, 491*n*3
Magnete, De (Gilbert), 76
magnetic circuits, 212, 231–32
magnetic field and electric field, 79, 83, 88–89, 99, 374
magnetic waves (lines of force), 91–92, 125, 231
magnetism
    compass needle and, 76, 82
    discovery of, 75–76
    electric circuit and magnetic field, 83
    overview, xi–xii
    and polarized light, 93
    rotating magnetic field, 315, 318
    See also electromagnetism
magnetite (lodestone), 75–76
magnets, 147, 231, 232
Man, Albon, 138, 155–56
Manhattan Elevated railway, 70, 247–48, 374
manufacturers, converting to central power stations, 426
Marconi, Guglielmo, 378–79, 379–80
marine-type triple-compound engines, 504*n*23
market fluctuations
    Baring Brothers failure, 277–78, 286, 293
    bears sabotage GE, 353–59
    financial panics, 191, 192, 326
    Great Crash, 437–38, 438–39
    money supply tightening leads to crashes, 386–87
    post-World War I money crunch, 427
    railway stocks plummet, 384
    rise in IUI stock, 436–37
    See also economic depressions
Marshall Field & Company, 433
Mars Works, 37
Martin, Thomas Commerford, 235
massing production concept, 414–15, 422, 430–31, 505*n*58
material abundance, industrialization and, xii–xiii
Mather, Robert, 392, 503*n*78
Maxim, Hiram S., 138, 154, 178, 179, 211–12, 474*n*54, 476*n*6. *See also* United States Electric Lighting Company
Maxwell, James Clerk "Daffy," 79, 90–95, 231, 467*n*63
McClelland, E. S., 312–13
McCollin, Thomas, 127–28, 129
McCormick, Cyrus, 66
McDonald, Forrest, 426, 434, 443
McElvaine, Charles, 324–25
McKeesport Light Company, 326
McKim, Mead & White, 349
McKinley, William, 379
McLennan, Donald R., 438, 439
Mead, Morris M., 342, 343, 347–48

mechanical meter, 164
mechanical poleboat, 42
Mellon, Andrew W., 430, 507n39
Menlo Park, New Jersey
  closing of, 176, 188
  Edison moves to NYC from, 166
  Edison's laboratory, 10–11, 133–34, 472n18
  Edison's manufacturing facility, 164–65, 174–75
  facility moved to Dearborn, MI, by Ford, 376
  funding for, 139–40, 142
  ohm as unit of electricity at, 472n22
  public displays of incandescent lighting, 151–54
  staff of, 143, 161, 162
  visitors to, 134, 160
Merchandise Creditors' Modified Plan, 390–91
mercury vacuum pump, 137, 163
meters
  accuracy issues, 231
  for AC system, 217
  demand meters, 406, 407
  Edison's electrolytic model, 164, 475n16
  interferometer, 96
  mechanical, 164
  for two-phase induction motor, 315
  volta-meter, 89
Michelson, Albert, 96
Michelson–Morley experiments, 467n72
Middle West Utilities Company, 424, 427, 435, 437, 438
Midland Utilities Company, 433–34, 507n48
Millard, Andre, 229
Miller, Mina, 203–4
Mills, D. O., 356, 369
mimeograph, 229
Miner's Friend, The (Savery), 20
Mississippi River, 40, 48–49, 58, 461n50
Mississippi Valley earthquake (1811), 49
Mitchell, George, 429
Mitchell, John J., 408, 410, 429
model electric house, 433
molecular currents, 83
Moll, Gerrit, 101
monocyclic system, 366–67
Morgan, J. P.
  as central bank for America, 387
  on Coffin's special trust plan, 355
  Drexel, Morgan firm, 139, 170–71, 192, 234, 254, 355, 484n45
  and Edison Electric, 254
  and J. P. Morgan & Company, 139–40, 192, 435
  on merging Thomson-Houston with Edison General Electric, 293, 294
  and Niagara Falls Power, 332
  and patent pooling fiasco, 369, 370
  and Tesla, 379–80, 501n42
  wiring in home of, 170–71, 173–74
Morgan, Junius S., 170
Morgan, Randall, 433
Morland, Sir Samuel, 19

Morley, Edward, 96
Morse, Samuel F. B., 104–8, 109, 468n20, 468n23, 469n28
Morse code, 104, 105, 204
Morton, Henry, 333
Mott, Henry A., 159–60
Mott, Samuel D., 143
Mullaney, Bernard J., 429
multiphase systems, 336–37, 339–40, 362
multiple-unit control scheme for electric railways, 255
multiplex telegraphy, 130
Munroe (Monroe), Charles A., 418, 421, 429, 506n5
Murdock, William, 25, 31, 54, 137
Musschenbroek, Pieter van, 464n18
mutual induction vs. self-induction, 468n13
Myers, L. E., 421

N
National Electric Light Association (NELA), 410
natural gas, 138, 195–96, 285, 330
Neumann, Franz Ernst, 94
Newcomen, Thomas, 20–22
Newcomen steam engine, 15, 21
New Orleans, Louisiana, 258
New Orleans (steamboat), 48–49
"New System of Alternate Current Motors and Transformers, A" (Tesla), 235–36
Newton, Isaac, 71, 77
New York Board of Electrical Control, 263, 264, 271, 272, 275
New York City, New York
  accidental deaths by electrocution, 270–71, 275
  deaths from various causes, 274
  electric wire panic, 271–75
  Goerck Street manufacturing facility, 168–69, 175, 217, 489n16
  Great Blizzard (1888), 262–63
  Medico-Legal Society of New York, 266
  power stations in, 116, 202, 361
  See also Pearl Street station
New York Herald
  on Board of Electrical Control, 272
  on changing from electricity to gas, 275
  on Corliss steam engine, 9
  on Edison's carbon filament lamp, 150–51
  on Edison's "etheric" force theory, 124
  and Edison's incandescent light, 142
  on electrocution, 265, 277
  power plant of, 169
New York Metropolitan Traction Company, 386
New York state, 276
New York Sun, 10, 151, 156–57, 269, 381
New York Times
  on Cockran's duplicity, 268
  on Edison, 152–53
  Edison's lights in offices of, 171

New York Times (continued)
　on electrocution, 266, 270
　on NY state's repeal of capital punishment, 276
　on soda water, 10
　on steam engine exhibits at Centennial
　　Exhibition, 11
　on Walter printing press, 5
New York Tribune, 6, 148, 272, 275
New York World, 264–65, 271, 276, 277
New York World's Fair (1939), 445–52
Niagara Falls, 330
Niagara Falls Power Company, 332, 349–51, 352.
　See also Adams, Edward Dean; Cataract
　　Construction
Niagara River, 330–31
Niagara River canal, 331
Niagara River Hydraulic Tunnel, Power, and
　　Sewer Company, 332
Niagara River tunnel, 331, 333
nitrogen, in carbon-electric lamp, 146
Nollet, Floris, 111
nonconductive substances, 78
North American Review, 273–74
North Shore Electric Company, 420, 422
Northwestern Gas Light and Coke, 422
Northwest Thomson-Houston Company, 355,
　358

O
Oak Park, Illinois, 433
Oersted, Hans Christian, 82, 84, 86, 89, 109, 110
Ohio River, 48–49
ohm, 472n22
Ohm, Georg Simon, 84
Ohm's law, 84, 144
oil-insulated transformers, 337
"Old Abe" (bald eagle), 11
Old Ironsides (locomotive), 59–60, 463n24
Olmsted, Frederick Law, 306–7
Omaha Herald, 52–53
"On the Mechanical Equivalent of Heat" (Joule),
　74
open-coil generators, 114, 128
open-end mortgages, 409–10
Opticks (Newton), 77
Ord, J. P., 490n47
Oregon (steamship), 217
Oruktor Amphibolos (Amphibious Digger), 37,
　40, 54, 460n18
Ott, Fred, 143
Ott, John, 143, 162, 205
Otto, Nikolaus August, 69–70
Overhead Conductor Electric Railway Company,
　252–53
overhead wires
　for Brush's arc light system, 166
　and deaths from electrocution, 270–72, 275
　public peril from, 263
　for telegraph, 107, 108, 109, 163, 263

　for telephones, 163
　torn down in NYC, 275
　underground vs., 163–64, 262–64
　Van Depoele overhead trolley, 245, 328, 367,
　　370

P
Pacinotti, Antonio, 112, 241
paddlewheels, 42, 43, 46. *See also* steamboats
Paine, Robert Treat, 359
Paine, Sidney B., 363
Palace of Fine Arts, 306, 307, 318
Pantaleoni, Guido, 213, 236–37
Papin, Denis, 20
parallel circuits, 144
parallel-connected transformers, 213, 214, 215
Paris Exposition (1889), 302, 305, 306
Paris Exposition (1900), 469n44
Paris International Electric Exposition (1893),
　169, 300
Parsons, Charles A., 372–73
"Passage to India" (Whitman), 62
Passer, Harold C., 283
patent disputes
　costliness of, 257, 367–68
　Edison against US Electric, 202, 326
　Edison lamp, 259, 260, 269–70, 283, 326
　Edison's feeder-and-main systems, 327
　and Edison's railway patents, 243
　and Evans, 35, 38–39
　and Morse, 109
　Niagara project and Adams' fear of, 340
　over carbon filament, 230
　over Edison lamp and Sawyer-Man patents,
　　259, 260, 269–70, 283, 313, 326
　Stanley's violations, 371
　and telephone, 134
　US Electric and Edison Electric, 326
　Westinghouse and Thomson-Houston over
　　transformer, 226
patent infringements, 178, 179, 190–91, 365
patent laws, 34, 39, 68, 494n8
Patent Office exhibit at Centennial Exhibition, 11
patent speculators, 157, 160, 178
Peabody, Frank S., 414–15, 427
Peabody Coal Company, 414–15
Pearl Street (New York, NY), 170
Pearl Street power station (New York, NY), 168,
　169, 170, 171–72, 201, 475n28, 479n2
Peck, Charles F., 218, 235
Pedro II of Brazil, Dom, 8, 9
Pennsylvania iron foundries, 64–65
Peoples Gas Light and Coke Company (Chicago,
　IL), 409, 411, 425–26, 428–31, 439, 507n38
Perisphere/Democracity exhibit, 445–46, 450–51
Peterson, Frederick, 266
petroleum, discovery of, 138
petroleum coke, 115, 469n47
Pevear, Henry A., 184, 185, 288

phase
  effect of competition, 366–67
  frequency and, 297–98
  for light circuits, 339
  multiphase systems, 336–37, 339–40, 362
  overview, 199–200
  polyphase, 296, 297
  single-, double-, and triple-, 235, 236, 297
  standards for, 298
  three-phase generators, 335
  three-phase system, 340, 341
  two-phase generators, 314
Philadelphia, Germantown & Norristown
    Railroad, 59–60
Philadelphia World's Fair. *See* Centennial
    Exhibition
Philips, Cynthia Owen, 48
phlogiston, 72
phonograph, 134
photophone, 304
*Physices elementa mathematica experimentis
    confirmata* ('sGravesande), 71–72
physics and electromagnetic fields, 93–94
Pickard, James, 26
Pirelli Manufacturing Company of Milan, 434
Pirelli underground cable, 434
pistons, 19–24, 26–27
Pittsburgh, Pennsylvania, 138, 359–61, 383, 392,
    489*n*16
Pittsburgh Reduction Company (Niagara, NY),
    350, 497*n*17
Pixii, Hippolyte, 110–11
Planck, Max, 95, 96–97
platinum filaments, 145, 147
Platt, Harold, 410, 412, 422, 503–4*n*21, 505*n*57
pneumatic power reverse gear, 61
polarization, 93
politics
  Bryan panic, 408, 504*n*38
  in Chicago, 408–9
  Insull's finesse with, 429
  presidential race (1876), 9–10
polyphase equipment, GE's display of, 315–16
polyphase generators, 340
polyphase motors, 296, 297
polyphase systems, 318–19, 337
polyphase transmission system, 315
Pope, Franklin L., 118, 214, 270
portable steam engines, 35–36, 66–67
Porter, Charles T., 68–69, 163
Porter, H. H., 409
Porter-Allen steam engines, 69, 168–69, 176
Portland, Oregon, 258, 335–36
Post, Robert C., 4
Potter, David M., xii
power stations. *See* central power stations
Preece, William H., 116, 141
preferred stocks, 281, 289, 437, 488*n*4, 489*n*28,
    503*n*78

Primary Heidelberg Electric Belt, 375
printing presses, 5, 10, 130
Prout, Henry, 290, 292, 297, 298, 318, 319, 375
Public Service Company of Northern Illinois
    (PSCNI), 421, 422
Public Service Investment Company, 433–34,
    507*n*48
Public Utility Commission of Illinois, 425–26
pulleys, 69, 242
Pulling, Ezra R., 160
pumping engine, 18–22, 25, 54
Pupin, Michael, 95

Q
quantity magnets, 101
Quarry Street, Chicago, IL, power station, 419

R
radio and radios, 376, 378–80, 432, 449
railroad industry
  and air brake, 4, 60, 193, 285, 329–30
  electric companies' inability to support, 385
  and electro-diesel locomotives, 61
  and farmers, 66
  growth in 1800s, 53, 56–57, 59–60
  high-pressure engines for, 54–55
  railroad cars, 57–58, 193
  regular train service in Charleston, SC, 53
  and United States interior, 39–40, 50, 57–58,
    61–62, 462*n*18
  See also electric railways; steam locomotives
Rankine, William B., 332
Rathenau, Emil, 420
RCA Building, NY World's Fair, 448–49
reaction water turbines, 372, 373
receiver for telegraph, 105–6
reciprocating steam engines, 19, 20–25, 36, 70,
    372, 374–75, 415
refrigeration industry, 431–32
refrigerators, 432
regional power network and growth of suburbs,
    422, 433
regulators
  for arc lighting, 115, 128–29, 183
  differential regulator, 127
  for raising AC voltage, 224
  in steam engines, 38
  Stillwell booster, 224
  thermal, 145, 149
  Thomson-Houston patent, 209
relay for boosting signal, 106
remote control, 378
"Reply to Mr. Edison, A" (Westinghouse), 273–74
resistance
  high-resistance lamps, 144, 145–46, 149
  high-resistance principle, 472*n*23
  internal resistance of generators, 147
  low-internal-resistance dynamo with
    high-resistance field, 242

resistance (*continued*)
  low-resistance armature generator, 242
  low-resistance lamps, 144, 145–46, 155, 164, 313
  Thomson's AC dynamo and, 210–11
Reynolds, Michael, 14
Rhode Island, 65
Rhode Island Locomotive Works, 244
Rice, Edwin W., Jr., 183, 226, 375, 376
Richmond Union Company (Richmond, VA),
  248–51, 483n28, 484n35
ring-clutch arc lamp, 114, 115
Rocket (locomotive), 55
Rock Island Railroad, 58
Roebuck, John, 23, 24
Roemer, Olaus, 96
Ronalds, Francis, 99
Roosevelt, Franklin, 451
Roosevelt, Nicholas, 48–49
Rose, Frank and Annie, 507n30
rotary converters, 298–99, 315, 318, 362, 405, 407
rotary engines, 373. *See also* steam turbines
rotary motors, Tesla's, 235–36
rotary steam engine, 194
rotative atmospheric engines, 25–26, 27, 29
Rowland, Henry A., 231, 338, 340, 341
royalties, 25, 237, 238
Rumford, Count (Benjamin Thompson), 72–73, 85
Rumsey, James, 42–45
Russell, Edward P., 424
Rust, H. B., 430

S
safety fuse, 162
safety instructions, customers' refusal to follow,
  228, 257
San Francisco, CA, central power station, 115–16
Sargent, Frederick "Fred," 414–15, 429, 505n57
saturation concept, 231
Savannah (steamship), 63, 463n31
Savery, Thomas, 20, 28
Sawyer, George, 270
Sawyer, William E., 138, 141, 146, 154–58, 160, 179,
  474n54, 474n60
Sawyer-Man incandescent lamps, 226
Saxton, Joseph, 111
Schenectady, New York
  Edison's Machine Works moves to, 205–6, 398
  GE complex in, 356, 357, 359, 369–70
  Kruesi in, 490n47
Schiff, Jacob, 391
Schilling, Baron Pavel, 99
Schmid, Albert, 224, 252, 290, 344
Schuyler, S. D., 178
science as community endeavor, 122
Scientific American, 69, 124, 180, 241, 242, 259
Scott, Charles F.
  on AC motors, 336–37
  on AC system reducing transformers, 335–36
  biographical information on, 299, 491n58

phase-changing transformer invention, 367
  on standards committee, 298
  and Tesla's motor, 290, 297, 339–40, 495n33
screw propellers, 63
second law of thermodynamics, 73, 75
Secrétan, Hyacinthe, 258
Security Investment Company, 386, 501n53,
  502n58
Seifer, Marc J., 501n42
self-excitation, 469n38
self-excited dynamoelectric machine, 112
self-exciting generators, 131
self-induction, 102
Sellers, Coleman, 333–34, 340, 341, 346, 348–49
series vs. parallel connections, 213, 480n35
sewing machines, 4
'sGravesande, Willem Jacob, 71–72
Shallenberger, Oliver B., 216, 223, 237
Sharlin, Harold I., 89, 91, 94, 469n35
Shreve, Henry M., 49
shunt-wound parallel-connected motor, 214
Siemens, Georg von, 335
Siemens, Sir William, 332, 469n38
Siemens, Werner von, 213, 241, 243
Siemens alternators, 213, 214
Siemens & Halske, 213, 232, 254, 288, 365, 371
signaling devices for railroads, 195
Simpson, James, 441
single-cylinder horizontal engine, 68
single-phase circuits, obsolescence of, 340
single-reduction No. 3 motor with gear casings,
  290–91, 292
single series circuit, 128
Sirius (steamship), 63
Skinner, Joseph J., 182
Slater, Samuel, 65
Smeaton, John, 28
Smithsonian Institution, 103
Society for the Prevention of Cruelty to Animals
  (SPCA), 259
Society for the Prevention of Cruelty to
  Children, 259–60
solar power, concept of, 37
solenoid, 83
South Side Elevated Railway Company of
  Chicago, 255
Southwick, Alfred P., 258, 260, 276
Special Theory of Relativity, 96–97
speed of light, 94, 96
spherical armature inside cylindrical magnetic
  field, 129
spinning thread, 17–18
Spitzka, Edward C., 276
Sprague, Frank J.
  and AC, 232
  biographical information on, 245–46
  and Edison, 242, 246–47, 483n13
  education of, 189
  electric elevator designs, 255

on Niagara Falls Power plans, 333
on reversible electric power, 239
and Thomson-Houston, 251–52, 254–55
and West End of Boston, 252
Richmond trolley system of, 248–51
Sprague Electric Elevator Company, 255
Sprague Electric Railway and Motor Company, 247–51, 253, 255
Sprengel, Hermann, 137
Sprengel pump, 147
stagecoach travel, 52–53, 57
Stahl, Georg Ernst, 72
stand-alone incandescent light systems, 162, 163, 169, 170–74, 176, 217
stand-alone power stations, 361, 363–64, 412–13, 426, 432
standardization of frequency, 366–67, 371
Standardizing Bureau, 230
Standard Oil, 280, 469n47
standards for phase and frequency, 298
Stanley, William
    on AC motors, 339
    AC power for arc-lighting system, 224
    AC system of, 212–15, 216, 237
    biographical information on, 211–12, 213–14, 375
    and GE, 364–65
    lamp system, 195, 227
    patent violation suits against, 371
    two-phase systems with condensers, 367
    and Westinghouse, 195, 364–65
Starr, J. W., 136, 148
static electricity, 79–80
stationary steam engines. See low-pressure engines
steam, 18–20
steamboats
    competition from stagecoach, 45–46
    eastern vs. western style, 50
    and Fitch, 40–47
    and Fulton, 40, 47–49, 55, 63
    invention of, 39–40, 48
    on lower Mississippi River, 48–49
    and Rumsey, 42–45
steam carriages or wagons, 31, 36, 37, 54, 55, 240–41
steam engines
    American dominance in, 11–12
    atmospheric, 21–23, 26, 31, 33
    as basis for electricity, xi
    at Centennial Exhibition, 2–4, 9, 11
    centrifugal governor, 27–28
    at Columbian Exposition, 302, 303, 307
    farmers' use of, 66–67
    Fitch's version of, 43
    indicator for, 28
    overview, 69–70
    portability of, 35–36, 66–67
    for powering Edison's dynamos, 152

for pumping water from mines, 18–22, 25
reciprocating, 19, 20–25, 36, 70, 372, 374–75, 415
refinement of, 67–69
replacing with gas engines, 220
rotary, 194
single-cylinder horizontal, 68
steam traction, 67
Watt's improvements, 15–16, 23–29
as weak leak in power generation, 371–72
See also high-pressure engines; low-pressure engines; steam turbines
steam locomotives
    American 4–4–0, 60, 61, 463n25
    Best Friend of Charleston, 55, 56
    cost of fuel and engine efficiency, 64
    dangers and discomforts of, 57–58
    development of, 59–61
    Evans' Oruktor, 37
    Old Ironsides, 59–60
    Stephenson's experimental design, 55
    Trevithick's work on, 55
steam power, golden age of, 69–70
steamships, 40, 62–64, 162, 163, 217, 463n31
steam traction engines, 67
steam turbines, 70, 372–75, 413, 414–16
Steele, Theophilus, 157–58
Steinheil, Karl A., 99
Steinmetz, Charles "Carl," 283, 363, 365, 366–67, 373
Stephens, Luther, 49
Stephenson, George, 54–55
Stephenson, Robert, 55
Stetson, Francis Lynde, 332, 334, 344
Stevens, John, 40, 51, 55–56, 63
Stevenson, Robert Louis, 113
Stiles, Aaron K., 244, 245
still coke, 469n47
Stillwell, Lewis B., 224, 298, 346–47
Stillwell booster, 224
stockholders in Insull's companies, 427
Stockly, George W., 114, 184
stopper lamp, 313, 326–27, 360
storage battery, 212
stream-boat, 42
Street Railway and Illuminating Properties, 356
street railways. See electric railways
Strouse, Jean, 387
Stuart, Harold L., 419–20, 428, 435, 439
Sturgeon, William, 100
Sturgeon's magnet, 100–101, 468n9
subdivision of light, 127, 137, 139, 141–42, 144
substations for universal power system, 327–28
suburban electric service, 257, 420–22, 433
sun-and-planet gear, 26
Sunbury, PA, power station, 190
superheater, 61
Swan, Joseph W., 137, 138, 179–80
Swan Electric Light Company, 195
switching devices for railroads, 195

synchronous alternators, 364
synchronous motors, 315, 495*n*35
Szigeti, Anthony, 199, 235

T
Tacoma, Washington, 258
Tamidine filaments, 180–81
Tate, Alfred O., 189–90, 229–30, 295–96, 479*n*15
technology, xi, xiii, 177, 299, 319, 328–29
telautograph, 316
teleautomaton, 378
telegraph
  battery for, 101, 106
  completion of, 107–8
  as conversions of mechanical and electrical
    energy, 89
  Edison's, 10–11, 134
  and electricity, 81, 98–103
  government funding for, 106–7, 108
  Morse and Morse code, 104–8, 109, 468*n*20,
    468*n*23, 469*n*28
  overhead wires for, 107, 108, 109, 263
  relay for boosting signal, 106
  Tesla's vision of global wireless system, 378
  wire examiner position, 130
telephone
  and Bell, 11, 96, 134, 203, 229, 316
  carbon transmitter for, 134, 203
  microphone relay for, 125–26
  at NY World's Fair, 448
television, 449
Telluride, Colorado, 336, 339
Terry, Charles A., 310, 478*n*54
Tesla, Nikola
  AC motors of, 197–200, 235–37, 252, 290,
    296–98, 336, 339–40
  arc-lighting system, 218
  and Batchelor, 199–200, 217–18
  biographical information on, 196–98, 218,
    377–78, 381, 478*n*58
  on communicating with Mars, 378, 501*n*37
  and Edison, 198, 200, 217–18
  on ether and electrical phenomena, 95
  on future of electricity, 71
  and Insull, 442–43
  laboratory in NYC, 234–35
  and Morgan, 379–80, 501*n*42
  multiphase system development, 336–37,
    339–40
  oscillators and fluorescent lamps, 380
  patents for polyphase motor, 340
  presentation at International Electrical
    Congress, 321–22
  Scott working on Tesla's motor, 290, 297,
    339–40, 495*n*33
  and Thomson, 236, 482*n*50
  turbine of, 380–81
  visualization skills of, 197, 198, 199, 378
  and Westinghouse, 221, 236–38, 237–38, 482*n*55

  and wireless communication, 378–80
  work display at Columbian Exposition, 315
Tesla Electric Company, 218
textile industry
  AC systems installed by GE, 363–64
  and carbon arc lights, 115
  carding technology, 34
  cost of fuel and engine efficiency, 64
  spinning thread and weaving cloth, 17–18
  and steam engines, 22–23, 28–29, 65
  water-based technologies for, 22
Thales, 76
thermal regulators, 145, 149
thermodynamics, 73–75
Thomas A. Edison Construction Department,
    189–90
Thompson, Benjamin (Count Rumford), 72–73,
    85
Thompson, Floyd E., 442
Thompson, Silvanus, 141
Thomson, Elihu
  and American Electric, 129–30, 182–84
  and arc-lighting system, 183
  biographical information on, 120–22, 130
  on Brown's use of Westinghouse equipment
    for electrocution of criminals, 268
  and Brush, 114–15, 128–29
  and Coffin, 185–86, 281–83
  constant-current transformer, 362–63
  and Edison, 124–25, 182, 226
  on electricity, 177, 319–21
  emphasis on safety, 225, 227–28
  funding sources, 281
  and General Electric, 295
  and Houston, 122–23, 470*n*11
  at International Electrical Congress, 321
  Model Room as private domain of, 225, 226
  retirement of, 376
  safety focus of, 225–26, 281
  and Tesla, 236, 482*n*50
  and Thomson-Houston, 207, 282–83
  at Universal Exposition in Paris, 126
  and Westinghouse, 226
  See also Thomson-Houston
Thomson, Frank, 242
Thomson, Mary Louise "Minnie," 207
Thomson, Sir William (Lord Kelvin), 74, 75,
    91–92, 338
Thomson-Houston Electric Company
  after safety disputes, 281–83
  arc-lighting system, 127–28, 182–83
  arc-lighting system of, 206–7
  Bentley-Knight order for equipment, 244
  and Brush AC system, 494*n*3
  and central power stations, 208–9, 257–58
  and Coffin, 207–9, 284, 292–95
  competition, 251–52, 254–55, 281–83
  and electric railways, 251–52
  establishment of, 185–87

and General Electric, 356–57
growth of, 186–87
intramural elevated railway at Columbian
    Exposition, 300, 304, 314
involvement in incandescent lighting, 196
merger negotiations with Edison General
    Electric, 292–95
preferred stock issue, 281
purchase of Bentley-Knight, 251
stock issues, 488n4
and Van Depoele, 245, 251
three-phase AC systems, 340, 341, 363, 366–67,
    367
three-phase generators, 335
threshing machines and thresherman, 66
Thurston, Robert H., 28, 32, 49, 201
Tilden, Samuel J., 10
Titusville, Pennsylvania, 138
Tocqueville, Alexis de, 50
Tomlinson, John, 203
Tom Thumb (locomotive), 56
Tone, Frank J., 350–51
Torricelli, Evangelista, 20
torsion balance, 79
Tower of Light, 305, 492n14
traction industry. See electric railways
transatlantic cable, 64
transformer motor, 364
transformers
    AC system reducing transformers, 335–36
    comparison of series and parallel connections,
        213, 480n35
    Gaulard-Gibbs model, 212–13, 216
    limitations of, 298
    oil-insulated, 337
    parallel-connected, 213, 214, 215
    phase-changing, 367
    Stanley's design for, 214–15, 216, 364
    in Thomson-Houston AC system, 226
    in two-phase induction motor, 315
    as voltage-changing device, 480n35
transmission
    amplifier to boost signals, 198
    Stanley's experiments with, 365
    three-phase AC transmission system, 363
    voltages for, 210
    Westinghouse's long-distance solutions,
        329–30
    See also distribution
transmitter for telegraph, 105–6
transportation industry
    capitalization of, 62, 279–80
    steamships, 40, 62–64, 162, 163, 217, 463n31
    See also railroad industry; steamboats
Treatise on Electricity and Magnetism
    (Maxwell), 92
Trevithick, Richard, 31, 32, 54, 65, 70
Triewald, Marten, 28
trolleys, 243. See also electric railways

Trylon, the, 445–46, 450–51, 509n3
tubular boilers, 63
turbines, 70, 372–75, 413, 414–16
turboalternators, 374–75
turbogenerators, 372–75, 415, 419
Tweed, William "Boss," 6
Twenty-seventh Street, Chicago, IL, power
    station, 407
two-gear motors, 252–53
Twombly, Hamilton McKown, 139–40, 293, 294,
    369
two-phase induction motor, 315
two-phase systems with condensers, 367
Tyndall, John, 118
typewriters, 11

U
underground conductors, 167–68, 434
underground wiring
    for avoiding weather, 163
    for Edison's central power station, 139, 141, 160,
        163, 165, 167–68
    insulators for, 165
    in NYC, 325
    overhead vs., 163–64, 262–64
    for telegraph, 107
Union Pacific 4-8-8-4 "Big Boy" locomotive, 61
Union Switch and Signal Company, 195
Union Traction Company, 419
United Edison Manufacturing Company, 206
United Electric Light and Power Company of
    New York, 361
United Gas Improvement Company
    (Philadelphia, PA), 220
United States
    anesthesia and attitude toward pain, 259
    credit shortage, 386–87
    in late 1930s, 445
    settling the interior of, 39–40, 50, 61–62
United States Electric Lighting Company, 178,
    179, 202, 224, 227, 289, 326
United States Illuminating Company, 263
United States Navy, 245–46
United States Rubber Company, 184
United States Steel, 379, 386, 505n54
Universal Exposition in Paris (1878), 126
universal power system
    at Columbian Exposition, 405
    development of, 335–38
    Edison's vision, 209, 261
    at Niagara Falls, 345–49
    overview, 319, 327–28
    Thomson-Houston polyphase system and,
        318–19
Upton, Francis
    and drum armature, 146–47
    on Eaton's plan to consolidate Edison plants,
        191
    and Edison, 133, 138

Upton, Francis (*continued*)
  on Edison Electric board, 192–93
  as lamp factory manager, 165, 175, 203, 204–5, 490*n*47
  mathematical skills of, 143, 144–45
  on ZBD AC system, 233
urban transportation
  cable cars, 240–41, 353, 418
  competition in market, 352–53
  Edison's beliefs about, 243
  as impetus for invention of motor, 240
  trolleys, 243
  See also electric railways; elevated electric railways
Usselman, Steven W., 292, 329–30
Utility Securities Company, 427

V

vacuum pumps, 147, 149, 163
vacuums (absence of matter), 20, 21, 33, 137, 149
Vail, Alfred, 106–7
valve gear, 8–9, 21, 31, 67–68, 463*n*43
Van Depoele, Charles J., 244–45, 251–52, 328, 370, 494*n*12
Van Depoele Electric Light Company, 244–45, 251
Van Depoele overhead trolley, 245, 328, 367, 370
Vanderbilt, William H., 174, 192
van Doren, Carl, 465*n*22
Varley, C. F., 469*n*38
velocity of electrostatic charges and electricity current, 94
vertical reciprocating engines, 36
Villard, Henry
  and *Columbia* project, 162
  consolidation of Edison's companies, 234, 253–55
  consolidation of major electricity companies, 284–85
  and Edison, 229
  on Edison Electric Light board, 192
  funding for electric railway, 243
  and merger of Edison General Electric with Thomson-Houston, 292–95
  and Westinghouse, 292
  See also Edison General Electric Company
VODER (Voice Operation Demonstration), 448
Voigt (Voight), Henry, 43, 45–46
Volta, Alessandro, 80
voltage for incandescent lighting system, 164, 210–15
voltaic batteries, 81, 82, 88, 101, 111, 465*n*29
volta-meter, 89

W

Wabash Plain Dealer, 98
Wagner, Richard, 8
Walker electric truck, 432

Walker Manufacturing Company of Cleveland, 365, 371, 499*n*62
Wallace, William, 131–32, 156
Wallace dynamos, 139, 146, 148
Wall Street Journal, 359, 361, 368, 387, 505*n*54
Walter (printing) press, 5
Wanamaker, John, 115
Wardenclyffe, New York, 379–80
Warning, A (Johnson and Edison Electric Light Company), 261–63, 485*n*14
Wasborough, Matthew, 25–26
Washington, Bushrod, 39
Washington, George, 42
Washington (steamboat), 49–50
water-commanding engine, 19
water frame, 22
water gas, 219, 428
Waterhouse Electric and Manufacturing Company, 224, 227
water-tube boilers, 67
water turbine for Niagara project, 341
waterwheels, 18, 22, 69, 305, 331–33, 335, 336
Watson, William, 77
Watt, James, 14–15, 15–16, 23–29, 73, 84
Waukegan, Illinois, 420
wave theory of light, 83, 91
weaving cloth, 17–18
Weber, Wilhelm, 94
Weed, Thurlow, 58
Welsbach, Carl Auer von, 220
Welsbach mantle, 220, 404, 411
Werdermann lamp, 152
West End Street Railway Company of Boston, 251–52, 353
Western Electric Company, 304–5
Western Union, 108, 134
Westinghouse, George, Jr.
  on AC for electrocution, 267
  air brake for railroad cars, 4, 60, 193, 285
  biographical information on, 193–96, 221–23, 290, 393–94
  on botched electrocution, 277
  capital stock holdings, 286, 501*n*53, 502*n*58
  and Coffin, 227, 287–88, 292, 368–71
  consolidation of firms owned by, 489*n*20
  creative genius of, 329–30
  on disputes within electric power business, 256
  and Edison, 232, 264–68, 267–68, 270, 352–54
  Edison compared to, 285–86, 291–92, 490*n*35
  and electric railways, 252
  and electric wire panic, 271–75
  employees' benefits from, 222–23, 392
  financial difficulties, 286–89, 381–91, 489*n*28, 502*n*68
  funding sources, 281
  and Gaulard-Gibbs transformers, 213, 216
  as impotent president, 391–93
  and Insull, 443
  and Mather, 392

multiple areas of development, 285
on Niagara project's request for ideas, 346
and NY state's repeal of capital punishment, 276
and Overhead Conductor Electric Railway, 252–53
and parallel-connected transformers, 214, 215
"A Reply to Mr. Edison," 273–74
retirement from electrical industry, 393
and Stanley, 195, 364–65
and Tesla, 221, 236–38, 237–38, 482n55
and Thomson-Houston AC system, 226
Westinghouse, George, Sr., 194
Westinghouse, Henry Herman, 194, 195
Westinghouse Building, NY World's Fair, 447–48
Westinghouse Electric and Manufacturing Company
    Allegheny Works, 313, 337–38
    buying smaller companies, 371
    and Columbian Exposition, 305, 309–12
    dividends paid by, 383, 393
    East Pittsburgh, PA, factory, 359–61, 383, 392
    financial report showing growth, 385
    investments in subsidiary companies, 383, 389
    Merchandise Creditors' Modified Plan, 390–91
    and Parson's turbine patent, 413
    pooling patents with GE, 368–71, 500n15
    purchase of Lackawanna securities, 385–86
    in receivership, 388–91
    theft of proposals and designs by GE, 342–44, 347–48
Westinghouse Electric reorganized as, 289
Westinghouse Electric Company
    AC distribution system patent, 225
    AC systems, 257, 335–36
    arc-lighting systems of, 227
    Brown's criticism of, 268, 269, 272–73
    central power stations installed by, 257–58
    and Consolidated Electric Light, 269–70, 289
    and deaths from electrocution, 271–75
    and depression in 1890s, 359
    display of AC-powered incandescent lighting system, 216
    generators from, for electrocution of criminals, 268
    growth of, 285
    and Niagara project, 348–50
    and Parson's steam turbine, 373
    polyphase transmission system, 315
    Sawyer-Man lamps of, 227
    and Stanley, 213–15
    voluntary reorganization of, 288–89
Westinghouse Machine Company, 194
Westinghouse railway motor, 252–53
Weston, Edward, 118–20, 146, 180–81, 206, 211–12

Weston Dynamo Electric Machine Company, 119
Weston Electric Light Company, 120
West Orange, NJ, laboratory (Edison), 206, 228–30, 233–34, 281
wet batteries, 130, 131
Wheatstone, Charles, 100, 469n38
Whig national convention, 107
White, John H., 61
Whitman, Walt, 62
Whitney, Henry M., 251–52
Whittier, John Greenleaf, 3–4
Wickes, Edward A., 332, 349
Wilcox, Stephen, 67
Wilde, Henry, 119, 469n38
Wilson, Woodrow, 425
wireless communication system, 378
wires and wiring
    batteries and, 78, 82–84, 86, 88–89
    copper wires, 101, 144, 263, 336, 341
    electric wire panic in NYC, 271–75
    in electromagnet, 100–101
    Evans's fine wire from bar iron, 34
    for filaments, 149–50
    heat from current flowing through, 144
    insulation for house wiring, 230
    lead tubing for, 107, 139, 141
    and magnetic fields, 92–93
    for Morgan's home, 170–71, 173–74
    platinum, 131, 136
    See also overhead wires; underground wiring
Wise, John, 26
Wollaston, William, 86
Woodbury, David O., 470n9, 470n11
Woodbury, Levi, 106
World's Fair. See Centennial Exhibition; Columbian Exposition; New York World's Fair
World Telegraphy Center, 379–80
World War I, 425
Wright, Arthur, 406, 407, 410
Wright, J. Hood, 139, 192, 294
Wrigley Building (Chicago, IL), 433, 507n46

X
X-rays, 376, 377, 447

Y
Yerkes, Charles T., 401, 408, 418, 504n40
Young, Owen D., 362, 440
Young Miller's Guide, The (Evans), 38
Young MillWright and Miller's Guide, The (Evans), 34–35

Z
ZBD AC system, 233.
Zipernowsky, Károly "Charles," 480n35

A NOTE ON THE AUTHOR

Maury Klein is the author of many books, including *The Life and Legend of Jay Gould*; *Days of Defiance: Sumter, Secession, and the Coming of the Civil War*; and *Rainbow's End: The Crash of 1929*. He is professor emeritus of history at the University of Rhode Island.